Few-Body Systems

Suppl. 9

Few-
Body
Systems

Editor: H. Mitter, Graz Managing Editor: W. Plessas, Graz

Supplementum 9

Mesons and Light Nuclei '95

Proceedings of the 6th International Conference,
Stráž pod Ralskem, July 3—7, 1995

Edited by
J. Adam, J. Dobeš, R. Mach, M. Sotona, and J. Dolejší

Springer-Verlag Wien New York

Dr. J. Adam
Dr. J. Dobeš
Dr. R. Mach
Dr. M. Sotona
Institute of Nuclear Physics
Academy of Sciences of the Czech Republic
Řež, Czech Republic

Dr. J. Dolejší
Faculty of Mathematics and Physics
Charles University
Prague, Czech Republic

Erziehungskooperation, Wien, Austria

Typesetting: Camera ready by the editors

Printed on acid-free and chlorine-free bleached paper

With 223 Figures

Die Deutsche Bibliothek – CIP-Einheitsaufnahme

Mesons and light nuclei '95: proceedings of the 6th international conference, Stráž pod Ralskem, July 3–7, 1995 / ed. by J. Adam ... – Wien ; New York : Springer, 1995
(Few body systems / Supplementum ; 9)
ISBN-13: 978-3-7091-9455-3 e-ISBN-13: 978-3-7091-9453-9
DOI: 10.1007/978-3-7091-9453-9
NE: Adam, J. [Hrsg.]; GT

ISSN 0177-8811
ISBN-13: 978-3-7091-9455-3

MESONS AND LIGHT NUCLEI

STRÁŽ POD RALSKEM, CZECH REPUBLIC

JULY 3 - JULY 7, 1995

INTERNATIONAL ADVISORY BOARD

LOCAL ORGANIZING COMMITTEE

Conference Address: Mesons and Light Nuclei,
Institute of Nuclear Physics, Academy of Sciences of the Czech Republic,
250 68 Řež, Czech Republic

Sponsored by

Austrian Ministry for Science and Research

Czech Ministry for Industry and Trade

ŠKODA PRAHA a.s

Organizers thank

Diamo s.p. Stráž pod Ralskem

for support

Preface

The International Conference **Mesons and Light Nuclei**, organized by the Institute of Nuclear Physics (INP), Řež, was held during July 2 - 7, 1995 in small north Bohemian town Stráž pod Ralskem. It was the sixth in a series of meetings which took place previously at Liblice 74 and 81, Bechyně 85 and 88, and Prague 91. The conferences gained already their firm position among intermediate energy nuclear physics activities. International nuclear physics community strongly supported our intention to continue the series. This year's venue for the conference was the accommodation and social area of the DIAMO company at Stráž.

The goal of the meeting was to summarize the present situation and the future perspectives concerning the experimental investigations and theoretical descriptions of light nuclei and their interactions with electromagnetic and hadronic probes, mainly at intermediate energies. The scientific program of the conference included the following areas of research: nuclear physics with pions and antiprotons, η-meson physics, baryonic systems with strangeness, relativistic few-body dynamics, and electroweak nuclear interaction. Representatives from many international groups working within different experimental facilities and with different theoretical methods were invited and asked to present their latest results and future research programs.

The Stráž conference, attended by 102 physicist from institutions in 22 countries, was sponsored by the Austrian Ministry for Science and Research, Czech Ministry for Industry and Trade, and by ŠKODA PRAHA a.s. Thanks to this sponsorship we could also invite several participants and students at essentially reduced cost.

Much of the success of the conference was due to the considerable help of our sponsors and the extensive efforts of many people. We would like to express our heartfelt thanks to the members of the International Advisory Commitee, for their advice on the program; to all session chairpersons, for their expertise and skillness; to Jaroslav Makovička, Director of DIAMO Company, for kind hospitality and support. We would like to thank particularly Professor Willibald Plessas for his advice and help in promoting and preparing the present Proceedings.

We are also grateful to Mrs. Renáta Novotná for her service as Conference Secretary. The program for accompanying persons was organized and successfully run by Miloš Tater.

Řež, September 1995 J. Adam, J. Dobeš, R. Mach, M. Sotona, J. Dolejší

SCIENTIFIC PROGRAM[1]

Monday morning, July 3, 1995

Opening Address: R. Mach, Director of INP, Řež

Chairperson: B.L. Berman
Speakers: D. Ernst (1), R.J. Peterson (17), J. Köhler (29)

Chairperson: T. Harada
Speakers: J.M. Nieves (36), B. Bassalleck (51), P. Bydžovský (61), J. Labarsouque (65)

Monday afternoon, July 3, 1995

Chairperson: S. Frullani
Speakers: O. Häusser (69), B.L. Berman (83), L. Canton (91)

Chairperson: E. Oset
Speakers: E. Friedman (97), S. Wycech (111), A. Cieplý (121)

Tuesday morning, July 4, 1995

Chairperson: S. Wycech
Speakers: J. Mareš (127), J. Dąbrowski (141), Y. Yamamoto (145), K. Miyagawa (150)

Chairperson: E. Friedman
Speakers: T. Motoba (495), T. Harada (155), Yu.A. Batusov (161), L. Majling (165)

Wednesday morning, July 5, 1995

Chairperson: R. Ceuleneer
Speakers: R.G.E. Timmermans (169), T. Ueda (177), J.L. Matthews (187)

Chairperson: J. Ahrens
Speakers: B.M.K. Nefkens (193), A. Švarc (203), L.Tiator (213)

---

[1] The number in brackets after the name of the speaker indicates the page of the proceedings where the contribution appears.
The following papers were not available for publication in the proceedings:
New Results from the NMS Spectrometer, by Ch. Morris;
Spin Observables in Medium Energy and Deep-Inelastic Electron Scattering, by J. van den Brand.

Wednesday afternoon, July 5, 1995

POSTER SESSION
S. Gmuca, S. Hirenzaki, D. Hüber, S.I. Kruglov, A.I. L'vov, H. Markum, J.L. Matthews, V.G. Nedorezov, Yu.L. Ratis, S. Shinmura, S. Simula, N.S. Topilskaya, S. Wycech, R. Yarmukhamedov

PARALLEL SESSION 1A
Chairperson: A. Švarc
Speakers: M. Batinić (219), J.-P. Didelez (223), V. Belyaev (227), F. Šimkovic (231), A. Krutenkova(237), M. Sadler(241)

PARALLEL SESSION 1B
Chairperson: E. Boschitz
Speakers: L. Mathelitsch (245), P. Cerello (249), V.P. Zavarzina (253), V. Kukulin (259), T. Wilbois (263), I. Gratch(267)

PARALLEL SESSION 2A
Chairperson: Y. Yamamoto
Speakers: D.E. Lanskoy (277), T. Yamada (281), E.Ya. Paryev (285), A. Parreño (293)

PARALLEL SESSION 2B
Chairperson: B.M.K. Nefkens
Speakers: S.S. Kamalov (297), R. Ceuleneer (303), N. Grion (307), J. Smejkal (311), J.R.M. Annand (315), P. Thörngren (319)

Thursday morning, July 6, 1995

Chairperson: A. Krutenkova
Speakers: V. Burkert (324), J. Ahrens (339), D. Moricciani (349)

Chairperson: L. Tiator
Speakers: R. Schumacher (355), T. Mart (369), S. Frullani (374), S. Shinmura (379)

Thursday afternoon, July 6, 1995

Chairperson: T. Ueda
Speakers: W. Glöckle (384), D. Hüber (399), A. Kievsky (405), R. Machleidt (410)

Chairperson: V. Kukulin
Speakers: J.W. Van Orden (415), W. Plessas (429), A.I. L'vov (439)

Friday morning, July 6, 1995

Chairperson: V. Belyaev
Speakers: Ch. Morris, U. van Kolck (444), L.V. Fil'kov (449), E. Oset (455)

Chairperson: J. Dąbrowski
Speakers: J. van den Brand, M. De Sanctis (461), S. Simula (466), Zh. Kurmanov (471)

Friday afternoon, July 6, 1995

Chairperson: B. Bassalleck
Speakers: C. Bennhold (475), A. Ramos (490)

CLOSING SESSION
Chairperson: W. Plessas
Speaker: R.J. Peterson (495)

Contents

Opening Address

Dear Colleagues, Dear Friends,

It is a great pleasure for me to welcome you on behalf of the Academy of Sciences of the Czech Republic, of the Institute of Nuclear Physics, and of the Organizing Committee to the sixth International Symposium **Mesons and Light Nuclei**.

Our symposium should help to summarize the recent progress reached in the three fields of nuclear intermediate energy physics, which are traditionally being discussed in this series of meetings.

Firstly, we shall learn about new development in the theory of the pion interaction with nuclei as well as we shall discuss some recent results obtained in the pion-nucleon scattering at low energies, in the physics of hadronic atoms, in the pion absorption, and in the pion double-charge reactions. A particular attention will be devoted to the pion interaction with few-body systems. The physics of eta-mesons will be covered at our symposium correspondingly.

The second big topic of our symposium is the hypernuclear physics. We shall deal with properties of hyperon-nucleon system, of strange hadronic systems, and with a large variety of nuclear systems containing strangeness. Strangeness production in reactions with different projectiles will be discused as well.

The third main topic of our symposium are the electro-weak interactions with nucleons and nuclei and manifestations of exchange currents. The problems to be discussed here are mostly associated with the physical program at CEBAF and at MAMI. New theoretical concepts will be presented based on the chiral models and the quark models.

In addition, some selected topics will be presented on the problems of nuclear structure, on the recent development in the physics of few-body systems, on the behaviour of two-pion systems, on polarization phenomena, etc.

Some discussions that start here will surely continue at the 8th Summer School on Intermediate Energy Physics organized by our younger colleagues as a satelite meeting of this Symposium. The School will be held in Prague since July 10 till July 15 and will mainly deal with the hadron dynamics at low and intermediate energies.

As most of you know, we intended to hold the Symposium at Třešt' castle which belongs to the Academy of Sciences. However, when the number of physicists wishing to attend the symposium exceeded the accommodation capacity of the castle, we had to solve the dilemma as to whether to introduce some regulations on the attendance or to change the venue of the symposium. We preferred, of course, the second option. Eventually, the DIAMO company offered us very favourable conditions. We appreciate it very much, and this is the reason why we hold the symposium here at Stráž.

I would like to mention that the activities of DIAMO company are in some sense associated with nuclear physics, since Diamo runs the uranium mines in the surroundings of Stráž. However, most of the mines have already been closed or will be closed soon. I think, it may be interesting for nuclear physicist to see how the uranium mines look like and how the mines affect the landscape.

One of our aims was to keep the costs of the symposium as low as possible and to make the attendance of the symposium affordable to students and to our young colleagues. We succeded in this effort also due to the contribution of the sponsors of the symposium. The sponsors are the Austrian Ministry for Science and Research, the Czech Ministry for Industry and Trade, and Škoda-Prague a.s.. Their contribution is highly appreciated.

It is a great honour for me to open this Symposium which, I hope, will be successful in fulfilling the aforementioned goals. We hope that this meeting will not be limited just to the scientific program but that it will contribute to a better understanding among the participants now as well as in the future. Our social program and the program for accompanying persons should help you to make an acquaintance with North Bohemia, the region where several chapters of the European history were written. I wish our Symposium great success. To our foreign guests, I extend once again warm welcome and hope you have a pleasant stay in our country.

R. Mach
Director of the Institute of Nuclear Physics,
Academy of Sciences of the Czech Republic

Few-Body Systems Suppl. 9, 1–16 (1995)

Few-Body Systems

Kaon, Pion, and Photon Interactions with the Nucleus Below 1 GeV

D. J. Ernst[1], M. F. Jiang[1], C. M. Chen[2], Mikkel B. Johnson[3]

[1] Department of Physics and Astronomy, Vanderbilt University, Nashville, TN 37235, USA

[2] Department of Physics, National Taiwan University, Taipei , Taiwan 10764, Republic of China

[3] Los Alamos National Laboratory, Los Alamos, NM 77845, USA

Abstract. Total cross sections for K^+ and high-energy pion scattering from nuclei are compared with contemporary momentum-space calculations. The first-order theory consistently under-predicts the measured cross sections indicating either substantial second-order corrections to the theory or a modification of the intrinsic properties of the nucleon by the nuclear medium. Medium broadening of the $D_{13}(1520)$ and the $F_{15}(1680)$ is determined phenomenologically from the total photo-reaction cross section and then used to examine the importance of this broadening for high-energy pion elastic scattering.

1 Introduction

Conventional nuclear physics assumes that the intrinsic properties of the nucleon remain unchanged when the nucleon is embedded in a nucleus. The success of many years of microscopic nuclear many-body theory precludes the possibility that the nucleon changes its properties by a large amount. However, there is not a clean demonstration that the nucleon does not change at the order of say 20%. The difficulties of a microscopic, parameter-free theory of the strong interaction prevent us from ruling out modifications at this level.

In addition to the nucleons already present in the nucleus, the same question can be asked of excited states of the nucleon when they are inserted into the nucleus. Do their masses and their widths change when they are embedded in nuclear matter? These questions have been studied for the nucleon and the $\Delta_{33}(1232)$, and, in the strange sector, for the $\Lambda(1115)$. For the Δ_{33}, a combination of medium effects [1-3] are required to reproduce pion- and photo-nucleus data. These include the binding of the nucleon, the mean-field acting on the Δ_{33}, Pauli blocking, Fermi averaging, pion true absorption, and correlation corrections. The total of these effects can be phenomenologically described [4] as shifts in the mass and width of the Δ_{33}. These modifications are all a result of the medium modifying the *effective* properties of the Δ_{33}. The question of whether the *intrinsic* structure of the hadron itself changes in the nuclear medium is more difficult. Only for the nucleon already existent in the nucleus do we have some hope of addressing this question.

In exploring the strong interaction, a variety of probes and reactions must be used. The theorist is always plagued by a lack of knowledge of the underlying strong interaction and by an ability to calculate without approximation only the lowest order of a theory whose higher-order corrections may not be totally understood. The situation is not different for electron or photon induced reactions. Even though we may know well the initial interaction, the question being addressed is inherently a question of the strong interaction.

This does not mean that the situation is impossible. It does mean that we must carefully choose those reactions and those phenomena which are least sensitive to a lack of knowledge and to theoretical limitations. Only when a variety of data can be brought into one coherent picture should we begin to believe that we have arrived at a new and basic understanding.

Elastic electron scattering is not the best candidate for studying modifications of the nucleon's properties when it is in the nuclear medium. The long-range character of the Coulomb interaction implies that one is measuring a folding of the proton charge distribution with the charge distribution of the nucleus. Small changes in the proton charge distribution would be masked by the dominant dependence on the nuclear charge distribution.

The best situation would be a short-ranged and weakly (in the sense of 'strength' not 'type') interacting projectile. In the limit of a very weak interaction, the total cross section on a nucleus would become just A times the spin-isospin averaged two-body total cross section. Deviations would imply a modification of the two-body interaction by the nuclear medium. The closest nature comes to this ideal situation is the K^+. The K^+–nucleon total cross section is in the 10 to 20 mb range. Pions above the Δ_{33} are the second weakest of the strongly interacting particles. Although there are broad resonance peaks, the total interaction is relatively independent of energy and has a typical total cross section in the range of 20 to 30 mb.

Working with a probe where the meson-nucleus total cross section is near A times the two-body cross section has a great advantage for the theorist. In this region, the second-order corrections, which are proportional to the square of the two-body amplitude, are small. Since the higher-order corrections are the largest uncertainty in the theory, it is best to keep them as small as possible. In addition, the second-order corrections contain an additional propagator so that going to high energies will also help to minimize the theoretical uncertainties. A simple estimate of the ratio of the second-order optical potential to the first-order is given by [5]

$$R = \frac{U^{(2)}}{U^{(1)}} = \sqrt{\sigma}\frac{\ell_c}{k}\rho \,, \tag{1}$$

where σ is the total two-body cross section, ℓ_c is the correlation length, and ρ is the density at which the projectile interacts. The factor $1/k$, where k is the incident pion momentum, arises from the extra propagator in the second-order potential. For pions on resonance, we find $R \approx 0.1$. R is less than one only because the density at which the interaction is taking place is small. For K^+, $R \approx .01$, and for pions at 500 MeV, $R \approx .04$, and at 1 GeV, $R \approx .02$. We thus expect second-order correlation corrections to be only a few percent. For other corrections, one can substitute the appropriate two-body cross section σ and the appropriate length scale ℓ to achieve a similar estimate.

Another advantage of a weak two-body interaction is that this will allow the projectile to penetrate further into the nucleus. Because the modifications to the nucleon should increase with nuclear density, it is important that the projectile see nucleons in a region where the density is substantial. A simple estimate of how far a projectile penetrates into the nucleus can be found from the arguments of [6]. This argument states that the radius of deepest penetration is equal to the impact parameter where the profile function of the target just equals one mean free path for the projectile. In [7] it is shown that the K^+ can just penetrate to the center of ^{40}Ca and to the half density point of ^{208}Pb. The pion at 700 MeV makes it to the center of ^{12}C but only to the half density point of ^{40}Ca.

Thus the high-energy pion and the K^+ are the best possible probes among the strongly interacting particles to investigate the possibility that the nucleon is modified in a fundamental way when it finds itself in the nuclear medium. To examine modifications of the properties of excited hadrons in the nuclear medium, the projectile must create these hadrons in the nucleus. These excited hadrons were discovered by their production in the pion-nucleon reaction. Thus high-energy pions are again a logical choice for the projectile. Electrons and photons can also be used.

In the next section, we review the properties of the momentum-space optical potential used to investigate meson-nucleus elastic scattering. The subsequent section contains recent results for K^+-nucleus scattering. Following that section, we examine pion and photon reactions where we must simultaneously address the questions of modifications of the nucleon in the target and modifications of the excited hadron created by the interaction. We end with a discussion of future work.

2 Formalism

To work with pions and kaons up to an incident energy of 1 GeV requires a covariant multiple scattering theory. A review of the formalism used can be found in [2]. Some important features include

1. Covariant kinematics. In order to incorporate covariant kinematics into a momentum-space multiple scattering theory requires only that one knows how to transform between reference frames. The Lorentz boost acting on a plane wave state is well defined. This observation combined with the

observation that the kinematic structure of the impulse approximation to the optical potential is that of a three-body problem (the projectile, the struck nucleon, and the A—1 residual nucleons) allows a straightforward construction of the necessary kinematics. In the calculations presented, we ignore the fact that the nucleons may be off their mass shell in treating the kinematic aspects of the theory. We have found that including such effects as shifting masses by binding energies has a totally negligible effect on the kinematics. Similarly, the Wigner precession of the nucleon spin, which is included in the theory, has no practical effect on our results. This approach to covariant kinematics can be generalized [8] to particles of arbitrary spin.

2. Computational technique. The computational technique employed makes use of the three-body structure of the impulse approximation. Two bases are defined. The first starts with the projectile-nucleon plane waves and boosts to a frame where the total three-momentum is zero. This defines the relative momentum between the projectile and the nucleon as the momentum of the projectile in this frame. The projectile-nucleon scattering amplitude is naturally expressed in this frame. The summed momentum of the projectile and nucleon is then combined with the momentum of the A—1 residual nucleus and a boost is performed to the frame where the total momentum of the system is zero, where the calculation is then performed. The second basis is generated by starting with the nucleon and the A—1 residual nucleons and boosting to a frame where the total three-momentum is zero. The momentum of the nucleon in this frame is then an appropriate definition of the relative momentum between the nucleon and the A—1 residual nucleus. This serves as the variable for the covariant definition of a wave function. The summed momentum of the nucleon and the A—1 residual nucleons is then combined with the momentum of the incident pion and a boost is performed to the frame where the total momentum of the system is zero. The two different orders of combining the momenta provide two distinct bases. If the plane wave states are angular momentum decomposed the overlap between the two bases are called [9] 'relativistic three-body recoupling coefficients.' These relativistic three-body recoupling coefficients lead to natural variables in which to perform the Fermi-averaging integral. What is termed 'Fermi-averaging integral' is also called 'full-folding' and in coordinate-space is called 'delta propagation.' In the resulting variables, the integrand falls smoothly and rapidly to zero, which allows the use of relatively few integration points. We find that it is important to perform this integration without approximation. Particularly for pions near the Δ_{33} resonance, the scale for the energy dependence of the pion-nucleon interaction is set by the half width of the resonance, 55 MeV. This scale is similar to typical nuclear target scales, i.e. is similar to the nuclear binding energy, the nuclear potential, or the average kinetic energy of a nucleon. The best approximation to the Fermi averaging integral is [10] the 'optimally fac-

torized approximation.' This approximation was originally proposed for the pion-nucleus problem, but it did not prove to be as useful [11] there as it has become [12] for the nucleon-nucleus problem.

3. Invariant amplitudes. As we are interested in mesons that will have kinetic energies much larger than their rest mass, it is important that one use invariant normalizations and phase-space factors. In order to incorporate invariant phase space factors and the nonlocality of the optical potential, it is necessary to work in momentum space. We also use the invariant amplitude [13] that is free of kinematic singularities. This guarantees that we do not mistakenly confuse off-shell behavior with implicit kinematic factors.

4. Two-body amplitude. For pions, the off-shell pion-nucleon t-matrix is taken from simple dynamic models. It was realized [14] quite early that even below inelastic threshold the effects of the virtual pion-production channels are important. Techniques were developed [14, 15] that allow the effects of the coupling of the inelastic channels to the elastic channels to be included in the models without explicit calculations of the inelastic channels. The approach uses dispersion theory and takes the ratio of the elastic to the total cross section in a spin-isospin channel from phase shift analyses and uses this as input to predict the phase shift in each channel. The method has been developed for separable potentials, the Chew-Low model, the Lee model, and for [16] models that contain a sum of terms for the basic interaction.

5. Formalism. The optical potential is defined [2] as the leading term in a systematic expansion. The diagrammatic approach used to derive the expansion is valid independently of the character of the underlying two-body interaction. Unlike conventional derivations, the formalism does not rely on a potential interaction between particles. The expansion is unconventional in that it treats Pauli corrections perturbatively. There are several advantages to this. First, the Pauli corrections when treated this way have the same structure as the other higher-order corrections. For pions, they are also necessarily of the opposite sign to the true absorption corrections. This can be seen from unitarity as the Pauli corrections are removing an extra channel included in the calculation but not in nature; true absorption is adding in a physical channel. Treating the Pauli correction perturbatively order by order has explicitly been shown [17] to be a convergent approach for pions near and below the resonance region. By not having a first-order optical potential defined in terms of a Pauli-blocked two-body t-matrix, we are able [13] to calculate the first-order optical potential and the resulting scattering without approximations.

6. Resonance-nucleus interaction. For the pion interaction in the resonance region, the interaction of the Δ_{33} with the remaining nucleons cannot be neglected. To incorporate a shell-model potential acting on the Δ_{33} would

require the solution of a three-body problem in the middle of constructing the optical potential. If, however, one expands [18] the potential out of the denominator and utilizes the fact that the mass of the nucleon is much larger than that of the pion, the correction for the potential is given by a matrix element which is target and energy dependent. We then insert this constant, called the 'mean-spectral energy,' back into the denominator to approximately sum the higher-order corrections. The K^+-nucleon amplitude, although weak, is a steadily rising function of energy. We have found [19] that the effect of the interaction of the intermediate nucleon and K^+ with the residual A—1 nucleons is not negligible.

The above ingredients have all been incorporated into a momentum-space optical model computer code [13] called ROMPIN for pions and ROMKAN for K^+. These codes are generally available for those who might wish to use them.

3 Results for Kaons

As noted in the introduction, the K^+ is the weakest of the strongly interacting particles. The weak two-body interaction makes it the most penetrating of the strongly interacting particles and makes the theoretical results the most reliable. There is, however, a limited amount of K^+-nucleus data. Elastic data [20] led to the first indications [21, 22] that theories would consistently under-predict the data. However, there is a 17% systematic error in the normalization of the data and a discrepancy of about twenty to 30% between experiment and theory. The short lifetime of the K^+ makes it very difficult to ascertain the absolute normalization of data. Ratios of cross sections are thus much more reliable. The total cross section for a nucleus divided by the total cross section for the deteuron is such a measurement. In Figs. 1 and 2, we present results for K^+ scattering from a variety of nuclei. The data are from ref. [23]. Presented is the ratio

$$r = \frac{\sigma_t(A)/A}{\sigma_t(D)/2} \; . \tag{2}$$

We see that there is an energy independent and nearly target independent discrepancy between the theory and the data. The discrepancy found is almost independent of the details of the theoretical calculation. Indeed, a calculation [24] based on the Kemmer-Duffin-Petiau equation also produces rather similar results. This can be understood by noticing that a simple eikonal model [19] produces results that are close to the full momentum space calculation. This implies that off-shell effects, the Fermi integration, and other features of the momentum-space calculation are not overwhelmingly important. This is supported by noting that the ratios of cross sections in Figs. 1 and 2 are all close to one. What the theorist is calculating is actually the difference from one. A constraint on a reasonable theoretical calculation is that the Born approximation to the optical potential gives for the total cross section nearly A times the spin-isopin averaged two-body total cross section (in the absence of

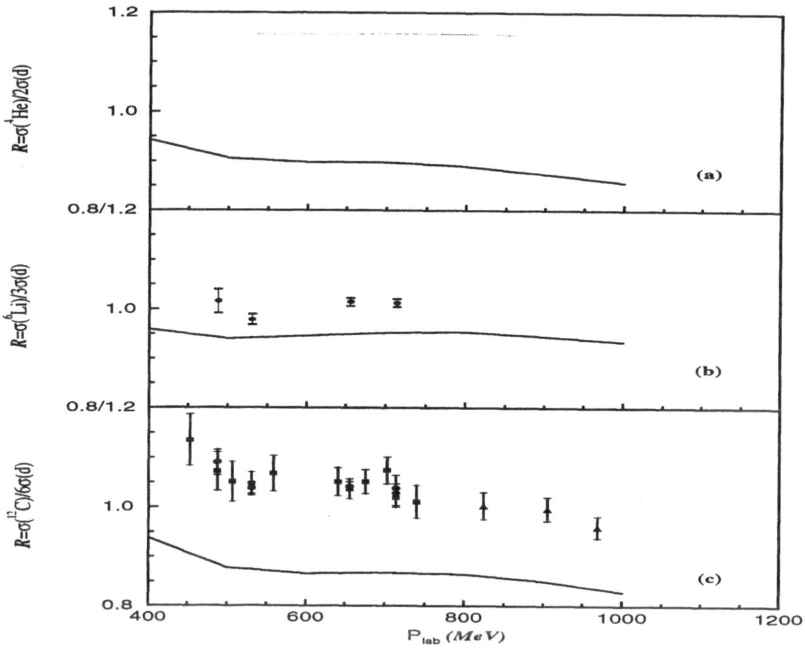

Figure 1. The total cross section ratio defined in Eq. (2) for ^{4}He, ^{6}Li, and ^{12}C as a function of the laboratory momentum of the K^+, P_{lab}. The data are from [23] and the curves are the results of the first-order optical potential utilizing ROMKAN.

Coulomb-nuclear interference). This is trivially true for the eikonal model. For the momentum-space approach the Born approximation to the optical potential contains a Fermi-averaging over the two-body total cross section. If the Fermi averaging is approximated [10] according to the 'optimal factorization' prescription, then the momentum-space results yield, in Born approximation, A times the two-body cross section evaluated at a small energy shift given by the difference between the average binding energy of the nucleons and the mean spectral energy. Given that the theories are only calculating these small differences from A times the two-body cross section and the shadowing that results from iterating the optical potential in an integral equation (only about a 5% effect), the agreement among theorists is not surprising.

The discrepancy which we find is about 20%. This may be viewed as caused by a modification of the nucleon when in the nuclear medium. Total cross section measurements cannot discriminate among models [22] which would simply increase the two-body scattering amplitude in the medium in an energy independent way.

Mesons exchange currents are a possible mechanism that could account for this discrepancy. The results of [25] would indicate that this is not the case. There it was found that exchange currents are both too small and would have a rapid energy dependence as the energy is increased above the threshold for pion

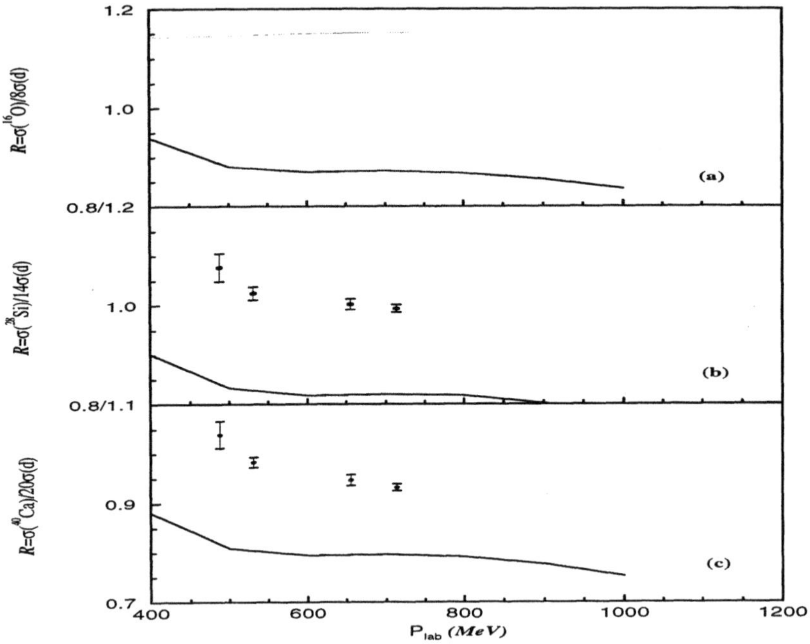

Figure 2. The same as Fig. 1 except the targets are ^{16}O, ^{28}Si, and ^{40}Ca.

production. Enhanced mesonic clouds resulting from reduced mesonic masses in the nucleus [26] would increase this contribution. In addition, a calculation which removes the static approximation in calculating the exchange currents and adds scattering from short and mid range correlations [27] yields a much larger effect, an effect that is of the order of the discrepancy between the theory and the data. Further investigation is needed to understand if the approach of [27] can produce results which are quantitatively in agreement with the energy and target dependence of the data.

Elastic differential cross sections would provide additional information on the origin of the observed discrepancy. It was the differential cross sections measured in [20] that lead to the original observation [21, 22] of a possible discrepancy. However, the systematic error in the data is given as 17% and the discrepancy [19] is nearly independent of angle. The possibility of a systematic normalization error somewhat larger than that quoted would resolve the problem. Thus, this data did not provide a clear and convincing argument for a discrepancy. The measurement of elastic differential cross sections with an absolute normalization of 5% would be most useful. However, the short lifetime of the K^+ has so far prevented measurements with absolute normalizations at this level.

Recent data [28] takes an interesting approach to this problem. The rms radius of ^{6}Li and ^{12}C are nearly equal, with ^{6}Li having the slightly larger radius. Roughly ^{12}C is twice as dense as ^{6}Li. In the forward direction the details of the density should make little difference. Thus one can take the ratio of elastic differential cross sections between these two nuclei and in the forward direction one would be examining the density dependence of the mechanism that is causing the discrepancy. In Fig. 1 we see that the less dense ^{6}Li appears to produce a smaller discrepancy for the total cross section than is seen in other more dense nuclei. The measurements of [28] confirm this observation. However, the theorists [19, 24] calculate ^{6}Li as if it were a spin zero target; it needs to be verified whether the incorrect treatment of the spin-flip part of the scattering in these calculations is causing a significant error in the theoretically predicted ratios of differential cross sections.

4 Results for Photons and Pions

As noted in the introduction, pions with kinetic energies above 400 MeV are the second weakest strongly interacting particle. Their interaction with the nucleon is predominantly through broad resonances. Thus their interaction with a nucleus necessarily involves two questions. One is whether they can be used as a complimentary probe to the K^+ for studying the possible modifications of the nucleon properties when it is in the nuclear medium. The second is can we study medium modifications of the resonances formed in the interaction?

Inclusive measurements with energetic photons [29, 30] can also be used to study the baryon resonances in nuclei. These photo-nuclear measurements find that the prominent peaks for the $D_{13}(1520)$ and $F_{15}(1680)$ resonances, present for the free nucleon, have essentially disappeared for the finite nucleus target. This result was interpreted in ref. [29] as the combined effect of Fermi averaging plus additional collision broadening that occurs when the baryon resonances propagate in nuclear matter. A somewhat more detailed model can be found in ref. [30]. In contrast to the photo-nuclear measurements, similar measurements with exclusive electron scattering [32] find prominent peaks for these two resonances with a nuclear target. The fate of the more massive resonances when embedded in a nucleus is thus not yet clear.

These resonances can also be produced by pions. A combined treatment of pion and photon scattering could elucidate the behavior of these resonances in the nucleus. As the coupling of the photon and the pion to the resonances differ, the details that govern the dynamics of photon and pion induced reactions will also differ. However, the same medium modifications of the resonances will be present in both cases and with sufficient data and theoretical thought a consistent picture could be developed.

We will assume that the pion-nucleus optical potential can be expressed in terms of a medium-modified pion-nucleon t-matrix. The free-space t-matrix is divided into resonance and background parts,

$$t = t_{\mathrm{nr}} + \sum t_{\mathrm{res},j}\,, \tag{3}$$

where t_{nr} is the nonresonant background amplitude and $t_{\mathrm{res},j}$ is the resonant amplitude for resonance j. We write the resonant amplitude in a Breit-Wigner form,

$$t_{\mathrm{res},j} = \frac{g^2_{\pi NN^*,j}}{(k+p)^2 - M_j^2 + \mathrm{i}M_j\Gamma_j} \tag{4}$$

characterized by a free-space elastic width (proportional to $g_{\pi NN^*,j}$), a free-space mass M_j, and a free-space total width Γ_j. (The free-space mass of the Δ_{33} is shifted [29] by 15 MeV to help produce a better fit to the data utilizing an energy-independent width). Here k is the pion and p is the nucleon four momentum.

In the model of Kondratyuk et al. [29], the resonant amplitude for forward Compton scattering on a nucleus is taken to be of the form of Eq. (3) with $g_{\pi NN^*,j} \rightarrow g_{\gamma NN^*,j}$. The resonance modifications in the nuclear medium are determined there from the total photon cross section on ^{238}U. To do this, the amplitude is first averaged over the motion of the nucleons and the free width is decreased by a Pauli blocking factor B_F. The masses of the resonances are left at their free values because the photon data do not exhibit any visible peaks from which they could be determined. The total free width of each resonance is then increased by adding a collision broadening contribution Γ^* until a fit to the photon total cross section for ^{238}U is obtained. The widths for eight resonances were modified in this fashion.

In [31] the model of ref. [29] was simplified by making use of the observation that the photo-nuclear data sensitively determines only the medium modifications of the the Δ_{33}, the $D_{13}(1520)$, and the $F_{15}(1680)$. In Fig. 3 we compare the original model [29] with all eight resonances altered by the medium (solid line) to our version in which just these three resonances are modified (dash line). The changes caused by the medium for the three-resonance model are exactly those prescribed in ref. [29]. The difference between the models is quite small, justifying our choice.

We next apply this same model to pion-nucleus total cross sections. We take the resonance medium modification from our version of the results of ref. [29]. We first adjust $g_{\pi NN^*,j}$ to fit the free pion-nucleon elastic amplitudes taken from [34]. However, since the energy-independent widths that are used in ref. [29] do not reproduce exactly the measured pion-nucleon total cross section, we add to the background amplitude in Eq. (3) the difference between the amplitude as determined from [34] and the amplitude that results from the sum of the three parameterized resonances. Our procedure of including the difference between these two curves as part of the background amplitude produces a model that, in the absence of any medium modifications, makes no error due to the lack of an exact fit to the two-body amplitude in the resonant channels. We extend the resonant channel amplitudes off the energy shell by utilizing a separable form. We use Gaussian functions for the form factor $v(\kappa)$ with a range of 1 GeV. We find that the predicted total cross sections for pion-nucleus scattering do not depend on the functional form chosen for the form factor nor do they depend on the range chosen for the form factor.

The pion-nucleus total cross sections is calculated utilizing the momentum-

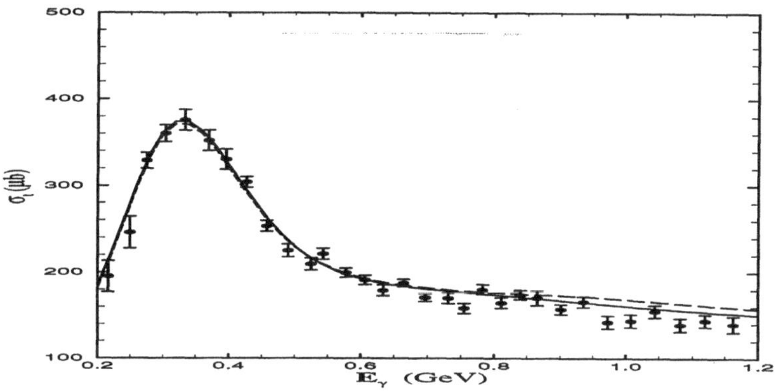

Figure 3. The total cross section for photon-^{238}U scattering as a function of photon energy. The data are from Kondratyuk et al. [29]. The solid line is the result of the model when eight resonances are assumed to have their widths modified in the medium; the dashed line assumes only three resonances have medium modified widths.

space optical potential approach developed in [13]. The theoretical calculations are presented in Fig. 4 for π^+ on ^{12}C. The medium-dash line is the result of the calculation in the absence of collision broadening. For purposes of comparison, the result for twelve free nucleons, $6(\sigma(\pi^+p)+\sigma(\pi^+n))$ is shown as the dot-dash line. We see that the combined effect of Fermi averaging (performed exactly by the computer code ROMPIN [13]) and multiple scattering decreases the cross section and, to a great extent, remove the bumps visible in the dot-dash line. The effect of Fermi-averaging alone is seen in the Born approximation to the momentum-space result, given as the short-dash curve in Fig. 4. By comparing this to the dot-dash curve, we see that the Fermi motion broadens the resonances and significantly smooths the energy-dependence of the cross section; effects of multiple scattering change the results from the short-dash curve to the medium-dash curve. Note that the momentum-space calculation falls below the data [35] by about 20%. A qualitatively similar discrepancy has been seen [1, 36, 37] in the elastic angular distribution [38] for 800 MeV/c pions on ^{12}C. A mechanism that uniformly increases the optical potential over the energy range examined will be able to approximately fit all the data that currently exists.

We now add the collision broadening of the three dominant resonances as determined from the photon total cross section. The physical origin of the broadening of the resonances is presumably the opening of additional decay channels when the resonance is created in the medium. Pion production and absorption are two examples of channels that can contribute to the decay of the resonances in the medium. Another possibility is that the heavier resonances, once produced in the nucleus, immediately decay because they find themselves in direct contact with many nucleons due to the increased radial size of these resonances [39].

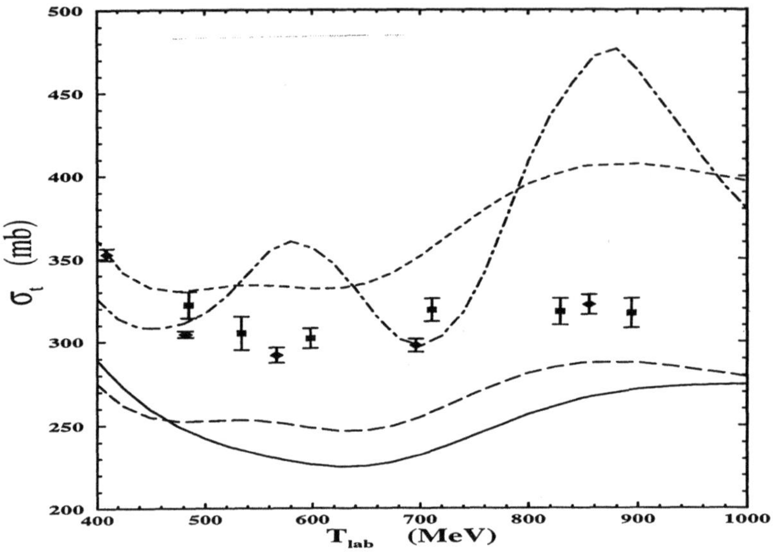

Figure 4. The total cross section for pion-^{12}C scattering as a function of pion kinetic energy. The data are from Crozon et al. and Clough et al. [35]. The dot-dash curve is the result of taking 12 times the spin-isospin average of the free two-body amplitude; the short-dash curve is the result of Fermi-averaging the free amplitude (the Born approximation to the optical potential); the medium-dash curve is the result of solving the optical potential in Klein-Gordon equation; the solid curve adds the additional collision broadening as determined in Kondratyuk et al. [29].

The result of collision broadening in pion scattering, as determined from the photo-nuclear data [29], is given as the solid curve in Fig. 4. We see that although the increase in widths has a noticeable effect on the predicted cross sections, the result is strikingly small and further reduces the theoretical predicted cross sections. Changes in the masses of the resonances will have very little additional effect since the resonances in our final model are quite broad. The most interesting feature is not the effects of the medium modifications *per se*, but rather the large discrepancy that stands out in comparing the solid curve in Fig. 4 with the data [35].

The discrepancy between the data and the complete theory including the collision broadening (solid curve) is about 20% and is approximately energy independent. If we go to the opposite extreme and neglect the additional broadening that we have taken from the fit to the photon total cross section [29], we still find a discrepancy of roughly 15%. Such a discrepancy could indicate a failure to include some important piece of the reaction mechanism. However, we expect that the reaction channels, such as true absorption and pion production, are dominated in this energy region by the resonances themselves, so that these effects are included automatically in our phenomenological description of collision broadening taken from photo-nuclear data.

A simple and possible explanation is that the coupling constant of the pion to excite resonances in the nucleus is modified in the medium. To simulate this, we find that a phenomenological increase of the pion-nucleon amplitude of 42% will reproduce the data. In a doorway model where the inelastic channels are coupled to the elastic channel only through the resonance channel, an increase in amplitude would be proportional to an increase in the square of the coupling constant. This model would then correspond to a 20% increase in all coupling constants, including those in the nonresonant channels. If we neglect the broadening of the resonances taken from the photon data, we find the data indicate a 30% increase in the in-medium amplitudes or a 15% increase in the all coupling constants.

The discrepancy found here is similar to the results for K^+ induced reactions [21, 19] presented in the previous section. There the theoretical cross sections are consistently, and of the order of 20% below the data. A possible explanation for the K^+ data [26] also could be an increase in the coupling of mesons to the nucleon. The role of two-body correlations, meson exchange currents, and true-absorption, although we believe them to be small for the pion-nucleus data, deserve additional investigation.

Although we have argued that reaction channels should be largely accounted for by the collision broadening, we have estimated what the effect of true absorption of the pion would be in the quasi-deuteron model [37] assuming that it is not resonance dominated. Our result is that it would contribute at most 5% to the total cross section, too small to account for the observed problem. Further investigations are needed to establish the actual contribution of such corrections to the observed discrepancy. A more dynamically motivated model for the photo-induced reaction is clearly needed to more cleanly separate the medium-modified resonance propagators from other effects such as the an enhanced photon-induced background [3]; such a model must, of course, maintain the established picture of the Δ_{33} that has come from microscopic models such as those in refs. [1, 2]. An enhanced photo-nuclear background would require less broadening, and the coupling constant enhancement we find would then lie between the 15% and 20% indicated above. In any case, we expect that the collision broadening would tend to *increase* the renormalization we have here found for energies above the Δ_{33}.

From a theoretical point of view, the modification of baryon properties in nuclear matter is an interesting and fundamental problem. One of the most promising developments in nuclear physics in recent years has been the realization that hadrons can be studied using methods of nonperturbative QCD [40], where masses and coupling constants can be related to the quark and gluon condensates [41]. Hadronic masses in nuclear matter have also been studied [42]. In the pioneering work of Drukarev and Levin [42], an estimate of the quark condensate in nuclear matter was given. Some of these condensates, particularly the four-quark condensates, are poorly known. The properties of the Δ_{33} in the nucleus are a promising source of information on these [43] condensates. Other sources of information, for example from higher-lying hadrons such as those discussed above, would be of great importance.

5 Conclusions

We have found that the first-order optical potential consistently under-predicts experimentally measured total cross sections for both K^+ and high-energy pions. A possible explanation could be found in [27] meson exchange currents and correlations or in modifications of the nucleon [22] by the nuclear medium.

A few new measurements of pion elastic scattering that could be of help are presently being analyzed at LAMPF and KEK. A number of additional measurements that would clarify the situation suggest themselves: (1) total and total reaction cross sections for pions on other nuclei to see whether the disagreement we find for carbon is a universal phenomenon; (2) measurement of elastic differential cross sections at sufficiently finely spaced intervals to identify structure in energy variation of the minima of the angular distributions; (3) measurement of single charge exchange (including (π, η) reactions), which would help separate isospin dependent effects. Other measurements, for example partial reaction cross sections would be very helpful for understanding the details of the reaction mechanism. Such measurements could be made in the low-energy region (pion kinetic energies up to 575 MeV) at LAMPF and at higher energies at KEK and at the AGS at Brookhaven.

We have also examined the effects of resonance broadening on high-energy pion-nucleus scattering. Although we find a visible effect, total cross sections are not the best measurement to investigate this phenomenon. Further work is required to understand which reaction would be the most sensitive and cleanest probe of the excited hadrons in the nuclear medium.

This work was supported, in part, by the US Department of Energy under Contract No. DE-FG05-87ER40376 and the National Science Council of the R.O.C.

References

1. C. M. Chen, D. J. Ernst, and M. B. Johnson: Phys. Rev. **C47**, R9 (1993)

2. M. B. Johnson and D. J. Ernst: Ann. Phys. (N.Y.) **219**, 266 (1992); Phys. Rev **C27**, 790 (1983)

3. R. C. Carrasco and E. Oset: Nucl. Phys. **A536**, 445 (1992)

4. L.S. Kisslinger and W. Wang: Phys. Rev. Lett. **30**, 1071 (1973); Ann. Phys. (N.Y.) **99**, 374 (1976)

5. D. J. Ernst, J. T. Londergan, G. A. Miller, and R. M. Thaler: Phys. Rev. **C16**, 537 (1977)

6. D. J. Ernst: Phys. Rev. **C19**, 896 (1979)

7. C. M. Chen, D. J. Ernst, and M. B. Johnson: Phys. Rev. **C48**, 841 (1993)

8. D. V. Ahluwalia and D. J. Ernst: Phys. Rev. **C45**, 3010 (1992)

9. D. R. Giebink: Phys. Rev. **C32**, 502 (1985); **C25**, 2133 (1982)

10. D. J. Ernst and G. A. Miller: Phys. Rev. **C21**, 1472 (1980); D. L. Weiss and D. J. Ernst: Phys. Rev. **C26**, 605 (1982)

11. D. J. Ernst, G. A. Miller, and D. L. Weiss: Phys. Rev. **C27**, 2733 (1983)

12. A. Picklesimer, P. C. Tandy, R. M. Thaler, and D. H. Wolfe: Phys. Rev. **C30**, 2225 (1984)

13. D. R. Giebink and D. J. Ernst: Comput. Phys. Commun. **48**, 407 (1988)

14. D. J. Ernst, J. T. Londergan, E. J. Moniz, and R. M. Thaler: Phys. Rev. **C10**, 1708 (1974); J. T. Londergan, K. W. McVoy, and E. J. Moniz: Ann. Phys. (N.Y.) **86**, 147 (1974); C. Corinis and R. H. Landau: Phys. Rev. **C24**, 605 (1981)

15. C. B. Dover, D. J. Ernst, R. A. Friedenberg, and R. M. Thaler: Phys. Rev. Lett. **33**, 728 (1974); K. K. Bajaj and Y. Nogami: *ibid.* **34**, 701 (1975); D. J. Ernst and M. B. Johnson: Phys. Rev. **C17**, 247 (1978); **C22**, 651 (1980); D. J. Ernst, G. E. Parnell, and C. Assad: Nucl. Phys. **A518**, 658 (1990); R. J. McLeod and D. J. Ernst: Phys. Rev. **C49**,1087 (1994)

16. R. J. McLeod and D. J. Ernst: Phys. Rev. **C29**, 906 (1984)

17. H. C. Chiang and M. B. Johnson: Phys. Rev. **C32**, 531 (1985)

18. D. J. Ernst and M. B. Johnson: Phys. Lett. **B237**, 164 (1990); Phys. Rev. **C32**, 940 (1985); D. J. Ernst, D. R. Giebink, and M. B. Johnson: Phys. Rev. **C32**, 940 (1985)

19. C. M. Chen and D. J.Ernst: Phys. Rev. **C45**, 2011 (1992); M. F. Jiang and D. J. Ernst: *ibid.* **C51**, 1037 (1995); M. F. Jiang, D. J. Ernst, and C. M. Chen: *ibid.* **C51**, 857 (1995)

20. D. Marlow et al.: Phys. Rev. **C25**, 2619 (1982)

21. M. J. Páez and R. H. Landau: Phys. Rev. **C24**, 1120 (1981)

22. P. B. Siegel, W. B. Kaufman, and W. R. Gibbs: Phys. Rev. **C30**, 1256 (1984); **C31**, 2184 (1985)

23. E. Mardor et al.: Phys. Rev. Lett. **65**, 2110 (1990); R. A. Krauss et al.: Phys. Rev. **C46**, 655 (1992); Sawafta et al.: Phys. Lett. **B307**, 293 (1993); R. Weiss et al.: Phys. Rev. **C49**, 2569 (1994)

24. B. C. Clark, S. Hama, G. R. Kälbermann, R. L. Mercer, and L. Ray: Phys. Rev. Lett. **55**, 592 (1985); L. Kurth and B. C. Clark: private communication

25. M. F. Jiang and D. S. Koltun: Phys. Rev. **C46**, 2462 (1992)

26. G. E. Brown, C. B. Dover, P. B. Siegel, and W. Weise: Phys. Rev. Lett. **26**, 2723 (1988)

27. C. García-Recio, J. Nieves, and E. Oset: Phys. Rev. **C51**, 237 (1995); U. G. Meissner, E. Oset, and A. Pich: preprint

28. R. Micjael et al.: submitted to Phys. Rev. C

29. L. A. Kondratyuk, M. I. Krivoruchenko, N. Bianchi, E. De Sanctis, and V. Muccifora: Nucl. Phys. **A579**, 453 (1994)

30. W. A. Alberico, G. Gervino, and A. Lavagno: Phys. Lett. **B321**, 177 (1994)

31. C. M. Chen, D. J. Ernst, M. F. Jiang, and M. B. Johnson: Phys. Rev. **C52** R485 (1995)

32. L. Elouadrhiri et al.: Phys. Rev. **C50**, R2266 (1994)

33. M. B. Johnson and H. A. Bethe: Comm. Nucl. Part. Phys. **8**, 75 (1978)

34. Program SAID (*Scattering Analysis Interactive Dial-in*), R. A. Arndt, Virginia Polytechnic Institute and State University

35. M. Crozon et al.: Nucl. Phys. **64**, 567 (1965); A. S. Clough: Nucl. Phys. **B76**, 15 (1974)

36. E. Oset and D. Strottman: Phys. Rev. **C44**, 468 (1991)

37. M. Arima, K. Masutani, and R. Seki: Phys. Rev. **C44**, 415 (1991); erratum *ibid.* **C48**, 2541 (1995)

38. D. Marlow et al.: Phys. Rev. **C30**, 1662 (1984)

39. N. Isgur and G. Karl: Phys. Rev. **D18**, 4178 (1978); **D19**, 2653 (1979); **D23**, 817 (1981)

40. M. A. Shifman, A. I. Vainshtein, and V. I. Zakharov: Nucl. Phys. **B147**, 385 (1979); 448 (1979)

41. V. M. Belyaev and B. L. Ioffe: Sov. Phys. JETP **56**, 493 (1982); L. J. Reinders, H. R. Rubinstein, and S. Yazaki: Nucl. Phys. **B213**, 109 (1983)

42. E. G. Drukarev and E. M. Levin: Pis'ma Zh. Eksp. Teor. Fiz. **48**, 307 (1988) [JETP Lett. **48**, 338 (1988); Nucl. Phys. **A511**, 679 (1990); Prog. Part. Nucl. Phys. **27**, 77 (1991); T. D. Cohen, R. J. Furnstahl and D. K. Griegel: Phys. Rev. Lett. **67**, 961 (1991); R. J. Furnstahl, D. K. Griegel and T. D. Cohen: Phys. Rev. **C46**, 1507 (1992); X. Jin, T. D. Cohen, R. J. Furnstahl and D. K. Griegel: *ibid.* **C47**, 2882 (1993); E. M. Henley and J. Pasupathy: Nucl. Phys. **A556**, 467 (1993)

43. M. B. Johnson and L. S. Kisslinger: Phys. Rev. **C52**, 1022 (1995)

Few-Body Systems Suppl. 9, 17–28 (1995)

Few-Body Systems

Pion-Nucleus Total and Reaction Cross Sections

R. J. Peterson*

Nuclear Physics Laboratory, University of Colorado, Boulder, CO 80309-0446, USA

Abstract. Data, including very recent results at low beam energies, for pion total and total reaction cross sections on complex nuclei are reviewed. The emphasis at low energies is on the coherent effects possible at long projectile wave lengths and accessible because of the long mean free path expected. At high beam energies the question of interest is the existence of nucleon isobars within complex nuclei. Comparison of existing pion data to eikonal calculations and with a comparison to recent analyses of photon cross sections agrees with the conclusions from the photon data that any N^* modes other than the delta are severely damped in a nucleus as heavy as carbon.

1 Introduction

Total and total reaction cross sections of projectiles on complex nuclei can be measured well and provide data of great value for reaction models. The special features of pion and photon beams are of particular interest since these are fields that can be absorbed, as well as scattered by the target nuclei. In terms of the partial wave scattering amplitudes η_ℓ for scattering by the nuclear forces these cross sections can be expressed as

$$\sigma_{\rm R} = \frac{\pi}{k^2}\Sigma(2\ell+1)[1-|\eta_\ell|^2] \tag{1}$$

$$\sigma_{\rm T} = \frac{2\pi}{k^2}\Sigma(2\ell+1)[1-{\rm Re}\eta_\ell]. \tag{2}$$

Coulomb effects for the scattering of a charged beam do not appear in these expressions. In the limit of a black disk of radius R, the reaction cross section is just πR^2 and the total cross section is $2\pi R^2$.

Pion reactions divide into three energy regions of interest for different reasons. Near 180 MeV beam energy, the prominent P_{33} resonance is observed in heavy nuclei as both total [1] and reaction [2] cross sections, prominent as a

**E-mail address:* peterson@spectr.colorado.edu

maximum for light nuclei but less evident for very heavy nuclei. Our understanding of this feature has become well-known through the delta-hole model [3, 4], and need not be reviewed here.

At low beam energies, the long wavelength of the beam and the long mean free path give access to coherent nuclear features within complex nuclei. New data will be presented in Section 2 to provide accurate insights into possible reaction mechanisms and their implications.

At high beam energies, the interesting question is on the presence of excited nucleons or N^* within complex nuclei, possibly expected by analogy to the survival of the delta, at least in light nuclei. The short wavelengths for the pions should allow use of a simple eikonal reaction model, and this will be tested in Section 3. Recent work with high energy photon beams on complex nuclei has also searched for N^* excitations in several nuclei, and these results will also be compared to the pion data base in Section 3.

Since standard methods to obtain total and reaction cross sections are difficult, other techniques have been attempted. Two of these will be reviewed and compared to standard data in Section 4.

Figure 1 presents an overview of the experimental situation for both pion and photon beams on a nuclear carbon target. The energy scale is taken to be the total energy available to the projectile in the laboratory frame. The curves will be described in Section 3.

Figure 1a shows both total and reaction cross sections for pions on carbon, with the total cross sections being notably larger than the reaction cross sections. This difference is just the nuclear total elastic scattering cross section. Data come from many sources, some quite ancient, especially at high energies. The small cross sections for low pion energies are evident, but on this scale the differences between pion signs can not be noted. No significant differences between positive and negative beams are seen at the high energies. All of the data shown here come from the standard 'good geometry transmission' method, described in Section 2 [14]. Figure 1b shows the photon total cross sections [15, 16, 17], with an energy scale to make easy comparison to the pion data. The pion total cross sections at their maximum are almost exactly 137 times greater than the maximum of the photon data for carbon.

2 Low Pion Beam Energies

For pion beam energies below 100 MeV, the wavelength of the projectile is comparable to the spacing between nucleons, and coherences among the nucleons can be important. This is most familiar from the second-order ρ^2 terms in pion-nucleon optical potentials [18, 19]. The small pion-nucleon cross sections lead to an expectation of access to the nuclear interior, where densities can be large enough for these ρ^2 terms to be important. Reaction cross sections measured at TRIUMF at 50 MeV for several light nuclei seemed to show a proportionality to the target mass A, as evidence for a transparency giving access to all nucleons [11]. We carried out a new experiment at TRIUMF designed

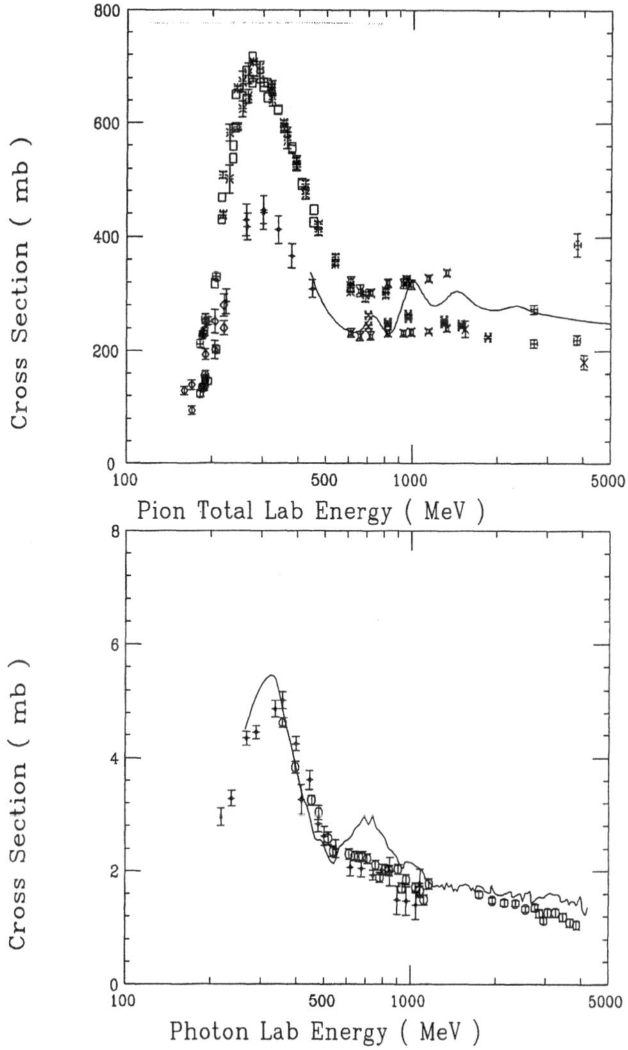

Figure 1. A survey of data for pion-nucleus total and total reaction cross section measurements, plotted against the pion total energy. Data are from refs.[1,2,5-13] The curve is from a calculation in the eikonal model, using free space pion-nucleon cross sections. Below are photon-cross sections, [15, 16], compared to six times those for deuterium [17].

to give small systematic uncertainties for pion beams of both signs, at several energies near 50 MeV, on $T = 0$ targets [12].

Our method was the 'classic transmission' geometry, where we count the beam particles incident upon the target, and the number that were not scattered beyond some cone of solid angle. The ratio of these (B/S) is compared

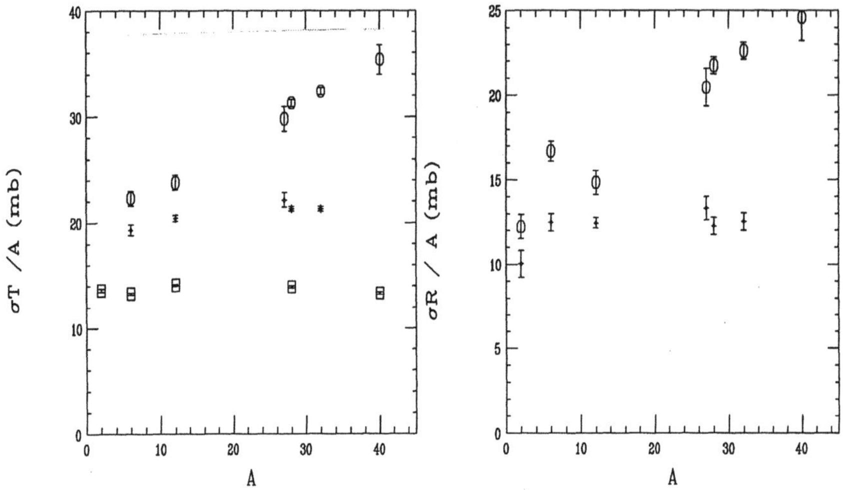

Figure 2. Pion-nucleus total and reaction cross sections per target nucleon at 50 MeV are plotted. Open points are for negative beam and solid points are for positive beam. Only statistical uncertainties are shown, and the data of Saunders have an overall systematic uncertainty not greater than 10%. The square total cross section points are from 531 MeV/c K^+ data [21].

to the ratio (B_0/S_0) without a target to give the raw attenuation cross section for that solid angle, with the number density N of target nuclei

$$\sigma_{\mathrm{att}}(\Omega) = \ln[S_0 B / S B_0]/N. \tag{3}$$

These cross sections are corrected at each solid angle for the integral of all elastic scattering beyond that solid angle. We parameterized the large body of elastic scattering data for our energies and targets, and used interpolations to treat each case needed for this elastic correction. We made various corrections, none large, for multiple scattering, decay and other experimental effects to obtain the reaction cross sections at each solid angle. These were extrapolated to zero solid angle to obtain the reaction cross sections.

Total cross sections were obtained for each solid angle by adding back to the reaction cross sections the nuclear elastic scattering. This was computed by our parameterized optical model [18, 19] for the scattering of neutral pions, with a small offset for the Coulomb energy. This method follows the suggestion of Kaufman and Gibbs [20], and is valid for penetrating particles. These total cross sections at ten solid angles were extrapolated back to zero degrees to give the total cross sections desired.

Only after much care can we assert that our systematic uncertainties do not exceed 10%. Our experience serves as a warning that the small systematic uncertainties reported for previous measurements of total and reaction cross sections are far too optimistic.

The beam energy with the greatest data base is 50 MeV. Figure 2 shows

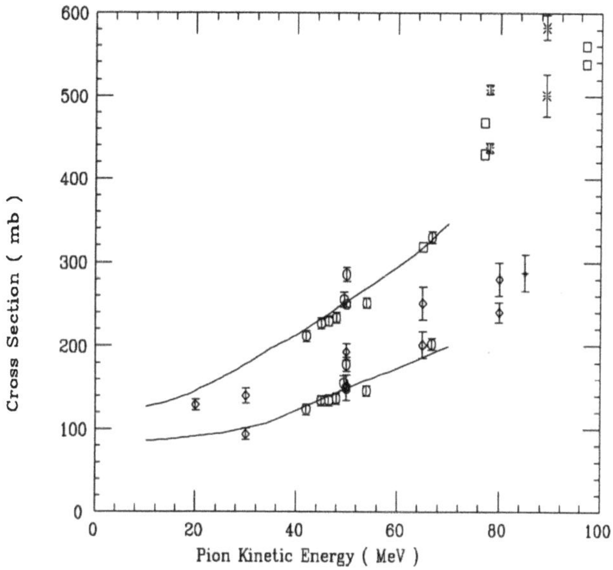

Figure 3. Pion-carbon total and reaction cross sections are plotted for beam kinetic energies below 100 MeV. The curves are from the unitary scattering theory for π^+ [22].

both the total and the reaction cross sections for both pion charge states on the $T = 0$ nuclei D, ^{6}Li, C, Si, S and Ca (and Al), together with previous data. These data have been divided by the target mass A. The rise for negative pions is presumably due to a Coulomb attraction drawing the cross sections towards the much larger values near the delta resonance. Otherwise, the data are remarkably constant, as evidence for complete transparency of the targets. The reaction cross sections for deuterium (using a different method for the elastic correction) are lower than the trend for the heavier targets, demonstrating a density or coherent effect.

Much has been made of the transparency of nuclei to K^+ mesons. Also shown in Fig.2 are total cross section data from a 531 MeV/c K^+ beam [21]. The pion data are just as constant as the K^+ results, and not much larger in magnitude. There is an important difference in the interpretation of the two experiments, however. The K^+ beam has a wavelength such that it can be said to be sensing nucleons within nuclei, but this is not so for the 50 MeV pions.

Pion-nucleon cross sections are decreasing for lower pion beam energies, and this should also influence pion-nucleus cross sections. Our TRIUMF experiment also measured total and reaction cross sections on carbon for energies between 42 and 65 MeV. These are shown in Fig.3, together with previous results. Results for negative beams are always above those for positive beams, from the expected Coulomb effect. The curves are from the unitary scattering theory for positive pions [22]. Agreement is generally quite good.

The next step in this analysis should be a comparison to pion-nucleus second-order optical model calculations, such as those carried through by Meirav [11]. This can not be done until a reanalysis of the first-order effects is completed. In the impulse approximation underlying all pion-nucleus optical models, the free space pion-nucleon amplitudes are used. These have been taken from broad-based phase shift analyses of data. Recently, however, there has been a significant change in the accepted values for pion-nucleon scattering observables at low energies, and these are certain to change the first-order optical model results in the range of beam energies considered here. We must be skeptical about the important second-order effects derived from pion-nucleus data until this new first-order analysis is reliably done.

3 Higher Pion Beam Energies

Above the delta resonance, the pion momenta give wave lengths small enough to sense nucleons within nuclei, and the cross sections are small enough to lead to enough penetration to allow interesting analyses. The most interesting question is on the existence of excited nucleons within complex nuclei. It is well-known that the delta can be found in small complex nuclei, but with modified properties. The strong N^* resonances found in free pion-nucleon scattering suggest a search for these in complex nuclei as well.

The lead in this class of study is now held by photon-nucleus cross sections. Several recent studies from Frascati and Mainz find smooth total cross sections for several nuclei. Data for carbon [15, 16] are shown in Fig. 1, and have been analyzed to determine the increased widths of the N^* [23]. Heavier nuclei have also been used for these studies [24]. Little if any evidence for discrete N^* peaks can be extracted, and the strength of the isobars excited on free nucleons is slightly lost in the smooth spectra found in heavy nuclei [16].

Pion-nucleon resonances are a yet better way to form N^* resonances, and we should examine pion-nucleus cross sections above the delta resonance to see if evidence for N^* can be found. With the strongly interacting beam, a reference total cross section is needed for comparison, since free scattering is not expected for complex nuclei. Such a reference can be taken from the eikonal or Glauber model, where nuclear total cross sections σ_{T} can be obtained from nucleon total cross sections σ_N by

$$\sigma_{\mathrm{T}} = 4\pi \int [1 - \exp(-\frac{1}{2}\sigma_N T(b))] b \, db$$

$$T(b) = \int_{-\infty}^{\infty} \rho(r) dz, \qquad r^2 = z^2 + b^2. \tag{4}$$

This model is most reliable for high energies, with small wavelengths. Calculations have been carried out using free-space pion-nucleon total cross sections [25, 26] and matter distributions taken from electron scattering. Data at 830 MeV [7] for reaction and total cross sections are shown in Fig.4, compared

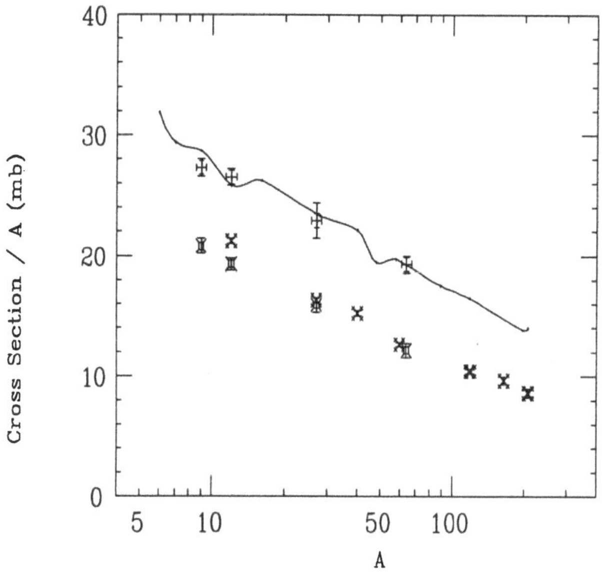

Figure 4. Pion-nucleus total and reaction cross sections for 830 MeV π^- are plotted for a range of target masses [7]. The solid line shows the total cross sections computed in the eikonal model. This agreement justifies uses of this model for high pion beam energies.

to eikonal calculations. Agreement for the total cross sections is sufficient to justify use of this simple model.

The energy dependence data set is reasonably complete only for reaction cross sections. These are shown in Fig.5 for carbon, calcium and lead (or bismuth), plotted per target nucleon. The smooth dependence can be compared to the somewhat structured total cross sections predicted from the eikonal model. Note that the structures are diminished for heavy nuclei in this calculation.

It is unfortunate that the pion-nucleus data base is so much sparser than the photon case, and almost no data for total cross sections are available for pions on heavy nuclei. The photon data show a particular sensitivity to the $D_{13}(1520)$ and $F_{15}(1680)$ N^* excitations, whereas the $D_{13}(1520)$, $D_{15}(1675)$ and $F_{37}(1950)$ excitations dominate the pion-nucleon spectrum. It is certainly now time to measure a systematic body of pion-nucleus total and reaction cross sections at energies across these resonances, with small enough uncertainties to permit a useful comparison to the widths of the free structures. The data now available do not seem to indicate strong N^* excitations in a nucleus as light as carbon. Measured total cross sections for carbon lie above the eikonal values in Fig.1a, indicating that resonance strength is not lost, as was the case for photons [16].

It might be naively expected that any higher N^* structures would be broadened in a complex nucleus, due to Fermi motion and collisional broadening. A

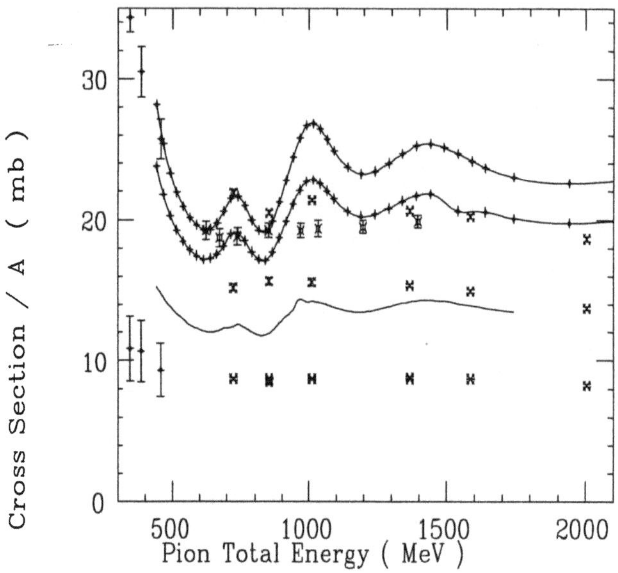

Figure 5. Measured pion-nucleus reaction cross sections per target nucleon for carbon, calcium and lead are plotted from top to bottom for a range of beam energies. The curves show the eikonal calculations for the total, not the reaction, cross sections. Nucleon internal momentum is not included in the calculations. Nevertheless, the pion-nucleon resonances still become less evident for heavier nuclei.

recent calculation, however, suggests that the $D_{13}(1520)$ excitation in a complex nucleus would be narrowed, not broadened, by medium effects [27]. Since this excitation decays largely into two pions of lower energy, the strong distortions of those pions have the effect of blocking the decay of the D_{13}. Figure 6 shows the calculations of Arima, with the free D_{13} and the smaller free S_{11} total cross sections shown by the dashed curve. The predicted narrowing of the D_{13} and S_{11} peaks in nuclear matter is shown by the solid curve. This should be readily noted in pion-nucleus total cross sections. Data for carbon are shown also in Fig.6, but so spaced in energy that the predicted narrowing could have been missed.

We have clear evidence that nucleons and deltas survive inside the strongly interacting nuclear medium, but higher excitations of N^* seem to be lost. Since the internal structure of these N^*'s require radial motion of the quarks, they can be expected to have larger radii. If this is comparable to nucleon spacings, perhaps their dissolution and deconfinement can be expected. The study of pion-nucleus reactions at high energies with good accuracy thus seems to lie at the center of our interest in strongly-interacting dynamics.

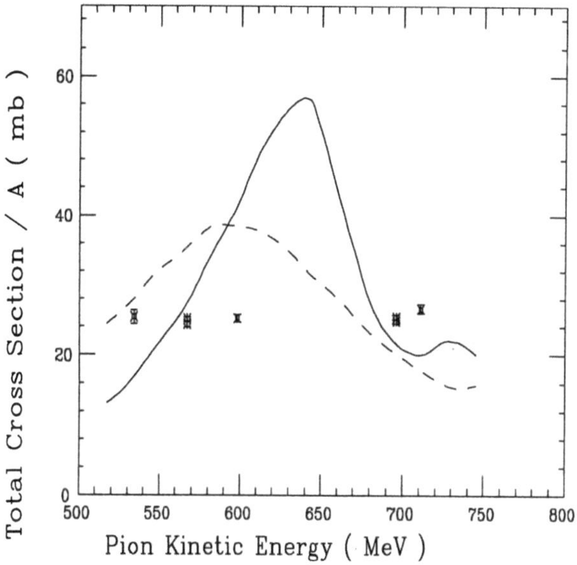

Figure 6. Total pion-carbon cross sections per nucleon are plotted, and compared to recent calculations predicting a drastic narrowing of the D_{13} resonance within nuclei [27]. The dashed curve shows the free D_{13} and S_{11} cross sections alone. This dashed curve becomes focussed to the solid curve in nuclear matter in the full calculation. The spacing of the data points may not rule out such a phenomenon.

4 Other Methods

The transmission method is the most direct means to measure total and reaction cross sections, but other methods have also been used, and their validity for pion-nucleus scattering needs to be evaluated.

The partial wave coefficients η_ℓ that enter the total and reaction cross section expressions also determine the nuclear elastic differential cross sections. For low energies and small energies, the number of partial waves in the sums is not large, and a fit to a full range of measured elastic scattering with the η_ℓ can perhaps also serve to determine the total cross sections. Total cross sections obtained by this means for ^{4}He below 250 MeV were compared to those measured directly, and good agreement was found [28]. We thus know that this idea can be made to work for some cases.

For larger nuclei and higher energies, where the need for the results is greater, the number of partial waves required is very great, and a demonstration of the validity of the method is necessary. The first such case has recently been carried out with data from KEK on pion-carbon elastic scattering at momenta from 610 to 895 MeV/c [29]. Their total and reaction cross sections fall below the data obtained from the direct transmission method, as shown in Fig.1. Since the differential elastic data extended only to about 50 degrees, the full range of needed partial wave coefficients was perhaps not determined by the

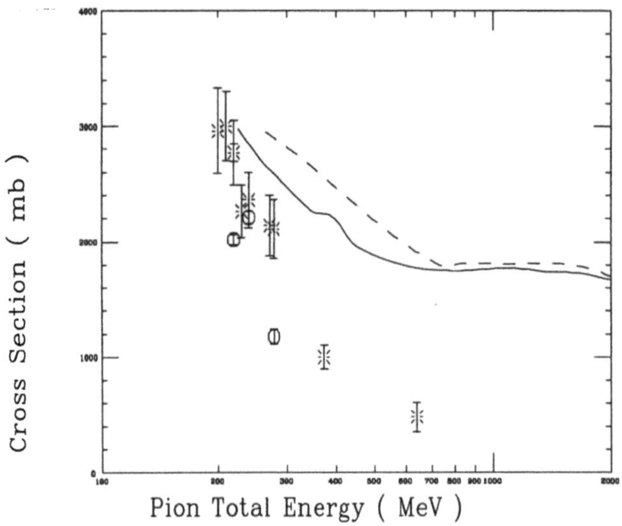

Figure 7. The energy dependence of pion-induced fission cross sections for uranium is compared to measured reaction cross sections on Pb or Bi [2, 5]. Stars show π^- and circles show π^+ fission data. The solid curve is for π^+, the dashed for π^-. It does not appear safe to assume that fission cross sections can be taken to equal reaction cross sections.

optical model method compared to the data by this group. For higher energies and heavier nuclei, this method must be yet more dangerous.

Reaction cross sections with photons have been difficult to measure by the transmission method, so an attempt was made to equate fission cross sections on sufficiently large nuclei with the total cross sections. The idea was that any reaction on a large, fragile nucleus would lead to fission, which is readily observed [30]. This gave rise to a 'universal curve' where the photon-nuclear cross sections per nucleon were the same for all nuclei [31, 32]. No N^* resonances appear in this curve for targets heavier than deuterium.

Comparison to directly measured photon reaction (or total) cross sections found this equality of fission and total cross sections to be true for uranium but that for thorium, only 90% of the reaction cross section went into fission [33]. The idea seems to work, but only for the most fragile nuclei.

We have measured pion-nucleus fission cross sections for uranium over a wide range of beam energies, with results shown in Fig. 7 [34]. The curves show measured reaction cross sections for Pb or Bi [2, 5]. The small fission cross sections at higher energies are strikingly below these curves indicating that the fission method is not suited to infer reaction cross sections for higher energy pions. Some reaction mechanism is operating that either does not result in excited nuclei (which would then fail to fission) or induces such a high nuclear excitation that prompt multifragmentation occurs, not yielding massive fission fragments. At resonance energies, a recent improvement to the data for actinide nuclei [35] finds good agreement with computed and measured reaction cross sections [36], so the equality does hold for the largest cross sections.

5 Overview

At low and resonance pion beam energies, the experimental coverage of total and reaction cross sections seems consistent and secure, with realistic experimental uncertainties. A re-evaluation of the pion-nucleus optical potential is needed at low energies, where the free pion-nucleon scattering data base has changed since the formulation of the optical potentials used.

Much more work is needed at higher pion energies, and this could be done well even with moderate beam intensities. Figures 1, 5 and 6 indicate that the existing data are not consistent enough and finely enough spaced to evaluate the role of N^* in complex nuclei. The recent work with photon beams provides guidance. Analysis of the pion data may rely on recent advances in optical models to give the elastic corrections. Before any fitting to search for N^* can be carried out, a realistic system of experimental uncertainties must be provided. This work is best carried out for light nuclei.

Cross sections at high pion energies on heavy nuclei should also be addressed. These data would constrain optical models, using the modern eikonal formulations that make calculations efficient [37]. Since the propagation of pions through nuclei is an important part of many reaction models for heavy ions, the total and reaction cross sections would also seem to form important inputs for those calculations. Another question for heavy nuclei is why the fission cross sections seem to be so much less than reaction cross sections at higher energies, in contrast to the situation with photons.

References

1. A.S. Carroll et al.: Phys. Rev. **C14**, 635 (1974)
2. D. Ashery et al.: Phys. Rev. **C23**, 2173 (1981)
3. E. Oset et al.: Phys. Rev. **83**, 281 (1982)
4. L.L. Salcedo et al.: Nucl. Phys. **A484**, 557 (1988)
5. B.W. Allardyce et al.: Nucl. Phys. **A209**, 1 (1973)
6. A.S. Clough et al.: Nucl. Phys. **B76**, 15 (1974)
7. M. Crozon et al.: Nucl. Phys. **64**, 567 (1965)
8. E. Friedman et al.: Phys. Lett. **B257**, 17 (1991)
9. M.J. Longo et al.: Phys. Rev. **125**, 701 (1962)
10. W.O. Lock and D. F. Measday: *Intermediate Energy Nuclear Physics*

11. O. Meirav et al.: Phys. Rev. **C40**, 843 (1989); **C36**, 1987 (1987)

12. A. Saunders et al.: (to be published)

13. C. Wilkin et al.: Nucl. Phys. **B62**, 61 (1973)

14. G. Giacomelli: Rep. Prog. Phys. **12**, 77 (1970)

15. M. Anghinolfi et al.: Phys. Rev. **C47**, 922 (1993)

16. N. Bianchi et al.: Phys. Lett. **B309**, 5 (1993); **B299**, 219 (1993)

17. T.A. Armstrong et al.: Nucl. Phys. **B41**, 445 (1972)

18. K. Stricker et al.: Phys. Rev. **C19**, 929 (1979); J.A. Carr et al.: Phys. Rev. **C25**, 952 (1982)

19. M.B. Johnson and E.R. Siciliano: Phys. Rev. **C27**, 1647 (1983)

20. W.S. Kaufmann and W.R. Gibbs: Phys. Rev. **C40**, 1729 (1989)

21. R.J. Weiss et al.: Phys. Rev. **C49**, 2569 (1994)

22. M.Kh. Khankhasayev: Nucl. Phys. **A505**, 717 (1989)

23. W. Alberico et al.: Phys. Lett. **B321**, 177 (1994)

24. L.A. Kondratyuk et al.: Nucl. Phys. **A579**, 453 (1994)

25. Scattering Analysis Interactive Dialin, a phase shift analysis maintained by R.A. Arndt. FA93 solution.

26. Particle Data Group: Phys. Rev. **D50**, 1 (1994)

27. M. Arima et al.: Phys. Rev **C51**, 285 (1995)

28. B. Brinkmoller and H.G. Schlaile: Phys. Rev. **C48**, 1973 (1993)

29. T. Takahashi et al.: Phys. Rev. **C51**, 2542 (1995)

30. Th. Frommhold et al.: Phys. Lett. **B295**, 28 (1992)

31. N. Bianchi et al.: Phys. Lett. **B325**, 333 (1994)

32. Th. Frommhold et al.: Z. Phys. **A350**, 249 (1994)

33. N. Bianchi et al.: Phys. Rev. **C48**, 1785 (1993)

34. R.J. Peterson et al.: Z. Phys. **A** (to appear)

35. S. deBarros et al.: Nucl. Phys. **A542**, 511 (1992)

36. S. deBarros et al.: Private Communication.

37. C.M. Chen et al.: Phys. Rev. **C48**, 841 (1993)

Few-Body Systems Suppl. 9, 29–35 (1995)

Few-Body Systems

Pion Absorption in Helium with the LADS Detector

J. Köhler for the LADS Collaboration*

Institut für Physik der Universität Basel, Basel, Switzerland

Abstract. Investigations on π–absorption on Helium reveal a substantial part of the absorption cross section to be multi–nucleon absorption. A careful search for different reaction mechanisms shows clear signatures of initial state interaction, whereas most of the multi–nucleon absorption follows a phase space like distribution. In comparable conditions the signatures look alike for either ^{3}He and ^{4}He. Fits of Monte Carlo simulations to the data give first results on the partial cross sections in ^{3}He.

1 Introduction

1.1 Motivation

π–absorption on nuclei is known to be a large fraction of the interaction cross section and therefore has been a field of interest for many years. Especially on light nuclei like ^{3}He and ^{4}He, there is the possibility to investigate the basic processes of π–absorption. The basic process seems to be understood, but earlier experiments have shown the existence of more complex mechanisms contributing to the absorption cross section [2, 3, 4, 5]. In this work, I'll concentrate on processes where more than two nucleons are involved.

1.2 Absorption Mechanisms

By energy and momentum conservation π–absorption is forbidden on a free nucleon and heavily suppressed on a single nucleon inside a nucleus. Therefore the simplest mechanism involves a nucleon pair on which the pion is absorbed. The remaining $(A-2)$–nucleus is acting as a spectator. This basic two nucleon absorption ($2N$A) is quite well understood and seems to be dominated by a rescattering process involving a formation of the Δ intermediate state. This basic mechanism can be seen in a momentum density distribution of the least energetic proton in the reaction ^{3}He(π^+, ppp) like Fig. 1. In the plot the $2N$A

*complete list see e.g. [1]

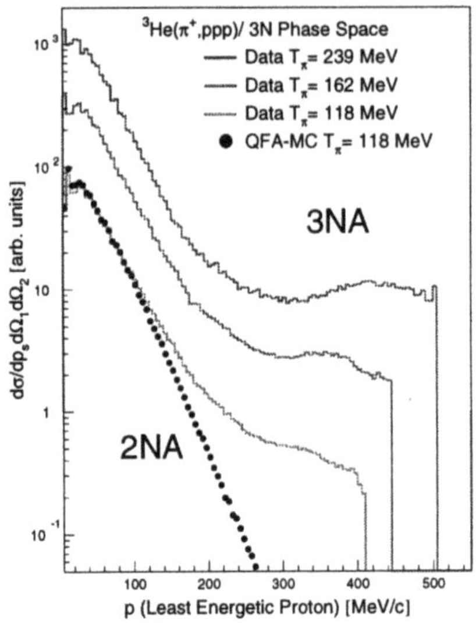

Figure 1. Acceptance Corrected Momentum Density Distribution of ^{3}He(π^+, ppp), T_π=239 MeV

also reported as Quasi Free Absorption (QFA) shows a fermi momentum distribution which rapidly decreases with increasing proton momentum. Note the y-axis is logarithmic. As a comparison at the lowest curve a QFA-simulation is plotted on top of the data. A clear deviation from the $2N$A picture is seen at higher proton momenta suggesting other mechanisms involving all nucleons. In ^{3}He those mechanisms will be referred to as three nucleon absorption ($3N$A), whereas in ^{4}He as multi–nucleon absorption (nNA). $3N$A (nNA) could be either multistep processes or some other process acting on all nucleons. Because of the phase space-like distribution at the low energy (118 MeV) it was suggested [2] that there exists a $3N$A process which can be described by a constant matrix element. But there is no conclusive theory available yet.

Earlier experiments did not identify multistep processes. The new data shows a structure at higher momenta, which depends on the pion energy. As the origin of this, we consider multistep processes where we distinguish in a simple approach two different mechanisms: Initial State Interaction (ISI), where the pion first scatters off one of the protons and is then absorbed on the remaining nucleon pair; or Final State Interaction (FSI) where the pion is absorbed first and one of the outgoing protons interacts with the spectator nucleon(s).

2 Experiment

To investigate multi–nucleon absorption, a 4π detector LADS (= Large Acceptance Detector System) has been built at PSI in Villigen Switzerland and experiments on different target nuclei (D, ^{3}He, ^{4}He, N, Ar, Xe) have been performed around the Δ-resonance energy. An extensive description of the detector will be published soon in NIM [6]. The advantage of LADS over previous experiments is the possibility of measuring multi–nucleon final states with complete information on all particles involved. Also good energy and angular resolution is obtained covering as much as 98% of the total solid angle. Having low proton thresholds gives an almost complete phase space coverage.

3 Analysis

3.1 Observables

One of the major points in the analysis is to give a convenient set of variables to describe multi–nucleon final states. In the case of ^{3}He it helps that the three protons form a plane in the CM-system. This plane and the orientation of the event in respect to this plane can be described by the Euler angles, where ξ describes the orientation of the plane relative to the incident pion and γ the rotation of the event in the plane. Because there is no polarization involved in our experiment, there is no dependence on the third Euler angle describing azimuthal properties. To describe the event completely it is sufficient to show in addition to ξ and γ two of the opening angles, or equivalently two of the particles' energies (Dalitz–plot).

In the case of the search for ISI, we have used the proposal of Salcedo et al. [7] to reconstruct the pseudo invariant mass of the intermediate pion. A contribution of ISI should show up as a peak in the pseudo invariant mass spectrum at the pion mass squared.

3.2 Monte Carlo Simulation

The simulations which were made to assist interpretation of the results used the CERN GEANT [8] package. All processes already discussed earlier were treated. The simulation also included the correct acceptance of the detector and all known losses due to reactions, particle misidentification, chamber efficiencies and reconstruction combinatorics. Tests in the $\pi d \rightarrow pp$ channel have shown good agreement with the data. All event generators including a $2N$A process (QFA,ISI,FSI) used a parametrization of the angular distribution by Ritchie et al. [9]. In addition the ISI-simulation used measured cross sections (SCATPI) [10] for πp knockout, whereas $N - N$ cross sections from SAID [11] were used for the FSI simulation.

In addition a three-body phase space simulation was done. The data shows a preference of multi–nucleon absorption to be in plane compared to phase space calculations. It can be shown that this can be linked to angular momentum in the same way as angular momentum is introduced by including a Legendre term in the angular distributions of the $2N$A [12]. Therefore, like in $2N$A, a second order Legendre term $P_2\{\cos(\xi)\}$ was introduced into the phase space simulation.

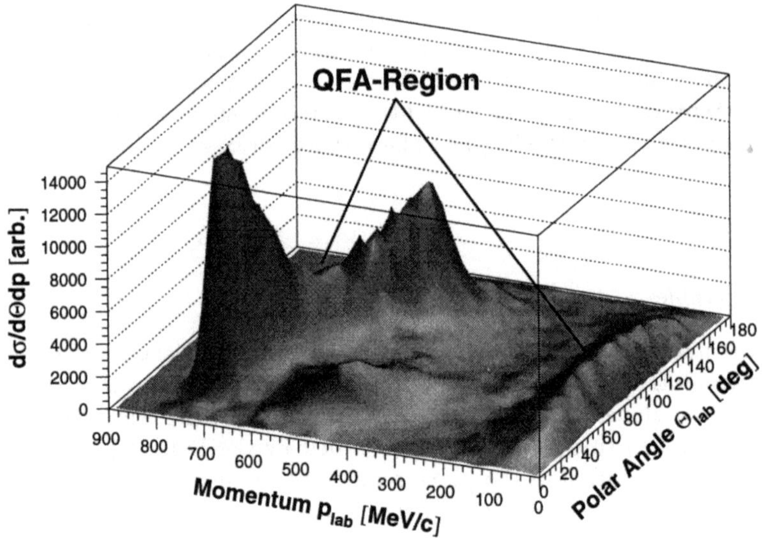

Figure 2. Θ_{lab} vs. p_{lab}, T_π=239 MeV

4 Results

4.1 ISI Signatures

A striking signature for ISI can be found in the correlation between the polar angle Θ and the momentum p in the lab–system. Figure 2 shows the Θ_{lab} vs. p_{lab} correlation of all protons in the reaction $^3\mathrm{He}(\pi^+, ppp)$ at T_π=239 MeV. One can clearly identify the regions of QFA, seen as two large peaks at high momenta showing the typical angular distribution as well as the broad bump of the spectator at low momenta. Beside these structures there is a band at forward angles visible which indicates πp knockout. Figure 3a) shows the same correlation but with a 30 MeV threshold on all protons to get rid of the QFA. On top of the distribution the free πp kinematics is drawn. As is seen the structure at forward angles agrees well with the kinematic line. Figure 3b) and c) show the correlation of the pseudo invariant mass squared m_x^2 versus p_{lab} and Θ_{lab} respectively. All possible combinations of m_x^2 are taken into account and plotted versus the remaining proton. For the ISI a peak is expected at the pion mass squared and indeed this structure is found in both plots. In addition a comparison of all plots shows that the structure in Fig. 3a) is reflected in the two other plots in the peaks at the pion mass squared. The simulation supports this picture by showing that this structure at forward angles can only be reproduced by the ISI simulation.

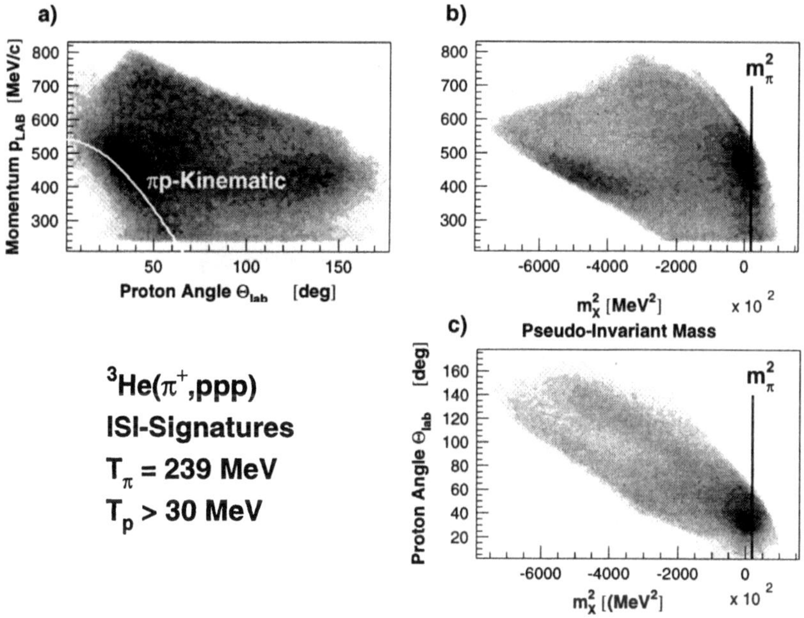

Figure 3. Signatures for ISI in ^{3}He(π^+, ppp), a) $\Theta_{\rm lab}$ vs. $p_{\rm lab}$, b) m_x^2 vs. $p_{\rm lab}$, c) m_x^2 vs. $\Theta_{\rm lab}$

4.2 Fit Results

Various fits have been performed, to get the partial and the total cross sections on ^{3}He(π^+, ppp). As an example, Fig.4 shows the results for $T_\pi = 239$ MeV on m_x^2. The numbers give the relative absorption cross sections not acceptance corrected. The fit agrees well and it is clear that only the ISI simulation is able to explain the strong peak at the right. The variables used in this fit are $\gamma, \xi, \Delta\Theta_{\rm max}, \Delta\Theta_{\rm min}$, hence the complete set of variables discussed in 3.1. Table 1 gives a breakup of the ^{3}He absorption cross section. The results of the total cross sections based on the analysis of the partial cross sections agree

Table 1. Partial and Total Cross Sections of ^{3}He(π^+, ppp)

T_π	[MeV]	118		162		239	
$\sigma_{\rm Total}$	[mb]	27.1 ± 0.8		23.9 ± 1.0		10.4 ± 0.7	
$\sigma_{\rm QFA}$	[mb]	19.4 ± 1.0		15.9 ± 1.3		6.2 ± 0.6	
σ_{3N}	[mb]	7.7 ± 1.0		8.0 ± 1.2		4.2 ± 0.4	
$\sigma_{3N}^{\rm PS}$	[mb]	4.9 ± 1.1	64%	5.6 ± 1.2	70%	2.7 ± 1.1	65%
$\sigma_{\rm ISI}$	[mb]	1.9 ± 0.6	25%	1.9 ± 0.8	24%	1.4 ± 0.4	33%
$\sigma_{\rm FSI}$	[mb]	0.9 ± 0.6	11%	0.4 ± 0.8	6%	0.1 ± 0.5	2%

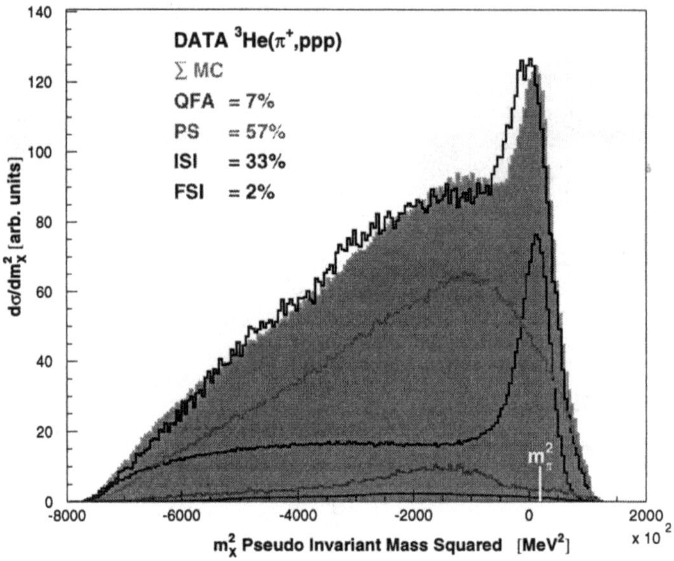

Figure 4. Fit of m_x^2 in $^3\text{He}(\pi^+, ppp)$, $T_\pi = 239$ MeV, $T_p > 30$MeV

within errors with the numbers from [1]. Deviations in the breakup of the cross sections into $2NA$ and $3NA$ can be explained by the number of models taken into account in the fit. The error bars reflect the uncertainties using different sets of variables in the fit, and do not include errors on model dependencies. All numbers in Tab. 1 are preliminary.

The simulation can be used to correct for acceptance losses, as in Fig. 1. In Figs. 2,3 we do not use the acceptance correction. That is reflected in a small angle cut–off for QFA and a strong suppression at high momenta and small angles. For the fit, no acceptance correction was applied to the data, but the acceptance losses were included in the simulation. For the calculation of the results in Tab. 1, the acceptance correction is taken into account by using the simulation with and without the acceptance correction.

4.3 Comparisons to ^{4}He

To compare ^{4}He to ^{3}He, the same conditions on the data set are required. This can be achieved by limiting the analysis to those events where three protons are measured, and the extra nucleon (here the neutron) has a momentum lower than 200 MeV/c and is therefore considered as a spectator. The system of the three outgoing protons is selected as the reference system. This choice allows us to describe the event with the set of variables described in 3.1.

There is a striking similarity of the different plots of both isotopes. Especially the Θ_{lab} vs. p_{lab} plot and the complete set of variables (γ, ξ, $\Delta\Theta_{\max}$,

$\Delta\Theta_{\text{min}}$) are almost identical. Only the peak in m_x^2 is not very pronounced in ^{4}He, but this can be explained by the broadening due to the extra nucleon, which is not completely at rest.

5 Summary

The present analysis shows for the first time a clear signature for ISI in the reactions $^3\text{He}(\pi^+, ppp)$ and $^4\text{He}(\pi^+, ppp)n$. In the case of ^{3}He it is possible to evaluate the partial cross sections using simulations of simple reaction models. The strength of the ISI is between 20 and 35 % of the $3N$A whereas FSI seems to be small. The remaining part of the $3N$A cross section shows a phase space like behaviour where we still lack theoretical explanations. There is a strong indication that taking into account the orbital angular momentum brought by the pion explains the data better. The similarity of ^{3}He and the special case in ^{4}He described in 4.3 suggests that the same $3N$A processes occur in ^{4}He as in ^{3}He.

The results of ^{3}He will be published soon [13], concentrating on the signatures of ISI, while theoretical calculations on the $3N$A will be performed in the near future. These calculations will help to answer some of the still open questions in π–absorption.

References

1. T. Alteholz et al.: Phys. Rev. Lett. **73**, 1334 (1994)
2. P. Weber et al.: Nucl. Phys. **A534**, 541 (1991)
3. S. Mukhopadhyay et al.: Phys. Rev. **C43**, 957 (1991)
4. H.J. Weyer: Phys. Rep. **195**, 295 (1990)
5. R.D. McKeown et al.: Phys. Rev. **C24**, 211 (1981)
6. D. Androić et al.: Nucl. Inst. Meth. (in preparation)
7. L. Salcedo, E. Oset and D. Strottmann: Phys. Lett. **B208**, 339 (1988)
8. Application Software Group: Comp. Net. Div., CERN Geneva, Schweiz
9. B. Ritchie: Phys. Rev. **C49**, 553 (1991)
10. J.B. Walter, G.A. Rebka Jr.: Techn. Rep. LA-7731-MS LAMPF 1979
11. R.A. Arndt et al.: Phys. Rev. **D32**, 1085 (1985)
12. N. Simičević, A. Mateos: Phys. Rev. **C51**, 797 (1995)
13. D. Androić et al.: Phys. Lett. (in preparation)

Few-Body Systems Suppl. 9, 36–50 (1995)

Few-Body Systems

Meson Exchange Contribution to K^+-Nucleus Scattering

J. Nieves [1] *, C.García-Recio [2†], E. Oset [1]

[1] Departamento de Física Teórica and Instituto de Física Corpuscular, Centro Mixto Universidad de Valencia-Consejo Superior de Investigaciones Científicas, E-46100 Burjassot, Valencia, Spain

[2] Departamento de Física Moderna, Facultad de Ciencias, Universidad de Granada, E-18071, Granada, Spain

Abstract. A microscopical many-body calculation of the K^+-nucleus optical-potential is carried out for kaon kinetic energies up to 500 MeV. In addition to the the impulse-approximation contribution, we have considered the contribution to the K^+-nucleus scattering from the interaction of the kaon with the virtual pion cloud. Previous approximations for the pionic cloud contribution, specially on the imaginary part, are shown to be inadequate. We also find new interaction mechanisms which provide appreciable corrections to the K^+-nucleus optical-potential. The inclusion of these new mechanisms in the inelastic part of the optical potential produces a significant improvement in the differential and total K^+ nuclear cross sections. Uncertainties in the real part of the optical potential are also discussed.

1 Introduction

Recent experimental [1-4] and theoretical work [6-8] on K^+-nucleus scattering has raised some puzzling questions. Since the K^+-nucleon system exhibits the weakest of the hadron-nucleon interactions at lower energies, with no resonances or bound states, it is expected that microscopic optical potentials should be calculable and reliable. With weaker interactions than the π, nucleon, or anti-nucleon, inelastic multiple-scattering should be considerably less important for the K^+. Thus, one would expect that optical model calculations, based on the impulse approximation (IA), give a reasonably accurate description of elastic and total K^+-nucleus cross sections.

* *Former address*: Physics Department, University of Southampton, SO17-1BJ, U.K.

† This work was sponsored by DGICYT (Spain) research Project No. PB92-0927 and CICYT (Spain) research project AEN93-1205.

However, discrepancies between experimental data and optical model calculations have persisted, in particular on the ratio of the K^+-nuclear to K^+-deuteron total cross section, and are not yet resolved. These discrepancies remain when a number of conventional nuclear corrections (Pauli blocking, nucleon-nucleon correlations, off-shell corrections, etc.) are taken into account.

As possible explanations of the discrepancies, there have been several proposals ([7],[9, 10]) related to the so-called nucleon *swelling* concept in the nuclear environment. An increase (10-20%) in the size of the nucleons in the nucleus over that in free space would lead to larger K^+- nucleus cross sections and would bring theoretical optical models predictions, based on the IA, in an acceptable agreement with the experimental data. A similar mechanism could be used to explain the European Muon Collaboration (EMC) effect in the deep inelastic lepton scattering from nuclei [11]. This led many authors to the conclusion that the K^+-nucleus scattering and the EMC effect reflect the same properties of nucleons in the nucleus.

In ref. [12] was suggested a different mechanism to account for these discrepancies: the interaction of the kaon with the modified pionic cloud surrounding the nucleons. In this reference, a qualitative estimate of the size of these effects is done by assuming that the K^+N cross section is increased by $\delta n_\pi \sigma(K^+\pi)$, where δn_π is the excess number of pions per nucleon in the nucleus. Jiang and Koltun ([13]) have analyzed more carefully the contribution of the excess of pions in the nucleus to the K^+-nucleus scattering. In that work a certain momentum distribution of pions in the nucleus and a $t_{K\pi}$- matrix depending on momentum are assumed, then the K^+-nucleus optical potential is obtained as a weighted integral (over three-momenta) of the $K^+\pi$ amplitude with the pion distribution in the nucleus. This is an improvement over the work of ref. [12], but still a static approximation is made for the $t_{K\pi}$- matrix to be able to factorize the pion three-momentum distribution.

In the present work we make a more rigorous evaluation of the pion cloud contribution to the K^+-nucleus optical potential. $K\pi$ scattering is not small at the relevant energies (K^+ mesons with a laboratory momentum of the order of 800 MeV) and we find that the effects of the mesonic cloud in K^+-nucleus scattering are relevant and of the right order of magnitude to account for the discrepancies of the conventional optical potential with the data.

We also show that the static approximation, used in [13], is not in general justified when the available phase space is properly taken into account and in the present case, it induces an error of about a factor 2 . On the other hand we find new mechanisms which enhance the contribution of the pion cloud to the K^+-nucleus optical potential by about a factor 2 or 3 with respect that of the standard mechanism considered in refs. [12] and [13].

2 Theoretical Framework: Standard MEC Contribution to K^+-nucleus Optical Potential

In this section we introduce and compare with previous works, our meson-exchange-current (MEC) model to account for the contribution to K^+-nucleus cross sections of K^+ scattering by virtual mesons, exchanged within the nuclear target[1]. We expect the dominant exchange to be that of π mesons, which have the lightest mass and longest range in the NN interaction. The model involves two important physical inputs: one is the pion propagator in the nuclear medium and the other is the isoscalar $K\pi$ scattering amplitude [2].

In the nuclear medium, the pion propagator is renormalised by allowing the pion to excite particle-hole (ph) and delta-hole (Δh) components in a correlated ground state; such a model provides a realistic description of the π-nucleus interaction and accounts for the basic components needed to produce realistic pion numbers in finite nuclei. More details can be found in [14].

For the $K\pi$ amplitude we use the model of ref. [13]. This model incorporates on-shell conditions and crossing symmetry. A detailed study is made in ref. [13] about uncertainties from the off-shell extrapolation, form factors, etc.

The calculations are done in infinite nuclear matter and the contributions to the K^+ self-energy[3] are obtained as a function of ρ_0, the nuclear matter density. By means of the local density approximation (carefully studied and justified in [15]), we obtain the MEC contribution to the K^+-nucleus optical potential as a function of $\rho(r)$. Our model for the K^+-nucleus potential is obtained by adding these new contributions (MEC) to the conventional ones from the impulse approximation and to the standard nuclear corrections (nuclear correlations, off-shell and binding effects, Pauli exclusion, etc., calculated in [3] and [7]). This new optical potential is then used to obtain the differential and total K^+-nucleus cross sections by solving numerically the Klein-Gordon equation.

The standard MEC contribution to the K^+-nucleus optical potential or selfenergy is shown in the diagram in Fig. 1(a). In an infinite spin-isospin symmetric nuclear medium is given by

$$-\mathrm{i}\Pi(k) = \int \frac{d^4q}{(2\pi)^4}\, \mathrm{i}D(q)\, (-\mathrm{i})\, \frac{1}{2}\, 3\, t^0(k,q;k,q), \tag{1}$$

where t^0 is the isoscalar $K^+\pi$ amplitude (average of $t_{K^+\pi^i}$ for the three charged pions) and the factor $\frac{1}{2}$ is a symmetry factor. However, as depicted in Figs. 1(b-e), the full propagator contains the free pion, one ph or Δh corrections and higher order corrections with $2ph$, $ph\ \Delta h$, $2\Delta h$, etc, excitations. The contribution from the free pion has to be subtracted because it corresponds to a

[1] Full details can be found in ref. [14].

[2] Since the pion is virtual, it is important to take into account properly the off-shell behaviour of the $K\pi$ amplitude

[3] Throughout this paper, we will refer indistinctly to the K^+ self-energy or to the K^+-nucleus optical potential, as they are related by $\Pi(k) = 2k^0 V_{\mathrm{opt}}(k)$.

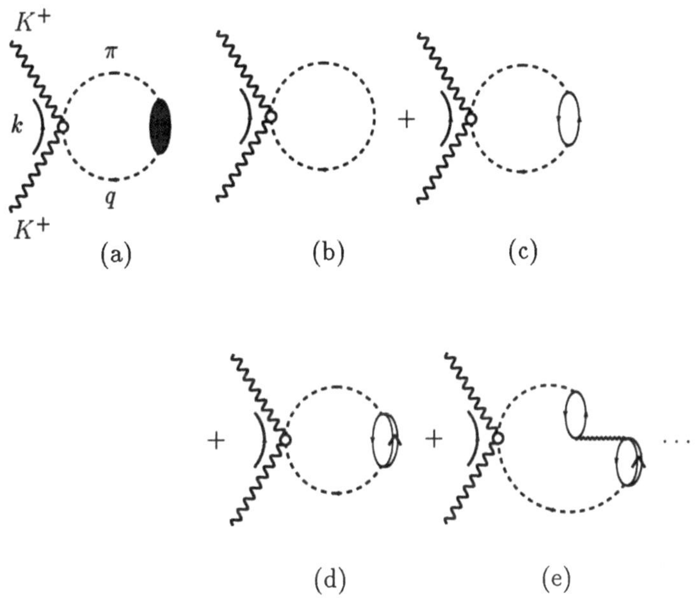

Figure 1. Standard K^+ self-energy diagrams due to the pion cloud. Our model for the propagation of pions, within the nuclear medium, accounts for the excitations of multiple particle-hole and delta-hole components, as well as short distance correlations among these.

piece in the free K^+ selfenergy. Analogously, once this subtraction is made, we will have terms in the selfenergy coming from $1ph$ or $1\Delta h$ excitations which are proportional to ρ. These terms must also be subtracted because they are implicitly accounted for in the IA selfenergy $\Pi^{\rm IA} = t_{KN}\,\rho$, where t_{KN} is the empirical KN t-matrix. Hence the genuine pion cloud contribution to the K^+ selfenergy is given by

$$\delta\Pi(k) = {\rm i}\int \frac{d^4q}{(2\pi)^4}\,\delta D(q)\,\frac{3}{2}\,t^0(k,q;k,q), \tag{2}$$

$$\delta D(q) = D(q) - D_0(q) - \rho\left(\frac{\partial D(q)}{\partial\rho}\right)_{\rho=0}. \tag{3}$$

In the static approximation the q^0 dependence in the t matrix is neglected. For instance in [13], q^0 is set to zero in $t^0(k,q;k,q)$. In this case one obtains

$$\delta\Pi_{\rm stat}(k) = \int \frac{d^3\boldsymbol{q}}{(2\pi)^3}\frac{\delta N(\boldsymbol{q})}{2\omega(\boldsymbol{q})}\,t^0(k,q;k,q)\big|_{q^0{\rm fix}} \tag{4}$$

where $\omega(\boldsymbol{q}) = \sqrt{m_\pi^2 + \boldsymbol{q}^2}$ and the "excess number of pions", $\delta N(\boldsymbol{q})$ is defined by

$$\delta N(q) = -6\omega(q) \int_0^\infty \frac{dq^0}{2\pi} \mathrm{Im}\,[\delta D(q)]. \tag{5}$$

$N(q)$ contains not only $n(q)$, the pion number distribution[4], but also the expectation values $\langle a_{\mathbf{q}\lambda}\, a_{-\mathbf{q}-\lambda}\rangle$ and $\langle a^+_{\mathbf{q}\lambda}\, a^+_{-\mathbf{q}-\lambda}\rangle$. These three factors correspond in fact to having the K^+ scattering with a pion, annihilating two pions from the ground state or creating two pions from the ground state. Our model, for the propagation of pions in the nuclear medium, leads to results for the momentum distribution of the excess number of pions in the nucleus ($\delta N(q)$), which agree qualitatively and quantitatively with calculations in finite nuclei [16].

Hence, in the static approximation of Eq. (4), $\Pi(k)$ comes as a weighted integral of the $K^+\pi$ amplitude with the pion distribution in the nucleus. This result looks intuitive . The approach of [13] corresponds to the static approximation of Eq. (4) with $\delta N(q) = 2 \times \delta n(q)$ and $\delta n(q)$ taken from [16]. The factor 2 accounts for the two pion creation ($\langle a^+_{\mathbf{q}\lambda}\, a^+_{-\mathbf{q}-\lambda}\rangle$) or annihilation ($\langle a_{\mathbf{q}\lambda}\, a_{-\mathbf{q}-\lambda}\rangle$) mechanisms as found in [17].

Our claim here is that the static approximation used in the approaches of references [12] and [13] is inaccurate, particularly for the imaginary part of $\Pi(k)$.

The imaginary part of $\Pi(k)$ corresponding to the standard MEC contribution, depicted in Fig. 1, can be calculated exactly without relying on the static approximation, by taking into account the analytical properties of the pion-propagator and the $t_{K\pi}$- matrix. In the model of ref. [13], the $t_{K\pi}$ amplitude satisfies crossing symmetry and has the following structure:

$$t^0(k,q;k,q) = \tilde{t}(s) + \tilde{t}(u) \tag{8}$$

$$s = (k+q)^2 \tag{9}$$

$$u = (k-q)^2 \tag{10}$$

where s and u are the usual Mandelstam variables ($t = 0$ in our case). Then, $\tilde{t}(x)$ verifies the following subtracted dispersion relation:

$$\tilde{t}(x) = P(x) + (x - x_0) \int_{x_0}^\infty \frac{dx'}{\pi} \frac{\mathrm{Im}\,\tilde{t}(x')}{(x - x' + i\epsilon)(x_0 - x' + i\epsilon)}, \tag{11}$$

[4] $n(\mathbf{q})$ the total number of pions is defined by

$$n(\mathbf{q}) \equiv \sum_\lambda n_\lambda(\mathbf{q}) = \sum_\lambda \langle a^+_{\mathbf{q}\lambda} a_{\mathbf{q}\lambda}\rangle, \tag{6}$$

where $n_\lambda(\mathbf{q})$ is the pion number distribution for a single class of pions, the symbol $\langle\ \rangle$ indicates the expectation value in the nuclear ground state and $a_{\mathbf{q}\lambda}$ the annihilation operator of a pion with momentum $\mathbf{q}$ and isospin λ. Thus,

$$\int \frac{d^3\mathbf{q}}{(2\pi)^3}\, n(\mathbf{q}) = \frac{n_\pi}{V} = \frac{n_\pi}{A}\rho, \tag{7}$$

with n_π/A the total number of pions per nucleon.

where $P(x)$ is a polynomial of real coefficients and $x_0 = (m_K + m_\pi)^2$. Using these explicit analytical structures in Eq. (2) and performing integrations over energies in the complex plane, we obtain an expression for the imaginary part of the kaon selfenergy only in terms of $\mathrm{Im}\, D_\rho$, $\mathrm{Im}\,\tilde{t}$ and P,

$$\mathrm{Im}\,\delta\Pi(k) \;=\; -\int \frac{d^3\boldsymbol{q}}{(2\pi)^3}\theta(q^0_{\mathrm{max}})\int_0^{q^0_{\mathrm{max}}} \frac{dq^0}{2\pi} 6\,\mathrm{Im}\,\tilde{t}(u)\mathrm{Im}\,\delta D(q). \qquad (12)$$

with $q^0_{\mathrm{max}} = k^0 - E(x_0)$ and $E(x_0) = [(\boldsymbol{k}-\boldsymbol{q})^2 + x_0]^{1/2}$. Note that for the imaginary part, *i.e.* the reaction channels of the kaon optical potential, the pion energy q^0 is restricted to the interval $0 < q^0 < k^0 - E(x^0)$, since $\mathrm{Im}\,\tilde{t}(u) = 0$ for $u < x_0$. The static approximation, by neglecting the q^0 dependence in $t_{K\pi}$, allows the q^0 integral to go up to ∞ (see Eqs. (4,5)), hence introducing unphysical contributions. Therefore, even if we make $\mathrm{Im}\,\tilde{t}(u)$ static in Eq. (12) in order to take it out of the q^0 integral, the pion excess number $\delta N(\boldsymbol{q})$ will not be generated, because the range $[0,\infty]$ in the q^0 integration is needed in Eq. (5). This occurs because the relevant magnitude is $\mathrm{Im}\, D(q)$, which provides the probability of finding a pion with momentum $\boldsymbol{q}$ and energy q^0. The probability of finding a pion of momentum $\boldsymbol{q}$ is an integral property obtained when one integrates over the energy of the pion from 0 to ∞. However, in decay processes the range of energies allowed is limited because of energy and momentum conservation, and the particle number can not be factored out. Note that the range of $\boldsymbol{q}$ is also restricted because $E(x_0) < k^0$. Then the whole phase space allowed is finite, as corresponds to the reaction channels accounted for by $\mathrm{Im}\,\delta\Pi(k)$. Under these circumstances one should not expect the static approximation to provide realistic results.

In Fig. 2 we present the imaginary part of the K^+ selfenergy as a function of its kinetic energy for nuclear matter density[5], $\rho = \rho_0$. As can be seen in this figure the static approximation used in [12] and [13] differ by an approximate factor of two from the exact calculation of Eq. (12). The leading part of the optical potential comes from the IA and the contribution of the standard mechanism of Fig. 1 is not enough to account for the discrepancies between the IA and the data.

In the next section we will discuss new mechanisms, which enhance considerably the total contribution of the pionic cloud to K^+-nucleus scattering. After the inclusion of these new mechanisms, it turns out that the total MEC-contribution (solid line in Fig. 2) becomes an important fraction of the IA optical potential. In the last section of the paper, we will see that these effects of the mesonic cloud are of the right order of magnitude to account for the discrepancies of the conventional optical potential with the data.

[5] Calculations with different values of ρ have shown that $\mathrm{Im}\,\delta\Pi$ behaves quadratically in density.

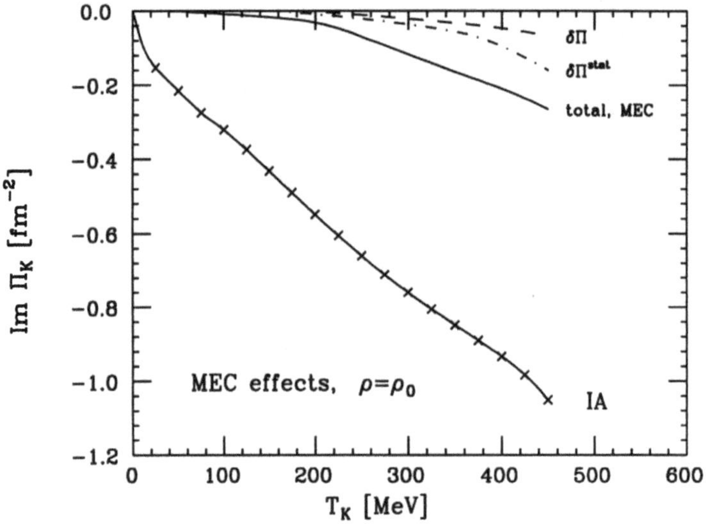

Figure 2. Imaginary part of the K^+ selfenergy for normal nuclear matter versus kinetic energy of the incoming kaon. The dot-dashed line is the result of the static approximation using q^0 fixed to zero. The dashed line displays the exact result calculated by means of Eq. (12). The IA contribution to the optical potential K^+-nucleus is displayed with a crossed solid line. The solid line stands for the total MEC contribution when the contribution of new mechanisms, discussed is the next section, is added to that of the standard mechanism of Fig. 1.

3 New MEC contributions to the K^+-nucleus Optical-Potential

The imaginary part of the kaon-selfenergy of Fig. 1 is due simultaneously to the reaction channels of the pion in nuclear matter (imaginary part of the pion propagator in Eq. (12)) and the reaction channels of the $K^+\pi$ scattering matrix ($\mathrm{Im}\,\tilde{t}(u)$ in Eq. (12)). For kaon kinetic energies up to 500 MeV, the only open channels in $t_{K\pi}$ are the elastic one or the charge exchange, $K^+\pi_i \to K^0\pi_j$, and thus $\mathrm{Im}\,t_{K\pi}$ is due only to the process $K\pi \to K\pi$. Then using the optical theorem in Eq. (12), we obtain the identity shown diagrammatically in Fig. 3.

This allows us to understand the processes to which $\mathrm{Im}\,\delta\Pi$ is due, these are: $K\,N \to K\,\pi\,N$, $K\,N \to K\,\pi\,\Delta$, where the interaction KN is renormalized in the medium. The incoming K^+ has to produce a kaon, a free pion and a nuclear excitation, so it is clear why the kinetic energy of the K^+ must be larger than the pion mass as shown in Fig. 2.

Looking again at Fig. 3 one realizes that not only one pion, but also the other pion and also both pions simultaneously must be modified by the nuclear medium. The imaginary part of the kaon selfenergy, Π^{MEC}, due to the pion cloud is then given by the diagram of Fig. 4, except that its linear part in

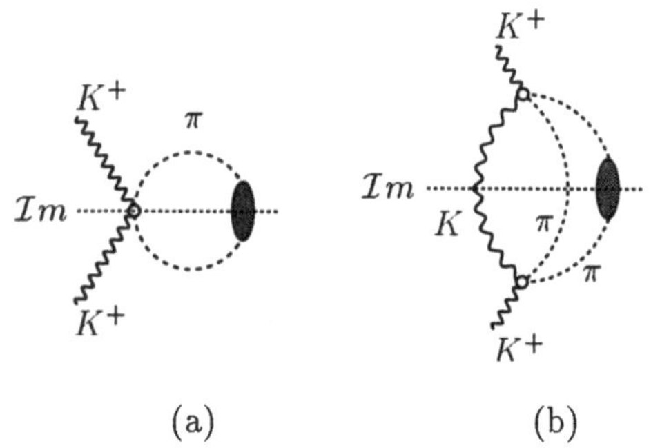

Figure 3. Feynman diagram (b) which imaginary part is related, through the optical theorem, to the imaginary part of the "standard" diagram of Fig. 1.

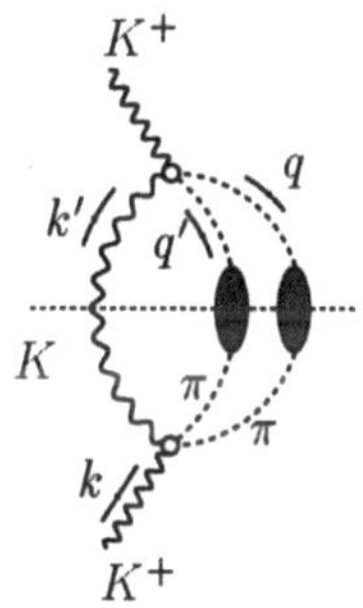

Figure 4. Diagram containing all the contributions to the imaginary part of the K^+-nucleus optical-potential up to second order in $t_{K\pi}$.

density is to be subtracted to eliminate selfenergy parts which are already included in the IA. By expanding the pion propagators in powers of the density, $D = D_0 + D_{(1)} + \delta D$ (being $D_{(1)}$ of order ρ, and δD the remaining terms of higher orders), using the symmetry $q \leftrightarrow q'$ (which holds due to the crossing symmetry of the amplitudes), applying Cutkosky rules and subtracting the terms constant and linear in ρ, we find that the imaginary part of the diagram of Fig. 4 is given by:

$$\begin{aligned}
\operatorname{Im} \Pi^{\text{MEC}}(k) = \; & 2 \int \frac{d^4q}{(2\pi)^4} \int \frac{d^4q'}{(2\pi)^4} \theta(q^{0\prime})\theta(q^0)\theta(k^{0\prime}) \times \\
& \operatorname{Im} D_K(k') \sum_\alpha |t^\alpha_{K\pi}(k', q'; k, -q)|^2 \Big|_{k'=k-q-q'} \times \qquad (13) \\
& \{ 2 \operatorname{Im} D_0(q') \operatorname{Im} \delta D(q) \qquad (\text{d1}) \\
+ \; & \operatorname{Im} D^{ph}_{(1)}(q') \operatorname{Im} D^{ph}_{(1)}(q) \qquad (\text{d2}) \\
+ \; & \operatorname{Im} D^{ph}_{(1)}(q') \operatorname{Im} D^{\Delta h}_{(1)}(q) + \operatorname{Im} D^{ph}_{(1)}(q) \operatorname{Im} D^{\Delta h}_{(1)}(q') \qquad (\text{d3}) \\
+ \; & \operatorname{Im} D^{\Delta h}_{(1)}(q') \operatorname{Im} D^{\Delta h}_{(1)}(q) \} \qquad (\text{d4})
\end{aligned}$$

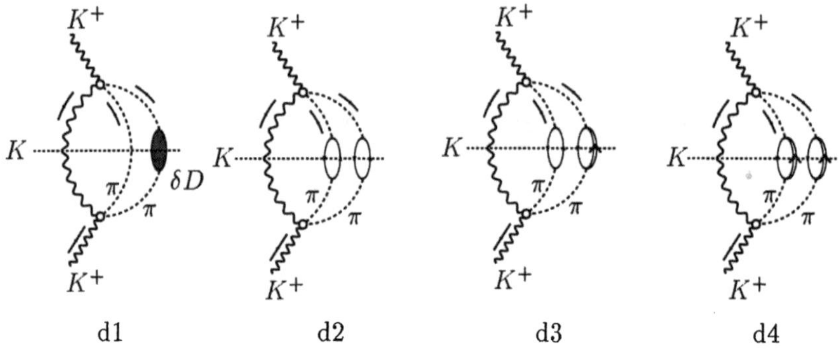

Figure 5. Diagrams contributing to $\mathrm{Im}\,\Pi^{\mathrm{MEC}}(k)$ up to second order in $t_{K\pi}$. The diagram d1 (standard contribution) corresponds to the process $K \rightarrow K\ \pi\ ph$, with threshold $T_K > m_\pi$. Diagrams d2, d3 and d4 are genuine new channels and correspond, respectively, to the processes: $K \rightarrow K\ ph\ ph$, $K \rightarrow K\ ph\ \Delta h$ and $K \rightarrow K\ \Delta h\ \Delta h$. Those processes have the following thresholds: $T_K \geq 222$ MeV $> m_\pi$ for diagram d1, $T_K > 0$ for diagram d2, $T_K \geq 181$ MeV $> m_\pi$ for diagram d3 and $T_K \geq 392$ MeV $> 2m_\pi$ for diagram d4.

where we have expanded the first medium correction to the pion propagator into ph and Δh components as $D_{(1)} = D_{(1)}^{ph} + D_{(1)}^{\Delta h}$. In Eq. (13), we have split $\mathrm{Im}\,\Pi^{\mathrm{MEC}}(k)$ into four different contributions, d1 (standard MEC contribution)[6], d2,d3, and d4, which correspond to the diagrams depicted in Fig. 5.

One could also think about effects from the renormalization of the intermediate kaon propagator in the diagram of Fig. 4. The impulse approximation $t_{KN}\ \rho$ provides the dominant part of the kaon selfenergy. From Fig. 2 one can see that $-\mathrm{Im}\,\Pi(k)/m_K^2 \sim 0.04$ is much smaller than $-\mathrm{Im}\,\Pi_\pi(k)/{m_\pi}^2$ and thus the corrections from this source can be estimated reasonably smaller than those obtained from pion renormalization.

The contributions of all MEC diagrams to the imaginary part of the K^+ selfenergy are approximately quadratic in density. In Fig. 6 the contribution of each of the diagrams[7] d1, d2, and d3 at density ρ_0 are shown for different kinetic energies of the kaon. Also the total of the MEC contributions, d1+d2+d3, is shown. For comparison the exact result for the "standard" calculation, $\mathrm{Im}\,\delta\Pi = \mathrm{Im}\,\Pi^{\mathrm{d1}}$, is depicted with dashed line. By itself it is much smaller than the total MEC contributions coming from diagrams d1, d2 plus d3. The diagram d2 is the most important MEC correction for low energies, but for higher energies the most relevant is d3, being more than half the total MEC

[6] By using the optical-theorem and after a little of algebra, one can prove [14] that $\mathrm{Im}\,\Pi^{\mathrm{d1}}$ is equal to the imaginary part of the standard (depicted in Fig. 1) MEC contribution, $\mathrm{Im}\,\delta\Pi(k)$, computed in Eq. (12).

[7] We do not evaluate the contribution of diagram d4 because its threshold is close to the highest energies we consider and we expect it to be small.

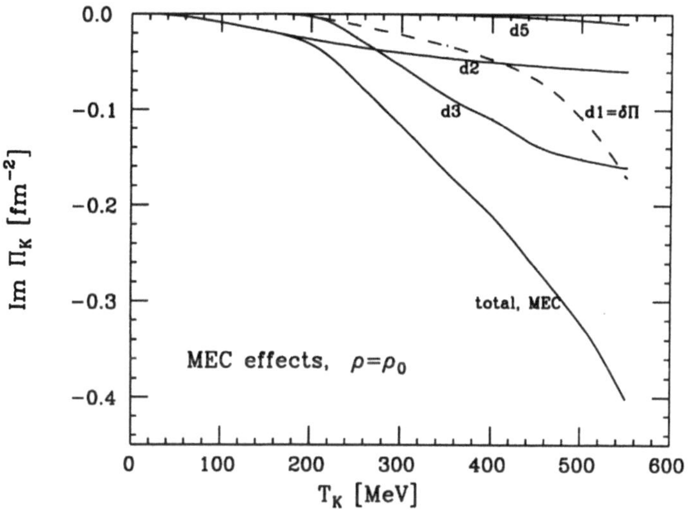

Figure 6. Imaginary part of the K^+ selfenergy for normal nuclear matter versus kinetic energy of the incoming kaon from the different mechanisms considered. Although not mentioned in the text, we also show the contribution (d5) of second order terms in density corresponding to a simultaneous excitation of two particles-two holes components by the pion in diagram d1 of Fig. 5. This contribution was computed in [14] and is negligible, in the range of energies considered, as seen in the figure.

effect for $T_K = 450$ MeV. In Fig. 2 we display the total MEC value together with the value of the IA. We see that the contribution due to the pionic cloud is sizable in relation to the dominant term which comes from the IA.

4 Real Part of the K^+-Nucleus Optical-Potential

We go back to Eq. (2) for $\delta\Pi(k)$ and substitute the isoscalar averaged t-matrix of Eq. (8) with its dispersion relation of Eq. (11). Then, we consider separately the contributions to $\delta\Pi$ coming from the analytical part of $\tilde{t}$ (this is P) and from its dispersive part (related to $\mathrm{Im}\,\tilde{t}$):

$$\delta\Pi \;=\; 3\,\mathrm{i}\int\frac{d^4q}{(2\pi)^4}\delta D(q)\left(P(s)+\int_{x_0}^{\infty}\frac{dx}{\pi}\frac{\mathrm{Im}\,\tilde{t}(x)}{x_0-x}\frac{s-x_0}{s-x+\mathrm{i}\epsilon}\right)\;, \quad (14)$$

where use has been made of crossing symmetry to cancel the factor 1/2. As $P(s)$ is a real polynomial in q^0 and, by doing a Wick rotation for q^0 , it can be proved that the first part (that going with $P(s)$) is real. The second part (going with $\mathrm{Im}\,\tilde{t}(x)$) is in general complex, its imaginary part being given by Eq. (12). This second part is linear in $\mathrm{Im}\,\tilde{t}(x)$. Due to the optical theorem $\mathrm{Im}\,\tilde{t}(x) \propto |\tilde{t}(x)|^2$. So, the first part is of order $\tilde{t}$ and the second one is of order $\tilde{t}^2$. We are keeping the leading order contribution to both, $\mathrm{Re}\,\delta\Pi$ and $\mathrm{Im}\,\delta\Pi$, this is: order $\tilde{t}$ for

$\text{Re}\,\delta\Pi$ and order $\tilde{t}^2$ for $\text{Im}\,\delta\Pi$. Within the same approximation , one should neglect for the real part the contribution of the diagrams d2, and d3. Note that to order $|\tilde{t}|^2$ there would be more diagrams besides these, which do not contribute, or contribute little, to the imaginary part, for instance d4.

With this approach of keeping the dominant order in $\tilde{t}$ and considering the incoming K^+ on-shell:

$$\text{Re}\,\delta\Pi(k) = -3\int \frac{d^4q}{(2\pi)^4}\text{Im}\,\delta D(q)\; P(s) \tag{15}$$

$$= -3\int \frac{d^3\boldsymbol{q}}{(2\pi)^3}\int_0^\infty \frac{dq^0}{\pi}\text{Im}\,\delta D(q)\left[\frac{\alpha_0}{2} + \beta_0(m_K^2 + {q^0}^2 - \boldsymbol{q}^2)\right] \tag{16}$$

where we have used $P(x) = \alpha_0/2 + \beta_0\, x$, which in the model of [13], is the dominant contribution to $P(x)$ and comes from the $s-$wave, the $p-$wave contribution has been neglected. The numerical values for the parameters α_0 and β_0 can be found in [13]. Direct experimental results for $K\pi$ scattering are not available. There are some indirect measurements which make use of the analysis of the $KN \to KN\pi$ reaction from which phenomenological on-shell $K\pi$ phase shifts can be derived. Unfortunately this information does not completely fix the parameters α_0 and β_0 of the $t_{K\pi}$ matrix: there are different pairs (α_0,β_0), compatible with the on-shell $K\pi$ phase shifts, which lead to totally different results[8] for $\text{Re}\,\delta\Pi(k)$ in Eq. (16). These uncertainties in $\delta\Pi$ result in large uncertainties in the K^+-nucleus cross section at small K^+ energies, while at large energies ($T_K^+ \sim 500-600$ MeV) these uncertainties are drastically reduced, as can be seen in Fig. 7, for the case $K^+ -\,^{12}C$ scattering.

However, in a recent work [18], Meißner, Oset and Pich have shown that, for symmetric nuclear matter, $\text{Re}\,\Pi^{\text{MEC}}$ is zero at lowest order in Chiral Perturbation Theory[9]. Obviously one may wonder what would happen if one goes to higher orders in the chiral expansion. The K^+ sector might be a privileged one for convergence of the chiral expansion, since one is far away from resonances in the $K^+\pi$ channel (the $K^*(892)$ has a negligible influence in the $K\pi$ amplitude in the range of energies considered in this work) and there are no resonances in the K^+N channel.

Thus, combining our results of the previous section for $\text{Im}\,\Pi^{\text{MEC}}$ with those of [18] for the real part, we find that in lowest order in the chiral expansion, and for symmetric nuclear matter, the contribution of the pionic cloud to the K^+-nucleus optical-potential is purely imaginary and is given by Eq. (13). In the next section we will solve the Klein-Gordon equation for a K^+ scattered by a nucleus and we will compare with the experimental data both the results obtained using an optical-potential given by the impulse approximation (IA)

[8] In references [13] and [14] was shown that different off-shell extrapolations give results for $\text{Re}\,\delta\Pi(k)$ which may differ in size up to an order of magnitude, and even in sign.

[9] In ref. [18], couplings of the type $NNKK\pi$ and $N\Delta KK\pi$ (which appear at the same order in the chiral expansion as the coupling $KK\pi\pi$, considered in the derivation of Eq. (16)) are also taken into account. The inclusion of new diagrams, derived from these couplings, results in multiple cancellations which finally lead to $\text{Re}\,\Pi^{\text{MEC}} = 0$.

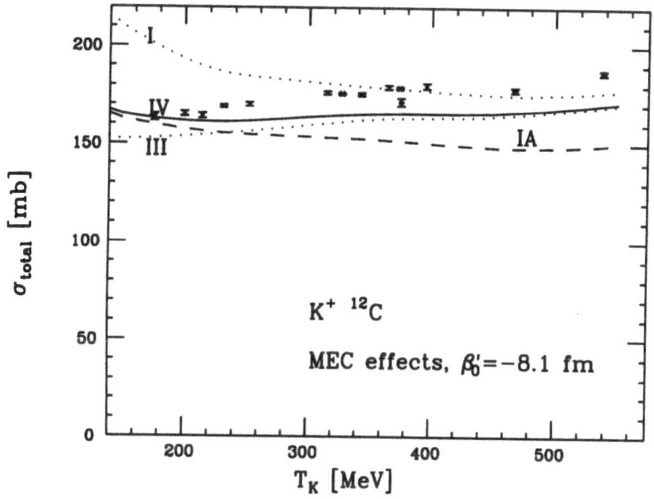

Figure 7. Total cross section for K^+ scattered by ^{12}C versus kinetic energy of the K^+. The experimental data with a cross are from ref. [4], the data with a diamond are from ref. [1], only the statistical errors are included, the systematic errors, not shown, are larger. The dashed line labeled IA corresponds to the impulse approximation. The curves labeled I, III, IV correspond to different values of Re $(\delta\Pi)$ coming from different $K^+\pi$ off-shell extrapolations and the same Im Π^{MEC}. The curve labeled IV (solid line) corresponds to Re $(\delta\Pi) = 0$. Note that Im Π^{MEC}, as computed in Sec. 3 does not depend on the parametrization.

and those obtained when the pionic cloud effects, Im Π^{MEC} given in Eq. (13) and Re Π^{MEC}=0 , are added to the IA-optical-potential (IA+MEC).

5 Results and Conclusions

In Fig. 8 we show the results for $d\sigma/d\Omega$ for K^+ with energy $T_K = 450$ MeV scattered by ^{40}Ca. Coulomb effects have not been included. Results with the impulse approximation are compared to those including pion cloud effects (with Re Π^{MEC}=0, parametrization IV in ref. [14]).

The total cross-section of K^+ scattered by ^{12}C versus kinetical energy of the K^+ is shown in Fig. 7. The curve labeled IV (solid line) corresponds to Re $\Pi^{MEC} = 0$.

Figures 7 and 8 show that the inclusion of MEC effects in the imaginary part significantly improves on the impulse approximation, bringing both the total and the elastic cross-sections closer to the experimental ones, and showing that MEC effects are large enough to have to be considered in this process.

Recent experiments measure the ratio of total cross section for $K^+ + {}^{12}C$ to that of $K^+ + d$ as a function of p_{lab} for a range $480 < p_{lab} < 740$ MeV/c [3, 4].

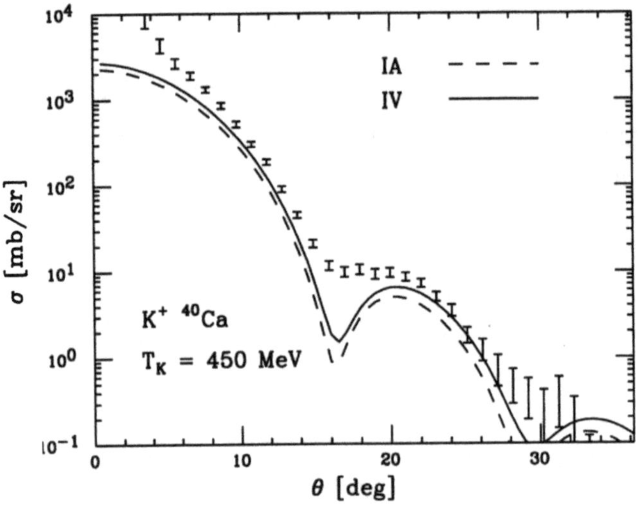

Figure 8. Differential cross section of a K^+ with kinetic energy of 450 MeV scattered by a ^{40}Ca nucleus. Dashed line depicts the results using the IA. Solid line includes IA plus the optical potential coming from the pionic cloud (with $\mathrm{Re}\,\Pi^{\mathrm{MEC}} = 0$, parametrization IV in ref. [14]). The experimental data are from ref. [3]. In computing the IA-optical potential, we have used the KN phase shifts of Arndt and co-workers [19].

An earlier experiment covered the higher momentum range $700 - 1000$ MeV/c [1]. In this ratio of cross sections, some uncertainties both theoretical (those related with the KN scattering amplitude used in the IA) and experimental (those related with the overall normalizations) partially cancel [7]. Data are plotted in Fig. 9 as $R = \sigma(K^+\ {}^{12}\mathrm{C})/6\sigma(K^+\ {}^{2}\mathrm{H})$, to emphasize the closeness of the ratio to unity. This would be the value in the limiting case that the K^+ scattering on these nuclear targets were given completely by free K^+N (isoscalar) single scattering, with no medium corrections.

In order to do a meaningful comparison of theoretical calculations with experiments, a more realistic K^+-nucleus standard optical potential (than the crude $t\rho$ impulse approximation used in Figs. 7 and 8) should be considered. There are corrections over this $t\rho$ which should be included like off-shell range, binding energy,... Also, there are nucleon-nucleon correlations which contribute to the optical potential in second order. These correlation effects are approximately ρ^2 by nature, which is the same form as the MEC contributions to the optical potential. Both kinds of corrections to the optical potential (and their effect on the K^+-nucleus cross-sections) have been calculated and/or estimated in the works of refs. [6, 7]. As result of their studies, they provide a band of uncertainty for the conventional calculation, the boundaries of the band are determined by varying the parameters in the theoretical model.

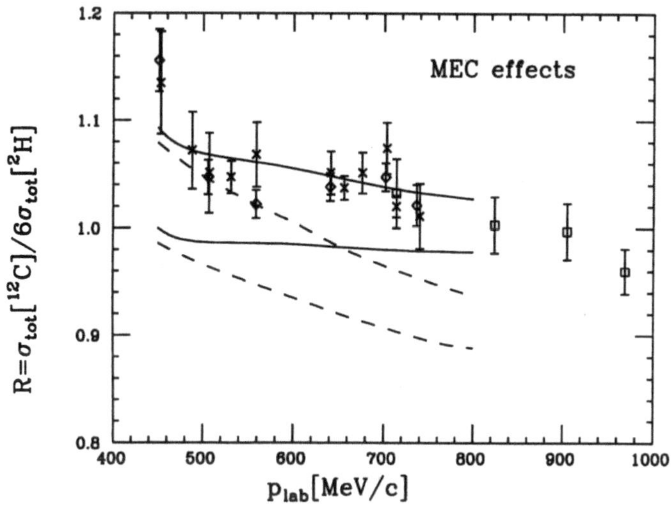

Figure 9. The ratio of cross-sections $R = \sigma(K^+\ ^{12}\mathrm{C})/6\sigma(K^+\ ^2\mathrm{H})$ is plotted against the lab momentum of the incoming K^+. The experimental data are: crosses from ref. [4], diamonds from ref. [3], squares from ref. [1]. The two dashed lines define the band of uncertainties of the theoretical results for R obtained with the conventional optical potential of ref. [7]. The two solid lines define the band of uncertainties of the theoretical results for R when the corrections due to MEC calculated here are added to the results obtained with the conventional optical potential of ref. [7].

In Fig. 9 the dashed-lines indicated the band of theoretical calculations for R with the conventional microscopic optical potential of ref. [7], for $p < 500$ MeV/c the results quoted are from ref. [3]. The solid-line band in Fig. 9 shows the effect of adding our MEC correction to the multiple scattering calculation of ref. [7].

We observe that the addition of MEC corrections to the conventional optical calculation of Siegel *et al.* provides a significative agreement of the calculation with the experiment for a large range of energies.

The main conclusion of this paper is that the effects of the mesonic cloud in K^+ nucleus scattering are relevant and of the right order of magnitude to account for the discrepancies of the conventional optical potential [7] with the data. However, there are uncertainties, particularly in the real part of the MEC, which do not allow us to draw stronger conclusions about the actual size of the corrections. The comparison of our results using $\mathrm{Re}\,\Pi^{\mathrm{MEC}} = 0$, as obtained in [18] at lowest order in the chiral expansion, with the data is reasonably good. On the other hand we have also observed that the direct relation of Π^{MEC} to the distribution of excess pions in the nucleus, as has been formerly assumed in refs. [12] and [13], is a consequence of a dangerous static approximation which should be avoided. In the present case it induced an error of a factor of two but in other cases it can induce errors of several orders of magnitude. We also found that in any case this contribution was only a small part of the total MEC corrections tied to the interaction of kaons with the nuclear pions.

Acknowledgement. J.N. wishes to acknowledge the hospitality of the Physics Department in Southampton, where part of this work was done. The Feynman graphs in this paper were produced with the Axodraw package by J.Vermaseren. We thank D.Burford for producing these graphs for us.

References

1. D. V. Bugg et al.: Phys. Rev. **168**, 1466 (1968)
2. D. Marlow et al.: Phys. Rev. **C25**, 2619 (1982)
3. Y. Mardor et al.: Phys. Rev. Lett. **65**, 2110 (1990)
4. R. A. Krauss et al.: Phys. Rev. **C46**, 2019 (1992)
5. J. Alster: Nucl. Phys. **A547**, 321c (1992)
6. P. B. Siegel, W. B. Kaufmann, W. R. Gibbs: Phys. Rev. **C30**, 1256 (1984)
7. P. B. Siegel, W. B. Kaufmann, W. R. Gibbs: Phys. Rev. **C31**, 2184 (1985)
8. C. M. Chen, D. J. Ernst: Phys. Rev. **C45**, 2019 (1992)
9. G. E. Brown, C. B. Dover, P. B. Siegel, W. Weise: Phys. Rev. Lett. **60**, 2723 (1988)
10. W. Weise: Nuovo Cimento **A102**, 265 (1989)
11. R. Jaffe et al.: Phys. Lett. **B134**, 449 (1984)
12. S. V. Akulinichev: Phys. Rev. Lett. **68**, 290 (1992)
13. M. F. Jiang D. S. Koltun: Phys. Rev. **C46**, 2462 (1992)
14. C. García-Recio, J. Nieves, E. Oset: Phys. Rev. **C51**, 237 (1995)
15. J. Nieves, E. Oset, C. García-Recio: Nucl. Phys. **A554**, 554 (1993)
16. E. L. Berger, F. Coester, R. B. Wiringa: Phys. Rev. **D29**, 398 (1984); E. L. Berger, F. Coester: Phys. Rev. **D32**, 1071 (1985); B. L. Friman, V. R. Pandharipande, R. B. Wiringa: Phys. Rev. Lett **51**, 763 (1983)
17. M. Ericson, M. Rosa-Clot: Phys. Lett. **B188**, 11 (1987)
18. U. Meißner, E. Oset, A. Pich: Phys. Lett. **B353**, 161 (1995)
19. R. A. Arndt, L. D. Roper P. II. Steinberg: Phys. Rev. **D18**, 3278 (1978); R. A. Arndt, L. D. Roper: Phys. Rev. **D31**, 2230 (1985)

Few-Body Systems Suppl. 9, 51–60 (1995)

Few-Body Systems

Searching for the H-Dibaryon at Brookhaven

AGS E813/E836 Collaboration

B. Bassalleck[12], M. Athanas[3*], P.D. Barnes[9], A. Berdoz[3], A. Biglan[3], J. Birchall[11], T. Bürger[4], M. Burger[4], R.E. Chrien[2], C. Davis[11], G.E. Diebold[14], H. En'yo[7], H. Fischer[12], G.B. Franklin[3], J. Franz[4], L. Gan[11], D. Gill[10], T. Iijima[7†], K. Imai[7], P. Koran[3], M. Landry[11], L. Lee[11], J. Lowe[1,12], R. Magahiz[3], A. Masaike[7], C.A. Meyer[3], R. McCrady[3], F. Merrill[3], J.M. Nelson[1], K. Okada[8], S. Page[11], P.H. Pile[2], B. Quinn[3], D. Ramsay[11], E. Rössle[4], A. Rusek[12], M. Rozon[3**], R. Sawafta[2], H. Schmitt[4], R.A. Schumacher[3], R.L. Stearns[13], R. Stotzer[12], R. Sukaton[3], V. Sum[11], R. Sutter[2], J.J. Szymanski[5], F. Takeutchi[8], W.T.H. van Oers[11], D.M. Wolfe[12], K. Yamamoto[7], M. Yosoi[7], V. Zeps[6], R. Zybert[1]

[1] U. of Birmingham, Dept. of Phys., Birmingham B15 2TT, UK
[2] Brookhaven National Lab, Upton, Long Island, NY 11973, USA
[3] Carnegie Mellon U., Dept. of Phys., Pittsburgh, PA 15213, USA
[4] U. of Freiburg, Fakultät f. Physik, D-79104 Freiburg, Germany
[5] Indiana U. Cyclotron Facility, Bloomington, IN 47405, USA
[6] U. of Kentucky, Dept. of Phys., Lexington, KY 40507-005, USA
[7] Kyoto U., Dept. of Phys., Sakyo-Ku, Kyoto 606, Japan
[8] Kyoto Sangyo U., Faculty of Science, Kyoto 603, Japan
[9] Los Alamos National Lab, Los Alamos, NM 87545, USA
[10] TRIUMF, Vancouver, BC V6T 2A3, Canada
[11] U. of Manitoba, Dept. of Phys., Winnipeg, Manitoba, R3T 2N2, Canada
[12] U. of New Mexico, Dept. of Phys., Albuquerque, NM 87131, USA
[13] Vassar College, Dept. of Phys., Poughkeepsie, NY 12601, USA
[14] Yale U., Phys. Dept., New Haven, CT 06511, USA

*now at U. of San Diego
†now at KEK
**now at U. of Alberta

Abstract. At the Brookhaven AGS several experiments are searching for the unique strangeness $S = -2$ H-dibaryon with the quark composition ($uuddss$). The E813/E836 collaboration, in particular, is using a high-intensity, separated 1.8 GeV/c K^- beam and two different target configurations. In E836 the reaction $K^- + {}^3\text{He} \rightarrow K^+ + H + n$ is used to search for a relatively deeply-bound H. Complementary to E836 the reactions $K^- + p \rightarrow \Xi^- + K^+$, followed by $(\Xi^-, d)_{\text{atom}} \rightarrow H + n$ are used to search near twice the Λ mass. The status of these two experiments is summarized, and other H-dibaryon searches are briefly reviewed.

1 Introduction

While QCD is the presently accepted theory of the strong interaction, its successes in the non-perturbative regime are clearly limited. One striking example of this are the difficulties in calculating the hadronic mass spectrum from fundamental principles of the theory. These difficulties have led to many QCD-inspired models like the MIT bag model. Apart from the spectroscopy of standard ($qqq, q\overline{q}$) hadrons the search (both experimentally and theoretically) for non-standard hadrons has been of fundamental importance for a long time. These non-standard hadrons ('exotics') contain more than the minimal number of quarks, and their properties correspond to quark configurations rather than hadronic molecules, like the deuteron.

Among the spectrum of six-quark (or dibaryon) states, the strangeness $S = -2$ sector plays a special role. Indeed, only a six-quark system containing $uuddss$ quarks can exist in an SU(3)-flavor singlet, a configuration which takes maximum advantage of the color-magnetic attraction. It is possible that such a six-quark system might be *stable* with respect to strong decay into all baryon-baryon channels (the lowest being $\Lambda\Lambda$). This was first noted by Jaffe [1], who predicted this ($uuddss$) state with $J^P = 0^+, I = 0$, which he called the H particle, to have a mass some 80 MeV/c^2 below the $\Lambda\Lambda$ threshold. Since Jaffe's bag-model prediction, many other calculations of the H-mass using many different models have appeared: variations of the bag model, quark cluster models, Skyrme models, lattice gauge calculations, instanton models, and QCD sum rules. Reference [2] contains a list of references with different H-mass calculations. These calculations give a very wide range of predicted H-masses, from much more deeply bound than in the original calculation to unbound relative to the $\Lambda\Lambda$ threshold. Although H-mass calculations are model dependent, the H does *not* appear to be an artifact of the bag model. Increased binding comes from the color-magnetic interaction, which gives the strongest attraction for the most symmetric color-spin representation, corresponding to the most antisymmetric SU(3)-flavor singlet representation. The significance of the H-dibaryon quark combination is inherent in QCD, hence the H appears in many different models.

Independent of calculational details, one expects that an object with the quantum numbers of the H is the most likely candidate for a stable six-quark bag state. If the H exists and is indeed stable, it will be a unique object in multiquark spectroscopy ($n_q > 3$). Given the theoretical uncertainties in the mass (and lifetime) of the H, its existence or non-existence clearly has to be settled by experiment.

2 H-Dibaryon Production Mechanisms

Experimentally the H can, in principle, be produced in many different reactions. The techniques that have been and/or are being implemented fall into two categories. Either a modern $S = -1$ secondary beam (K^- or Σ^-) is used in a strangeness exchange reaction like $K^- + X \rightarrow K^+ + H + Y$, or a non-strange beam (protons or relativistic heavy ions) is used in a double associated production like $p+p \rightarrow K^+ + K^+ + H$. An ideal H-search experiment needs to encompass the following goals: good sensitivity over as wide a range in H mass as possible, no dependence on the (unknown) lifetime nor the decay modes of the H, simple and experimentally clean triggers and reactions, and using low momentum transfer in order to enhance the chances of the six quarks actually forming the H.

We will focus on two recent Brookhaven experiments employing a new high-intensity separated K^- beam. They were designed to cover complementary H-mass ranges, and both are completely independent of the decay modes and the lifetime of the H. Several other H-searches will be briefly summarized at the end.

3 AGS E813 and E836 Experiments

Both experiments employ the (K^-, K^+) double strangeness exchange reaction in order to transfer two units of strangeness to a nuclear target, and thus end up with a $S = -2$ system. In E813 1.8 GeV/c K^- are used to produce Ξ^- via $K^-p \rightarrow \Xi^- K^+$. The K^- momentum is determined in the final stage of a high-intensity and high-purity kaon beamline, which is described in ref.[3]. The K^+ is analyzed under small forward angles in a single, large-aperture dipole spectrometer. The detector setup for both E813 and E836 is shown in Fig.1.

The relatively low-energy Ξ^- from the above reaction are slowed down in W degraders, and a fraction of them will range out in a liquid deuterium target. The geometry of this combined LH_2/LD_2 target of E813 was optimized using a Monte Carlo simulation with the purpose of maximizing the stopping fraction of Ξ^-, see Fig.2. Nevertheless the majority of Ξ^- decay in flight. Si detectors between the hydrogen and deuterium vessels are used to tag the very slow Ξ^-, i.e. the ones with maximum stopping probability. Finally H-formation is searched for via $(\Xi^- d)_{\text{atom}} \rightarrow Hn$, where a monoenergetic neutron would carry the information on the mass of the H. The branching ratio for this last reaction is of course unknown. However, it has been calculated by Aerts and Dover [4]. Especially for a lightly-bound H this branching ratio has been predicted to be quite large, thus maximizing our sensitivity near $\Lambda\Lambda$-threshold. In both experiments momentum, time-of-flight, and aerogel Cerenkov detectors

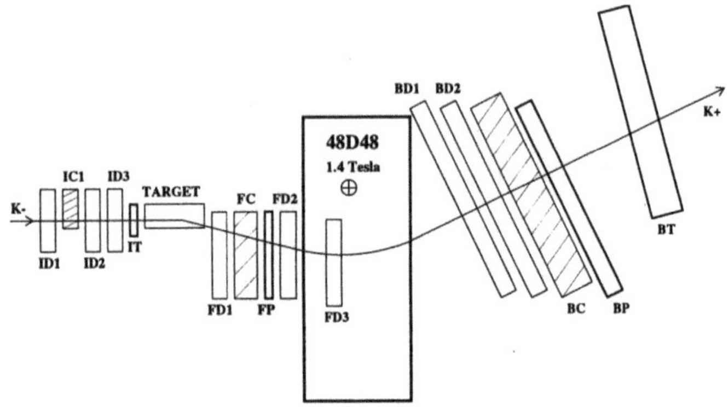

Figure 1. Side view of E813/E836 detector setup. ID, FD, and BD are drift chambers for tracking and momentum measurement. IT, FP, BP, and BT are plastic scintillator hodoscopes for triggering and timing. IC, FC, and BC are aerogel Cerenkov detectors. Not shown are the plastic scintillator neutron time-of-flight detectors, mounted on the left and right of the target.

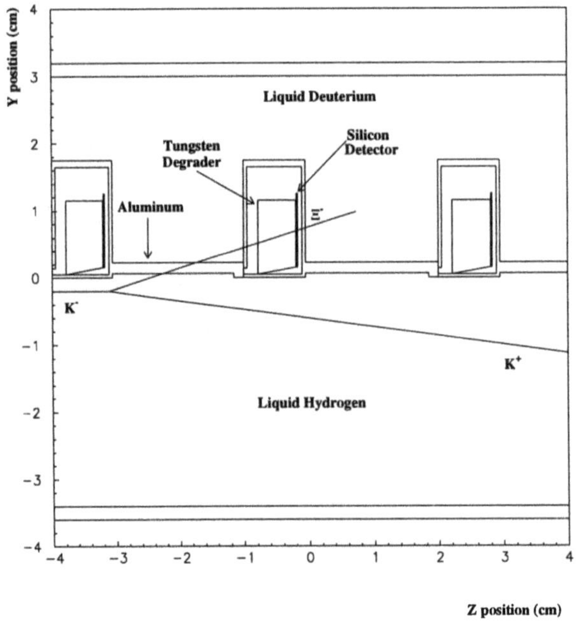

Figure 2. E813 target geometry. Shown are three of the twenty cells.

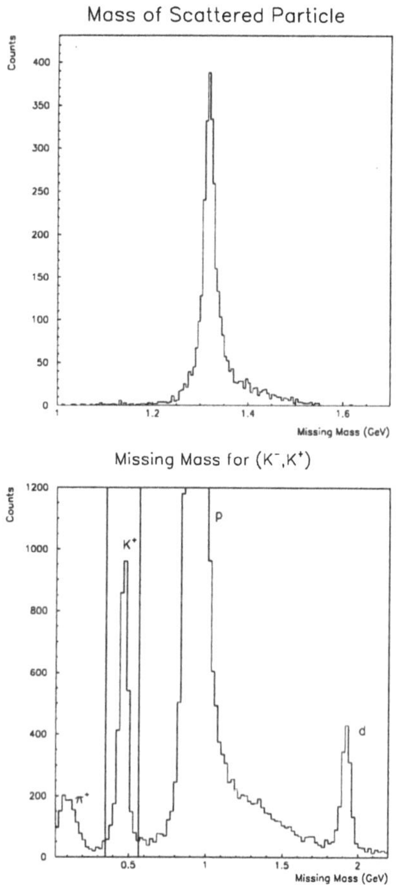

Figure 3. Examples of particle identification in E813: reconstructed mass of outgoing particle (top), and reconstructed missing mass peak corresponding to Ξ^-.

are used for particle identification, see Fig.3.

Even though only a small fraction of the Ξ^- stop in the deuterium, the stopping probability per event is maximized in the analysis by carefully chosen cuts on the outgoing K^+ angle and cuts on the Si detector pulse height [5]. This 'tagging' of the slow hyperons was tested by producing Σ^- via $\pi^- p \rightarrow \Sigma^- K^+$, and stopping the Σ^- in hydrogen (upper target vessel filled with LH_2 instead of LD_2). The kinematics of this reaction is very similar to $K^- p \rightarrow \Xi^- K^+$. With about 50% branching ratio the $(\Sigma^- p)_{atom}$ will decay into $\Lambda + n$, resulting in a monoenergetic 43 MeV neutron. This reaction chain therefore tests our tagging ability as well as the neutron detection. Fig.4. shows a preliminary result from this Σ^- test, establishing the feasibility of the overall technique of E813.

The vast majority of the E813 (K^-, K^+) data for the H-search were taken

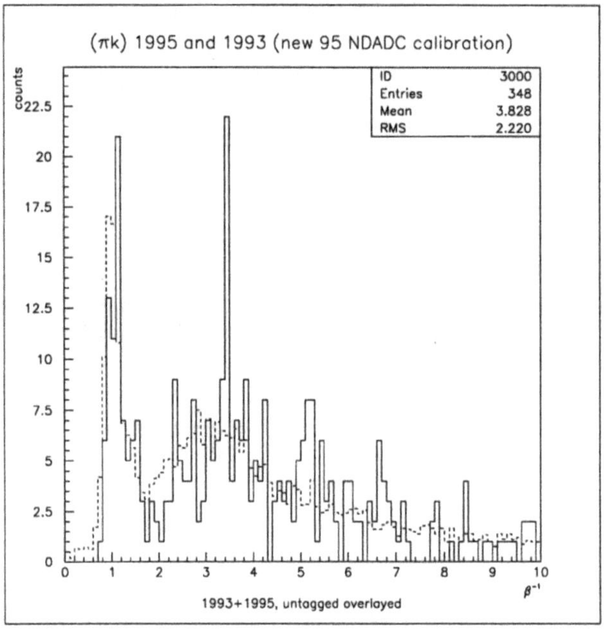

Figure 4. *Preliminary* neutron time-of-flight spectrum in units of inverse beta for tagged Σ^- events. The peak at $\beta^{-1} = 1$ corresponds to prompt gammas, whereas the peak around $\beta^{-1} = 3.5$ is the expected signal (43 MeV neutron) from $(\Sigma^- p)_{\text{atom}} \rightarrow \Lambda + n$. The dashed line represents the measured untagged background, scaled by the measured tagging fraction.

in 1993 and 1995. Only the 1993 data are analyzed at this point, and therefore only preliminary results are available. Some representative neutron time-of-flight spectra for tagged and untagged events from the 1993 (K^-, K^+) data are shown in Fig.5. It is worth pointing out that the untagged spectrum is *not* the result of a fit, but is directly measured. Scaled by the measured tagging fraction it obviously describes the overall shape of the tagged spectrum very well. A detailed and elaborate statistical analysis of the 1993 data has been performed [5]. It is applied to a search for potential peaks (including establishing their statistical significance) as well as to the setting of upper limits on the branching ratio for $(\Xi^-, d)_{\text{atom}} \rightarrow H + n$. Within the limited statistics of the presently analyzed data we do not see any statistically significant structure associated with H-production. For example, studies of the expected statistical fluctuations have shown that the probability of a fluctuation anywhere in the spectrum as large as or larger than that seen at $\beta^{-1} \sim 3.0$ is 0.08. An example of fitting a spectrum with background shape plus a Gaussian of known width is shown in Fig.6. The width in β^{-1} of any signal peak is known from various neutron detector resolution studies. Within such a peak width the number of background neutrons is around 8. The results of Fig.6 thus show that even

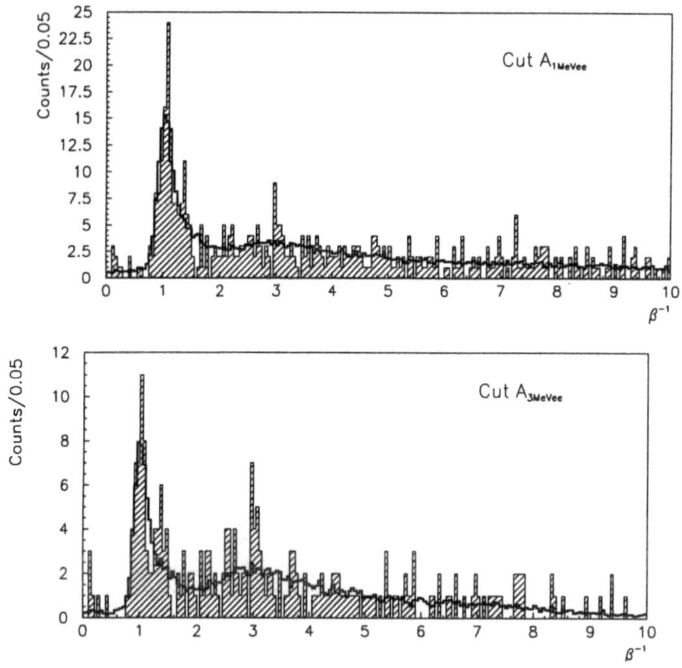

Figure 5. *Preliminary* tagged neutron time-of-flight spectra (in units of inverse beta) from the 1993 E813 data for 1. MeVee (top) and 3. MeVee (bottom) neutron pulse height threshold. The untagged spectrum, scaled by the tagging fraction, is superimposed as a solid line.

for the most significant deviation in the spectrum the signal to background ratio is only $\sim 1.4 : 1$. Addition of the 1995 data will result in a significant improvement of the overall statistics. Assuming that no signal will be found, we should be able to set upper limit on the above-mentioned branching ratio clearly below the theoretical predictions of ref.[4] for a lightly-bound H.

In E836 the same (K^-, K^+) reaction was used on a single nuclear target, with data existing for ^{3}He, ^{6}Li, and ^{12}C. Effectively a *pp*-pair is converted into an H. The $^3\text{He}(K^-, K^+n)H$ reaction is particularly interesting since ^{3}He is a light nucleus with well-known wavefunction, and a detailed theoretical calculation exists for comparison [6]. In addition, our experimental setup allows the measurement of the same three momenta as in E813 (K^-, K^+, n), for a kinematically complete determination. However, as pointed out and calculated by Aerts and Dover [6], a relatively deeply-bound H will manifest itself as a well-separated, narrow peak in the K^+ momentum spectrum above the region of quasi-free Ξ^- production ($K^- + ^3\text{He} \rightarrow K^+ + \Xi^- + p + n$). Therefore

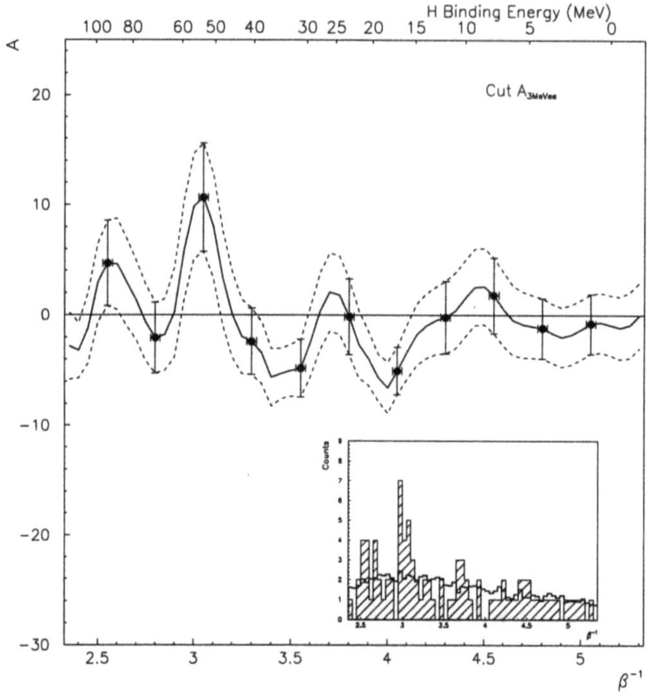

Figure 6. Results of fitting the bottom spectrum of Fig.5 (inlayed histogram) with background shape plus a Gaussian of known width. The central line is the maximum likelihood number of neutrons in the peak resulting from the fit. The upper and lower lines are the one σ limits.

our main thrust is a search for such a structure above the endpoint of the quasi-free region. Data for E836 were taken in 1994, and the analysis is ongoing. Preliminary results are shown in Fig.7, indicating no significant structure. E836 requires excellent particle identification, especially π/K separation. The reason is that both incoming π^- (via $\pi^- p \to \Sigma^- K^+$) and outgoing π^+ (via $K^- p \to \Sigma^- \pi^+$) can produce events above the Ξ^- quasi-free endpoint. We are therefore analyzing the reponses of our time-of-flight system and of the Cerenkov detectors in great detail. Preliminary results are very encouraging, as shown in Fig.7. The tail of the quasi-free Ξ^- production limits the sensitivity of E836 for a lightly-bound H, i.e. near $\Lambda\Lambda$-threshold. Therefore we are optimizing the momentum resolution in order to enhance the sensitivity in the threshold region as much as possible. E836 is characterized by acceptance over a wide range of H-masses, from a few tens of MeV of binding energy down to hundreds of MeV, and is thus complementary to E813. We also note (see Fig.7 bottom) that we should be able to set upper limits against H-production

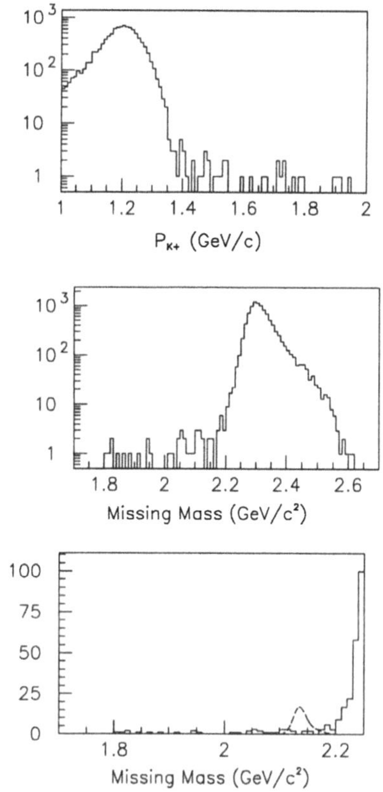

Figure 7. *Preliminary* results from E836, $^3\mathrm{He}(K^-, K^+n)H$. The top shows the reconstructed outgoing K^+ momentum spectrum, dominated by quasi-free Ξ^- production with an endpoint near 1.4 GeV/c. The middle and bottom show a "missing mass" spectrum, calculated assuming a di-proton target, corrected for binding energy in ^{3}He. The dashed line in the bottom spectrum is the expected signal for an H-mass of 2.13 GeV/c^2, using the theoretical calculation of ref.[6] plus folding in the experimental momentum resolution.

that are significantly below theoretical expectations. Therefore the final E836 results will represent one of *the* most sensitive H-searches. Significant results from ^{6}Li will also be available.

4 Other H-Searches at Brookhaven

The first BNL H-search experiment [7], using $pp \to K^+K^+X$ and resulting in lower sensitivity than theoretically needed, will be repeated at KEK as E248.

E888, using a setup from a rare K-decay experiment, used 24 GeV/c protons in $p+\mathrm{Cu/Pt} \to H + X$ collisions [8]. Large P_t Λ's from $H \to \Lambda n$ were searched for via $\Lambda \to p\pi^-$. In addition, H signature was looked for via $Hp \to \Lambda\Lambda p$ in an active dissociator. The analysis of E888 is presently being finished.

Several relativistic heavy ion experiments also focus, at least partly, on H-searches. E810 (14.6 AGeV/c Si + Pb collisions) is searching for evidence for $H \to \Sigma^- p$ decays in a TPC. They report some interesting candidate events [9], but remain cautious at this stage and continue to process more data. E864 is taking data now searching for long-lived H's [10], while E896 will be a search for short-lived H's [11]. Both use the relativistic AGS Au beam of 11 AGeV/c.

References

1. R. L. Jaffe: Phys. Rev. Lett. **38**, 195 (1977); **38**, 1617(E) (1977)
2. M.A. Moinester et al.: Phys. Rev. **C46**, 1082 (1992)
3. P.H. Pile et al.: Nucl. Instr. and Meth. **A321**, 48 (1992)
4. A.T.M. Aerts and C.B. Dover: Phys. Rev. **D29**, 433 (1984)
5. F. Merrill: Ph.D. dissertation. Carnegie Mellon University 1995
6. A.T.M. Aerts and C.B. Dover: Phys. Rev. **D28**, 450 (1983)
7. A.S. Carroll et al.: Phys. Rev. Lett. **41**, 777 (1978)
8. A.J. Schwartz: SLAC-Report-444, p.393 (1993)
9. R. Longacre: Proc. of Quark Matter '95
10. J. Sandweiss et al.: AGS E864
11. H. Crawford et al.: AGS E896

Few-Body Systems Suppl. 9, 61–64 (1995)

Few-Body Systems

Elastic Scattering of K^+ from Light Nuclei

P. Bydžovský, M. Sotona

Institute of Nuclear Physics, Řež, Czech Republic

Abstract. Sensitivity of the total cross section for ^{12}C and d to the relativistic corrections is investigated in the multiple scattering formalism. The relativistic treatment of the dynamics of the K^+-nucleus (K^+A) and the K^+-nucleon (K^+N) systems is found to be important. Results of calculations with a coupled-channel separable model for the K^+N amplitude are presented for the $\sigma^{\rm tot}$ of the K^+N and K^+A scattering.

The K^+-meson is supposed to be a good tool for studies of the nuclear interior due to its relatively weak interaction with nucleons. The long free path of K^+ inside the nucleus suggests that a single and double K^+N scattering dominates the K^+A collision [1] so that the first-order optical potential approximation can give a reasonable description of the process.

The optical model calculations of the ratio $R_{\rm T} = \sigma^{\rm tot}(^{12}{\rm C})/6\sigma^{\rm tot}(d)$, however, underestimate the experimental data by ≈10% [1]. Attempts to explain this discrepancy are based on a modification of the K^+N amplitude inside the nucleus [1, 2] introducing density-dependent effective interaction.

The conventional medium effects such as Pauli blocking, binding and recoil effects were found to give a small (≈1%) contribution to the differential cross section [1]. In ref. [1] an off-shell extension of the K^+N amplitude was introduced. The differential cross section was found to be sensitive to variations in the range of the form factor in the first minimum and the second maximum.

Our aim is to show how the results of the model calculations depend on the relativistic treatment of Green's functions, transformation of momenta, energy and the K^+N t–matrix and on variations in the off-shell form of the K^+N t–matrix. In the following we first briefly outline the main points of the formalism which was previously applied to the pion-nucleus scattering [3] and than discuss the results.

The full-folding integral is simplified by the "optimal factorization approximation" (OFA). The optical potential than reads

$$< \boldsymbol{Q}'|U^{(1)}(E)|\boldsymbol{Q} >= (A-1)t(\varepsilon,\boldsymbol{p}',\boldsymbol{p})\mathcal{F}(\boldsymbol{q}), \tag{1}$$

where $\boldsymbol{Q}(\boldsymbol{Q}')$ and $\boldsymbol{p}(\boldsymbol{p}')$ are initial (final) kaon momenta in the K^+A (Ac.m.) and the K^+N (2c.m.) center of mass frames, respectively, and $\boldsymbol{q} = \boldsymbol{Q}' - \boldsymbol{Q}$. The effective nucleon momentum in Ac.m. is then set to $\boldsymbol{k}_e = \boldsymbol{q}/2 - (\boldsymbol{Q}' + \boldsymbol{Q})/2A$ in OFA.

In a semi-relativistic model (R) Ac.m. $\rightarrow$ 2c.m. transformation of kaon momenta is chosen as

$$\boldsymbol{p} = \frac{\epsilon_2}{\omega}\boldsymbol{Q} - \frac{\epsilon_1}{\omega}\boldsymbol{k}_e, \tag{2}$$

with $\epsilon_{1,2}$ being the relativistic energies of K^+ and N in Ac.m. and ω being the total K^+N energy in 2c.m. The effective two body energy in 2c.m. is postulated

$$\varepsilon = E - W_{KN} - E_{\mathrm{cor}}, \tag{3}$$

where W_{KN} stands for the K^+N center of mass kinetic energy and E_{cor} is the total energy of the A-1 nucleons (a core). The K^+N t-matrix is evaluated in 2c.m. and transformed into Ac.m. Assuming also a relativistic treatment of the K^+N scattering a relation between the K^+N t-matrix in Ac.m. and the amplitude in 2c.m. reads as follows

$$t_{A\mathrm{cm}}(\varepsilon, \boldsymbol{p}', \boldsymbol{p}) = -2\pi\sqrt{\frac{\omega\omega'}{\epsilon_1\epsilon_1'\epsilon_2\epsilon_2'}} f_{2\mathrm{cm}}(\varepsilon, \boldsymbol{p}', \boldsymbol{p}). \tag{4}$$

The K^+A Green's function in the Lippman-Schwinger type equation was replaced by the relativistic form

$$< \boldsymbol{\kappa}|G_0(E)|\boldsymbol{\kappa} >= [E - E(\boldsymbol{\kappa}) + \mathrm{i}\eta]^{-1}, \tag{5}$$

with $E(\boldsymbol{\kappa})$ the total relativistic K^+A energy in the intermediate state.

The calculations were performed with an on-shell K^+N amplitude of Martin [4] and with an off-shell separable one. The latter one was developed from a two-channel separable model of the elastic K^+N scattering which accounts for an inelasticity via coupling to an effective $\tilde{K}N$ channel with an effective pseudoscalar strange meson $\tilde{\mathrm{K}}$. The s, p, d and f partial waves were included.

The nuclear form factor $\mathcal{F}(\boldsymbol{q})$ in Eq. (1) was taken to be a Fourier transform of the Fermi density for ^{12}C. The deuteron f.f. corresponds to the Bonn OBEPQ wave function where the d-wave part was omitted. The Coulomb static K^+A interaction was included.

In Fig. 1 we show a prediction of the semi-relativistic model R with the on-shell Martin amplitude for the total cross section for d (1a) and ^{12}C (1b) and the ratio R_{T} (1c). Curves show various approximations to the full solution of the integral equation for a K^+A T-matrix. The long dashed curve displays PWIA ($T = U^{(1)}$), whereas the short dashed one shows K-matrix approximation ($T = U^{(1)} - \mathrm{i}\pi U^{(1)}\delta(E - H_0)T$).

The full solution for $\sigma^{\mathrm{tot}}(d)$ is in a very good agreement with the data whereas $\sigma^{\mathrm{tot}}(^{12}\mathrm{C})$ underestimates the data which results in a $\approx$5% discrepancy between the calculations and the data for R_{T}. This, however, means an improvement of the description of the data, especially for $p_{\mathrm{lab}} > 600$ MeV/c in a comparison with the result of Siegel et al. [1].

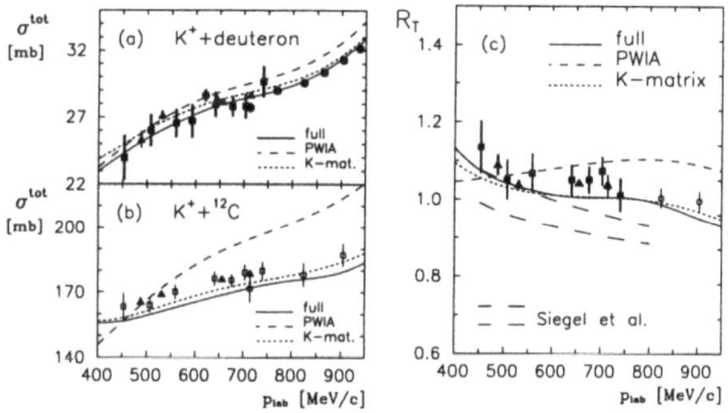

Figure 1. Prediction of the model R with the on-shell Martin's amplitude for $\sigma^{\text{tot}}(d)$ (a), $\sigma^{\text{tot}}(^{12}\text{C})$ (b) and their ratio R_{T} (c). In part (c) a band indicated by long dashed curves represents the range of the conventional calculations by Siegel et al.[1].

The K-matrix approximation fits the data on ^{12}C better than the full solution but it makes the result worse for d. It suggests that accounting for the term $U^{(1)}\frac{\mathcal{P}}{E-H_0}T$ in the K^+A equation gives undesirable contribution to the $\sigma^{\text{tot}}(^{12}\text{C})$. The PWIA approximation fails in the case of ^{12}C but it is reasonable for d. The PWIA result for ^{12}C can be improved if the off-shell extension of the K^+N amplitude is assumed, as it was done in ref. [1] with α=300 MeV/c.

The model R gives good results also for an elastic differential cross section, deviations from the data, however, occur in the vicinity of the first minimum where the medium corrections start to be important.

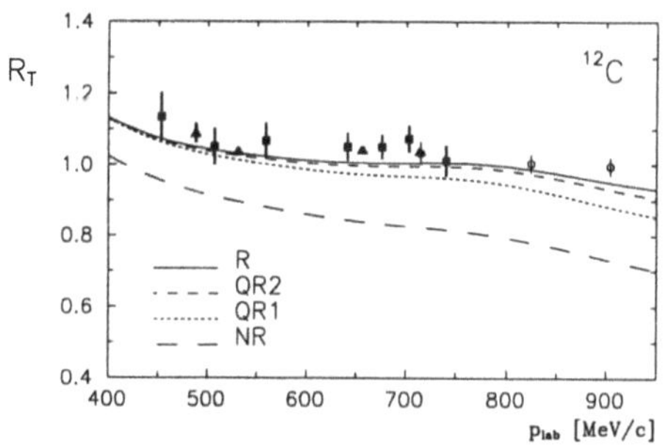

Figure 2. The ratio R_{T} is ploted as in Fig. 1. The different curves mean a various level of the relativistic corrections included (see the text for details).

In Fig. 2 results for R_{T} are plotted when the relativistic corrections are switched off step by step in the following way: QR2: momentum $\boldsymbol{p}$ and energy ε are given by a non-relativistic expression (Eq. (2) and (3)) and the K^+N t-matrix is supposed to be Galilean invariant; QR1: in addition to the case QR2 the non-relativistic relation between the K^+N t-matrix and the amplitude is assumed; NR: non-relativistic analogs of Eqs. (2-4) are used.

A comparison between NR and QR2 indicates, that the relativistic form of the K^+A Green's function (Eq. (5)) and the relativistic relation between the K^+N t-matrix and the amplitude is very important for the OFA model to fit the data on R_{T}. The relativistic transformation of momenta, energy and the

K^+N t-matrix (R versus QR2) is less important here. However, only QR2 gives good results for σ^{tot}, too. QR1 predicts σ^{tot} systematically shifted above the data for both ^{12}C and d but the discrepancy is canceled in the ratio R_T.

In Figs. 3a and 3b two separable models of the K^+N amplitude are compared with the analyses by Martin [4] and Hyslop [5]. The set of parameters sep2 possesses a stronger coupling to the inelastic channel in S_{01}, P_{13} and D_{03} waves and a shorter range of the form factors in S_{10}, S_{11} and D_{03} waves than the set sep1. The strong coupling causes significant improvement of K^+N σ^{tot} for $p_{lab} > 900$ MeV/c in both isospin channels. The results are still not satisfactory, an improvement is in progress.

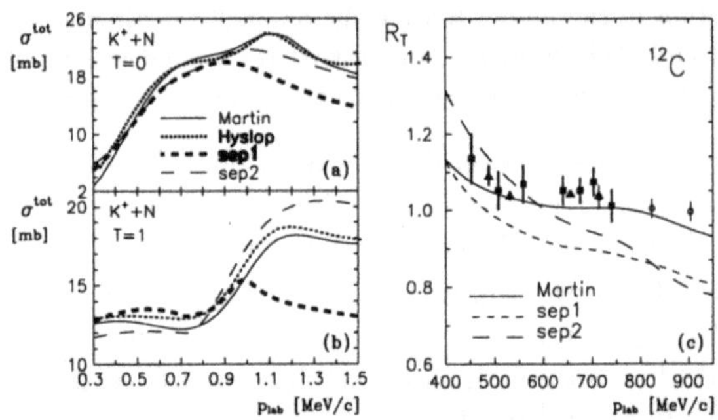

Figure 3. The total cross section for K^+N scattering is displayed for two sets of parameters of the separable K^+N model in 3a and 3b. The result is compared with analyses of Martin[4] and Hyslop[5]. In Fig. 3c the ratio R_T is ploted for the two types of the K^+N amplitudes.

The calculations of the ratio R_T for ^{12}C and d with the fully-off-shell K^+N amplitudes sep1 and sep2 are displayed in Fig. 3c. The discrepancy with the data is caused mainly by the deviation of the on-shell values of the K^+N amplitudes from those of Martin and Hyslop (see Fig. 3a,b). The ratio R_T, however, exhibits a significant sensitivity to the off-shell form of the K^+N amplitude which should be taken into account in the OFA model.

References

1. P.B. Siegel, W.B. Kaufmann, W.R. Gibbs: Phys. Rev. **C30**, 1256 (1984); Phys. Rev. **C31**, 2184 (1985)
2. G.E. Brown, C.B. Dover, P.B. Siegel, W. Weise: Phys. Rev. Lett. **60**, 2723 (1988); J.C. Caillon, J. Labarsouque: Phys. Lett. **B295**, 21 (1992)
3. M. Gmitro, J. Kvasil, R. Mach: Phys. Rev. **C31**, 1349 (1985)
4. B.R. Martin: Nucl. Phys. **B94**, 413 (1975)
5. J.S. Hyslop, R.A. Arndt, L.D. Roper, R.L. Workman: Phys. Rev. **D46**, 961 (1992)

Few-Body Systems Suppl. 9, 65–68 (1995)

Few-Body Systems

K^+ Scattering on Light $N = Z$ Nuclei

J.C. Caillon, J. Labarsouque

Centre d'Etudes Nucléaires de Bordeaux-Gradignan, Université Bordeaux I, rue du Solarium, F-33175 Gradignan Cedex, France

Abstract. In the calculation of the K^+-nucleus total cross sections, the coupling of the mesons exchanged between the K^+ and the target nucleons to the polarization of the Fermi and Dirac seas has been taken into account in the RPA approximation. The agreement with experiment is improved.

The scattering of K^+ mesons from nuclei holds a very peculiar and exciting position as the weakest of the strong-interaction processes which can be used to obtain information on the nuclear interior [1]. Since the K^+ mesons penetrate deeply inside the nucleus they should be sensitive, in particular, to the in-medium properties of the nucleons and mesons. Especially, they should bring information about the in-medium mesons which are known to couple strongly to the polarization of nuclear matter.

In this work, in the calculation of the K^+-nucleus cross sections, we have taken into account the coupling of the mesons exchanged between the K^+ and the target nucleons to the excitations of the Fermi and Dirac seas.

The scattering amplitude has been calculated by solving a relativistic Lippman-Schwinger type equation in momentum space [2].

$$\langle \boldsymbol{k}'|T(E)|\boldsymbol{k}\rangle = \langle \boldsymbol{k}'|U(E)|\boldsymbol{k}\rangle + 2\frac{E_A E_K}{E}\int \frac{\langle \boldsymbol{k}'|U(E)|\boldsymbol{k}"\rangle\langle \boldsymbol{k}"|T(E)|\boldsymbol{k}\rangle}{\boldsymbol{k}_i{}^2 - \boldsymbol{k}"^2 + i\epsilon} d^3k" \quad (1)$$

where $E_A = \sqrt{\boldsymbol{k}_i{}^2 + M_A^2}$ and $E_K = \sqrt{\boldsymbol{k}_i{}^2 + m_K^2}$ are the nucleus and K^+ energies, $\boldsymbol{k}_i$ is the initial momentum and $E = E_A + E_K$. This equation is solved by partial wave decomposition, discretization and matrix inversion.

The optical potential describing the elastic scattering of K^+ mesons from nuclei has been constructed by folding the nuclear proton and neutron densities, ρ_p and ρ_n, with the density-dependent K^+-nucleon amplitude assumed to be the same as in infinite nuclear matter at the same density.

$$\begin{aligned}\langle \boldsymbol{k}'|U|\boldsymbol{k}\rangle &= \int e^{i(\boldsymbol{k}'-\boldsymbol{k})\cdot\boldsymbol{r}}[Z\langle \boldsymbol{k}'|t_{Kp}(\rho_p(\boldsymbol{r}))|\boldsymbol{k}\rangle\rho_p(\boldsymbol{r}) \\ &\quad + N\langle \boldsymbol{k}'|t_{Kn}(\rho_n(\boldsymbol{r}))|\boldsymbol{k}\rangle\rho_n(\boldsymbol{r})]d^3r \end{aligned} \quad (2)$$

The total Coulomb-free K^+-nucleus cross section has been obtained from the forward scattering amplitude using the optical theorem. The nuclear point-like proton distributions required in the present analysis are that deduced, after the proton finite-size correction has been made, from the electron-scattering charge densities of Li et al. [3] for ^{6}Li, Sick and McCarthy [4] for ^{12}C, Li, Sick and Yearian [5] for ^{28}Si, Frosch et al. [6] for ^{40}Ca and we have chosen equal n and p-distributions.

For the K^+-nucleon amplitude in free space, we have used here the full Bonn boson exchange model [7] which is actually one of the more elaborate descriptions of the KN interaction. The solution B1, which provides the best agreement with experimental data, will be used here. We think that this agreement, though not perfect, is good enough to expect that the main part of the physics has been taken into account. In nuclear matter, this amplitude has been modified in order to take into account the coupling of the σ and ω mesons, whose exchange provides the dominant part of the medium-range KN interaction, to the polarization of the Fermi and Dirac seas. This polarization, calculated in the one-loop approximation, has been summed up to all orders in the meson propagators (RPA approximation). The heavier particles exchanged lead to very short-ranged processes less influenced by the nuclear environment. A Pauli blocking for the nucleon intermediate states has been introduced in the integral equation generating the in-medium KN amplitude from the boson-exchange kernel.

The expressions for the Fermi-sea part of the polarization in the σ, $\sigma\omega$ and ω channels can be found in ref. [8]. For the Dirac-sea part, in order to eliminate divergent terms, a renormalization procedure has to be applied. It has been done as usual [9] by requiring that, in free space, the three- and four-σ vertices vanish at zero momenta and the corrections to the σ and ω masses and wave functions are zero at $q^2 = \mu^2$ where μ is a renormalization scale. Moreover, in order to recover the Bonn KN amplitude in free space for any energy-momentum transfer, we have subtracted to the polarization its value in free-space. Doing this the renormalization scale μ disappears. This leads, for the σ channel, to:

$$\begin{aligned} \Pi^V_{\sigma\sigma}(q) &= \frac{3g^2_{\sigma N}}{2\pi^2}[3({M^*}^2 - M^2) - 4(M^* - M)M \\ &\quad -({M^*}^2 - M^2)\int_0^1 \ln\frac{{M^*}^2 - x(1-x)q^2}{M^2}dx \\ &\quad - \int_0^1 (M^2 - x(1-x)q^2)\ln\frac{{M^*}^2 - x(1-x)q^2}{M^2 - x(1-x)q^2}dx] \end{aligned} \tag{3}$$

and for the ω channel, to:

$$\Pi^{\omega V}_q(q) = \frac{g^2_{\omega N}q^2}{\pi^2}\int_0^1 x(1-x)\ln\frac{{M^*}^2 - x(1-x)q^2}{M^2 - x(1-x)q^2}dx \tag{4}$$

The coupling constants $g_{\sigma N}$ and $g_{\omega N}$ have been determined at each density in order to obtain a realistic saturation curve of nuclear matter and realistic values for the proton-nucleus optical potential in the Relativistic Hartree Approximation (RHA) of the $\sigma - \omega$ model [10] with density-dependent coupling constants. We have used here the saturation curve obtained by Brockmann and Machleidt from a Dirac-Brueckner calculation [11] and the proton-nucleus optical potentials obtained by Hama et al. [12] from experimental data over a wide range of energy and nuclei.

The K^+-nucleus total cross sections obtained when the polarization of the nuclear medium is taken into account as indicated above, are shown in Fig.1, (full lines). The experimental points are taken from ref. [13] (squares), from ref. [14] (triangles) and from ref. [15] (circles).

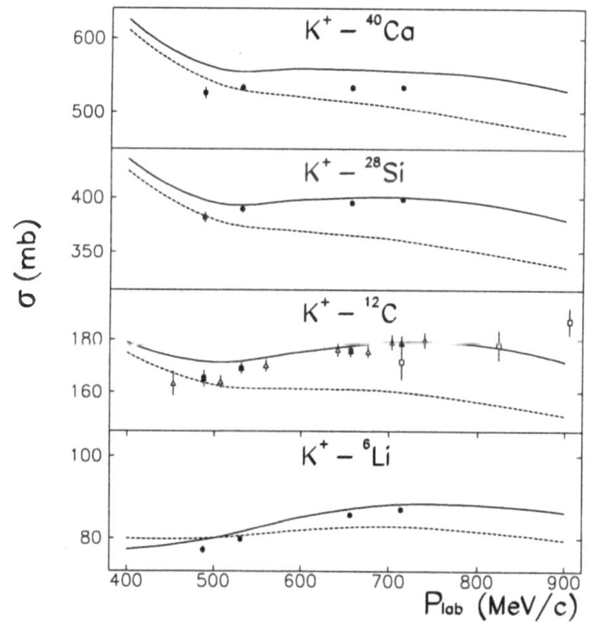

Figure 1. Total K^+-nucleus cross sections as a function of $p_{\rm lab}$ calculated, with the free-space K^+N interaction (dashed line) and (full line) with a density-dependent K^+N interaction taking into account the coupling of the σ and ω mesons with the polarization of nuclear matter (see text).

As we can see, the coupling of the mesons exchanged between the K^+ and the target nucleons to the polarization of the Fermi and Dirac seas, leads to an enhancement of the K^+-nucleus cross-section from the values obtained with the free-space KN interaction (dashed lines), above 500 MeV/c.

The agreement with experiment is considerably improved for ^{6}Li, ^{12}C and ^{28}Si. For ^{40}Ca , where the values obtained with the free-space KN interaction were not very far from the experimental results, the dressing of the KN interaction leads to better shaped but overestimated cross sections.

Thus, in this work we have shown that a dressing of the mesons exchanged between the K^+ and the target nucleons by the polarization of the Dirac and Fermi seas, improves significantly the agreement with experimental total K^+-nucleus cross sections. Even if other medium effects (scattering on the pionic cloud, Fermi averaging, correlations other than RPA, ...) should also be taken into account, we believe that such a global agreement with experiment clearly shows that an important part of the physics has been understood.

References

1. C.B. Dover and G.E. Walker: Phys. Reports **89**, 1 (1982) and references therein
2. I. T. Todorov: Phys. Rev. **D3**, 2351 (1971)
3. G. C. Li et al.: Nucl. Phys. **A162**, 583 (1971)
4. I. Sick and J. S. McCarthy: Nucl. Phys. **A150**, 631 (1970)
5. G. C. Li, I. Sick and M. R. Yearian: Phys. Lett. **B37**, 282 (1971)
6. R. F. Frosch et al.: Phys. Rev. **174**, 1380 (1968)
7. R. Büttgen et al.: Nucl. Phys. **A506**, 586 (1990)
8. L. S. Celenza, A. Pantziris and C. M. Shakin: Phys. Rev. **C45**, 205 (1992)
9. S. A. Chin: Ann. Phys. **108**, 301 (1977)
10. B. D. Serot and J. D. Walecka: Advances in Nuclear Physics **16**, 1 (1986)
11. R. Brockmann and R. Machleidt: Phys. Rev. **C42**, 1965 (1990)
12. S. Hama et al.: Phys. Rev. **C41**, 2737 (1990)
13. D. Bugg et al.: Phys. Rev. **168**, 1466 (1968)
14. R. A. Krauss et al.: Phys. Rev. **C46**, 655 (1992)
15. R. Sawafta et al.: Phys. Lett. **B307**, 293 (1993) ; R. Weiss et al., Phys. Rev. **C49**, 2569 (1994)

Few-Body Systems Suppl. 9, 69–82 (1995)

Few-Body Systems

Experiments with Polarized ^{3}He and Muonic ^{3}He: Pion Elastic Scattering and Muon Capture

O. Häusser*†

TRIUMF, 4004 Wesbrook Mall, Vancouver, B.C. Canada, V6T 2A3, and Simon Fraser University, Burnaby, B.C. Canada, V5A 1S6

Abstract. Novel asymmetry measurements on the A=3 system using secondary beams of pions and muons have recently become feasible. Asymmetries A_y for π^+ elastic scattering have been measured for the first time across the P_{33} resonance region. The asymmetries are large and change sign between 180 and 256 MeV. The data disagree in detail with calculations that use Faddeev wavefunctions for ^{3}He and first-order optical potentials. Inclusion of a Δ-neutron spin-spin interaction greatly improves the agreement with the data.

Muonic ^{3}He has been polarized on the time scale of the muon lifetime by direct spin exchange with optically pumped Rb vapor. Asymmetries for 1.9 MeV tritons from muon capture have been observed for the first time. With future improvements of the experiment the vector analyzing power for the tritons may provide an accurate and nearly model independent value for the induced pseudoscalar coupling g_P of the weak interaction.

1 Introduction

The study of few-body nuclear systems is presently receiving new impulses through the measurement of polarization-dependent observables. These observables add qualitatively new information to spin-averaged cross sections and rates, and help to probe the underlying physics at a deeper level. Although the basic optical pumping techniques for producing polarized ^{3}He gas targets

For the E1267 Collaboration: S.P. Blanchard, E.J. Brash, B. Brinkmöller, G.R. Burleson, W.J. Cummings, B.J. Davis, D. Dehnhard, P.P.J. Delheij, C.M. Edwards, M.A. Espy, R. Henderson, B.K. Jennings, M.K. Jones, B.A. Lail, J.L. Langenbrunner, B. Larson, W. Lorenzon, K. Maeda, C.L. Morris, B. Nelson, J.M. O'Donnell, M.A. Palarczyk, B.K. Park, S.I. Penttilä, D.R. Swenson, D. Thiessen, D. Tupa, Q. Zhao.

†*For the E683 Collaboration:* J. Behr, P. Bogorad, E.J. Brash, G.D. Cates, W.J. Cummings, A. Gorelov, M.D. Hasinoff, R. Henderson, K. Hicks, R. Holmes, J.C. Huang, K. Kumar, B. Larson, W. Lorenzon, G.M. Marshall, J. McCracken, E. Saettler, P.A. Souder, D. Swenson, D. Tupa, X. Wang, A. Young.

had been established since the early 1960's too few nuclei could be polarized to allow extensive measurements. With the advent of powerful lasers in the infrared, and the development of gas compression techniques which retain high polarization, many new measurements have become feasible during the past few years. They range from the application as spin filters for thermal neutrons, to the study of reaction mechanisms and spin effects in hadronic reactions, to the measurement of electromagnetic nucleon form factors and nucleon spin structure functions, and to the determination of fundamental induced couplings of the weak interaction.

We present here data obtained with secondary beams of pions and muons at the meson factories LAMPF and TRIUMF. The targets at pressures of 6-12 amagat (atm at 273 K) are polarized by spin exchange collisions with optically pumped Rb vapor [1]. In the first part of the paper new measurements of asymmetries in pion elastic scattering are discussed which, together with existing cross section data, probe the π-nucleus reaction mechanism for a target ($A > 2$) with well understood nuclear structure. We then present first observations of the vector analyzing power in muon capture, ^{3}He + μ^- $\rightarrow$ ^{3}H + ν_μ, from polarized muonic ^{3}He. With further improvements this experiment could provide in the future a new competitive experimental value for the induced pseudoscalar coupling constant g_P of the weak interaction.

2 Elastic π^+ Scattering from Polarized ^{3}He

Previous work on pion elastic scattering from polarized 1p-shell nuclei of spin I=1/2 had revealed large discrepancies between theoretical predictions and experiment. The observed asymmetries A_y for ^{15}N [2] and ^{13}C [3,4] are generally small, in contradiction with theory [5,6]. Calculations have found [6,7] a strong dependence of A_y on the assumed nuclear spin densities. It is not clear at present whether the discrepancies between experiment and theory signal deficiencies in the π-nucleus interaction and specifically its spin-dependent part, or in the nuclear wave functions of these nuclei.

^{3}He is ideal for a study of π-nucleus reaction models because the ^{3}He wave function can be calculated with good accuracy from the Faddeev equations using realistic NN potentials as input [8-10]. Therefore, for ^{3}He the nuclear structure uncertainty is almost negligible compared to p-shell nuclei.

Pion-eering scattering experiments on polarized ^{3}He were carried out with 100 MeV $\pi^\pm$ beams at TRIUMF [11], where a high-density, optically pumped ^{3}He gas target had been employed successfully since 1989. In the TRIUMF experiment the scattering vertex for each event was reconstructed using position information from wire chambers in the beam and at the entrance of the magnetic spectrometer. Very large values of A_y were found in π^+ elastic scattering at 100 MeV. At this energy the asymmetry is rather insensitive to the nuclear wave function and shows only a slight dependence on the reaction model [12]. However, for energies at and above the P_{33} resonance, the asymmetries were predicted [12] to become increasingly sensitive to details of the wavefunction

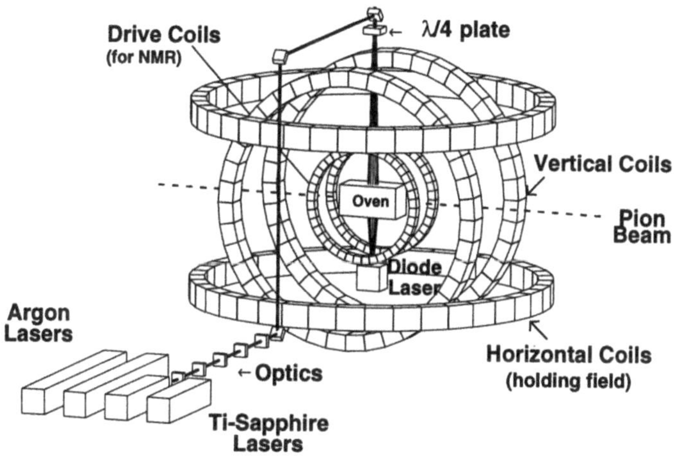

Figure 1. The TRIUMF Polarized ^{3}He Target used at LAMPF during E1267.

and the reaction model.

Measurements at energies near and above the P_{33} resonance could not be done at TRIUMF because of the lack of a suitable spectrometer and the relatively low beam fluxes at the higher energies. Therefore, the E1267 collaboration was formed and the TRIUMF target was set up in the P^{3}E area at LAMPF. The target setup is shown schematically in Fig. 1.

The ^{3}He gas was contained in cylindrical glass cells, 4.8 cm in diameter and 6.5 cm long. The cells were made of quartz glass that was 1.5 mm thick at the cylindrical cell walls and 0.4 mm thick at the hemispherical endcaps where the pion beam entered and exited the cell. The cells were filled with 6-7 amagat of ^{3}He gas, a trace of Rb, and ~100 Torr of N_2 which served as buffer gas to quench fluorescence in Rb. At an oven temperature of 450 K and a Rb density of 4×10^{14} cm^{-3} ~ Watts of circularly polarized laser light at 795 nm were needed for efficient optically pumping of the Rb. Spin exchange occurs during collisions of ^{3}He and Rb via the Fermi contact interaction, with a typical time constant of 8 hours. When the glass cell was hot, small amounts of ^{3}He leaked from the cell. The pressure in the cell was monitored periodically using the fact that the linewidth for absorption of resonant light is proportional to the ^{3}He pressure in the cell [1]. By recording the time periods when the cells were hot a correction for the pressure loss could be made. During the experiment a diode laser array system [13] was added which illuminated the target from below (see Fig. 1). The diode laser array added to the optical pumping power and significantly increased the polarization after one of the Ar lasers had failed. The ^{3}He polarization was in the range between 35 and 50%.

The helicity of the laser light, and thus the direction of the ^{3}He polarization, was determined from the orientation of the fast axis of the quarterwave plate and, independently, with a liquid crystal which transmits only left-hand

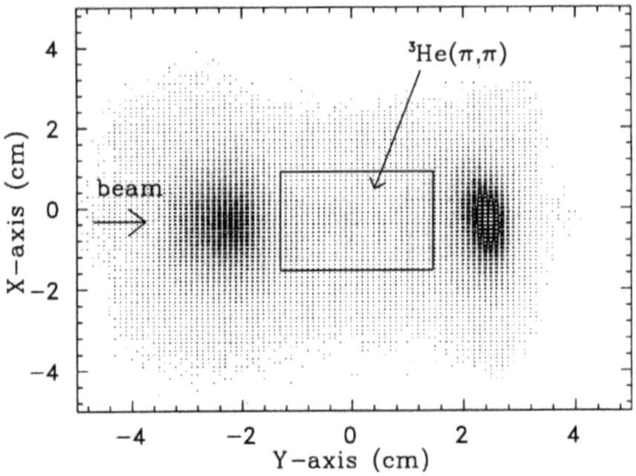

Figure 2. Intersection of particle tracks with a vertical plane perpendicular to the central ray of the spectrometer and centered at the target.

circularly polarized light. The horizontal coils provide a holding field for the polarization. Together with the vertical coils they can be used to adiabatically reverse the direction of the target polarization at intervals of a few minutes, with negligible loss of polarization [14]. The magnitude of the polarization was measured using the nuclear magnetic resonance (NMR) technique of adiabatic fast passage, with absolute normalization factors obtained by comparson with NMR signals from the weak signals from protons in a water-filled cell of the same dimension [1].

Because of the low duty cycle of LAMPF and the large instantaneous pion flux, in-beam detectors could not be used. Instead, a lead beam collimator, which was tapered according to the convergence of the beam, prevented excessive beam halo from hitting the walls of the glass vessel at the target centre. The scattered pions were detected with the Large Acceptance Spectrometer (LAS). The LAS uses a magnetic quadrupole doublet, a magnetic dipole, and several sets of two-dimensional wire chambers to identify the scattered pions and to measure their momenta. The front wire chambers, located between the quadrupole doublet and the dipole, allow traceback of the scattered particle trajectories to a plane that intersects the center of the target. This plane is perpendicular to the central ray of the LAS. The projections of the reaction vertices onto this x-y plane were used to discriminate between events from ^{3}He and the glass in the endcaps (Fig. 2) whereas events from the beam halo striking the cylindrical cell walls were largely eliminated by the conical lead collimator.

2.1 Asymmetry results and discussion

Measurements were taken at incident energies $T_\pi = 100, 142, 180$, and 256 MeV and laboratory scattering angles ranging from 40° to 100°. The top and center panels of Fig. 3 show missing mass spectra measured at 180 MeV and $\theta_{\rm lab} =$

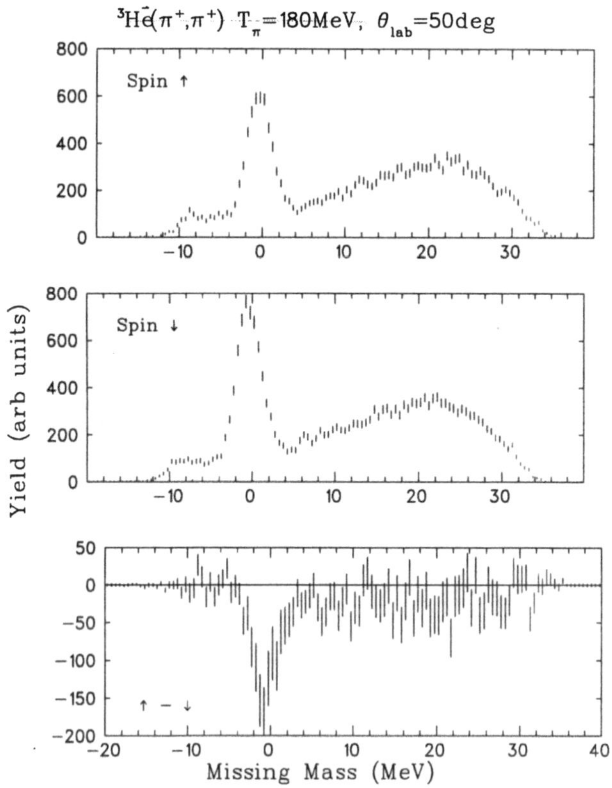

Figure 3. Typical normalized yield spectra for π^+ scattering from the target at $T_\pi = 180$ MeV and $\theta_{\text{lab}} = 50°$. The top two panels are for target spin "up" (↑) and "down" (↓), the difference spectrum is shown in the bottom panel.

50°, normalized to integrated beam flux, for the two target spin orientations normal to the reaction plane. The ^{3}He elastic peak has a width of about 4 MeV (FWHM). The difference spectrum is shown in the bottom panel of Fig. 3. The asymmetries were obtained using the expression

$$A_y = \frac{N_\uparrow - N_\downarrow}{N_\uparrow + N_\downarrow} \frac{1}{p}. \tag{1}$$

Here $N_\uparrow$ and $N_\downarrow$ are the normalized number of counts in the ^{3}He elastic peak for the ^{3}He target spin orientations "up" and "down", respectively, and p is the target polarization. A large negative asymmetry is apparent in the region of the elastic peak from ^{3}He (Fig. 3, bottom panel, Missing Mass = 0).

Experimental and theoretical angular distributions of A_y for π^+ elastic scattering are presented in Fig. 4. The 100 MeV data are from previous work at TRIUMF [11], the others are from the present work. The theoretical curves

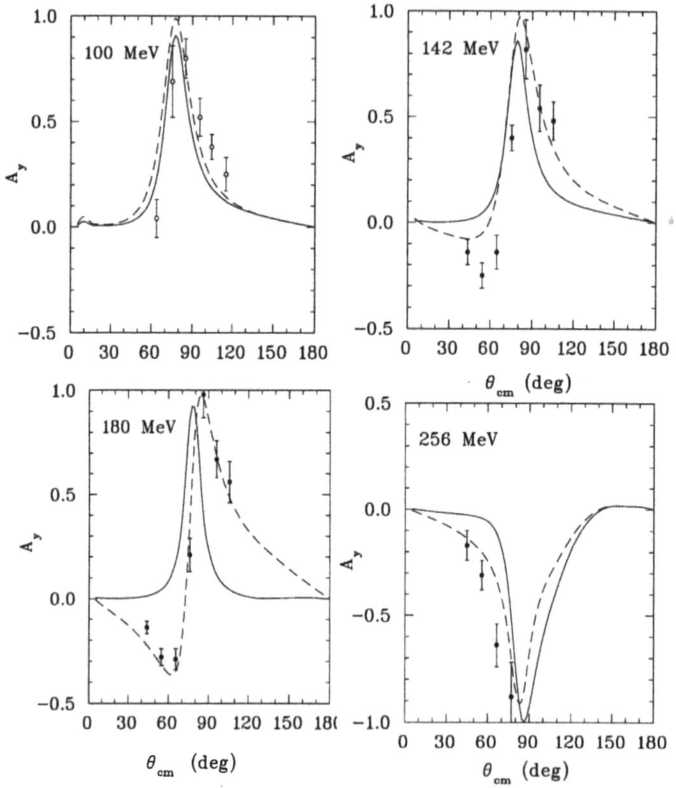

Figure 4. Asymmetry angular distributions for elastic π^+ scattering from polarized ^{3}He at T_π = 100, 142, 180 and 256 MeV. The 100 MeV results are from ref. [11], the others from E1267. The solid lines correspond to first-order calculations of ref. [12]. The dashed lines correspond to calculations which include a $\Delta - N$ spin-spin interaction.

(solid lines in Fig. 4) were obtained using the distorted wave impulse approximation (DWIA) and state-of-the-art Faddeev wave functions [8-10] for ^{3}He.

Neither the first-order calculations of ref. [12] which include multiple scattering (Fig. 4, solid lines), nor those of refs. [15] and [16] (not shown) give a satisfactory description of the data. At both 142 and 180 MeV we observe large asymmetries as predicted by the first-order calculations, however, the maxima are shifted towards larger angles. The negative values of A_y near 60° were unexpected since all conventional model calculations predict positive asymmetries between 100 and 180 MeV. At 256 MeV the asymmetries are negative at the measured angles, probably a result of multiple scattering which becomes more important at higher energies.

In pion scattering from a spin $I = 1/2$ nucleus the asymmetry can be written as $A_y = 2\,\mathrm{Im}(F \times G^*) \,/\, (|F|^2 + |G|^2)$, where F and G are the complex spin-independent and spin-dependent scattering amplitudes, respectively. When the P_{33} resonance dominates the elementary π-nucleon interaction, isospin coupling coefficients imply much larger scattering amplitudes, F and G, for π^+ elastic scattering from protons than from neutrons. In the impulse approximation, and for the predominant $(1s)^3$ component of the ^{3}He wave function, the π^+ interacting with one of the paired-off protons cannot lead to a spin-dependent amplitude. G results then only from scattering from the unpaired neutron. F has a large component from scattering from the protons and a small one from scattering from the neutron. Nevertheless, the A_y can be large near the minimum of F, where G can have a relative maximum. Qualitatively, this simple model gives asymmetries similar to those of the first-order optical model calculations (Fig. 4, solid lines).

If the intermediate Δ^{++}, generated with very high probability in π^+ scattering on one of the two protons, interacts with the neutron, a large second-order contribution to G can arise [17]. The magnitude of this second-order contribution to G has been investigated using a meson exchange model for the Δ-neutron interaction. The π, ρ, ω, σ and η mesons were included. For the π and ρ mesons the Pauli exchange diagrams are also important. The meson-Δ couplings were obtained from the meson-nucleon couplings by use of SU(3) symmetry. Gaussian single particle wave functions were used. Correlations were included by multiplying the single particle wave function by 1 minus a Gaussian that depends on the relative distance of the interacting particles. The rms radius of the density was kept fixed at the experimental ^{3}He value, and the width of the Gaussian was varied. The best fit for 180 MeV was obtained with a range of the correlation Gaussian that was 1/4 that of the single particle Gaussian. The contributions from the direct ρ and ω exchange largely cancel. The remaining ρ and ω contribution is very sensitive to short-range correlations. The η exchange is not very significant.

This model (dashed lines in Fig. 4) provides an excellent fit to the data at 180 MeV: the negative asymmetries near 60° and the shift of the positive maximum towards larger angles are reproduced. The agreement with the data at 142 and 256 MeV is improved. There still remain small discrepancies: at 142 MeV and 256 MeV the experimental data near 60° are more negative than predicted by the model. Further theoretical and experimental work could be helpful to confirm the conclusions suggested by the phenomenological model. On the experimental front asymmetry measurements for π^- elastic scattering scattering would be particularly decisive. For π^- the first-order term is much larger than the second-order term and the effect from the Δ-nucleon interaction on the asymmetry is predicted to be negligible.

3 Muon Capture in Polarized Muonic ^{3}He

Information on the structure of the nucleon as probed by weak charged currents is contained in F_A and F_P, the axial and the induced pseudoscalar form factors. The induced pseudoscalar coupling constant,

$$g_P = \frac{M_\mu}{2M} F_P(t = -0.88 M_\mu^2) \tag{1}$$

can be investigated by ordinary muon capture (OMC) in hydrogen, $p + \mu^- \rightarrow n + \nu_\mu$. Theoretically g_P is dominated by the pion pole contribution and can be estimated by combining the PCAC hypothesis and the Goldberger-Treiman relation at $t = 0$. With QCD corrections using chiral Ward identities a value accurate to a few percent ($g_P = 8.44 \pm 0.23$) has recently been calculated [18].

Experimentally the situation is less satisfactory. Because of molecular difficulties [19] g_P determinations from OMC in hydrogen are of limited accuracy ($\geq 40\%$). Radiative muon capture (RMC) in hydrogen, although having an enhanced sensitivity to g_P, suffers from an unfavorable RMC/OMC ratio of 10^{-5}. The result for g_P from RMC in hydrogen by the E452 collaboration at TRIUMF is expected later this year [20].

Muon capture from nuclei containing at least two protons has the advantage that difficulties due to molecule formation are absent. Furthermore, detection of the charged recoil nucleus is easier. Muon capture on ^{3}He has the added advantage that the $^3\text{He} \rightarrow {}^3\text{H}$ transition involves the same spin, isopin and parity as the $p \rightarrow n$ transition. In the "elementary particle" model (EPM) of Kim and Primakoff [21] the weak current of this transition is parametrized by four form factors $F_{\text{V,M,A,P}}$ which can be determined from data on beta decay, magnetic moments and electron scattering form factors of the $A = 3$ system using CVC and assuming the validity of the PCAC hypothesis [22]. The partial capture rate has been measured recently at PSI [23] and was found to be in excellent agreement with the EPM prediction of $\Lambda_c = 1497 \pm 11\ \text{s}^{-1}$ [22]. Extraction of g_P from the partial capture rate depends on the relative importance of meson exchange current effects in electromagnetic and semileptonic weak interactions at the relevant momentum transfer. The importance of MEC effects in muon capture can be inferred from impulse approximation calculations in which the contributions of individual nucleons are added using elaborate $A = 3$ Faddeev wavefunctions. The value obtained, $\Lambda_c = 1304\ \text{s}^{-1}$, is substantially lower than the EPM one which implicitly includes the MEC effects.

The observation of spin dependence in the muon capture reaction $\mu^- + {}^3\text{He} \rightarrow {}^3\text{H} + \nu_\mu$ (0.32 % branch) can imply a significant enhancement in sensitivity to g_P compared to the partial capture rate. The angular distribution of the tritons is

$$\frac{d\Gamma}{d\Omega} = \frac{\Gamma_0}{4\pi}[1 + P_v A_v P_1(\cos\theta) + P_t A_t P_2(\cos\theta) + \Delta A_\Delta], \quad (2)$$

where Γ_0 is the unpolarized rate, θ is the angle between the polarization axis and the triton direction, and P_1 and P_2 are Legendre polynomials. The quantities P_v (vector polarization), P_t (tensor polarization), and Δ are defined by:

$$P_v = N(1,1) - N(1,-1) \quad (3)$$

$$P_t = N(1,1) + N(1,-1) - 2N(1,0) \quad (4)$$

$$\Delta = N(1,1) + N(1,0) + N(1,-1) - 3N(0,0), \quad (5)$$

where $N(F, M)$ are population densities for finding the muonic ^{3}He atom in a given hyperfine state. A_v, A_t, and A_Δ are the analyzing powers for the above

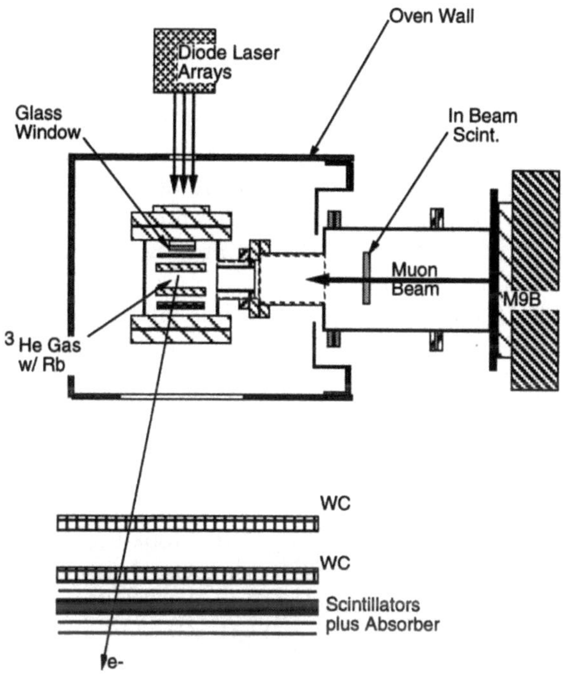

Figure 5. Schematic top view of the apparatus used for E683

quantities. For the vector analyzing power, A_{v}, Congleton and Fearing [22] find a 3.5 times larger sensitivity to g_{P} than for the partial capture rate.

Earlier attempts to observe spin dependence in muon capture from ^{3}He either using polarized muons [24] or polarized ^{3}He [25] where hampered by low muon polarizations of only a few percent. With a new technique [26] in which neither the muon beam nor the ^{3}He target are polarized, but where the polarization rises on the time scale of the muon lifetime through collisions with laser optically pumped Rb, high muon polarizations can be obtained. The $(\mu^-\mathrm{He})^+$ ion produced after the muon cascade in the ^{3}He atom rapidly forms a molecular ion $\mathrm{He}(\mu^-\mathrm{He})^+$. This ion is dissociated and polarized by the reaction

$$\mathrm{Rb}\uparrow + \mathrm{He}(\mu^-\mathrm{He})^+ \rightarrow \mathrm{Rb}^+ + \mathrm{He} + (\mu^-\mathrm{He})^+ e^- \uparrow\ , \tag{6}$$

where the arrows indicate the spin polarization of the transferred electron. For neutral atoms $(\mu^-\mathrm{He})^+\mathrm{e}^-$ polarization occurs during spin exchange collisions with Rb

$$\mathrm{Rb}\uparrow + (\mu^-\mathrm{He})^+ e^- \downarrow \rightarrow \mathrm{Rb}\downarrow + (\mu^-\mathrm{He})^+ e^- \uparrow\ . \tag{7}$$

With 8 amagat ^{3}He contained in glass cells of 8 cm^3 volume at 200°C average muon polarizations of 26% were determined from asymmetries for the decay electrons [26].

3.1 Experimental Setup and Preliminary Results

A measurement of the vector analyzing power of tritons from the $^3\mathrm{He} + \mu^- \rightarrow$ $^3\mathrm{H} + \nu_\mu$ reaction has to satisfy the following requirements: a) a large fraction of the muons should be stopped in ^{3}He gas of several amagats pressure; b) Rb vapor of several 10^{14} atoms cm^{-3} must be created by heating the entire apparatus to about 200 °C; the Rb has to be optically pumped with many Watts of circularly polarized laser light of 795 nm; c) the vector polarization P_v must be measured by observing the asymmetry of decay electrons; d) the energy of the triton and the direction of the track must be measured using a drift chamber which converts position into time information.

After initial attempts at LAMPF (E1231) and TRIUMF (E683) the setup shown in Fig. 1 has been used at TRIUMF in the fall of 1994. The stopping time of the muon was obtained from a thin in-beam scintillator. The reaction vessel was made of stainless steel; it contained 8 amagat of ^{3}He and 2% of N_2 and was operated at 200 °C. With a vertical probe laser whose wavelength was scanned over several nm, the Rb number density could be estimated roughly from the width of the absorption dip. It was found that it took from many hours to days until an equilibrum value of a few 10^{14} atoms cm^{-3} was attained, whereas it takes only minutes when glass vessels are used. The Rb was polarized using two diode laser arrays [13] assembled at TRIUMF which delivered a total of 28 W of circularly polarized light with a linewidth of about 2 nm FWHM at 795 nm over an area of about 20 cm^2. A holding field of about 1 mT was applied parallel to the laser beam direction. The detection system for decay electrons consisted of two wire chambers, followed by an absorber to eliminate the low energy electrons, and a scintillator telescope. Preliminary results from more

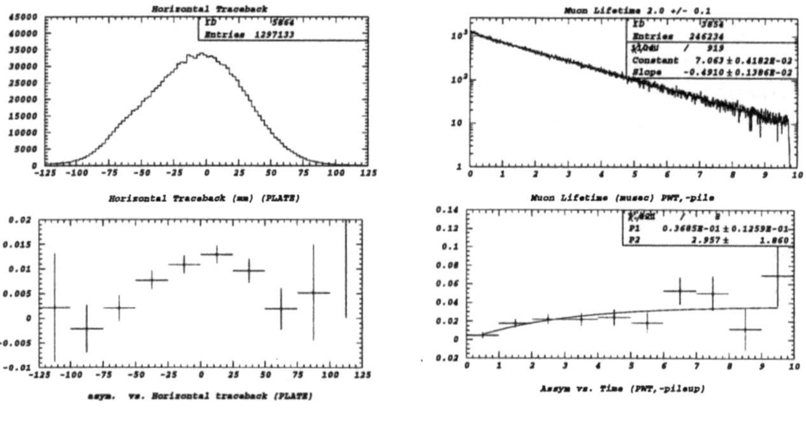

Figure 6. Top left: horizontal traceback position for decay electrons. Top right: logarithmic plot of muonic ^{3}He lifetime from decay electrons. Bottom left: electron asymmetry versus horizontal position. Bottom right: electron asymmetry versus decay time.

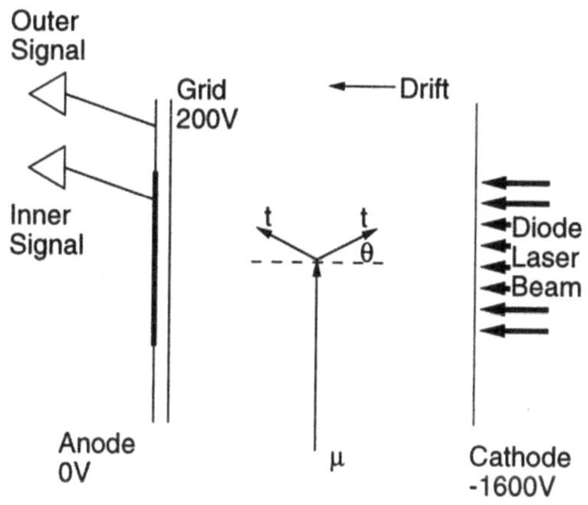

Figure 7. Ionization chamber driftcell configuration. Field shaping electrodes between Frisch grid and cathode grid are not shown.

than 4×10^6 electron events collected are shown in Fig. 6. From the traceback to the horizontal position shown in the top left panel of Fig. 6 complex sources of multiple scattering have already been eliminated by applying appropriate cuts. The decay time relative to the in-beam scintillator shows an exponential falloff compatible with the decay constant for muonic ^{3}He (top right panel). The average decay electron asymmetry, shown in the bottom left panel versus the horizontal traceback position, reaches a value of 1.40 ± 0.13 % at the target center. The time differential asymmetry for the central region has a risetime of about 3 μs (bottom left panel).

For a measurement of energy and direction of recoil tritons an ionization chamber developed at Princeton operated reliably at elevated temperatures and in the presence of Rb vapor[1]. The drift cell configuration is shown in Fig. 7. The addition of two equipotential rings between Frisch grid and cathode grid was necessary to ensure reasonable drift field uniformity. The angle of the triton track can be inferred from the pulse shape at the anode. Tritons emitted in the direction of the laser beam towards the anode (called "downers") show a Bragg peak at early times, those emitted away from the anode (called "uppers") show a Bragg peak at late times. The pulse width is proportional to the cosine of the emission angle θ. Tritons whose tracks are not confined in the region of the inner anode deposit less charge at the inner anode and can be vetoed by the signal at the outer anode. The charge signal from the muon which may have stopped several μs before triton emission is superimposed on the triton charge distribution making it necessary to fit individual pulse shapes. In Fig. 8 three examples of "uppers" (left panel) and "downers" (right panel) are shown. The contributions of muon and triton track are shown separately. The largest

[1] A gas scintillation chamber developed at TRIUMF worked well up to 180°C but became unreliable when Rb vapor was introduced

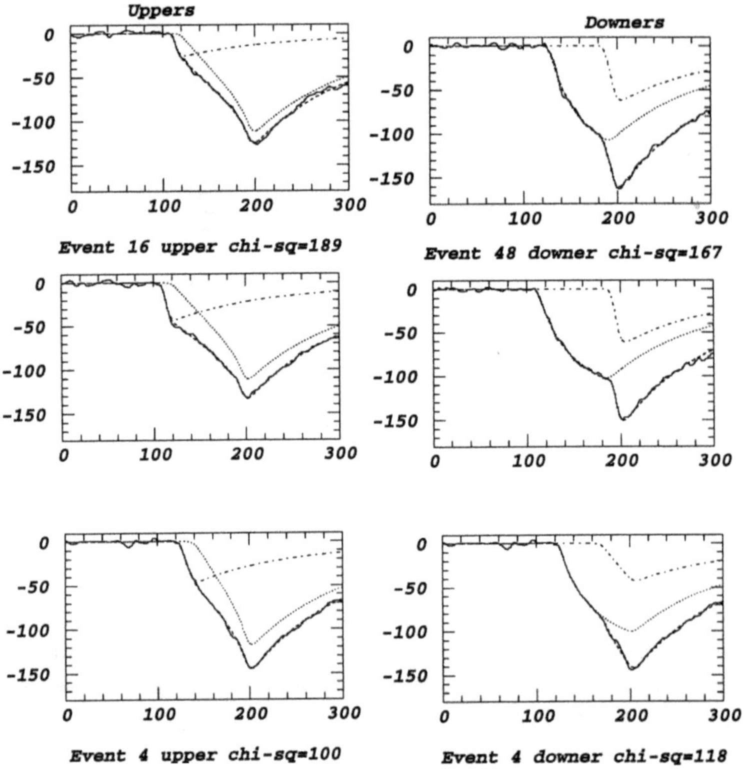

Figure 8. Examples of pulse shapes for tritons emitted away from the anode ("uppers", shown on the left) and emitted towards the anode ("downers", shown on the right). The separate contributions of muon (dot-dashed lines) and triton (dotted lines) pulse shapes are also shown.

differences in χ^2 for "uppers" and "downers" occur for extreme values of cosθ which have the largest sensitivity to the vector analyzing power.

Systematic effects arising from differences in the acceptances for "uppers" and "downers" can be cancelled by forming a "super asymmetry" which utilizes the fact that no change occurs when flipping simultaneously the drift angle by 180° and the helicity of the light:

$$A_s = \frac{\sqrt{N_{\mathrm{CCW}}^{\mathrm{uppers}} N_{CW}^{\mathrm{downers}}} - \sqrt{N_{\mathrm{CCW}}^{\mathrm{downers}} N_{\mathrm{CW}}^{\mathrm{uppers}}}}{\sqrt{N_{\mathrm{CCW}}^{\mathrm{uppers}} N_{\mathrm{CW}}^{\mathrm{downers}}} + \sqrt{N_{\mathrm{CCW}}^{\mathrm{downers}} N_{\mathrm{CW}}^{\mathrm{uppers}}}} \tag{8}$$

The "super asymmetry" versus the pulse width of the triton pulse, and thus $\cos\theta$, as obtained from fits of the type shown in Fig. 8, is shown in Fig. 9. The sample includes only data from four days of running and better statistical accuracy is expected when all data will be analyzed. The fit in Fig. 9 corresponds to a very preliminary value of the triton asymmetry of 5.5 ± 0.5 %.

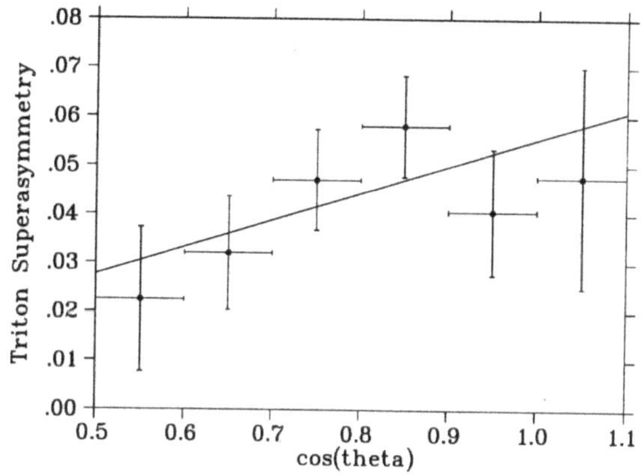

Figure 9. Super asymmetry ratio (see definition in the text) versus the triton pulse width which is proportional to the cosine of the triton emission angle.

For a meaningful extraction of the vector analyzing power it is essential that the triton superasymmetry A_s and the decay electron asymmetry refer to the same fiducial volumes in the target. This can only be ensured with careful Monte Carlo simulations which include the accurate geometry of the experiment. Such simulations are presently being carried out and are necessary before a value for g_P can be quoted.

3.2 Summary and Outlook

Future improvements in the determination of the vector analysing power A_v are highly desirable, especially after the precise determination of the partial capture rate Λ_c at PSI [23]. A_v and Λ_c depend in different ways on the renormalization of g_P and g_A which occur in the $A = 3$ system due to the existence of meson exchange effects. Thus accurate values for both A_v and Λ_c are needed to determine these renormalizations from experiment.

Several improvements of the present experiment appear feasible. Diode laser arrays are in a state of rapid development such that considerably more laser power is already available. One can then operate at higher Rb density causing the muon polarization to rise more rapidly. With more laser power a larger volume can be uniformly polarized, with less dependence of the muon polarization on the chosen volume. Further segmentation of the anode signal would also be helpful to eliminate triton tracks at the edge of the polarized volume.

References

1. B. Larson et al. : Phys. Rev. **A44**, 3108 (1991)

2. R. Tacik et al. : Phys. Rev. Lett. **63**, 1784 (1989)

3. Yi-Fen Yen et al. : Phys. Rev. Lett. **66**, 1959 (1991)

4. J.T. Brack et al. : Phys. Rev. **C45**, 698 (1992)

5. R. Mach and S.S. Kamalov: Nucl. Phys. **A511**, 601 (1990)

6. P.B. Siegel and W.R. Gibbs: Phys. Rev. **C48**, 1939 (1993); Private Communication

7. Yi-Fen Yen et al. : Phys. Rev. **C50**, 897 (1994)

8. R.A. Brandenburg, Y.E. Kim, and A. Tubis: Phys. Rev. **C12**, 1368 (1975)

9. J.L. Friar et al. : Phys. Rev. **C34**, 1463 (1986)

10. D.R. Tilley, H.R. Weller, and H.H. Hasan: Nucl. Phys. **A474**, 2 (1987)

11. B. Larson et al. : Phys. Rev. Lett. **67**, 3356 (1991); Phys. Rev. **C49**, 2045 (1994)

12. S.S. Kamalov, L. Tiator, and C. Bennhold, Phys. Rev. **C47**, 941 (1993); C. Bennhold, B.K. Jennings, L. Tiator, and S.S. Kamalov, Nucl. Phys. **A540**, 621 (1992)

13. W.J. Cummings et al. : Phys. Rev. **A** (in press)

14. W.J. Cummings et al. : Nucl. Instr. Meth. **A346**, 52 (1994)

15. P.B. Siegel and W.R. Gibbs: Phys. Rev. **C48**, 1939 (1993); Private Communication

16. S. Chakravarti, C.M. Edwards, D. Dehnhard, and M.A. Franey: Few-Body Systems, Suppl. **5**, 267 (1992)

17. B.K. Jennings: Private Communication

18. V. Bernard, N. Kaiser and Ulf-G. Meissner: Phys. Rev. **D50**, 6899 (1994)

19. G. Bardin et al. : Nucl. Phys. **352**, 365 (1981)

20. M. Hasinoff: Private Communication

21. C.W. Kim and H. Primakoff: Phys. Rev. **B140**, 566 (1965)

22. J.G. Congleton and H. Fearing: Nucl. Phys. **A552**, 534 (1993)

23. C. Petitjean and J. Deutsch: Private Communication

24. P.A. Souder et al. : Phys. Rev. **A22**, 33 (1980)

25. N.R. Newbury et al. : Phys. Rev. Lett. **69**, 391 (1992)

26. A. Barton et al. : Phys. Rev. Lett. **70**, 758 (1993)

Few-Body Systems Suppl. 9, 83–90 (1995)

Elastic and Inelastic Pion Scattering on ^{3}H and ^{3}He

B.L. Berman[1], K.S. Dhuga[1], W.J. Briscoe[1], S.K. Matthews[1*], D.B. Barlow[2†], B.M.K. Nefkens[2], C. Pillai[2], L.D. Isenhower[3], M.E. Sadler[3], S.J. Greene[4], I. Slaus[5]

[1] Center for Nuclear Studies, Department of Physics, The George Washington University, Washington, DC 20052, USA

[2] Department of Physics, University of California at Los Angeles, Los Angeles, CA 90024, USA

[3] Department of Physics, Abilene Christian University, Abilene, TX 79699, USA

[4] Los Alamos National Laboratory, Los Alamos, NM 87545, USA

[5] Rudjer Boskovic Institute, Zagreb, Croatia

Abstract.
Over a period of several years, the UCLA-GWU-ACU-LAMPF Collaboration has carried out an extensive and systematic program of pion-scattering measurements on the $A = 3$ nuclei. By means of these measurements, we have probed the matter distributions of nucleons in these few-body nuclei, particularly the neutron distribution in ^{3}H (which cannot be accessed with electron scattering), and have investigated the extent to which charge symmetry is broken in the strong interaction.

We have measured the differential cross sections and the ratios of the scattering yields for pion scattering from ^{3}H and ^{3}He. The measurements were performed at the EPICS facility at LAMPF, over the energy range spanning the Δ resonance. The data set includes, in addition to our phase–1 cross-section measurements at several energies and at forward angles [1, 2, 3], phase–2 measurements in the non-spin-flip (NSF) dip region, a large-angle excitation function, and an angular distribution for backward angles (from 114° to 168°) near the peak of the Δ at 180 MeV [4, 5, 6]. The analysis of the phase–2 data is now complete; we provide here a summary of the more notable features that have emerged.

*Present address: Catholic University of America, Washington, DC 20064
†Present address: Los Alamos National Laboratory, Los Alamos, NM 87545

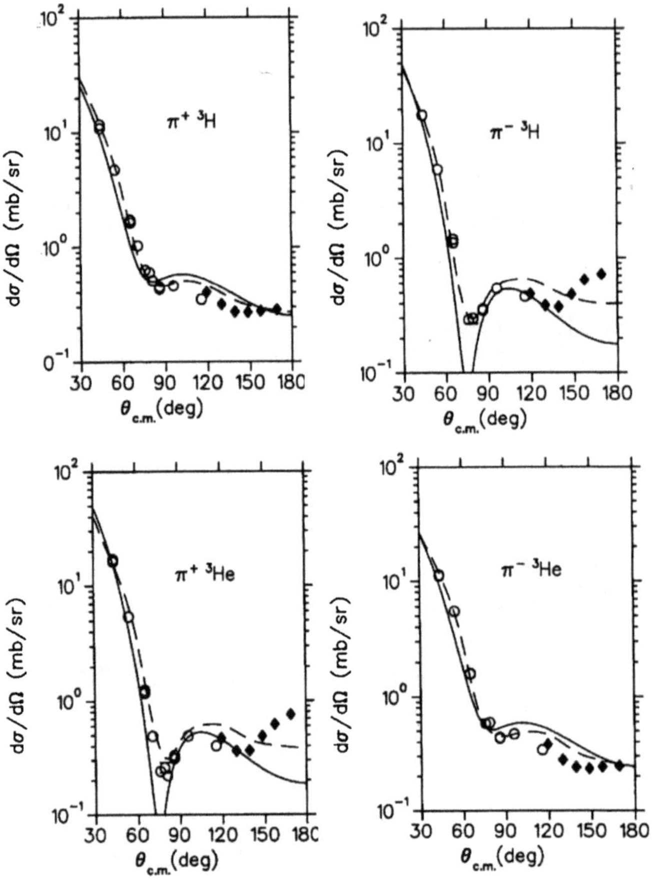

Figure 1. Elastic cross sections for 180–MeV pions

One of the most interesting features of our new results is the observation of a steep rise in the cross sections (for angles greater than 140°) for the reactions $^3He(\pi^+,\pi^+)^3He$ and $^3H(\pi^-,\pi^-)^3H$, i.e. for the *even-nucleon* cases, as shown in Fig.1 [4]. No such rise is seen in the reactions for the *odd-nucleon* cases, $^3He(\pi^-,\pi^-)^3He$ and $^3H(\pi^+,\pi^+)^3H$. We note that both $\pi^+ + ^3He$ and $\pi^- + ^3H$ couple to form isospin 3/2, whereas both $\pi^- + ^3He$ and $\pi^+ + ^3H$ couple to produce both isospin 1/2 and 3/2. In a recent calculation [7] (based on multiple-scattering theory) of pion elastic-scattering cross sections for the trinucleon system, the authors employed correlated three-body wave functions obtained from the solution of the Faddeev equations with the use of the Reid potential for the NN interaction. Also included in the calculation is a phenomenological second-order term (a ρ^2 scalar term obtained by fitting angular-distribution data for heavy nuclei) to allow for absorption and other higher-order processes.

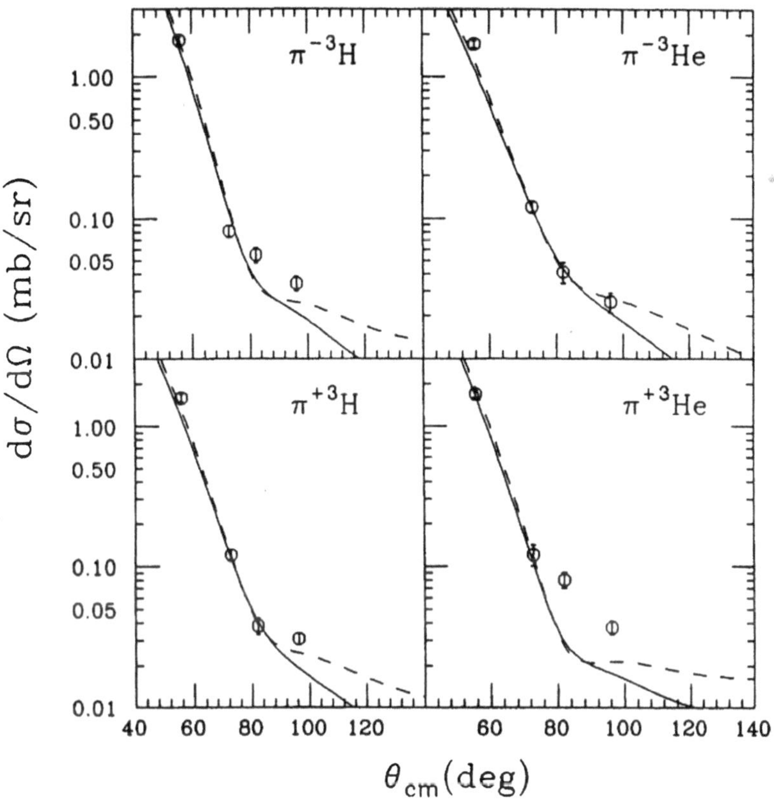

Figure 2. Elastic cross sections for 256–MeV pions

This calculation (the solid line in Fig.1) does a good job of describing the angular distributions for all four (pion + trinucleon) cases up to about 110°, the extent of the phase–1 data [1, 2]. For the new measurements at backward angles, however, the *odd-nucleon* cross sections are reasonably well described, but the *even-nucleon* cross sections are much larger than predicted. Another calculation [8], performed primarily to describe our ratio measurements (see below) shows similar behavior(the dashed line in Fig.1). This enhancement of the large-angle pion-nucleus scattering cross section has been noted before [9]; as far as we are aware, it has yet to receive a satisfactory explanation.

Our elastic cross-section results for 256 MeV [5] are shown in Fig.2, together with the recent theoretical predictions of Kamalov et al.[7]. The solid lines in the figure have the same meaning as in Fig.1, i.e., including both first-order and second-order (ρ^2) corrections. Here we see that the calculated results fall below the data even at the middle angles (around 90°). On the other hand, the calculated results *without* the ρ^2 term match the data better, although they still

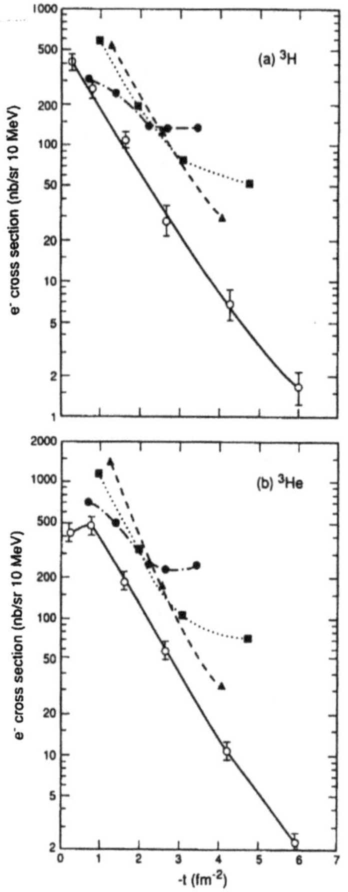

Figure 3. Inelastic cross sections for e^- and π^+ scattering

fall below the data at the larger angles. This calls into question the very idea of using a nucleon-density term fitted to heavy nuclei for the modest-density configuration of the three-body nuclei.

For inelastic pion scattering at 180 MeV, our results [3] can be compared with electron-scattering data [10], where both sets of data have been obtained for the 10–MeV excitation energy region just above the breakup thresholds. These data are shown in Fig.3 as a function of the four-momentum transfer. Since the charge and proton-matter form factors should have the same t dependence, we anticipate that these barely inelastic cross sections for pions (solid symbols) and electrons (open circles) will have a similar t dependence in the appropriate kinematic region. This appears to be the case at pion kinetic energy of 220 MeV (triangles) but not so at 180 MeV (squares) in the backward direction. At 142 MeV (circles) the cross-section shapes are totally different.

Perhaps this should not be surprising, since the barely inelastic cross-sections are different as well. One factor that probably contributes to the lack of correspondence in shape is that the pion data should exhibit interference between s- and p-wave scattering while the electron data do not. Diffraction minima for both would occur only at a much larger momentum transfer.

We now discuss the results in terms of certain charge-symmetric ratios, namely, r_1, r_2, and the superratio R, which is the product of r_1 and r_2. These ratios are defined as

$$r_1 = d\sigma(\pi^+ \, {}^3\mathrm{H})/d\sigma(\pi^- \, {}^3\mathrm{He})$$

$$r_2 = d\sigma(\pi^- \, {}^3\mathrm{H})/d\sigma(\pi^+ \, {}^3\mathrm{He})$$

$$R = r_1 \times r_2 \, .$$

The cross sections and ratios are determined from the measured pion-nucleus elastic-scattering yields. The charge-symmetric prediction for each of these ratios is unity. Thus, a sizable deviation from unity of any one of these ratios (particularly of R, which is independent of the systematic uncertainties in the monitoring of the pion beam) as a function of energy and angle can be interpreted as evidence for charge-symmetry breaking in the strong interaction. Such deviations also can be thought of as arising from (small) differences in the neutron and proton distributions in these mirror nuclei. Of course, the Coulomb interaction must be taken into account properly.

For resonance-energy pions, single-scattering models based on the impulse approximation predict that the scattering in and around the NSF dip is dominated by the spin-flip part of the total pion-nucleon amplitude, and that the ratio r_1 is primarily proportional to the squared ratio of the *odd-nucleon* matter form factors in the $A = 3$ nuclei (i.e., the proton in ^{3}H and the neutron in ^{3}He). These same models show that the scattering at angles approaching 180° is dominated by the non-spin-flip amplitude, and that r_2 is proportional to the squared ratio of the *even-nucleon* matter form factors of the $A = 3$ nuclei (i.e., the neutrons in ^{3}H and the protons in ^{3}He). Thus, pion-scattering measurements, under the favorable kinematic conditions described, are a very sensitive way of studying neutron and proton distributions in these nuclei, in addition to allowing one to isolate the contributions of the spin-flip and the non-spin-flip amplitudes in pion-nucleus interactions, without the need for polarized targets.

At the peak of the Δ resonance, our phase–1 data [1, 2] show the superratio R for elastic scattering to be significantly greater than unity in the angular region spanning the NSF dip, as shown in Fig.4a. These same data show that r_1, on the other hand, is consistent with unity throughout this angular region, and that the observed magnitude and the angular behavior of R is a reflection of the behavior of the ratio r_2, since $R = r_1 \times r_2$.

A detailed theoretical analysis [8] of the phase–1 data concluded that the magnitude and the angular behavior of these ratios, particularly that of the superratio R, is indicative of a significant charge-symmetry-breaking effect above and beyond that which is expected from the Coulomb interaction. It also allows one to determine the differences between the odd- and even-nucleon matter

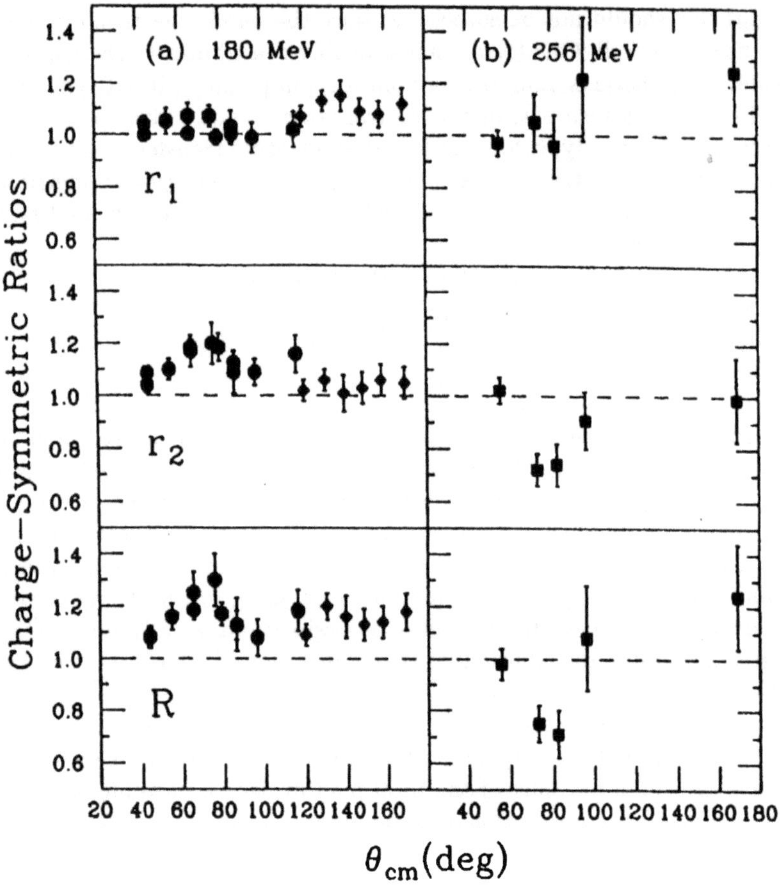

Figure 4. Charge-symmetric ratios for pion energies of (a) 180 and (b) 256 MeV

radii in ^{3}H and ^{3}He (quantities equal to 0.030-0.040 fm) with a precision of a few times 10^{-18} m.

Results of our latest angular-distribution measurements, at 180 MeV, also show that the superratio R is greater than unity for all the measured angles in the range 114° to 168°, as is also shown in Fig.4a [6]. However, we find that a remarkable role-reversal occurs for the ratios r_1 and r_2 in this angular range; r_2 is now consistent with unity, and it is r_1 that deviates significantly from this value. Perhaps even more remarkable, this role-reversal was predicted by the authors of ref.[8] solely from their analysis of the forward-angle data at this energy.

We now turn to our results for 256-MeV incident pions, shown in Fig.4b [5]. These measurements also yielded a very surprising result: for the first time, the measured superratio turned out to be substantially *smaller* than unity, in complete contrast to be results at 180 MeV. Interestingly, the behavior of the individual ratios r_1 and r_2 at the two energies is qualitatively the same; r_1 is consistent with unity across the NSF dip, and r_2 deviates quite significantly from unity, leading to the observed behavior in R.

By now, several models have appeared in the literature [7, 8, 11, 12] that attempt to explain the behavior of the superratio. For example, the authors of ref.[12] use as input for their (optical-model) calculation the measured electromagnetic form factors of the three-body nuclei, along with the argument that the Coulomb force distorts the nuclear force sufficiently to cause the observed deviation of R from unity. However, all preliminary calculations with this model indicate that it is going to be very difficult, if not impossible with the limited physics contained in the model thus far, to obtain a good simultaneous description of both the cross sections and the superratio at 180 and 256 MeV. Indeed, even the models of both refs.[7] and [8] fail to reproduce even qualitatively the behavior of R at 256 MeV. We strongly suspect that the radically different behavior of the superratio at 180 and 256 MeV is an indication of scattering and/or absorption processes that have not yet been implemented in the models.

Our 180-MeV data establish that R is greater than unity, but the 256-MeV measurements show R to be substantially less than unity in the NSF-dip region. Clearly, a transition must occur as a function of energy. In addition to our earlier results at 142 MeV, we also have obtained data (in the NSF-dip

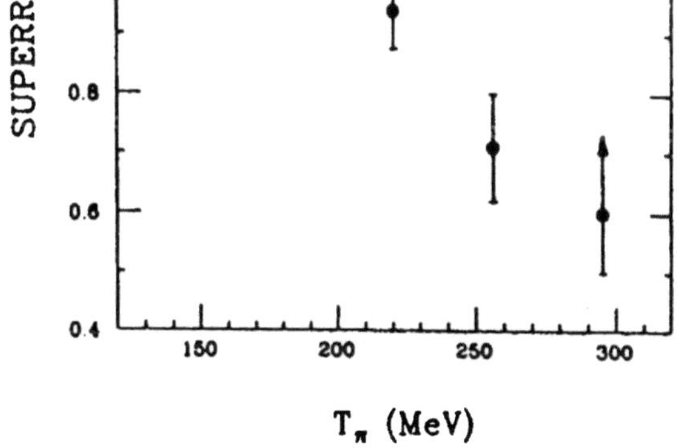

Figure 5. The superratio as a function of pion energy

region) at 220 MeV and 295 MeV, in order to locate the transition energy and to shed more light on our results at 256 MeV. These data, shown in Fig.5, confirm the transition in behavior of the superratio in which it is substantially larger than unity for the lower energies and significantly smaller than unity for the higher energies, with the transition occuring at around 210 MeV. For the 295 MeV point, we were only able to establish a limit on the superratio because of the very small $\pi^- \, ^3$He cross section at this energy. However, the downward trend of the superratio as a function of incident pion energy is evident. The reproduction of this behavior of the superratio is a major challenge to all the theoretical models considered.

References

1. B.M.K. Nefkens et al.: Phys. Rev. **C41**, 2770 (1990)
2. C. Pillai et al.: Phys. Rev. **C43**, 1838 (1991)
3. B.L. Berman et al.: Phys. Rev. **C51**, 1882 (1995)
4. S.K. Matthews et al.: Phys. Rev. **C51**, 2534 (1995)
5. K.S. Dhuga et al.: Phys. Rev. **C** (to be submitted)
6. W.J. Briscoe et al.: Phys. Rev. **C** (to be submitted)
7. S.S. Kamalov, L. Tiator, C. Bennhold: Phys. Rev. **C47**, 941 (1993) and Private Communication
8. W.R. Gibbs and B.F. Gibson: Phys. Rev. **C43**, 1012 (1991)
9. K.S. Dhuga et al.: Phys. Rev. **C35**, 1148 (1987) and references therein
10. G.A. Retzlaff et al.: Phys. Rev. **C49**, 1263 (1994)
11. Y.E. Kim, M. Krell, L. Tiator: Phys. Lett. **B172**, 287 (1986)
12. K.T. Kim, Y.E. Kim, R.H. Landau: Phys. Rev. **C36**, 2155 (1987)

Few-Body Systems Suppl. 9, 91–95 (1995)

Few-Body Systems

Pion Absorption on Few–Nucleon Systems

L. Canton, G. Cattapan

Istituto Nazionale di Fisica Nucleare e Dipartimento di Fisica dell'Universitá, Via F. Marzolo n.8, I-35131, Padova, Italia

Abstract. Pion absorption is incorporated in a set of four-body equations describing the πNNN dynamics. We outline the general chain-labelled structure of the equations and the implications for a complete calculation of the absorption amplitude.

1 Introduction

From the mid–eighties, much attention has been addressed to pion absorption on A=3 nuclei, and more generally on few–nucleon systems. This interest was triggered by the wealth of experimental results obtained in recent years at the meson facilities (LAMPF, PSI, and TRIUMF) where the employment of coincidence techniques, together with detector system of large angular acceptance (see [1] and references therein contained) allowed to establish that, in the energy region around the Δ resonance, three-nucleon absorption mechanisms play a significant rôle. These mechanisms contribute up to 30% of the total absorption cross section, while the dominant absorption process is on isoscalar nucleon pairs, with the third nucleon acting as a mere spectator (Quasi–Free Absorption, QFA).

While there have been extensive theoretical investigations about QFA on isoscalar and isovector nucleon pairs, multinucleon absorption on light nuclei still awaits for a proper theoretical treatment. An inherent theoretical difficulty relies on the fact that, in the kinematical regions where more than two nucleons share the energy of the incoming pion, the effects due to initial–state interactions (ISI) and final–state interactions (FSI) have to be disentangled: this requires a simultaneous and "consistent" treatment of the three–nucleon and four–particle (πNNN) continua, a task of formidable difficulty. The issue then is how to extend the Yakubovsky–Grassberger–Sandhas (Y–GS) quantum mechanical formulation of multiparticle scattering [2] to the case of πNNN systems, where obviously absorption processes may occur.

Clearly, there is an outstanding difficulty related to the fact that production/absorption processes force the theory to deal with an infinite number of degrees of freedom, and the recasting of the theory into a four–body–like approach unavoidably implies that higher–order diagrams with intermediate multipion states have to be neglected. The same problem appeared already in the related question of how to extend the Faddeev theory to the case of the πNN system. Here, there have been a number of theoretical formulations leading to the reduction of the field–theoretic πNN problem into an effective connected–kernel three–body problem, coupling all the relevant channels of two and three–body scattering, absorption, and production. Recent reviews on these approaches are given in refs. [3] and for a critical review of the plagues due to the underling field–theoretic truncation, see also ref. [4]. With respect to this point, the progress made recently with the introduction of field–theoretic convolution techniques [5] allowed to overcome a serious problem with the nucleon renormalization which was plaguing earlier πNN theories, and at the same time allowed to include certain diagrams with intermediate two–pion states previously missing in the theory, but known to be important [6]. In our opinion, this advancement put new hopes in the use of non covariant unitary theories to describe the π–multinucleon dynamics.

Recently, we have obtained new connected–kernel equations [7] starting from disconnected Afnan–Blankleider–like (AB) equations for the πNNN–NNN system. It is the purpose of this paper to illustrate how one can obtain connected–kernel equations for the πNNN system starting from the general structure implied by the πNN–NN theories. The method relies only on the combinatorial properties of the partitions of the system into clusters, with a crucial rôle played by the treatment of the production/absorption vertices, which unavoidably introduce further disconnected contributions with respect to standard multiparticle scattering theory. The resulting equations are similar in structure to the four–body Y-GS chain–labelled equations, but contain in addition the complete couplings to the absorption channels.

2 Connected–Kernel πNNN–NNN Equations

πNN theories lead to equations having the general structure

$$U_{ab} = G_0^{-1}\bar{\delta}_{ab} + \sum_c \bar{\delta}_{ac} t_c G_0 U_{cb} + F_a g_0 U_b^\dagger, \tag{1}$$

$$U_a^\dagger = F_a^\dagger + \mathcal{V} g_0 U_a^\dagger + \sum_c F_c^\dagger G_0 t_c G_0 U_{ca}, \tag{2}$$

$$U_a = F_a + \sum_c \bar{\delta}_{ac} t_c G_0 U_c + F_a g_0 U, \tag{3}$$

$$U = \mathcal{V} + \mathcal{V} g_0 U + \sum_c F_c^\dagger G_0 t_c G_0 U_c, \tag{4}$$

($\bar{\delta}_{ab} = 1 - \delta_{ab}$). The very same equations (though disconnected) can be derived for the πNNN system [8, 9], but in such a case the U_{ab} represent the scattering amplitudes connecting all three–cluster partitions one can have with four particles. The indices a, b represent one of the six possible three–cluster partitions of type $(NN)+N+\pi$ or $(\pi N)+N+N$. In this last case, the partition a may be identified also by the i-th nucleon coupled to the pion, $i \equiv a$.

The single amplitude U refers to the unclusterized $N+N+N \leftarrow N+N+N$ scattering process in the NNN sector. The remaining 6+6 operators U_a and $U_a^\dagger$ represent production and absorption operators for processes of type $N+N+N \leftrightarrow a$. The free Green's functions in the NNN and πNNN sectors are denoted by g_0 and G_0 respectively, while $\mathcal{V}$ represents the sum of the pair OPE potentials in the NNN sector, and may include higher order contributions. t_a is the two–body t–matrix in the four–body sector, (in the case of πN t–matrices, only the non–polar term is retained). The production/absorption vertex operators F_a, $F_a^\dagger$ are defined in terms of the elementary vertices $f(i)$, $f^\dagger(i)$ by

$$F_a = \sum_{i=1}^{3} \bar{\delta}_{ia} f(i), \qquad F_a^\dagger = \sum_{i=1}^{3} \bar{\delta}_{ia} f^\dagger(i). \tag{5}$$

We now introduce the partitions of the system into two clusters, and denote such partitions in the four–body sector by a'. For a' we distinguish three classes: the $(NNN)+\pi$ partition is called of type `III`, while the three $(\pi NN)+N$ partitions are of type `I`. The remaining three partitions $(\pi N)+(NN)$ are denoted as type `II`. For each partition of the system into two clusters we have to define the corresponding sub–amplitudes, which satisfy independent three–body–like equations.

type I

$$(u_{a'})_{ab} = G_0^{-1}\bar{\delta}_{ab} + \sum_c \bar{\delta}_{ac} t_c G_0 (u_{a'})_{cb} + (f_{a'})_a g_0 (u_{a'}^\dagger)_b, \tag{6}$$

$$(u_{a'}^\dagger)_a = (f_{a'}^\dagger)_a + \mathcal{V}_{a_1} g_0 (u_{a'}^\dagger)_a + \sum_c (f_{a'}^\dagger)_c G_0 t_c G_0 (u_{a'})_{ca}, \tag{7}$$

$$(u_{a'})_a = (f_{a'})_a + \sum_c \bar{\delta}_{ac} t_c G_0 (u_{a'})_c + (f_{a'})_a g_0 (u_{a'}), \tag{8}$$

$$(u_{a'}) = \mathcal{V}_{a_1} + \mathcal{V}_{a_1} g_0 (u_{a'}) + \sum_c (f_{a'}^\dagger)_c G_0 t_c G_0 (u_{a'})_c. \tag{9}$$

type II

$$(u_{a'})_{ab} = G_0^{-1}\bar{\delta}_{ab} + \sum_c \bar{\delta}_{ac} t_c G_0 (u_{a'})_{cb} + (f_{a'})_a g_0 (u_{a'}^\dagger)_b, \tag{10}$$

$$(u_{a'}^\dagger)_a = (f_{a'}^\dagger)_a + \sum_c (f_{a'}^\dagger)_c G_0 t_c G_0 (u_{a'})_{ca}, \tag{11}$$

$$(u_{a'})_a = (f_{a'})_a + \sum_c \bar{\delta}_{ac} t_c G_0 (u_{a'})_c + (f_{a'})_a g_0 (u_{a'}), \tag{12}$$

$$(u_{a'}) = \sum_c (f_{a'}^\dagger)_c G_0 t_c G_0 (u_{a'})_c. \tag{13}$$

type III

$$(u_{a'})_{ab} = G_0^{-1}\bar{\delta}_{ab} + \sum_c \bar{\delta}_{ac} t_c G_0 (u_{a'})_{cb}. \tag{14}$$

Here, for any type of partition a', it is intended that the sub-amplitudes have components related to each sub-partition of the system into three clusters (namely, $a, b, c \subset a'$). The 6 sub-amplitudes of type I and II have also one component in the three-nucleon space, the component being related to one well defined partition into two clusters of the *NNN* sector, denoted as a_1.

A crucial rôle is played by the definition of the new production/absorption operators $(f_{a'})_a$ and $(f^\dagger_{a'})_a$, which we call *internal* vertices and are the inputs for the sub-amplitude equations. These vertices are defined as

$$(f_{a'})_a = \sum_{i=1}^{3} \bar{\delta}_{ia} \delta_{i,a\subset a'} f(i), \tag{15}$$

and satisfy the two-cluster sum rule

$$F_a = \sum_{a'} (f_{a'})_a, \qquad F^\dagger_a = \sum_{a'} (f^\dagger_{a'})_a. \tag{16}$$

The vertex F_a can be viewed as a 6×1 rectangular matrix connecting the 6 three-cluster partitions of the four-body space to the single three-cluster partition of the three-nucleon space. Instead, $(f_{a'})_a$ is interpreted as a rectangular matrix connecting all chains in the four-body sector (a chain is a pair of partitions $\{a', a\}$ with $a \subset a'$) to new, nonstandard chains in the absorption space (defined as a pair of partitions $\{a', a_1\}$ where a_1 follows from the absorption of the pion *inside* the partition a').

By means of the above defined sub-amplitudes, we have obtained a set of equations describing scattering and absorption [7] with a kernel that is connected after iteration [10].

$$\begin{aligned}
U_{a'ab'b} &= \bar{\delta}_{a'b'}\delta_{ab}(G_0 t_a G_0)^{-1} + \sum_{c'c} \bar{\delta}_{a'c'}(u_{c'})_{ac} G_0 t_c G_0 U_{c'cb'b} \\
&\quad + \sum_{c'(\in \mathrm{I,II})} \bar{\delta}_{a'c'}(u_{c'})_a g_0 U^\dagger_{c'c_1b'b}, \qquad\qquad (17)\\
U^\dagger_{a'a_1b'b} &= \sum_{c'(\in \mathrm{I,II})c} \bar{\delta}_{a'c'}(u^\dagger_{c'})_c G_0 t_c G_0 U_{c'cb'b} + \sum_{c'(\in \mathrm{I,II})} \bar{\delta}_{a'c'} u_{c'} g_0 U^\dagger_{c'c_1b'b}, \\
U_{a'ab'b_1} &= \sum_{c'c} \bar{\delta}_{a'c'}(u_{c'})_{ac} G_0 t_c G_0 U_{c'cb'b_1} + \sum_{c'(\in \mathrm{I,II})} \bar{\delta}_{a'c'}(u_{c'})_a g_0 U_{c'c_1b'b_1}, \\
U_{a'a_1b'b_1} &= g_0^{-1}\bar{\delta}_{a'b'} + \sum_{c'(\in \mathrm{I,II})c} \bar{\delta}_{a'c'}(u^\dagger_{c'})_c G_0 t_c G_0 U_{c'cb'b_1} \\
&\quad + \sum_{c'(\in \mathrm{I,II})} \bar{\delta}_{a'c'} u_{c'} g_0 U_{c'c_1b'b_1}.
\end{aligned}$$

For pion absorption, we have used the residue method to obtain the complete expression for the amplitude

$$<\phi_0|\mathcal{A}|\Psi>=\sum_{a',a,b}<\phi_0|(u^\dagger_{a'})_a G_0 t_a G_0 U_{a'ab'b}|\Psi_b>+\sum_{a',b}<\phi_0|(u_{a'})g_0 U^\dagger_{a'a_1b'b}|\Psi_b>.$$

The final and initial asymptotic states, $<\phi_0|$ and $|\Psi>$, represent the three–nucleon plane wave and the π–trinucleon plane wave, respectively ($|\Psi_b>$ are the Faddeev components of the *NNN* bound state). The partition b' is obviously of type III, while a' refers only to partitions of type I and II.

Whereas the present formalism has been developed with reference to the AB approach, extensions to other descriptions of π–multinucleon dynamics can be envisaged. Thus, one could start from models where the πNN vertex is not explicitly taken into account, and coupling of the no–pion sector with the πNNN space is accomplished through an effective $\pi N\Delta$ vertex [11, 12]. In such a case, apart from the obvious absence of terms explicitly referring to the πNN vertex, one gets equations where the coupling of scattering and absorption/emission channels is structurally the same in the two cases, which is enough to guarantee the applicability of our method. A more ambitious goal would be the extension of the present πNNN–NNN formalism to a convolution approach. It is worth to stress that, convolution equations can be written in a form quite similar to the AB scheme, plus an integration over an additional energy variable, for amplitudes defined in the *NN* space [5].

References

1. T. Alteholz et al.: Phys. Rev. Lett. **73**, 1336 (1994)
2. O.A. Yakubovsky: Sov. J. Nucl. Phys. **5**, 937 (1967); P. Grassberger and W. Sandhas: Nucl. Phys. **B2**, 181 (1967)
3. H. Garcilazo and T. Mizutani: *πNN Systems*. Singapore: World Scientific 1990; B. Blankleider: Nucl. Phys. **A543**, 163c (1992)
4. D.R. Phillips and I.R. Afnan: preprint FIAS/R/223 Flinders Uni. 1993
5. A.N. Kvinikhidze and B. Blankleider: Phys. Lett. **B307**, 7 (1993)
6. B.K. Jennings: Phys. Lett. **B205**, 187 (1988)
7. L. Canton and G. Cattapan: Phys. Rev. **C50**, 2761 (1994)
8. Y. Avishai and T. Mizutani: Nucl. Phys. **A393**, 429 (1983)
9. G. Cattapan, L. Canton, J.P. Svenne: Nuovo Cim. **A106**, 1229 (1993)
10. G. Cattapan and L. Canton: Few–Body Sys. **17**, 163 (1994)
11. M. Betz and T.S.H. Lee: Phys. Rev. **C23**, 375 (1981); A. Matsuyama and T.S.H. Lee: Phys. Rev. **C32**, 516 (1985)
12. H. Pöpping, P.U. Sauer, Zhang Xi–Zhen: Nucl. Phys. **A474**, 557 (1987)

Few-Body Systems Suppl. 9, 97–110 (1995)

Few-Body Systems

Strong Interaction Physics from Hadronic Atoms

E. Friedman* †

Racah Institute of Physics, The Hebrew University, Jerusalem 91904, Israel

Abstract.
Strong interaction effects in kaonic, sigma and antiprotonic atoms are analyzed with density dependent (DD) $t_{\rm eff}$ in a $t_{\rm eff}\rho$ approach to the optical potential. Different radial sensitivities are observed and the importance of realistic nuclear density distributions is demonstrated. For kaonic atoms the phenomenological DD potential can be related to the propagation of the Λ (1405) in the nuclear medium. For Σ^- atoms the potentials can be related to a relativistic mean field (RMF) approach which places constraints on various meson-Σ hyperon coupling constant ratios. For $\bar{p}$ atoms DD potentials enable reasonable p-wave terms to be accommodated. In all cases the low density limit can be respected with the DD potentials.

1 Introduction and Outline

Hadronic atoms have long been recognized as a source of unique information on the strong interaction of various particles with nuclei at zero kinetic energy. Such atoms are formed when a negatively charged particle (e.g. π^-, K^-, Σ^- and $\bar{p}$) is captured in an atom. While it cascades down the atomic levels, Auger electrons and x-rays are emitted. The latter have characteristic line spectra that terminate with a strong-interaction broadened and shifted line, compared to that expected for a pure electromagnetic interaction. The energy shifts and level widths thus provide information on the strong interaction with the nucleus. The very first treatment of strong interaction effects in pionic atoms was that of Deser et al. [1] in 1954, which was confined to 1s states in very light atoms. Later it became clear that in general these effects are non-perturbative, and with the accumulation of experimental results it was realized that the interaction of a particle with a nucleus in a hadronic atom is not simply related to the corresponding interaction with a free nucleon [2]. Fairly simple optical

**Alternative address (1994-5):* TRIUMF, 4004 Wesbrook Mall, Vancouver, B.C., Canada V6T 2A3

†*E-mail address:* elifried@vms.huji.ac.il

potentials, however, have been shown to describe pretty well the experimental results and were able to shed some light on the interactions themselves.

The strong interaction in a hadronic atom involves the overlap between an atomic wavefunction and the nucleus. Because of the different strength of the nuclear absorption process for the various particles, the last observed state in the atomic cascade will have widely different overlap with nuclei, thus exhibiting a whole range of medium-modification phenomena of the strong interaction. The present talk will focus on these effects.

In the second section we will show how the radial sensitivity of various hadronic atoms can be determined almost quantitatively, and the radial sensitivity will be presented for hadronic atoms of pions, kaons, sigma hyperons and antiprotons. Three kinds of hadronic atoms will be discussed in the subsequent sections, where emphasis will be placed on medium modification or density dependence of the interaction.

2 Radial Sensitivity of Hadronic Atom Data

In the present section it is assumed that for each kind of hadronic atoms an optical potential giving a good fit to the data can be established. We then proceed to check what radial regions of the potential are indeed determined by the data. By the nature of hadronic atoms data one cannot readily apply "model independent" methods [3], but an alternative is possible by introducing a local perturbation or a "notch" into the potential and examining its effects on the fit to the data.

The sensitivity of the data to the potential around the radius $R_{\rm N}$ is obtained by multiplying the best-fit potential by a factor:

$$f = 1 - d\ \exp[-(\frac{r - R_{\rm N}}{a_{\rm N}})^2] \tag{1}$$

representing a "notch" in the potential around $r = R_{\rm N}$ spread approximately over $\pm a_{\rm N}$, whose relative depth is d. By varying the depth d and observing the changes in χ^2, the sensitivity (defined as the values of d that cause χ^2 to increase by one unit) can be determined,

$$\chi^2(R_{\rm N}, d^{\pm}) = \chi^2_{\rm min} + 1 \tag{2}$$

$$\frac{\delta V(R_{\rm N})}{V(R_{\rm N})} = \frac{1}{2}(d^+ - d^-). \tag{3}$$

For $a_{\rm N}$ the value of 0.5 fm was chosen. By scanning over $R_{\rm N}$, the radial region where the fit to the data is sensitive to the potential is determined. $R_{\rm N}$ is scanned in steps of a nuclear diffuseness a_0, i.e., $R_{\rm N} = R_0 + a_0\Delta$, with $R_0 = 1.1A^{1/3}$ fm and $a_0 = 0.5$ fm. It should be emphasized that this notch is applied to all the data at the same time. Therefore the results should be largely free from local variations. Figure 1 is an example of such a notch test, applied to the best fit potential for kaonic atoms [4,5]. It is seen that in this case the

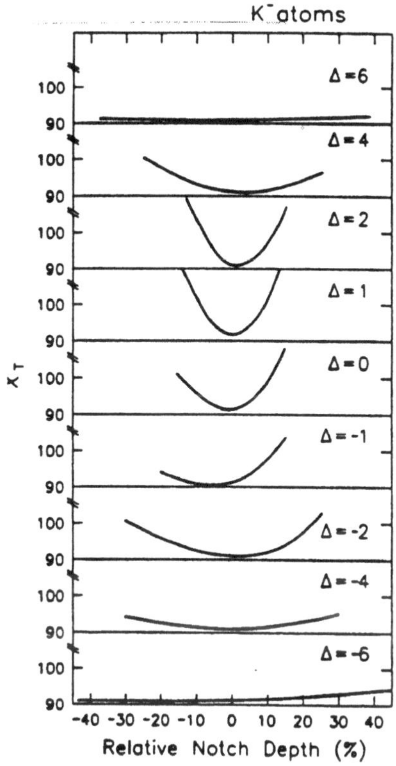

Figure 1. Total χ^2 values as a function of the relative notch depth d (cf. Eq.(1)) superimposed on the best-fit DD optical potential for kaonic atoms [4,5] for several positions R_N ($R_N = R_0 + a_0\Delta$) expressed via steps Δ of a nuclear diffuseness, with a radius $R_0 = 1.1A^{1/3}$ fm and a diffuseness $a_0 = 0.5$ fm.

radial region which is "sampled" by the data is quite broad, extending from well within the nuclear volume to outside of the nuclear surface.

Figure 2 is a collection of results of notch tests, applied to all the types of hadronic atoms for which reasonably accurate data exist. The results are shown as the relative uncertainty in the potential as a function of the notch position. Also shown is a reference nuclear density, parameterized as a Fermi distribution, which enables one to relate a notch position to the average local density. It is seen that there are large differences between hadronic atoms. Whereas pionic and kaonic atoms are quite sensitive to the potentials in the nuclear interior, sigma and antiprotonic atoms are sensitive only to the potential outside of the nuclear surface. The level of uncertainty also varies considerably, presumably due to the widely different accuracy of data, for the various types of hadronic atoms.

The above results raise another question, namely, that of the nuclear densities to be used in the optical potentials (see below). For sigma and antiprotonic

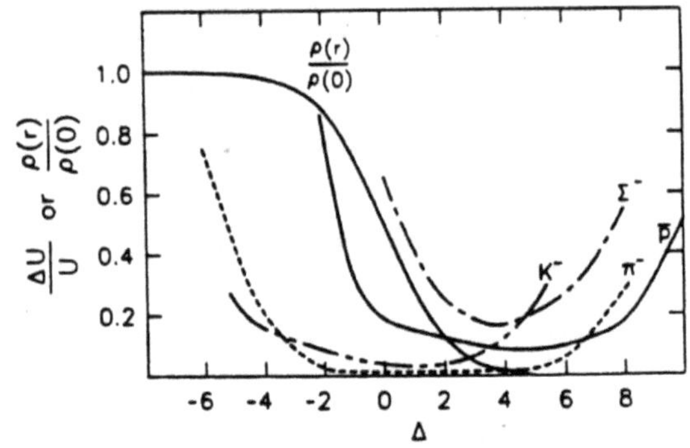

Figure 2. Relative accuracies of optical potentials for different types of hadronic atoms as determined from "notch tests". Positions are given in units of a nuclear diffuseness parameter, relative to the nuclear half-density point. Also shown is a reference nuclear density distribution.

atoms one needs densities that will be most appropriate at the extreme nuclear surface whereas for pionic and kaonic atoms the interior is also important. A convenient "macroscopic" (MAC) parameterization of nuclear densities is in terms of a 2- or 3- parameter Fermi function or, for light nuclei, in terms of a modified harmonic oscillator density. Experimental information on nuclear charge densities is often presented [6] using these functions, and the proton densities may be obtained by properly unfolding the finite size of the proton. Such densities are adequate around the nuclear surface where the *rms* radius is determined. However, such distributions cannot be adequate far outside of the nucleus because they do not possess the correct exponential fall-off, as determined by the binding energy of the least bound particles. In that region it is more appropriate to use singe-particle (SP) densities or better still densities obtained from a more realistic Hartree-Fock calculation.

The method chosen in the present work for generating "microscopic" or SP nuclear densities was to fill in single particle levels in Woods-Saxon potentials separately for protons and neutrons. The radius parameter of the potential was adjusted to reproduce the *rms* radius of the charge distribution (after folding in the finite size of the proton). The binding energy of the least bound particle was set equal to the corresponding separation energy [7]. For $N > Z$ nuclei the *rms* radius of the neutron density distribution was chosen to be slightly larger than that of the protons. By using this method, rather realistic density distributions are obtained in the region of the exponential fall-off.

Figure 3 shows the two types of densities for Ca where small differences are observed both in the nuclear interior and at the extreme tail of the distributions. Also shown is an example of the radial wavefunction squared for the 1h state of Σ^- atoms. It is the overlap between this and the density that determines the strong interaction effects and it is clear from the figure how small differences

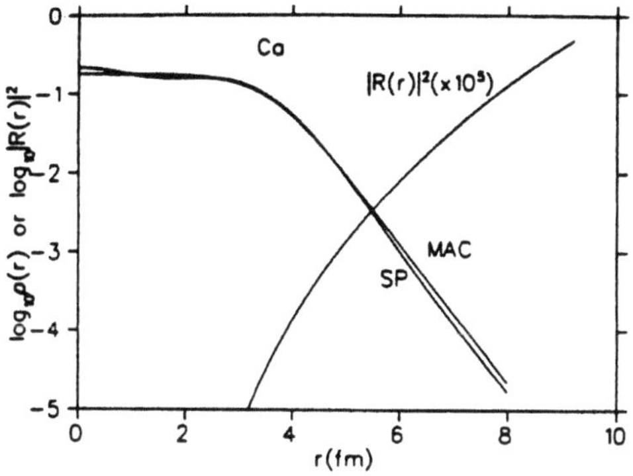

Figure 3. Nuclear density distributions for ^{40}Ca from a macroscopic and single particle formulations. Also shown is the absolute value squared of the radial wavefunction for the $1h$ state of Σ^- atom.

between densities at large radii may cause significant effects in hadronic atoms.

3 Kaonic Atoms

The simplest approach to the optical potential is $V(r)=t_{\text{eff}}\rho(r)$, where $\rho(r)$ is the nuclear density distribution normalized to the total number of nucleons A. Within this approach strong interaction effects in kaonic atoms are known to be non-perturbative, as evidenced by the fact that although the level shifts are repulsive, the real part of the empirical potential invariably comes out attractive ($\text{Re}t_{\text{eff}} < 0$). This is a direct consequence of the (absorptive) imaginary part of t_{eff} being comparable in magnitude to or larger than its real part. A purely attractive potential poses a problem in the limit of low density, where one expects [8-10] $V_{\text{opt}} \to t_{\bar{K}N}\rho$, with $t_{\bar{K}N}$ denoting the $\bar{K}N$ t matrix at threshold for which $\text{Re}t_{\bar{K}N} > 0$, hence a repulsive potential.

The clue to the low-energy K^- nuclear interactions apparently lies with the nuclear dynamics of $\Lambda(1405)$, which is commonly considered an unstable $\bar{K}N$ $I = 0$ bound state, some 27 MeV below the K^-p threshold. Normally, $t_{\bar{K}N}$ appears repulsive due to this bound state. One then argues [11] that the energy appropriate to a $\bar{K}N$ collision is shifted downward in the nuclear medium, into a regime where $t_{\bar{K}N}$, for $I = 0$, appears attractive. Several model calculations [12-15] have been published in the last two decades to show how this mechanism forces the K^- nuclear optical potential at threshold to become attractive, contrary to what a naive extrapolation of the low density limit into normal nuclear densities would give. This kind of situation, where $\text{Re}V_{\text{opt}}(r)$ apparently contains both attractive and repulsive components, naturally suggests a non linear density dependence of the optical potential. Kaonic atoms were therefore the first to be analyzed in the recent series of studies of density dependent (DD) effects in hadronic atoms [4,5,16-18].

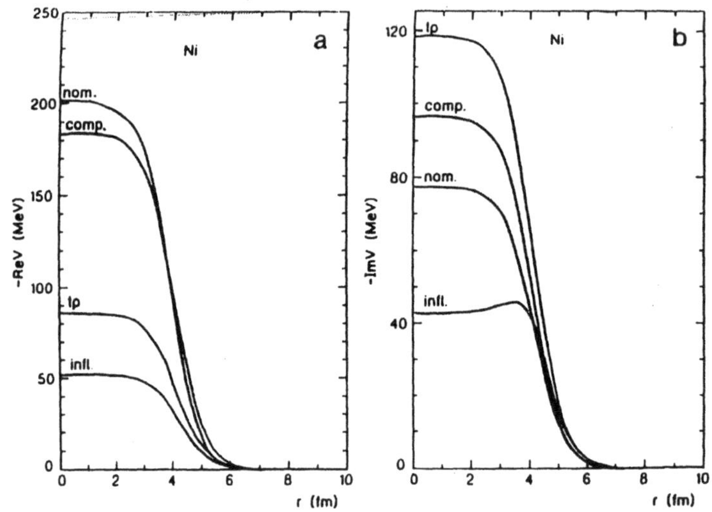

Figure 4. K^- nucleus potentials for Ni for several families of solutions. (see ref.[5])

For the DD optical potential we choose the form

$$2\mu V_{\rm opt}(r) = -4\pi(1+\frac{\mu}{m})\left\{\left[b_0 + B_0(\frac{\rho(r)}{\rho(0)})^\alpha\right]\rho(r) + \left[b_1 + B_1(\frac{\rho(r)}{\rho(0)})^\alpha\right]\delta\rho(r)\right\} \quad (4)$$

where μ is the kaon-nucleus reduced mass, m is the mass of the nucleon and $\rho(r) = \rho_n(r) + \rho_p(r)$ is the nuclear density distribution normalized to the number of nucleons A. An isovector term is also included in the above general expression although it was not used in the majority of fits to kaonic atom data consisting of 65 data points. The isovector density is given by $\delta\rho(r) = \rho_n(r) - \rho_p(r)$. The parameters b and B are given in units of fm and are obtained from fits to the kaonic atom data. Note that for $B = 0$, Eq.(4) reduces to the standard $t_{\rm eff}\rho$ parameterization of the optical potential in terms of a density independent $t_{\rm eff}$. The complex coefficients b and B, and the exponent α, can be determined by a least squares fit to the K^- atom data. Whilst the values of these complex constants change considerably when gridding on α, we found that certain combinations of these parameters or integral quantities such as the volume integrals or *rms* radii of the real and imaginary potentials displayed remarkably smooth and weak dependence on α. The various results could therefore be grouped into "families" of solutions with significantly lower χ^2 values than for $B = 0$.

Figure 4 shows some of the resulting potentials for the case of Ni. Examining this figure together with the accuracies displayed in Fig. 2, it is noted that the differences between the various potentials are larger than the uncertainties throughout most of the nuclear interior where the densities and potentials saturate. In particular we note that the depth of the real potential is around

180 MeV for typical nuclear densities, a result which might have astrophysical consequences [19]. From the radial sensitivity displayed in Fig. 2 it is expected that the results will not depend sensitively on the type of nuclear density used in the analysis. This is indeed found to be the case as can be seen from Table 3.1. These results show very little difference between MAC- and SP- based fits, both for a $t_{\rm eff}\rho$ and DD potentials, the latter based on the "nominal" solution of ref.[5] which respects the low density limit for the isoscalar component. It is also evident that the isovector term is not determined at all by the data, and again this conclusion is insensitive to the type of densities used.

Phenomenological results such as presented above are most useful if they can serve as a guide towards a more physical approach. In the case of kaonic atoms the first candidate for the mechanism of this density dependence is the Λ (1405) resonance just below the K^-p threshold. Some successes and some difficulties of this mechanism have been discussed in [5] in connection with these DD phenomenological potential.

Table 3.1. Results for kaonic atoms

	MAC	SP	SP	MAC	SP	SP
Reb_0	0.63	0.81	0.85	$-\underline{0.15}$	$-\underline{0.15}$	$-\underline{0.15}$
	±0.06	±0.07	±0.07			
Imb_0	0.89	0.89	0.87	$\underline{0.62}$	$\underline{0.62}$	$\underline{0.62}$
	±0.05	±0.06	±0.08			
Reb_1			−0.32			−0.18
			±0.34			±0.34
Imb_1			0.06			−0.04
			±0.31			±0.25
ReB_0				1.65	1.59	1.58
				±0.06	±0.08	±0.1
ImB_0				−0.06	0.10	0.02
				±0.06	±0.07	±0.08
α				0.23	0.18	0.17
				±0.03	±0.04	±0.04
χ^2	134.5	130.5	127.1	109.2	112.5	112.2

b and B are given in units of fm. Underlined quantities were held fixed during the fits. The top row indicates the type of nuclear density used.

4 Σ^- Atoms

A recent DD analysis of almost all available data on Σ^- atoms [16,17] showed that much better fits to the data are possible compared to those obtained with a $t_{\rm eff}\rho$ potential. A surprizing result was that the DD real potential had a small attractive pocket just outside of the nuclear surface, turning into repulsion towards the nuclear interior. Those results were based on the use of MAC density distributions. As the radial sensitive region of Σ^- atom data is well outside the nucleus, the analysis has been repeated using SP densities, as explained in Sec.2. The potential of Eq.(4) was then used with the only difference being that the imaginary potential was based on proton densities, because absorption of Σ^- is assumed to occur only by the $\Sigma^- p \longrightarrow \Lambda n$ process. Table 4.1 summarizes results of fits based on the two sets of nuclear densities. The results shown are for α=0.4 as a typical value for the respective "family" of solutions, as discussed in Sec. 3. It is seen that in the $t_{\text{eff}}\rho$ approach the χ^2 values are around 48 (for 23 data points) and the real isovector term is not determined by the data. Applying DD to the MAC densities the χ^2 goes down to 22 but the isovector term is still not determined. Only by using SP densities and including DD terms does the χ^2 go down to 17 (or 0.9 per degree of freedom) and then a repulsive isovector component emerges. This is a demonstration of the importance of using as realistic nuclear densities as possible in the relevant region. By comparing ReB_0 with Reb_0 it is realized that the potential indeed becomes repulsive in the nuclear interior. Figure 5a shows the DD potential for the case of Si.

Table 4.1. Results for sigma atoms

	MAC	MAC	MAC	MAC	SP	SP	SP	SP
Reb_0	0.31	0.38	1.5	1.3	0.33	0.31	1.1	2.4
	±0.04	±0.09	±0.6	±0.7	±0.05	±0.09	±0.5	±1.2
Imb_0	0.21	0.20	0.37	0.35	0.24	0.24	0.43	0.49
	±0.03	±0.03	±0.07	±0.08	±0.04	±0.05	±0.11	±0.12
Reb_1		−0.3		−0.2		0.03		−1.0
		±0.3		±0.3		±0.25		±0.5
ReB_0			−2.5	−2.0			−1.7	−4.3
			±1.5	±2.2			±1.2	±3.4
α			$\underline{0.4}$	$\underline{0.4}$			$\underline{0.4}$	$\underline{0.4}$
χ^2	47.7	45.4	21.9	21.8	47.9	47.9	29.2	16.7

b and B are in units of fm. Underlined quantities were held fixed during the fits. Absorption is on protons only, i.e. Imb_1=−Imb_0. The top row indicates the type of nuclear density used.

The relativistic mean field (RMF) approximation [20] is an interesting alternative to more conventional nuclear models. Recent calculations of hypernuclei have demonstrated that it can be successfully extended to more general baryon-nucleus systems [21,22]. It is thus topical to apply the RMF approach to constructing the Σ-nucleus optical potential and to establish constraints imposed on the Σ hyperon couplings by fitting the Σ^- atomic data. We therefore used the RMF formalism for developing the Σ^--nucleus optical potential [23].

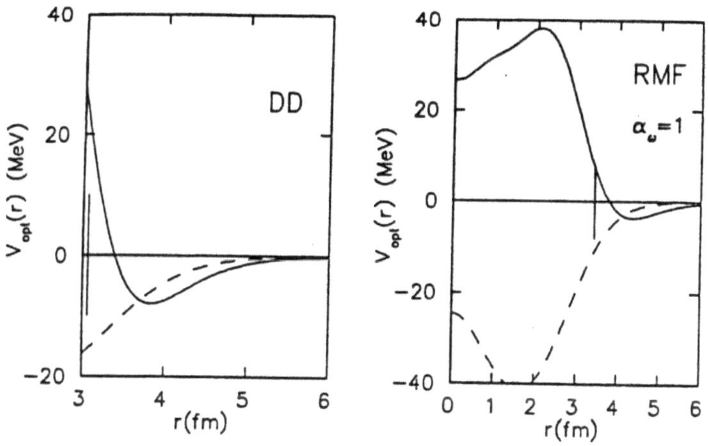

Figure 5. $\mathrm{Re}V_{\mathrm{opt}}^{\Sigma}$ (solid lines) and $\mathrm{Im}V_{\mathrm{opt}}^{\Sigma}$ (dashed lines) in Si as functions of r for (**a**) the best-fit DD optical potential, (**b**) the RMF Σ^- optical potential. The vertical bar indicates the position of the corresponding nuclear *rms* radius.

For any particular nucleus we calculated the scalar (σ) and vector (ω, ρ) mean (meson) fields and constructed the Schrödinger equivalent (SE) Σ-nucleus potential:

$$\mathrm{Re}V_{\mathrm{opt}}^{\Sigma}(r) \equiv V_{\mathrm{SE}}(r) = S(r) + \frac{EV(r)}{M_\Sigma} + \frac{(S^2(r) - V^2(r))}{2M_\Sigma} \quad , \tag{5}$$

$$\text{where} \quad S(r) = g_{\sigma\Sigma}\sigma(r) \quad \text{and} \quad V(r) = g_{\omega\Sigma}\omega(r) - g_{\rho\Sigma}\rho_0(r) \ . \tag{6}$$

The SE potential was used as a real part of the optical potential in the calculations of Σ^- atoms.

The imaginary part describing the conversion $\Sigma^- p \longrightarrow \Lambda n$ was used in the form:

$$\mathrm{Im}V_{\mathrm{opt}}^{\Sigma}(r) = t\rho_p(r) \ . \tag{7}$$

Here, the proton density ρ_p was calculated in the RMF model and t was taken as a parameter to be determined by fitting the atomic data.

In the case of the vector meson coupling ratio α_ω ($\alpha_i = g_{i\Sigma}/g_{iN}$, $i = \sigma$, ω, ρ) three values, namely 1/3, 2/3 and 1, were used. The choice of $\alpha_\omega = 1/3$ was inspired by early hypernuclear calculations, the 2/3 ratio follows from the constituent quark model, and finally, the equality $g_{\omega\Sigma} = g_{\omega N}$ is motivated by a QCD sum rule evaluation [24]. For each of the above choices of α_ω the ratio α_σ together with t from Eq.(7) were fitted to the experimental atomic shift and width in Si. The isovector part of the optical potential was then included and the ratio α_ρ was determined by fitting the shift and width in Pb while holding the isoscalar parametrization fixed. Typical results for Si are illustrated in Fig. 5b. Having determined the isoscalar as well as isovector parameters of $V_{\mathrm{opt}}^{\Sigma}$ by

Table 4.2. Parameters of the RMF Σ^- nucleus optical potentials. The linear parametrization from ref.[25] was used for the nucleonic sector. $\alpha_i = g_{i\Sigma}/g_{iN}$ ($i = \sigma$, ω, ρ), and t is defined in Eq.(7).

potential	α_ω	α_σ	α_ρ	t [MeV fm^3]	χ^2
L1	1.0	0.770	2/3	-400	18.1
L2	2/3	0.544	2/3	-300	23.9
L3	1/3	0.313	0.65	-265	26.5

fitting the Si and Pb data we constructed optical potentials for all nuclei for which Σ^- atomic data exist.

The Σ^- atomic data appear to be sufficient to significantly constrain the RMF parametrization of the Σ nucleus optical potential. Not only that the value of the ρ-Σ coupling ratio $\alpha_\rho \approx 2/3$ holds unambiguously for all the parametrizations used, but in addition, a weak ω-Σ coupling is almost certainly ruled out by fitting to the Σ^- atom data. The application of the RMF calculational scheme to Σ^- atom data in terms of three coupling-constant ratios $(\alpha_\omega, \alpha_\sigma, \alpha_\rho)$ showed that very good quality fits, reproducing these data, can be made. The larger α_ω is (in the range 0 to 1), the better is the fit, as demonstrated in Table 4.2. Of these fits the best ones, in the range $\alpha_\omega \simeq 2/3$ to 1, indeed produce SE $V_{\rm opt}^{\Sigma}$ with a volume repulsion in the nuclear interior.

It is gratifying to conclude that both approaches, the phenomenological one and the RMF approach, agree with each other, for a comparable degree of fit, in producing an isoscalar repulsion in the nuclear interior and, for Σ^-, an added isovector repulsion. We stress that this isovector repulsion, with $\alpha_\rho \simeq 2/3$, was derived independently of the values assumed by the isoscalar coupling constants ratios. The potentials obtained will generally not bind Σ hyperons in nuclei, in agreement with recent experimental results [26].

5 Antiprotonic Atoms

Data of particularly high quality from the LEAR facility have become available in the last decade. Previous data have also been quite useful in providing information on the $\bar{p}$ interaction with nuclei at zero energy. In this case the optical potential is dominated by its imaginary term, which originates in the annihilation of antiprotons on nucleons. Under such circumstances it is not clear what information can be expected on the real potential and in particular how it is related to the free $\bar{p}N$ interaction. Batty [27] showed that a $t_{\rm eff}\rho$ potential gave good fit to the data and its real part corresponded to about 100 MeV attractive potential inside nuclei. However, in the low density limit one expects Re$V_{\rm opt}$ to be repulsive and of the same magnitude [18]. Antiprotonic atoms are therefore another case where DD effects are expected to be important.

The very strong absorption of antiprotons by nuclei suggests that the radial

Table 5.1. Results for antiprotonic atoms for $t_{\rm eff}\rho$ and the LESS data

	Reb_0	Imb_0	Reb_1	Imb_1	χ^2	χ^2/F
"free"	−0.9	0.9	≤0.2	≤0.05		
MAC	2.6±0.3	2.8±0.4			112.	2.8
MAC	3.8±0.3	2.3±0.5	−12±2	4.2±2.6	59.5	1.6
SP	2.5±0.3	3.4±0.3			49.9	1.3
SP	2.7±0.3	3.1±0.4	−1.2±1.6	1.6±1.3	47.7	1.3

parameter values are in units of fm

Table 5.2. Results for antiprotonic atoms for $t_{\rm eff}\rho$ and the LESS data

	Reb_0	Imb_0	Rec_0	Imc_0	χ^2	χ^2/F
"free"	−0.9	0.9	0.8	1.9		
SP	3.1±2.0	1.3±1.5	−2.2±4.7	5.2±3.9	49.0	1.3
SP	4.5±0.7	4.5±0.5	−4.0±1.4	−2.4±0.7	43.5	1.1

b values are in units of fm, c values are in fm^3.

region which is determined by the data is outside the nuclear surface, and this is indeed observed in Fig. 2. For this reason it is expected that the use of SP density distributions will be important. In order to partly separate effects due to the surface densities from isospin effects three data sets have been used [18] in the analysis: (1)all 48 data points available (denoted as ALL), (2)all the data except the isotopes of ^{16,18}O and ^{92,98}Mo (denoted by LESS) and (3)the above mentioned isotopes. Table 5.1 shows results for a $t_{\rm eff}\rho$ potential with or without an isovector term (b_1) for the MAC and SP densities. Also shown are the scattering lengths for the free $\bar{p}N$ interaction. Several points are readily obsereved: (1) better fits are obtained with the SP densities, (2) the Reb_1 value for the MAC densities is unreasonably large, (3) the b_1 values for the different densities are totally inconsistent and (4) in all cases the values of b_0 are very different from the free $\bar{p}N$ values. However, an effective isospin-dependent potential based on the SP densities is capable of producing good fits to the data.

Table 5.2 shows the effect of introducing a p-wave term into the potential, in analogy with pionic atoms [2], using the SP densities. Good fits to the data are possible but with no resemblance at all between the effective parameters and those expected for the free $\bar{p}N$ interaction. If one constrains the values of b_0 and c_0 to the free values of Table 5.2, then Table 5.3 shows that with DD and isospin dependent terms reasonably good fits are possible for the three data sets and the parameters obtained in the fits are consistent within their respective errors.

Table 5.3. Results for antiprotonic atoms for DD potentials and constrained b_0 and c_0

Data	α	ReB_0	Reb_1	Imb_1	χ^2	χ^2/F
LESS	0.18±0.06	8.4±1.9	−1.2±1.7	1.7±1.4	52.7	1.4
ISO	0.19±0.12	7.7±3.7	−0.4±2.6	2.5±2.0	17.8	2.2
ALL	0.19±0.06	8.6±2.1	−0.6±1.5	1.6±1.1	69.8	1.6

B values are in units of fm.

In conclusion it is necessary to use SP densities in the analysis of $\overline{p}$ atoms in order to get internal consistency with isotopic data. Equally good fits are possible with either an isospin dependent s-wave or an s- plus p-wave isospin-independent $t_{\text{eff}}\rho$ potential. However, in order to respect the low density limit by having for these isoscalar terms the free $\overline{p}N$ interaction, one must introduce DD into the potential. Within such a potential a reasonable p-wave strength can definitely be accommodated.

6 Summary and Discussion

Density dependence (DD) of t_{eff} in the $t_{\text{eff}}\rho$ description of the hadron-nucleus optical potential is clearly revealed in phenomenological analyses of extensive hadronic atom data. In kaonic atoms the results predict that inside nuclei the potentials are twice as deep as the potential obtained from density-independent analyses. This may have important consequences for the evolution of strangeness in high-density stars. The origin of the DD is traced to the propagation of the Λ (1405) in nuclei, which had been suggested long ago.

For Σ^- atoms a very small attractive pocket is found just outside the nuclear surface, accompanied by a repulsive potential in the nuclear interior. A repulsive isospin term is determined for the first time. Similar results are obtained from a RMF approach which enables one to place some constraints on various meson-Σ hyperon couplings. It is concluded that Σ hyperons will generally not bind in nuclei.

For antiprotons the low density limit can be respected if DD is introduced into the potential. That allows also for the inclusion of a reasonable p-wave term. About equally good fits can be obtained with an isoscalar s- plus p- wave potential or with an s-wave isovector potential. Internal consistency is possible only when SP densities are used.

It has been demonstrated that the use of realistic nuclear densities is essential for analyzing hadronic atoms data. Together with a phenomenological density dependence the analysis is capable of revealing new effects that could be related to more fundamental aspects of the hadron-nucleon interaction. Undoubtedly more extensive data, particularly for Σ^- atoms, will be most welcome.

Acknowledgements. The results reported above have been obtained in various collaborations with C.J. Batty, A. Gal, B.K. Jennings and J. Mareš who have provided numerous suggestions and ideas.

References

1. S. Deser et al.: Phys. Rev. **96**, 774 (1954)
2. M. Ericson, T.E.O. Ericson: Ann. Phys. (NY) **36**, 323 (1966)
3. C.J. Batty et al.: In:*Experimental Methods for Studying Nuclear Density Distributions* (Adv. Nucl. Phys. **19**), p.1. New York-London: Plenum Press 1989
4. E. Friedman, A. Gal, C.J. Batty: Phys. Lett. **B308**, 6 (1993)
5. E. Friedman, A. Gal, C.J. Batty: Nucl. Phys. **A579**, 518 (1994)
6. H. de Vries, C.W. de Jager, C. de Vries: At. Data Nucl. Data Tables **36**, 495 (1987)
7. D.J. Millener, P.E. Hodgson: Nucl. Phys. **A209**, 59 (1973); C.J. Batty, G.W. Greenlees: Nucl. Phys. **A133**, 673 (1969)
8. C.B. Dover, J. Hüfner, R.H. Lemmer: Ann. Phys. (NY) **66**, 248 (1971)
9. J. Hüfner, C. Mahaux: Ann. Phys. (NY) **73**, 525 (1972)
10. M. Lutz, A. Steiner, W. Weise: Nucl. Phys. **A574**, 755 (1994)
11. W.A. Bardeen, E.W. Torigoe: Phys. Rev. **C3**, 1785 (1971); Phys. Lett. **B38**, 135 (1972)
12. M. Alberg, E.M. Henley, L. Wilets: Phys. Rev. Lett. **30**, 255 (1973); Ann. Phys. (NY) **96**, 43 (1976)
13. M. Thies: Phys. Lett. **B70**, 401 (1977); Nucl. Phys. **A298**, 344 (1978)
14. W. Weise, L. Tauscher: Phys. Lett. **B64**, 424 (1976); R. Brockmann, W. Weise, L. Tauscher: Nucl. Phys. **A308**, 365 (1978)
15. M. Mizoguchi, S. Hirenzaki, H. Toki: Nucl. Phys. **A567**, 893 (1994)
16. C.J. Batty, E. Friedman, A. Gal: Prog. Theoretical Phys. Suppl. **117**, 227 (1994)
17. C.J. Batty, E. Friedman, A. Gal: Phys. Lett. **B335**, 273 (1994)
18. C.J. Batty, E. Friedman, A. Gal: Nucl. Phys. **A** (in print)
19. G.E. Brown et al.: Phys. Lett. **B291**, 355 (1992)

20. B.D. Serot, J. D. Walecka: Adv. Nucl. Phys. **16**, 1 (1986)

21. J. Mareš, B.K. Jennings: Phys. Rev. **C49**, 2472 (1994)

22. J. Schaffner et al.: Phys. Rev. Lett. **71**, 1328 (1993); Ann. Phys. **235**, 35 (1994)

23. J. Mareš et al.: Nucl. Phys. (submitted)

24. X. Jin, M. Nielsen: Phys. Rev. **C51**, 347 (1995)

25. C.J. Horowitz, B.D. Serot: Nucl. Phys. **A368**, 503 (1981)

26. R. Sawafta: Nucl. Phys. **A585**, 103c (1995)

27. C.J. Batty: Nucl. Phys. **A372**, 433 (1981)

Few-Body Systems Suppl. 9, 111–120 (1995)

Few-Body Systems

Nuclear Neutron Haloes as Seen by Antiprotons

S. Wycech*, R. Smolańczuk†

Sołtan Institute for Nuclear Studies, Hoża 69, PL-00-681 Warsaw, Poland

Abstract. Nuclear interactions of antiprotons in atomic states are discussed. The total as well as partial widths for single nucleon capture events are calculated. These are compared to the X-ray and recent single nucleon capture data. The rates of the neutron or proton captures test nuclear density distributions at the extreme nuclear surface. Recently found cases of neutron and proton haloes are analysed.

1 Introduction

It has been known for a long time that hadronic atoms are a way to test the nuclear surface region: the tail of nuclear density distribution and its isospin structure and correlations [1]. There are two methods to learn these properties:

1) *Measurements of the X ray cascade in hadronic atoms that provide atomic levels and widths.* Some fractions of the level energies are due to nuclear interactions and some parts of the widths are due to nuclear captures of the hadron. For highly excited atomic states, the nuclear effects are small. For low levels, the nuclear capture probability increases rapidly with decreasing orbital radii and the cascade terminates suddenly. These natural limitations allow to measure one level width per atom. Only in some special cases two widths and one level shift may be obtained. The levels in question are of large angular momenta that locate the nuclear interactions on the nuclear surface. At first, the atomic data are used to learn the strength and form of hadron optical potentials. Next, some of the level widths may be used to test the nuclear density tail. The shifts are usually difficult to interpret and provide a check on the optical potential.

2) *Measurements of the nuclear capture products.* A unique determination of the emitted particles may discriminate captures on protons from captures on neutrons and signal nuclear correlations. Many experiments using various detection techniques have presented data, in principle, more informative than the X-ray data. Unfortunately, these are also more difficult to interpret as the

**Internet address:* wycech@fuw.edu.pl

†*Internet address:* smolan@fuw.edu.pl

initial atomic states are not known and the final state interactions, in particular the charge exchange reactions, are uncertain.

Some related results in the $\bar{p}$ and other atoms are discussed in reviews [2,9] and other talks at this meeting. This paper discusses the neutron density distributions tested by recent antiprotonic CERN experiments [3] that are of the second kind and follow previous antiprotonic [4] and kaonic [5] studies. However, the separation of $\bar{p}p$ and $\bar{p}n$ annihilation modes is done in a different way. Instead of the final state mesons, the final nucleus is detected by radiochemical methods. The rates of reactions $\bar{p}(N,Z) \rightarrow mesons(N-1,Z)$ and $\bar{p}(N,Z) \rightarrow mesons(N,Z-1)$ are thus found. The ratio of these rates allows to detect the number of neutrons relative to the number of protons in the region of nuclei where the $\bar{p}$ capture occurs. The measurements done in nuclei where the method is applicable yield results that differ widely. For some nuclei (^{58}Ni,^{96}Ru) one finds the n/p ratio about unity, in heavy nuclei (^{176}Yb,^{232}Th,^{238}U) it is as large as 5 or 8 while in ^{144}Sm it happens to be significantly less than unity. Once the final and initial states are known and the reaction mechanism is understood, one can determine where the capture occurs and what nuclear region is tested by these experiments. Then one can interpret the n/p ratio in terms of "neutron halo" or "neutron skin " and give more precise meaning to these terms. The purpose of this work is the presentation of the basic elements of such an analysis [6].

The known difficulties: uncertainty of the capture state and the necessity to describe the final state interactions are still present. In particular, the radiochemistry is selective and only cold final nuclei are seen. A proper description of the final states becomes a question. Fortunately, an additional constraint follows from the measurement itself. It is given by the ratio of two capture rates: the rate for single nucleon captures that end with cold residual nuclei and the total capture rate. The former, cold captures, ammount to 10-20 percent of the total and are almost independent on the nucleus. If the capture occurs from a definite atomic state, the total rate is given by the level width and is in principle measurable by the X-rays. Such clean experiments are not yet feasible. The chances of nuclear capture from various initial $\bar{p}$ states have to be calculated with an optical potential derived from other atomic data. An experimental check is expected with forthcoming CERN experiments that hopefuly will supply also the transition probabilities per single stopped antiproton [6].

2 Nuclear Interactions of Atomic Antiprotons

Antiprotons captured into atomic orbits cascade down, emitting Auger electrons and X-rays, to be finally absorbed by the nucleus. This happens in atomic states of high angular momenta, presumably in circular orbits. Such a scenario is formed by the atomic cascade that tends to populate states of high angular momenta l. Because of the centrifugal barriers and large $N\bar{N}$ absorptive cross sections, the nuclear interaction of $\bar{p}$ is rather well localised at distances as large as twice the nuclear radius. An important aspect of such peripherality is that it allows low density approximations in the theoretical description: quasi-free interactions and a single-particle (s.p.) picture of the nucleus. It also facilitates the description of final mesons, an important issue in understanding of the absorption experiments. On the other hand, the surface studies are complicated by the sensitivity of results to an uncertain range of hadron-nucleon forces.

Here, we present a phenomenological description of the antiproton absorption by nuclei. The level widths are calculated in terms of $\bar{p}$ optical potentials. The simplest one, linear in nuclear density, is of the form [7]

$$V^{\rm opt}(\boldsymbol{R}) = \frac{2\pi}{\mu_{N\bar{N}}} t_{N\bar{N}} \rho(\boldsymbol{R}) \tag{1}$$

where $\mu_{N\bar{N}}$ is the reduced mass, $\rho(\boldsymbol{R})$ is a nuclear density, and $t_{N\bar{N}}$ is a complex scattering length. The density $\rho(\boldsymbol{R})$ in Eq.(1) is not the "bare" nucleon density $\rho_0(\boldsymbol{R})$ but a folded one

$$\rho(\boldsymbol{R}) = \int d\boldsymbol{u} \rho_0(\boldsymbol{R} - \boldsymbol{u}) v(\boldsymbol{u}) \tag{2}$$

where v is a formfactor that describes the $N\bar{N}$ force range. The length $t_{N\bar{N}}$ in Eq.(1) is extracted from the most precise X-ray measurements done in oxygen isotopes [7]. The best fit to these data yields $t_{N\bar{N}}$ of about $-1.5 - i2.5$ fm [7,8]. The corresponding potential is deep and black. For the central densities, Im$V^{\rm opt}$ is 200 MeV strong and the mean free path is well below 1 fm. However, both the form and the strength of $V^{\rm opt}$ are tested only in the surface region. Thus, Im$V^{\rm opt}$ is determined by the atomic level widths, given by

$$\Gamma = 4\frac{\pi}{\mu_{N\bar{N}}} {\rm Im} t_{N\bar{N}} \int d\boldsymbol{R} \rho(\boldsymbol{R}) \mid \Psi_{\bar{N}}(\boldsymbol{R}) \mid^2 \tag{3}$$

where $\Psi_{\bar{N}}(\boldsymbol{R})$ is the atomic wave function. Since $\Psi_{\bar{N}} \approx R^l$ is determined essentially by the angular momentum l and only high momentum states are available, the absorption strength is peaked at the surface.

A typical nuclear interaction region in ^{58}Ni is shown in Fig.1, where the absorption density $W = \rho \mid \Psi \mid^2 R^2$ is plotted. There are two special atomic states in the capture process. One is called an "upper" level which usually is the last one that can be detected by the X rays before the cascading down $\bar{p}$ is absorbed. Its width is obtained from the intensity loss of X-ray transitions. In ^{58}Ni and in many other nuclei, the absorption is most likely to happen from this level. The next circular state below it, "the lower state", may be reached in some nuclei. In such a case, one measures the shape of X-ray line and extracts the level width and shift. Nuclear absorptions may happen also in higher atomic orbits in a way that is not detected by X-ray studies. The chances for the $\bar{p}$ to reach the low levels in question are not known well.

The peripherality of capture depends on the range of $N\bar{N}$ forces. The range parameters in Eq.(2) may be adjusted to fit the atomic and low energy scattering data. Gaussian profile formfactors $\exp(-(r/r_0)^2)$ have been used [8], and

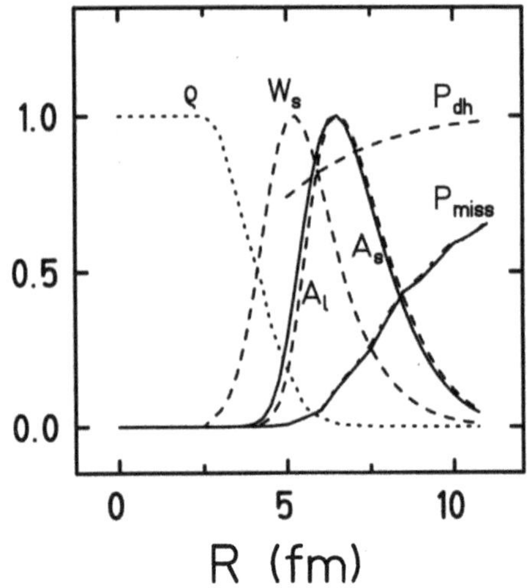

Figure 1. The antiproton absorption densities from n=6, l=5 (upper level) in ^{58}Ni: W_s for the $N\bar{N}$ annihilation range r_0 = .75 fm, ρ is a "bare" neutron density. The cold antiproton absorption densities on a neutron (integrand in Eq.(4)): A_l for the $N\bar{N}$ annihilation range r_0 = 1 fm and A_s for the range r_0 = .75 fm. Normalisations are arbitrary. Missing probabilities (left scale): P_{miss} continuous is due to phase space alone, P_{miss} dash-dotted is calculated with corrections for the experimental pion momentum distribution. The flat dashed curve is P_{dh} from the HF model.

typical best fit values are: $r_{0i} \approx 1$ fm (for Im V) and $r_{0r} \approx 1.5$ fm (for Re V). On the other hand, calculations based on the $N\bar{N}$ potentials yield values r_{0i} of 0.7 fm up to 1.5 fm [11], for different $N\bar{N}$ states and different ways to go off-shell. The latter values are probably the upper and lower limits of r_{0i}, while the best fit number is located in-between. An effect of the range uncertainty is shown in Fig.1 for a partial decay width. The effect on the full width is essentially the same.

Optical model calculations based on the $N\bar{N}$ interaction potentials [10,11] show that the lengths $t_{N\bar{N}}$ are not the $N\bar{N}$ S-wave scattering lengths. The latter are smaller and repulsive (positive). The effective Re$t_{N\bar{N}}$ is density dependent and has a complicated structure. At the nuclear surface, it reflects a long attractive tail of the pion exchange forces. At distances around the nuclear radius, it may turn to repulsion due to repulsive scattering lengths, and it is rather uncertain at the nuclear matter densities. On the other hand, the phenomenological best fit Im$t_{N\bar{N}}$ represents a cumulative effect of the S and P wave absorptive amplitudes and can well be understood in terms of the free Im$t_{N\bar{N}}$. Theoretical optical potentials are more complicated than formula (1) but do not reproduce the data as accurately as the latter with the best fit parameters.

The level widths reflect all modes of nuclear absorption. The initial stage of this process, an elementary $N\bar{N}$ annihilation, generates ≈ 2 GeV energy that turns mostly into kinetic energy of the final state mesons. These mesons excite residual nuclei by inelastic processes. To calculate the widths one sums over unobserved nuclear excited states. The large energy release allows a closure over these states and the effective $\mathrm{Im}t^{N\bar{N}}$ is close to the absorptive part of the free $\mathrm{Im}t^{N\bar{N}}$. This is not true when final nuclear states are limited by the measurements. Thus, the radiochemical method detects only cold nuclei, i.e. nuclei either in the ground states or excited less than the neutron separation threshold [3]. The effect of this limitation is discussed in the next section.

3 Nuclear $N\bar{N}$ Annihilation and Final State Interactions

The aim of this section is to calculate the rate of nuclear $\bar{p}$ annihilations that lead to cold final nuclei. This is done in few steps:

1) An amplitude for the $N\bar{N}$ annihilation into mesons $t_{N\bar{N}\to M}$ is assumed and introduced into the nuclear transition amplitude in the impulse approximation.

2) The emission probabilities are calculated and summed over mesonic and nuclear final states. For an isolated $N\bar{N}$ annihilation, this procedure would produce the absorptive cross section and via unitarity condition the absorptive amplitude $\mathrm{Im}t_{N\bar{N}}$. For nuclear captures leading to cold nuclei, we limit the summation over final states to the states of elastic meson nucleus scattering. This limited summation generates the $\mathrm{Im}t_{N\bar{N}}$ again, but now it is folded over nuclear final state interaction factors, $P_{\mathrm{miss}}(\boldsymbol{R})$, which describe the probability that the annihilation mesons born at point $\boldsymbol{R}$ miss the residual nucleus.

3) Let a nucleon N occupy a single particle level α with a wave function $\varphi_\alpha(\boldsymbol{X})$. Then, the $\bar{p}N$ annihilation process limited by P_{miss} leads to a single hole final nuclear state. The experimental condition allows only those initial levels α that end in cold final nuclei. A factor $P_{\mathrm{dh}}(\boldsymbol{X}) = \sum_\alpha^{\mathrm{ltd}} \varphi_\alpha^2 / \sum_\alpha \varphi_\alpha^2$ accounts for this condition.

4) The finite range effects are described by separable potentials as done in [11].

The result for a partial width corresponding to a cold capture on a nucleon $s =$ (proton or neutron) is now

$$\Gamma_s(\mathrm{cold}) = 4\frac{\pi}{\mu_{N\bar{N}}}\mathrm{Im}t^s_{N\bar{N}} \int \mid \Psi_{\bar{N}}(\boldsymbol{Y}) \mid^2 v(\boldsymbol{Y}-\boldsymbol{X})\rho_s(\boldsymbol{X})P_{\mathrm{dh}}(\boldsymbol{X})P_{\mathrm{miss}}(\boldsymbol{R}) \quad (4)$$

where $t^s_{N\bar{N}}$ is an effective length and $\boldsymbol{R}$ is the birthplace of the mesons. The assumption, justified later, is that all the mesons are emitted from the central point of the annihilation region $\boldsymbol{R} = \frac{1}{2}(\boldsymbol{X}+\boldsymbol{Y})$.

Equation (4) is the result. Now, we explain briefly the final state interactions that determine P_{miss} and in the next section we turn to calculations of the nuclear densities.

The spectrum of mesons consists essentially of pions correlated in a sizable fraction into ρ and ω. These heavy mesons propagate some 1 fm and then turn into pions. The pion multiplicities range from 2 to 8 with an average 4–5 and an average momentum as large as 400 MeV. Nuclear interactions of these pions may be absorptive, inelastic, or elastic. All the absorptive and almost all the inelastic processes would not leave the residual nuclei in cold states and so the production rate for cold nuclei is given essentially by the elastic scattering. This allows an optical potential description. In addition, in the bulk of phase space, the pions are fast enough to allow also an eikonal description. Following this, the wave function for each pion is taken in the form

$$\bar{\psi}^{(-)}(\boldsymbol{p}\boldsymbol{\xi}) = \exp(\mathrm{i}\boldsymbol{p}\boldsymbol{\xi} - \mathrm{i}S(\boldsymbol{p}, \boldsymbol{\xi})) \tag{5}$$

where $\boldsymbol{p}$ is a momentum, $\boldsymbol{\xi}$ is a coordinate and S is expressed in terms of the pion-nucleus optical potential

$$S(\boldsymbol{p}, \boldsymbol{\xi}) = \int_0^\infty ds(\sqrt{(p^2 - U^{\mathrm{opt}}(\boldsymbol{\xi} + \hat{\boldsymbol{p}}s)} - p)\,. \tag{6}$$

The S is calculated by integrating the local momentum over the straight line trajectory. Due to nuclear excitations and pion absorptions, this wave is damped with a rate described by ImS generated by absorptive part of the pionic optical potential ImU^{opt}. This damping follows the whole path but the main effect comes from regions of large nuclear densities and not the region around the birth place $\boldsymbol{\xi}$. We assume that all functions $S(\boldsymbol{p}, \boldsymbol{\xi})$ are related to the central point of annihilation $\boldsymbol{R}$ which is the $N\bar{N}$ CM coordinate. With the mesonic wave functions (5), the dependence on the total momentum of mesons $\boldsymbol{P}$ factorizes approximately to a plane wave form. One consequence is that the $N\bar{N}$ CM "conservation" $\delta(\boldsymbol{R} - \boldsymbol{R}')$ arises in the transition probabilities integrated over final momenta. Now, the final state pion interaction factors may be collected into a probability distribution

$$P_{\mathrm{miss}}(\boldsymbol{R}) =< \prod \mid \exp(-S(\boldsymbol{p}_i, \boldsymbol{R})) \mid^2 > \tag{7}$$

which is a product of the eikonal factors averaged over the number and phase space of final pions with some allowance for an unknown momentum dependence generated by $t_{N\bar{N}\to M}$.

The calculations of P_{miss} are performed in a Monte Carlo procedure. The optical potential for pions must cover wide momentum range from the threshold up to 0.9 GeV but the phase space favours a region just above Δ resonance. This potential is related to the pion nucleon forward scattering amplitudes and in this way to the pion nucleon cross sections. That method is established around the Δ [13]. Here, it is extended to higher N^*_{11}, N^*_{13} resonances which are described by the Breit-Wigner amplitudes. The two nucleon absorption mode is taken in a phenomenological form [14]. Performing these calculations one finds that: high energy expansion of the square root in Eq.(6) is satisfactory, higher resonances cannot be neglected, and the black sphere limit is a good approximation in dense regions. The result is close to a pure geometrical estimate that relates $P_{\mathrm{miss}}(\boldsymbol{R})$ to the solid angle of the nucleus viewed from the point $\boldsymbol{R}$ [9].

Table 1. Column 2 contains main quantum number n for the lower and upper circular states. The weigths give probabilities for nuclear capture calculated under the assumption that the circular atomic state $n = n_{\mathrm{upper}} + 1$ is fully occupied. The absorption widths: total Γ^{t} and cold Γ^{c} are calculated with the AD model, $R_{np} = .63$.

ELEMENT	n	weight	$\Gamma^{\mathrm{c}}/\Gamma^{\mathrm{t}}$	Γ_n/Γ_p
^{58}Ni	4	0	.095	.69
	5	.16	.097	.69
	6	.83	.11	.70
	7	.01	.15	.71
	8	0	.22	.71
^{90}Zr	6	.24	.106	4.67
	7	.72	.128	5.30
^{154}Sm	7	.01	.087	3.65
	8	.75	.099	3.98
^{238}U	9	.29	.106	6.55
	10	.71	.138	8.24

In the surface region of interest, the P_{miss} is a linear function of the radius. This is very fortunate as it makes calculations fairly independent on the size of the $N\bar{N}$ annihilation region and on the uncertain range of the heavy meson propagation.

Examples of P_{miss}, P_{dh}, and the density A for cold absorption generated by Eq.(4) are given in Fig.1. The latter is seen to be more peripheral than the total absorption density W. The uncertainty due to the $N\bar{N}$ force range is rather small, being moderated by effects of the strong absorption. The uncertainty of the initial atomic state is shown in Table 1. The ratio of cold to total capture rates $\Gamma^{\mathrm{c}}/\Gamma^{\mathrm{t}}$ raises quickly with the increasing angular momentum. The experimental results, given in Table 2, indicate that a significant contribution of high l states is unlikely. On the other hand, some participation of lower l states is possible. However, due to the atomic cascade properties, that seems to be unlikely. A definitive answer requires experimental and theoretical studies which are being undertaken [6].

The absorbtion widths are given by superpositions of high moments of the nuclear density distributions. We find the $2l-2$ moment as the dominant one for the total widths and the $2l$ moment to dominate the cold capture width. The "neutron halo " is thus understood as a ratio of these high moments of neutron and proton density distributions. The "neutron skin " is related the mean squared radius or other low moments.

4 Nuclear Densities

As the simplest estimate, we use an asymptotic density (AD) model. It follows the Bethe-Siemens approach [12] although the larger input includes: charge density distribution, neutron and proton separation energies, and difference of the rms radii of proton and neutron density distributions. At central densities, the Fermi gas of protons and neutrons is assumed. The Fermi momenta are determined by the densities and the Fermi energies are fixed by the separation energies. This gives depth of the potential well which in the surface region is extrapolated down in the Woods-Saxon form. The densities are given by the exponential damping of the nucleon wave functions due to the potential barriers. For protons, the Coulomb barrier is added and potential parameters (c, t) are fitted to reproduce the experimental charge density down to 5 percent of the central density. For neutrons, the same t is used but c is chosen to obtain the rms radius equal (or larger by 0.05 fm in the heaviest nuclei) to the proton density rms radius. This model misses shell effects and correlations and is expected to generate average level densities.

A second model used to determine neutron and proton densities is the Hartree-Fock (HF) and the Hartree-Fock-Bogolyubov (HFB) scheme with the effective two-body Skyrme-type interaction. Our aim in using HF and HFB methods to find nucleon densities at the extreme tails of the nuclear matter distribution (at distances of 8−15 fm from the center) is rather unusual. The necessary practical condition is the use of a HF code not restricting the asymptotic form of s.p. wave functions. This excludes e.g. all codes using the harmonic oscillator basis. In the present work, we have applied the code solving HF equations on the spatial mesh, in which all fields and densities are expressed in the coordinate representation.

The HF method disregards pairing correlations completely. To account for them, we used the HFB theory [16,17] which unifies the self-consistent description of nuclear orbitals, characteristic of HF method, and the mean field treatment of residual pairing interaction into a single variational theory. The effective force is the ten-parameter Skyrme SkP interaction described in ref.[18]. It has a virtue that the pairing matrix elements are determined by the force itself, contrary to other Skyrme-type interactions which define only the particle-hole channel. The paired HFB ground-state has not the BCS form. So, there is no simple pairing gap parameter although a kind of average gap can be defined as the pairing potential average over occupied states.

The most severe restriction of the presented results is the imposed spherical symmetry, both in the HF and HFB codes. It allows enormous simplification of solutions. In particular, the HFB equation takes the form of two coupled differential equations in the radial variable for each value of s.p. angular momentum.

The density matrix is obtained by summing contributions from the lowest s.p. orbits. The degeneracy of the spherical subshells is handled by taking contribution of the last orbit in the filling approximation, i.e., an appropriate occupation probability smaller than one is, if necessary, associated with this

Table 2. Experimental and calculated results for Γ^c/Γ^t and Γ_n/Γ_p. Calculations are averaged over few atomic orbitals weighted as in Table 1, with $R_{np} = .82$.

	Γ^c/Γ^t	Γ_n/Γ_p	Γ^c/Γ^t	Γ_n/Γ_p	Γ^c/Γ^t	Γ_n/Γ_p
	Exp. [3]		Asympt.		HF	
^{58}Ni	.098(8)	0.9(1)	.11	.90	.110	.785
^{90}Zr	.161(22)	2.6(3)	.12	4.9	.125	2.54
^{96}Ru	.113(17)	0.8(3)	.10	1.7	.099	.944
^{130}Te	.184(36)	4.1(1)	.12	2.6	.124	3.14
^{144}Sm	.117(20)	$\leq .4$	.09	1.9	.094	1.38
^{154}Sm	.121(20)	2.0(3)	.10	5.1	.110	3.34
^{176}Yb	.241(40)	8.1(7)	.12	4.8	.111	3.23
^{232}Th	.095(14)	5.4(8)	.12	7.6	.087	3.80
^{238}U	.114(9)	6.0(8)	.13	10	.092	4.09

orbit.

Some results are collected in Table 2. To find Γ_n/Γ_p a ratio of $\bar{p}n$ and $\bar{p}p$ absorptive amplitudes, R_{np} is needed. One number $R_{np} = .63$ follows from mesonic studies in ^{12}C [4]. It depends on uncertain final charge exchange processes and we use the deuteron value $R_{np} = .82$ [15]. The latter coincides with a good fit to Γ_n/Γ_p in ^{58}Ni.

Our conclusions are:

1) The AD model overestimates the neutron haloes. Those are not given by the binding energies and Coulomb barriers alone. The shell effects (angular momentum) are essential. Correlations of HFB type have rather small effect. These conclusions are similar to the results of subcoulomb neutron pickup studies [19].

2) Even at these nuclear peripheries, at least 2-3 nucleon orbitals are involved in the capture.

3) There are two anomalous cases: Yb and Te. The anomalies are apparently related to a strong E2 mixing in those atoms.

4) An interesting case of proton halo is seen in ^{144}Sm. It is not yet understood.

5) The antiprotonic study of nuclear surface is a promising method despite some uncertainties.

Acknowledgement. The authors acknowledge support by KBN Grants 2 P302 010 07 and Pb2 0956 91 01

References

1. D.H. Wilkinson: Phil. Mag. **4**, 215 (1959)

2. C.J. Batty: Rep. Prog. Phys. **52**, 1165 (1989); C.J. Batty et al.: Adv. Nucl. Phys. **19**, 1 (1989)

3. J. Jastrzębski et al.: Nucl. Phys. **A558**, 405c (1993); P. Lubiński et al.: Phys. Rev. Lett **73**, 3199 (1994)

4. W.M. Bugg et al.: Phys. Rev. Lett. **31**, 475 (1973); M. Leon and R. Seki: Phys. Lett. **48B**, 173 (1974)

5. D.H. Davis et al.: Nucl. Phys. **B1**, 434 (1967); E.H.S. Burhop: Nucl. Phys. **B1**, 438 (1967); Nucl. Phys. **B44**, 445 (1972)

6. The CERN Proposal, Experiment PS 209; S.Wycech et al.: to be published

7. Th. Kohler et al.: Phys. Lett. **B176**, 327 (1986); D. Rohmann et al.: Z. Phys. **A325**, 261 (1986); C.J. Batty: Nucl. Phys. **A372**, 433 (1981)

8. C.J. Batty: Phys. Lett. **B189**, 393 (1987); E. Friedman and J. Lichtenstadt: Nucl. Phys. **A455**, 573 (1986)

9. J. Cugnon and J. Vandermeulen: Ann. Phys. (Paris) **14**, 49 (1989)

10. T. Suzuki and H. Narumi: Nucl. Phys. **A426**, 413 (1984); S. Adachi and H.V. von Geramb: Acta Phys. Aust. **XXXVII** 627 (1985); Nucl. Phys. **A470**, 461 (1987); O.Dumbrajs et al.: Nucl. Phys. **A457**, 491 (1986)

11. A.M. Green and S. Wycech: Nucl. Phys. **A467**, 744 (1987); Nucl. Phys. **A377**, 441 (1982)

12. H. Bethe and Ph. Siemens: Nucl. Phys. **B21**, 587 (1970)

13. L.C. Liu and C.M. Shakin: Phys. Lett. **B78**, 389 (1978)

14. J.N. Ginocchio: Phys. Rev. **C17**, 195 (1978)

15. R. Bizzari et.al.: Nuovo Cim. **22A**, 225 (1974)

16. N.N. Bogolyubov: JETP (Sov. Phys.) **7**, 41 (1958); Sov. Phys. Usp. **2**, 236 (1959); Usp. Fiz. Nauk **67**, 549 (1959)

17. P. Ring and P. Schuck: *The Nuclear Many-body Problem.* New York: Springer 1980

18. J. Dobaczewski, H. Flocard and J. Treiner: Nucl. Phys. **A422**, 103 (1984)

19. H.J. Körner and J.P. Schiffer: Phys. Rev. Lett. **27**, 1457 (1971)

·Body Systems Suppl. 9, 121–126 (1995)

Few-Body Systems

Nuclear Structure Effects in Light Pionic Atoms

A. Cieplý*

Racah Institute of Physics, The Hebrew University, Jerusalem 91904, Israel

Abstract. We provide an analysis of the $1s$ and $2p$-level characteristics of light pionic atoms using a momentum space formulation of the optical potential constructed in the framework of multiple scattering theory. This potential accounts for non-local effects extending the standard parametrization given in terms of nuclear density and its powers. Particularly, the effects due to center-of-mass NN correlations and the spin-orbital Fermi motion correction are discussed in some detail.

In recent years, a bulk of new very precise experimental data on π-atomic levels has revealed the limitations in our understanding of the pion - nucleus interaction at its zero energy limit. In spite of enormous efforts undertaken when fitting the data with the refined versions of the standard Ericson - Ericson optical potential [1] and also with the other newly developed optical potentials [2, 3], the problems persist. It well may be that the so called anomalous data represent effects not involved in the framework of present theories. In our recent work we stressed the need of going beyond the parametrization of the optical potential in terms of nuclear density and its powers only [4]. In our opinion, other nuclear ingredients are needed to enter the potential as well. The long-range NN correlations and the spin-orbital interaction in the nuclear valence shell are only two examples of effects that influence the characteristics of pionic $1s$ and $2p$ levels analysed thoroughly in ref. [4].

The detailed description of the pion - nuclear optical potential used in our calculations can be found in [4]. The first order is a coherent sum of the elementary πN amplitudes

$$\langle \mathbf{Q}' \,|\, V_N^{(1)}(E) \,|\, \mathbf{Q} \rangle = A \frac{\mathcal{M}}{\mu} \left[a_{00} + \frac{2}{A} \boldsymbol{t} \cdot \boldsymbol{T} \, a_{01} \right] F(\boldsymbol{q}) \tag{1}$$

where a_{00} and a_{01} denote their isoscalar and isovector combinations, $\mathcal{M}$ (μ) stands for the pion - nucleus (nucleon) reduced mass, and $\boldsymbol{T}$ ($\boldsymbol{t}$) is the nuclear (pion) isospin vector. The term $V_N^{(1)}(E)$ is linear in nuclear density, or by other

---
* *also at:* Nuclear Physics Institute, CZ-250 68 Řež near Prague, Czech Republic

words proportional to its Fourier transform $F(\boldsymbol{q}) = \mathcal{F}\rho \equiv \int d^3r\, e^{-\mathrm{i}\boldsymbol{q}\cdot\boldsymbol{r}}\rho(\boldsymbol{r})$, $\boldsymbol{q} = \mathbf{Q} - \mathbf{Q}'$. The Lorentz - Lorenz renormalization and the rescattering corrections are expected to contribute effectively to the second order term in our model.

It was also shown in ref. [4], that the terms arising from the spin-flip part of the elementary πN amplitude can be neglected in the s-wave. However, when the pion-nuclear p-wave (or higher orbital momenta) is taken into account one has to consider the spin-orbital term that reflects the Fermi motion of nucleons inside the nucleus. This correction to the first order optical potential can be written in the form [4]

$$\langle \mathbf{Q}' \,|\, V_N^{(\sigma l)}(E) \,|\, \mathbf{Q} \rangle = \frac{A-1}{A}\frac{\mathcal{M}}{M} q^2 \langle \Psi_0 \mid \frac{j_1(qr)}{qr} f_1(\theta)\, (\boldsymbol{\sigma}\cdot\boldsymbol{l}) \mid \Psi_0 \rangle \tag{2}$$

where M is nucleon mass, $f_1(\theta)$ stands for the spin-flip part of the elementary πN amplitude, $j_1(x)$ is the spherical Bessel function, and the matrix elements of the $(\boldsymbol{\sigma}\cdot\boldsymbol{l})$ operator are to be evaluated in the nuclear ground state Ψ_0. Since the πN amplitude is real at zero energy, the term $V_N^{(\sigma l)}(E)$ affects mainly energy shifts leaving the level widths almost intact. Its matrix elements are equal to zero for magic nuclei (^{4}He, ^{16}O, ^{40}Ca, ...) reaching maximal magnitude for nuclei with half-filled valence shells.

Processes associated with pion annihilation are beyond the scope of the nonrelativistic multiple scattering theories and are usually taken into account on a phenomenological level. The leading mechanism involves at least two nucleons and its contribution to the pion - nucleus optical potential is accounted for by the second order term

$$\langle \mathbf{Q}' \,|\, V_N^{(2)}(E) \,|\, \mathbf{Q} \rangle = A\,(A-1) \left[dB_0 + \frac{C_0}{d}\, \mathbf{Q}'\cdot\mathbf{Q} \right] G(\boldsymbol{q}) \tag{3}$$

where B_0 and C_0 are the s-wave and p-wave parameters, that characterize pion interaction with a correlated nucleon pair, and $d = 1 + m/2M$ (m is the pion mass) is a kinematical transformation factor commonly used in the coordinate - space calculations. In our model, the function $G(\boldsymbol{q})$ is constructed from the two-particle nuclear density and includes the long-range NN correlations associated with the nuclear recoil. When harmonic oscillator wave function (with the harmonic oscillator parametr a) are used to characterize the nucleonic states and the centre of mass system (CMS) motion, the function $G(\boldsymbol{q})$ reads as

$$G(\boldsymbol{q}) = \frac{1}{(2\pi)^3} \int d^3p\, F(\boldsymbol{p})\, F(\boldsymbol{p}-\boldsymbol{q})\, e^{\mathrm{i}\frac{a^2}{2A}(\boldsymbol{p}\cdot\boldsymbol{q} - p^2)} \tag{4}$$

In the limit of heavy nuclei ($A \to \infty$), Eq.(4) converges to the Fourier transform of nuclear density squared, $G(\boldsymbol{q}) \to \mathcal{F}\rho^2$, and the second-order optical potential (3) becomes the momentum-space version of the standard ρ^2 term.

The pion - nucleus optical potential defined by Eqs.(1-4) assumes equal distributions of protons and neutrons inside the nucleus. It is well known, however, that the correct description of heavier pionic atoms requires an introduction of neutron densities different from the proton ones. The corresponding generalization of the first-order optical potential is straightforward and can be found elsewhere [3]. In the second-order term we simply assume that the pion interacts dominantly with deuteron-like NN pairs which leads to the $\rho_n \rho_p$ dependence in the integral of Eq.(4). Another parametrization of the second-order term was suggested in Ref.[3] where the authors considered separately the interaction with both the pp and np pairs. We have preferred the standard and rather conservative approach to keep the changes of our free parameters B_0 and C_0 under better control. Since the effective number of deuteron-like pairs is approximately equal to $A(A-1)/4$, we can absorb the 1/4 factor into the parameters B_0 and C_0 keeping the A-dependence of the second-order optical potential essentially unchanged. The neutron densities used in the present work were generated in the Hartree-Fock calculations performed by Angeli et al. [5] while the standard set of charge densities [6] was used to get the distributions of proton matter.

To complete the specification of the optical potential the parameters B_0 and C_0 were fixed to reproduce the characteristics of pionic ^{16}O ($1s$ and $2p$ levels) and ^{40}Ca ($2p$ level). In our calculations, we use three different parametrizations of the optical potential:

A $\rho_n(r) = \rho_p(r)$ assumed for all the nuclei and the CMS correlations neglected in the second-order optical potential [i.e., the standard parametrization $G(q) = \mathcal{F}\rho^2$ is used instead of Eq.(4)]

B $\rho_n(r) = \rho_p(r)$ assumed for all the nuclei and the CMS correlations taken fully into account

C different neutron densities $\rho_n(r) \neq \rho_p(r)$ incorporated for $A > 16$ and the CMS correlations neglected in the second-order optical potential

The corresponding parameters B_0 and C_0 are listed in the Table 1.

The results of our recent analysis of the π-atomic $1s$ levels are presented in Fig. 1 for both the energy shifts due to strong interaction (ΔE_N) and the absorption widths (Γ_{ABS}). The experimental data were listed in refs. [7, 8]. As one can see, the description of the data is generally good and similar to the results

Table 1. Parameters B_0 and C_0 (in the units of m^{-4} and m^{-6}, respectively) corresponding to different versions of the optical potential used in our calculations.

choice	Re B_0	Im B_0	Re C_0	Im C_0
A	−0.096	0.048	−0.170	0.080
B	−0.102	0.051	−0.155	0.075
C	−0.096	0.048	−0.178	0.078

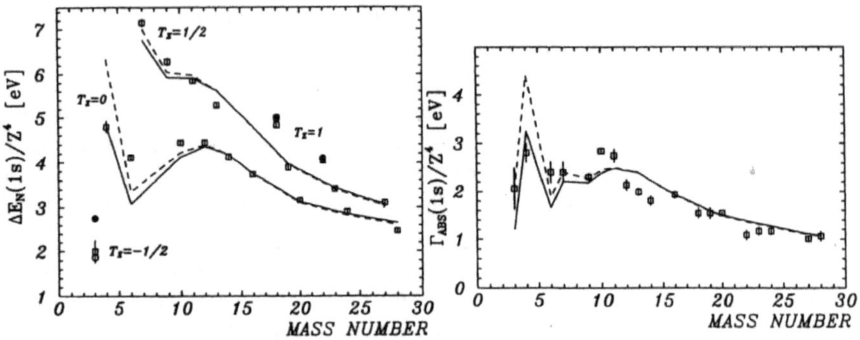

Figure 1. The calculated strong interaction shifts (left) and absorption widths (right) of the pionic $1s$ levels are connected by the full and dashed lines (or denoted by full and empty circles) that correspond to the use of the optical potentials B and A, respectively. The available experimental data are given by squares. The shift data for nuclei with different isospin are displayed separately. In the case of pionic ^{3}He, the shift is presented with the opposite sign.

of other analyses. Particularly, the isospin dependence of the shifts $\Delta E_N(1s)$ is nicely reproduced in our model. One should notice, that this achievement is not a result of the fitting procedure, because the isospin structure of our optical potential fully reflects the isospin structure of the elementary πN amplitude. The discrepances observed for $\Delta E_N(1s)$ in the region of very light nuclei require a more detailed investigation.

As we expected, the long - range center-of-masss correlations influence significantly just the problematic region of very light π-atoms ($A \lesssim 10$) leaving the characteristics of heavier ones almost intact. Unfortunately, the beautiful agreement with experimental data achieved for ^{4}He is payed for by much worse description of the characteristics of ^{3}He and some other pionic atoms (e.g. ^{6}Li). Leaving apart the magic nucleus of ^{4}He the deviations indicate that some repulsion is missed in the optical potential and the corresponding correction should be of $1/A$ type dying out for heavier nuclei. Many years ago Tauscher and Schneider [7] studied so called spin - isospin $1/A$ corrections to the second order optical potential following the prescription given by Ericson and Ericson. Unfortunately, their conclusions have proved to be inaccurate. In fact, if the antisymmetry properties of the nuclear wave function are correctly accounted for, the spin - isospin corrections to the isoscalar part of the second-order optical potential appear to be almost A-independent and no $1/A$ effects can be explained in this way [9, 4].

The influence of the $(\boldsymbol{\sigma}\cdot\boldsymbol{l})$ term (2) on the characteristics of the $1s$ level is negligible [4]. Also the widths of the π-atomic $2p$ levels are not affected by this correction. On the other hand, the $2p$-level shifts are rather sensitive to the inclusion of the Fermi motion correction as one can see in Fig. 2. In the region of deformed $s-d$ shell nuclei ($24 \leq A \leq 32$), the effect represents some

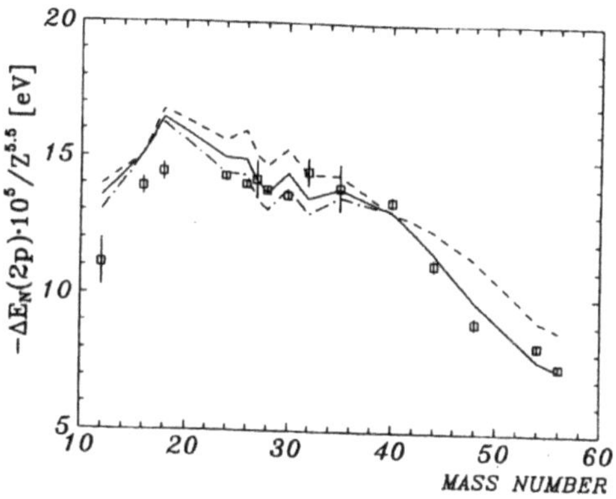

Figure 2. Comparison of the calculated and measured energy shifts $\Delta E_N(2p)$ The full (dashed) line connects the results obtained when the $(\sigma \cdot l)$ term was included (omitted) in our calculations. The dot-dashed line corresponds to the calculations with the $(\sigma \cdot l)$ correction evaluated using the simple model of closed nuclear subshells.

$5-7\,\%$ of the total strong-interaction shift. The effect is even larger $(15-20\,\%$ of the shift) for nuclei filling the valence $1f\,7/2$ subshell. Since the energy shifts $\Delta E_N(2p)$ change their sign in the region around $A \sim 90$, we expect much more profound effects for heavier nuclei with nucleons filling the nuclear subshells of much larger orbital momenta. Unfortunately, we lack reliable nuclear wave function (of nuclei beyond the $1f\,7/2$ subshell) to calculate the matrix elements of the $(\boldsymbol{\sigma} \cdot \boldsymbol{l})$ operator and to go further in our analysis of the $2p$ and higher levels observed in π-atomic spectra.

Finally, in Fig. 3 we show the sensitivity of our results to the introduction of realistic neutron densities. Since the characteristics of the π-atomic $1s$ levels are practically the same for both the A and C versions of the optical potential (mostly due to our assumpion of the same A-dependence of the second-order optical potential and due to equal distributions of neutrons and protons assumed for $A \leq 16$), we present only the results obtained for the $2p$ levels. The dashed line in Fig. 3(a) corresponds to the full line shown in the previous Fig.2. It is well known that the introduction of neutron densities improves the description of differences observed for various isotopes of the same pionic atom. This effect is nicely observed especially in the case of the absorption widths of the calcium isotopes (A = 40, 44, 48), and also for oxygen (A = 16, 18) and silicon (A = 28, 30). Unfortunately, we have no explanation for rather large deviations observed for absorption widths of the iron isotopes ($A = 54$, 56).

The introduction of realistic neutron densities improves also the description of the strong-interaction energy shifts $\Delta E_N(2p)$, especially in the region of relatively light nuclei. In the case of energy shifts, the observed isotopic effects are caused not only by different distributions of protons and neutrons but also by variations in the structure of the valence subshells of different isotopes (and correspondingly different magnitude of the correction due to the spin-orbital

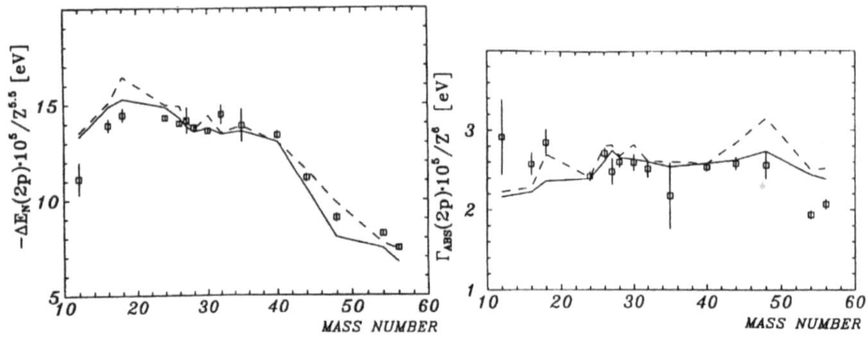

Figure 3. The strong interaction shifts (left) and absorption widths (right) of the pionic $2p$ levels calculated using the optical potentials A (dashed line) and C (full line) are compared with the available experimental data.

Fermi motion term). The existence of isovector term in the first-order optical potential plays an important role mainly in the description of the $1s$-level shifts. However, it affects the isotopic differences observed for higher levels as well. Our results show that all these effects have to be taken into account to achieve a reasonable description of the experimental data.

The momentum - space formulation of the optical potential presented here is very suitable for studying the nonlocal features of the pion - nucleus interaction and the influence of nuclear structure on the π-atomic bound states. The description of the pionic $1s$ and $2p$ levels is generally good in spite that some problems remain especially in the region of very light nuclei.

The author would like to thank R.Mach and A.Gal for many useful discussions and comments on the presented contribution.

References

1. M. Ericson and T.E.O. Ericson: Ann. Phys. (N.Y.) **36**, 323 (1966)
2. B.L. Birbrair et al.: Journ. of Phys. **G9**, 1473 (1983); **G11**, 471 (1985)
3. J. Nieves, E. Oset, C. Garcia-Recio: Nucl. Phys. **A554**, 509 (1993)
4. A. Cieplý and R. Mach: Phys. Rev. **C49**, 1454 (1994)
5. I. Angeli et al.: J. Phys. G - Nucl. Part. Phys. **6**, 303 (1980)
6. C.W. De Jager, H. De Vries, C. De Vries: Atom. Nucl. Data Tables **14**, 479 (1974); **36**, 495 (1987)
7. L. Tauscher and W. Schneider: Z. Phys. **271**, 409 (1974)
8. J. Konijn et al.: Nucl. Phys. **A519**, 773 (1990)
9. J.F. Germond and R.J. Lombard: Nucl. Phys. **A526**, 722 (1991)

Few-Body Systems Suppl. 9, 127–140 (1995)

Strange Baryonic Systems

J. Mareš*†, B. K. Jennings

TRIUMF, 4004 Wesbrook Mall, Vancouver, B.C., Canada V6T 2A3

Abstract. Strange baryonic systems are investigated by using the relativistic mean field (RMF) approach. Present discussion ranges from ordinary hypernuclei to multiply strange objects. Application to the shell model as well as optical model calculations allows to study various aspects of hyperon-nucleus interactions.

1 Introduction

This contribution is concerned with recent efforts to describe hyperon-nucleus interactions within the relativistic mean field theory [1, 2, 3]. This approach has been extremely successful in nuclear structure as well as nuclear dynamics calculations [4, 5]. Attempts for its extrapolation to more general baryonic systems are thus justified.

Hypernuclei, as bound objects of different types of baryons, represent a sound generalization of traditional nuclear matter to a many-body baryonic system. Their study provides a direct test of various models for the baryon-baryon and baryon-nucleus interactions. Recent investigations taking into account the Lorentz-tensor coupling of the ω meson to the Λ hyperon [6, 7, 1] have proved that a consistent description of both nuclear and hypernuclear systems can be achieved within the RMF model. This is regarded a great success of the Dirac approach.

Once the RMF model accounts for the hypernuclear data it is rather straightforward to extend considerations to more "strange" objects – multiply strange baryonic systems [8, 9, 10, 11]. Strange matter, either quark or baryonic, is of much interest in astrophysics [12] and in the physics of relativistic heavy ion collisions [13].

Present calculations of Σ atoms not only revealed that the RMF model is capable of high quality fits to the data but also demonstrated that the data are sufficient to constrain the couplings of mesons to a Σ hyperon [3]. This has important consequences for the spectroscopy of Σ hypernuclei.

**Permanent address:* Nuclear Physics Institute, 250 68 Řež, Czech Republic
†*E-mail addresses:* mares@reg.triumf.ca, jennings@triumf.ca, mares@ujf.cas.cz

Hyperon-nucleus scattering [2] is still a rather unexplored field. Currently there is no experimental data available, however, this may change soon [14]. Recent calculations [2] thus should be considered just a first qualitative estimate of the hyperon-nucleus scattering observables. The RMF model predicts a large effect of the Lorentz-tensor coupling on the predictions for the analyzing power.

There are other applications of the RMF approach to the physics of baryonic systems with strange particles which could not be discussed here due to limited space. Hypernuclear magnetic moments and currents [15], and decays of hypernuclei [16] should be mentioned among others. Interested reader will find relevant information and references in the contributions of Bennhold [17] and Ramos [18] to this conference and in the recent review [19].

In the next section, we introduce the underlying RMF model (subsect.2.1) and discuss the choice of parameters (subsect.2.2). Application to the spectroscopy of hypernuclei is presented in Sect.3. A brief summary of the RMF predictions for multi-strange baryonic systems follows in Sect.4. In Sect. 5, we apply Dirac phenomenology to costructing the optical potential for hyperons. First, we make use of the Sigma atom data to extract some information about the Σ-nucleus interaction and discuss implications for Σ hypernuclei. Then, we apply the RMF model to description of a hyperon-nucleus scattering. Conclusions are drawn in Sect.6.

2 Relativistic Mean Field Model

2.1 Lagrangian

The RMF formalism describes baryons as Dirac spinors interacting via meson fields in the mean field approximation [4]. The underlying Lagrangian density appropriate for strange baryonic systems can be expressed as:

$$\mathcal{L} = \mathcal{L}_N + \mathcal{L}_Y \;, \quad Y = \Lambda,\, \Sigma,\, \Xi, \tag{1}$$

$$\mathcal{L}_Y = \bar{\Psi}_Y \left[\,\mathrm{i}\,\gamma_\mu \partial^\mu - g_{\omega Y}\,\gamma_\mu V^\mu - (M_Y + g_{\sigma Y}\phi)\right] \Psi_Y + \mathcal{L}_{\rho Y} + \mathcal{L}_{AY} + \mathcal{L}_{\mathrm{T}} \;. \tag{2}$$

The detailed form of the standard nuclear part $\mathcal{L}_N$ can be found elsewhere [4]. In general, the Lagrangian density $\mathcal{L}_Y$ includes interactions of a hyperon with the isoscalar (σ, ω) meson fields, as well as contributions from the ρ meson ($\mathcal{L}_{\rho Y}$) and Coulomb ($\mathcal{L}_{AY}$) fields. For a particular hyperon $\mathcal{L}_{\rho Y}$ and $\mathcal{L}_{AY}$ acquire the following form:

$$\begin{aligned} \mathcal{L}_{\rho\Lambda} = \mathcal{L}_{A\Lambda} &= 0 \;, & (3)\\ \mathcal{L}_{\rho\Xi} + \mathcal{L}_{A\Xi} &= -\bar{\Psi}_\Xi \left[\frac{1}{2} g_{\rho\Xi}\, \boldsymbol{\tau}_\Xi \cdot \boldsymbol{b}^\mu \gamma_\mu + \frac{e}{2}(\tau_{3,\Xi} - 1)\gamma_\mu A^\mu\right] \Psi_\Xi \;, & (4)\\ \mathcal{L}_{\rho\Sigma} + \mathcal{L}_{A\Sigma} &= -\bar{\Sigma}_{ij} \left(\frac{g_{\rho\Sigma}}{2}\gamma_\mu \Theta^\mu_{jk} + \frac{e}{2}\gamma_\mu A^\mu (\tau_3)_{jk}\right) \Sigma_{ki} \;, & (5) \end{aligned}$$

where

$$\Sigma = \begin{pmatrix} \Psi_{\Sigma^0} & \sqrt{2}\Psi_{\Sigma^+} \\ \sqrt{2}\Psi_{\Sigma^-} & -\Psi_{\Sigma^0} \end{pmatrix} \;, \tag{6}$$

and

$$\Theta^{\mu} = \begin{pmatrix} \rho_0^{\mu} & \sqrt{2}\rho_+^{\mu} \\ \sqrt{2}\rho_-^{\mu} & -\rho_0^{\mu} \end{pmatrix} . \tag{7}$$

Finally, the Lagrangian density $\mathcal{L}_{\mathrm{T}}$ describes the ω-Y anomalous coupling,

$$\mathcal{L}_{\mathrm{T}} = \frac{f_{\omega Y}}{2M_Y} \bar{\Psi}_Y \, \sigma^{\mu\nu} \partial_{\nu} V_{\mu} \Psi_Y \, . \tag{8}$$

It is to be noted that similar tensor coupling term for nucleons is omitted since the coupling constant $f_{\omega N}$ is small. We neglect the ρYY and ρNN coupling terms, as well. There effect was found to be negligible [1].

The Euler-Lagrange equations lead to a system of equations of motion for both baryon and meson fields which was solved fully self-consistently (for more details see [1]). This appeared important particularly for calculating the hypernuclear magnetic moments [20] and for evaluation of the isovector ρ contribution to the hyperon binding energies [1].

2.2 *Parametrization*

The parameters of the nucleonic part $\mathcal{L}_N$ of the Lagrangian from Eq.(1) are usually fitted to the properties of spherical nuclei. Here, we adopted parameter sets of Sharma et al. [21], Horowitz and Serot [22] and Reinhard [23].

The Λ couplings (characterized via the ratios $\alpha_{i\Lambda} = g_{i\Lambda}/g_{iN}$, $i = \sigma,\ \omega$) that describe the experimental single particle spectra reasonably well are located along a line in the $(\alpha_{\sigma\Lambda},\ \alpha_{\omega\Lambda})$ plane [9]. This linear dependence results from the constraint to obtain the correct well depth $U_{\Lambda} \approx 28$ MeV in nuclear matter [11]:

$$U_{\Lambda} = -(\alpha_{\sigma\Lambda} S_N + \alpha_{\omega\Lambda} V_N) = M_N \left(1 - \frac{M_N^*}{M_N}\right)\alpha_{\sigma\Lambda} - \frac{g_{\omega N}^2}{m_{\omega}^2 \rho_0}\alpha_{\omega\Lambda} \ ,$$

where S_N and V_N are the scalar and vector potentials for nucleon, respectively.

If the tensor coupling contribution $\mathcal{L}_{\mathrm{T}}$ (Eq.(8)) is not considered, the empirically known small spin orbit splitting is achieved at a price of an extremely weak Λ coupling ($\alpha_{i\Lambda} \approx 1/3$) [8]. This value, however, is in contradiction with the constituent quark model predictions, as well as, with the bounds obtained from neutron star mass calculations [12], namely $\alpha_{i\Lambda} \geq 0.5$. In this contribution we demonstrate that a consistent description of hypernuclei can be achieved even with the quark model values of the hyperon-meson couplings provided the tensor coupling term $\mathcal{L}_{\mathrm{T}}$ is introduced in the Lagrangian $\mathcal{L}_Y$ (Eq.(2)) [6, 7, 1].

We used the constituent quark model to determine the values of the coupling constants $g_{\omega Y}$, $f_{\omega Y}$ and $g_{\rho Y}$. The $g_{\sigma\Lambda}$ couplings were fitted to the experimental Λ hypernuclear data. For Ξ hypernuclei, we chose $g_{\sigma\Xi}$ to roughly obtain the potential well depth $U_{\Xi} \approx 20 - 25$ MeV which is compatible with a scarce data from Ξ hypernuclei [24]. For Σ hyperon, we used quark model and/or QCD sum rules predictions [25] for $\alpha_{\omega\Sigma}$. Corresponding $\alpha_{\sigma\Sigma}$ as well as $\alpha_{\rho\Sigma}$ were then fitted to Σ^- atomic level shifts and widths [3]. In view of new results

Table 1. The couplings used for hyperonic sector. The ratios of hyperon to nucleon couplings for $Y = \Lambda$, Σ and Ξ are presented here. For α_{TY} the values of Nijmegen models F [26] ($\star$), D [27] ($\dagger$), and the values from Dover and Gal [28] ($\ddagger$) are used for comparison.

	$\alpha_{\sigma Y}$	$\alpha_{\omega Y}$	$\alpha_{\rho Y}$	α_{TY}		
Λ	0.621	2/3	0.0	0.0	-0.541*	-1.0
Σ	0.544 0.77	2/3 1.0	2/3 2/3	0.0	0.76 ‡	1.0
Ξ	0.375	1/3	1.0	0.0	-0.4 †	-2.0

from Σ^- atoms (presented in Sect.5.1), the case of Σ hypernuclei deserves more attention and will be discussed in separate subsection 5.1.1.

The values of $\alpha_{\sigma Y} = g_{\sigma Y}/g_{\sigma N}$, $\alpha_{\omega Y} = g_{\omega Y}/g_{\omega N}$, $\alpha_{\rho Y} = g_{\rho Y}/g_{\rho N}$ and $\alpha_{TY} = f_{\omega Y}/g_{\omega Y}$ used in the following sections are presented in Table 1.

3 Hypernuclear Shell Model

The RMF model introduced in Sect.2 yields reasonable description of known hypernuclear characteristics – hyperon binding in nuclear matter, spin orbit interaction, single particle spectra [1]. In this section, we select only a few examples illustrating the role of different terms in the Lagrangian $\mathcal{L}_Y$ from Eq.(2).

Figure 1 demonstrates the competing effects of the Coulomb and isovector (ρ meson) interactions in Ξ hypernuclei. The attractive Coulomb potential leads to a considerably stronger binding of Ξ^- in the nuclear medium when compared with neutral hyperons (Λ, Ξ^0). The ρ meson field significantly reduces the effect of the Coulomb interaction in nuclei with a neutron excess. In Zr and Pb the ρ meson contribution is repulsive (attractive) for Ξ^- (Ξ^0) and thus tends to compensate the contribution from the Coulomb potential. The same role is played by the above fields in the $\Sigma^{\pm}$-nucleus interaction.

The tensor coupling (Eq.(9)) is crucial for evaluating the hypernuclear spin orbit interaction. In Fig.2, ${}^{17}_{Y}$O serves as an example of a quite different contribution of the tensor coupling term $\mathcal{L}_T$ to the spin orbit splitting in a case

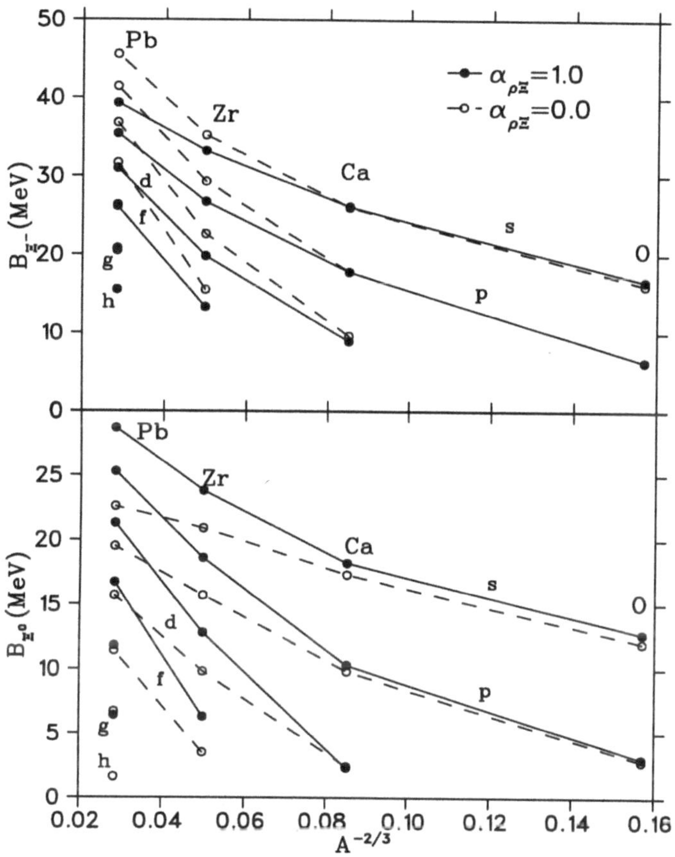

Figure 1. Comparison of the Ξ^- and Ξ^0 single-particle levels in hypernuclei for $\alpha_{\rho\Xi} = 0$ and $\alpha_{\rho\Xi} = 1$.

of the three kinds of hyperons ($Y = \Lambda,\ \Sigma^0,\ \Xi^0$). It is to be stressed that for sake of comparison we adopted here the well depth of Σ-nucleus potential of the same size as that for Λ, $U_\Sigma \approx U_\Lambda$.

The results of Fig. 2 become apparent from Schrödinger equivalent spin orbit potential:

$$V_{ls}^Y\, \boldsymbol{l}\cdot\boldsymbol{s} = \frac{1}{2M_{\rm eff}^2}\left[\frac{1}{r}\left(g_{\omega Y}V_0' - g_{\sigma Y}\phi' + 2f_{\omega Y}\frac{M_{\rm eff}}{M_Y}V_0'\right)\right]\boldsymbol{l}\cdot\boldsymbol{s}\ ,$$

$$M_{\rm eff} = M_Y - \frac{1}{2}\left(g_{\omega Y}V_0 - g_{\sigma Y}\phi\right)\ .$$

For $f_{\omega Y} = 0$ the spin-orbit splitting for hyperons is reduced when compared to the nuclear case due to a larger mass $M_{\rm eff}$ in the denominator and due to smaller couplings to σ and ω mesons. The quark model values of $f_{\omega Y}$ for Λ, Σ and Ξ hyperons differ in their strengths and signs. Consequently, the tensor coupling contribution is comparable in magnitude with the original $\sigma - \omega$ part and is negative (positive) for Λ (Σ). It is negative and even larger than $\sigma - \omega$

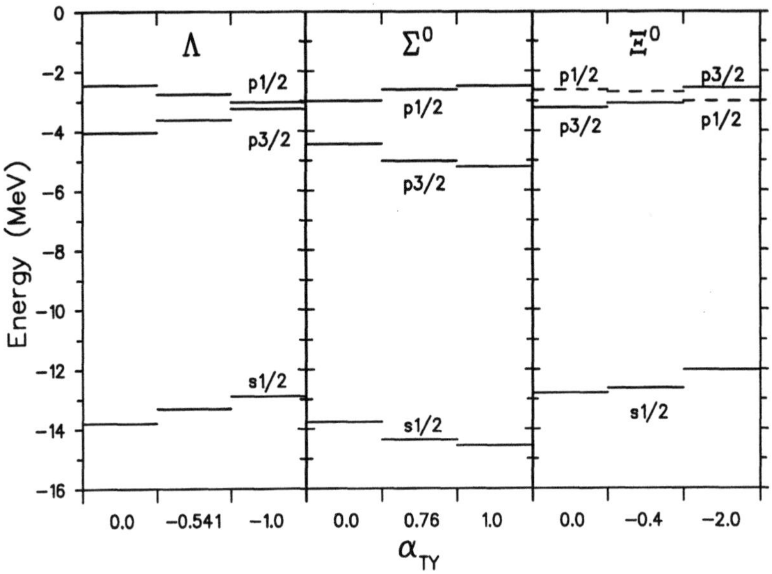

Figure 2. The hyperon single-particle levels in ${}^{17}_{Y}$O as a function of $\alpha_{TY} = f_{\omega Y}/g_{\omega Y}$.

term for Ξ. The resulting spin orbit interaction for Λ nearly vanishes, it is almost doubled for Σ, and changes sign for Ξ.

It should be noted, however, that although the effects of tensor coupling are relatively large, the absolute shifts of energy levels are probably still beyond the reach of contemporary experimental resolution.

4 Multiply Strange Baryonic Systems

Baryonic objects with larger contents of strangeness are expected to be produced in the relativistic heavy ion collisions [13]. In addition, certain form of strange (baryonic) matter is predicted to be present in the cores of neutron stars [29, 12]. Investigations of these multi-strange systems are expected to provide unique information about the baryon-baryon interaction. This could help to distinguish between various models more or less equivalent in describing the YN scattering or single hypernuclei. Unfortunately, our empirical knowledge is limited to a few events of double-Λ hypernuclei: ${}^{6}_{\Lambda\Lambda}$He, ${}^{10}_{\Lambda\Lambda}$Be, ${}^{13}_{\Lambda\Lambda}$B. The calculations revealed that the RMF model, as formulated in section 2, has to be improved by additional meson exchanges (scalar f_0 and vector ϕ) acting exclusively between hyperons in order to get stronger YY interaction in agreement with the $\Lambda\Lambda$ hypernuclear data and predictions of a one-boson exchange potential [10, 11].

The RMF models predict a possibility of forming bound systems with an appreciable number of hyperons. The dependence of the binding energies per particle, density distributions and rms radii of such systems on a number of hyperons n_Y, results from a delicate interplay between the effect of Pauli blocking (hyperon is distinguishable from N) and the weaker YN interaction compared to the NN one.

First studies of multi-strange objects have limited themselves to calculations of ΛN systems[8, 9]. However, the considerations can not be restricted to nucleons and Λ particles only as far as Ξ^0 and Ξ^- are necessary constituents of the strange baryonic matter [10]. This is because the process $\Lambda\Lambda \longrightarrow \Xi N$ becomes energetically favourable due to the Pauli blocking of Λ's in some multi-Λ hypernuclei. This greatly increases the possible amount of strangeness in the bound systems (stable against strong decay). Figure 3 adopted from ref.[11] serves as an example. Here, the binding energies per particle (E_B/A) for ^{56}Ni + Y ($Y = \Lambda$, Ξ) are calculated.

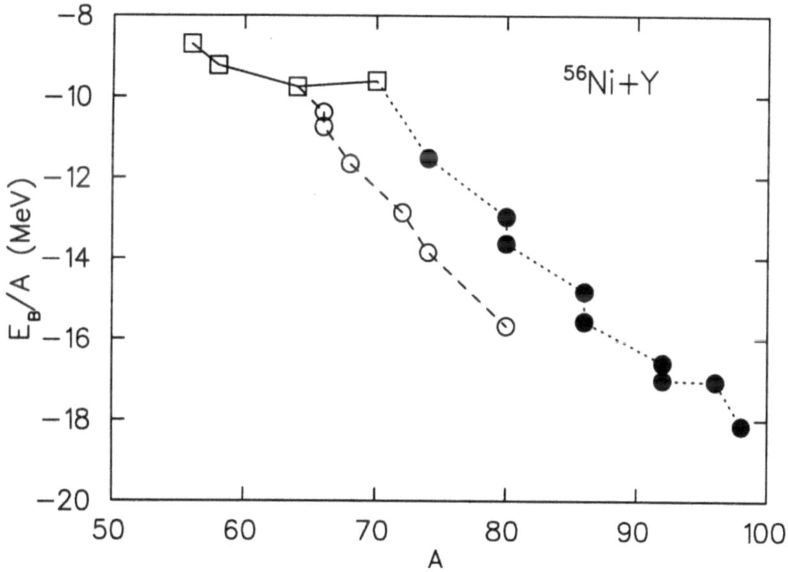

Figure 3. The binding energy per particle (E_B/A) in ^{56}Ni + Y as a function of A. The squares correspond to $Y = \Lambda$ only, the circles are obtained by adding Ξ's to a system. (The figure is adopted from ref.[11].)

The RMF calculations yield baryonic (N, Λ, Ξ) systems with densities $\rho \approx (2-3)\rho_0$, $|S|/A \approx 1$ and $|Z|/A \ll 1$. These values resemble the speculations about droplets of strange quark matter "strangelets" [29]. However, the baryonic objects are more loosely bound, $|E_B/A| \approx 10 - 20$ MeV. Since it is much less than the $\Lambda - N$ mass difference ($\approx$ 177 MeV), these objects will decay by weak interaction with lifetimes $\approx 10^{-10}$ s.

An excellent review on the (N, Λ, Ξ) systems can be found in ref.[11].

5 Optical Potential for Hyperons

So far we have discussed the RMF calculations of bound nuclear systems, either hypernuclei or strange hadronic matter. However, the Dirac approach is a valuable tool to also describe the optical model as we illustrate here on two examples. First, we apply the RMF approach to constructing the Σ-nucleus optical potential to be used in calculations of Σ-atoms. Then, we extend further our considerations to higher energies and apply Dirac phenomenology to Λ and Σ scattering off nuclei.

5.1 What We Have Learnt from Σ^- Atoms

Recent phenomenological analyses of level shifts and widths in Σ^- atoms by Batty et al. [30] suggest that the real part of the Σ-nucleus potential $\mathrm{Re}V^{\Sigma}_{\mathrm{opt}}$ is attractive only at the nuclear surface, changing into a repulsive potential as density increases in the interior (for more details see contribution of E. Friedman). The shallow attractive pocket of such potential does not provide sufficient binding to form Σ hypernuclei. These conclusions are seemingly in contradiction with previous analyses [31, 32], yielding $-\mathrm{ReV}^{\Sigma}_{\mathrm{opt}} \approx 25-30$ MeV. Since calculations of Σ hypernuclei have often been based on the above analyses it is desirable to apply the RMF approach directly to determining the Σ-nucleus optical potential by fitting to Σ^- atom data.

In our work [3] we used as a real part of the Σ-nucleus potential the Schrödinger equivalent potential constructed out of the scalar (σ) and vector (ω, ρ) meson mean fields. A purely phenomenological imaginary part of the form $\mathrm{Im}V^{\Sigma}_{\mathrm{opt}} = t\rho_p(r)$ was chosen in order to account for the conversion $\Sigma^- p \to \Lambda n$. While the proton density $\rho_p(r)$ was calculated within the RMF model, the parameter t was fitted to the atomic data. The other free parameters of the model were the scalar meson coupling ratio α_σ and isovector meson coupling ratio α_ρ. The values of the coupling ratio α_ω were adopted from constituent quark model [6] ($\alpha_{\omega\Sigma} = 2/3$) and QCD sum rules evaluations [25] (($\alpha_{\omega\Sigma} = 1$). Since a rather detailed description of our analysis can be found in the contribution of E. Friedman to this conference and in ref.[3] we limit ourselves here to mere summary of the results:

1) The RMF approach is capable of very good quality fits to the Σ atomic data.
2) The best fits were obtained for larger values of α_ω ($2/3 \leq \alpha_\omega \leq 1$).
3) The ρ-Σ coupling ratio $\alpha_\rho \approx 2/3$ holds unambiguously for all the parametrizations used.
4) The RMF model yields $\mathrm{ReV}^{\Sigma}_{\mathrm{opt}}$ with a volume repulsion in the nuclear interior and a shallow attractive pocket at the surface in agreement with the latest density dependent phenomenological analyses [30]. For illustration, we show in Fig.4 one of the Σ nucleus potentials compatible with the data.

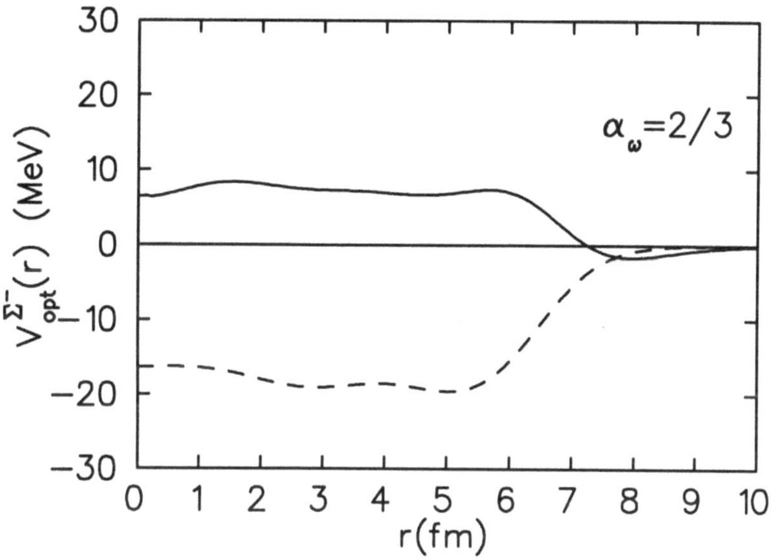

Figure 4. Re$V^{\Sigma}_{\rm opt}$ (solid line) and Im$V^{\Sigma}_{\rm opt}$ (dashed line) as a function of r for the Σ^- optical potential in Pb. Linear RMF model with $\alpha_\omega = 2/3$ was used.

5.1.1 Do Σ Hypernuclei exist?

Up to now, no Σ hypernuclear bound state has been clearly established [33, 34], except perhaps a $J^\pi = 0^+$, $I = 1/2$ ${}^4_\Sigma$He bound state [35].

The volume repulsion obtained in the previous section for the isoscalar component of $V^{\Sigma}_{\rm opt}$ in fact precludes binding for Σ^0 hypernuclei.

For Σ^+, the attractive isovector part of $V^{\Sigma}_{\rm opt}$ generally does not overcome the repulsion from the isoscalar and Coulomb interactions. It is thus unlikely to bind a Σ^+ in nuclei. It is to be noted, that above considerations could be modified in very light hypernuclei (${}^4_\Sigma$He) where isovector potential dominates over the Coulomb interaction [36, 37].

For Σ^- and high Z nuclear cores, the attractive Coulomb potential gives rise to Σ^- bound states that might be called nuclear as the corresponding wave-functions are located within the nucleus or at its surface. When $V^{\Sigma}_{\rm opt}$ is included these Coulomb levels are shifted to lower binding energies and acquire strong interaction widths. These states may be considered an example of "Coulomb assisted states" [37]. Their location could provide a valuable information about $V^{\Sigma}_{\rm opt}$. However, since these states are "pushed" outside the nucleus, the cross sections to excite them by a nuclear reaction are not expected to be sizeable.

The chances of establishing a meaningful Σ hypernuclear spectroscopy are thus vanishingly small at present.

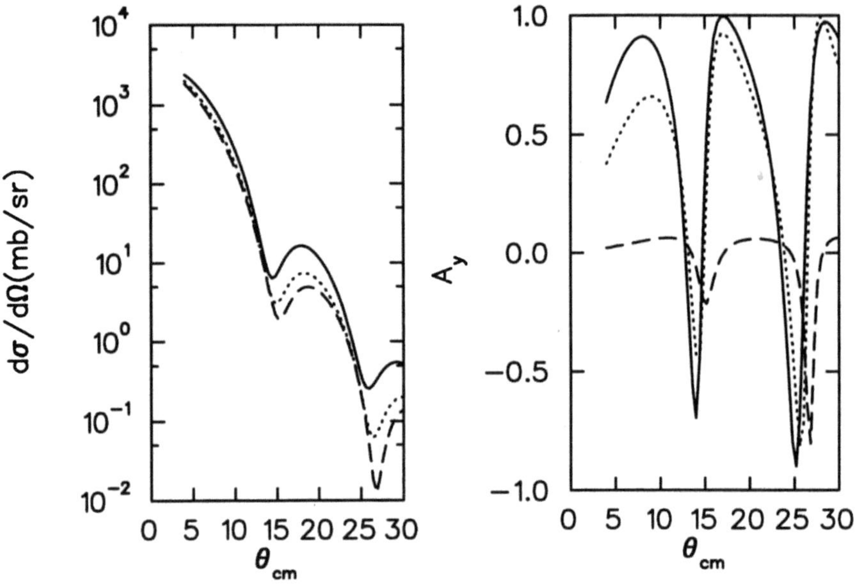

Figure 5. The effect of the tensor coupling on the observables for 300 MeV Λ's elastically scattered from ^{40}Ca. The predictions for $\alpha_T = -1$ (dashed line), $\alpha_T = 0$ (dotted line) and $\alpha_T = 1$(solid line) are compared.

5.2 Hyperon-Nucleus Scattering

The Dirac approach yields remarkably accurate results for proton-nucleus scattering observables at medium energies [5]. Similar investigations of a hyperon optical potential have been missing till recently as there are no data on hyperon scattering off nuclei. The situation could change, however, as an experiment at KEK [14] may soon provide data on Σ^+-^{12}C. In addition, knowledge of the hyperon-nucleus optical potential will become necessary for description of quasi-free hyperon production ($\gamma K^+ \Lambda$) proposed at CEBAF [38].

With this motivation we have developed an optical potential describing hyperon-nucleus interaction at medium energies. We used the global optical model for nucleon-nucleus scattering [39] to get the shape and energy dependence of the potential. The strength of the potential was determined according the information coming from quark model, Λ hypernuclei and Σ atom data [3].

We made several assumptions which remain to be verified:
1) the energy dependence of the Y-nucleus interaction is that of nucleon- nucleus potential;
2) the imaginary strengths of the potentials for hyperon are obtained from the imaginary parts for protons by multiplying the latter by the squares of the ratios for the real potentials;
3) the absorption due to the $\Sigma N \rightarrow \Lambda N$ conversion is energy independent.

The above optical potential was used in calculations of the differential cross

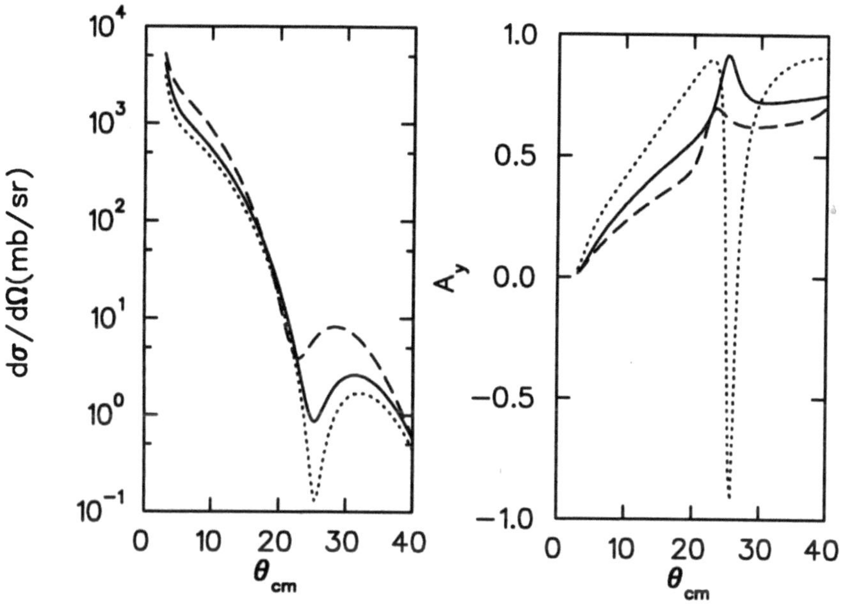

Figure 6. The observables for 200 MeV Σ^+ elastically scattered from ^{12}C calculated for potentials compatible with the Σ atom data. The predictions for $\alpha_{\omega\Sigma} = 1$ (dashed line) and $\alpha_{\omega\Sigma} = 2/3$ (solid line) are compared. The prediction for $\alpha_{i\Sigma} = \alpha_{i\Lambda}$ (dotted line) is included for comparison.

sections and analyzing powers for Λ and Σ scattering off nuclei.

Figure 5 illustrates the dominant role of the tensor coupling in predictions of the analyzing power A_y. Scattering of Λ from ^{40}Ca at 300 MeV serves here as an example. The cross sections are qualitatively similar for different values of the tensor coupling ratio α_T ($\alpha_T = -1, 0, +1$) with maxima and minima at roughly same angles. On the contrary, the predictions for A_y strongly depend on the value of the tensor coupling. Particularly $\alpha_T = f/g = -1$ (quark model value for Λ) yields considerably smaller values of analyzing power; A_y is close to zero for forward angles. This result suggests that measurements of A_y would give information about the tensor coupling of the vector meson to a hyperon.

In order to get closer to experiment [14] we have calculated the scattering of Σ^+ from ^{12}C at 200 MeV using coupling ratios determined by fitting the Σ atom data [3]. The results presented in Fig.6 demonstrate the sensitivity of A_y to the strengths of Σ-meson couplings. For comparison we have included also results for the case of $U_\Sigma = U_\Lambda$, which was assumed in our previous calculations [2].

In view of the assumptions made the present calculations should be considered just a first qualitative estimate of the hyperon-scattering observables. There is still a lot of theoretical as well as experimental work to be done.

6 Concluding Remarks

This contribution aimed to demonstrate that the relativistic mean field approach provides a natural description of hyperon-nucleus interactions:

With reasonable (quark model inspired) values of the meson-hyperon couplings it is possible to reproduce the hypernuclear data.

Extrapolation of the RMF theory from ordinary hypernuclei to multi-strange systems predicts rather weakly bound stable objects composed of N, Λ, Ξ^0 and Ξ^- baryons, of arbitrarily large A, high strangeness content and small charge.

The Σ^- atom data are sufficient to significantly constrain the possible hyperon-nucleus couplings. The analysis yields potentials with a repulsive real part in the nuclear interior. Consequently, the chances of forming bound nuclear states of Σ (except perhaps the lightest systems) are very limited.

Hyperon-nucleus scattering experiments could become another important source of information about hyperon-nucleon interaction. In particular, we predict a strong sensitivity of the analyzing power A_y to the tensor coupling.

Experimental programs in BNL, KEK, CEBAF, DAΦNE promise new perspectives in a study of hyperon-nucleus interactions.

Acknowledgement. Calculations and investigations which formed basis for this contribution were done under fruitful and stimulating collaboration with E.D. Cooper, A. Gal and E. Friedman. To them belongs our gratitude. B.K.J. would like to thank the Natural Sciences and Engineering Research Council of Canada for financial support.

References

1. J. Mareš and B.K. Jennings: Phys. Rev. **C49**, 2472 (1994)
2. E.D. Cooper, B.K. Jennings, J. Mareš: Nucl. Phys. **A580**, 419 (1994)
3. J. Mareš et al.: Preprint TRI-PP-95-15, TRIUMF, 1995; submitted Nucl. Phys. A
4. B.D. Serot and J.D. Walecka: Adv. Nucl. Phys. **16**, 1 (1986)
5. S.J. Wallace: Ann. Rev. Nucl. Part. Sci. **37**, 267 (1987)
6. B.K. Jennings: Phys. Lett. **B246**, 325 (1990)
7. J. Cohen and H.J. Weber: Phys. Rev. **C44**, 1181 (1991)
8. M. Rufa et al.: J. Phys. G **13**, 143 (1987); Phys. Rev. **C42**, 2469 (1990)
9. J. Mareš and J. Žofka: Z. Phys. **A333**, 209 (1989); **A345**, 47 (1993)
10. J. Schaffner et al.: Phys. Rev. Lett **71**, 1328 (1993)

11. J. Schaffner et al.: Ann. Phys. **235**, 35 (1994)

12. N.K. Glendenning: Z. Phys. **A326**, 57 (1987); N.K. Glendenning, S.A. Moszkowski: Phys. Rev. Lett. **67**, 2414 (1991)

13. M. Wakai, H. Bandō, M. Sano: Phys. Rev. **C38**, 748 (1988); H. Bandō: Il Nuovo Cimento **102A**, 627 (1989)

14. M. Ieiri, D.R. Gill: Private Communication

15. J. Cohen, J.V. Noble: Phys. Rev. **C46**, 801 (1992)

16. A. Ramos et al.: Nucl. Phys. **A544**, 703 (1992)

17. C. Bennhold: Contribution to this conference

18. A. Ramos: Contribution to this conference

19. J. Mareš, B.K. Jennings, E.D. Cooper: Progr. Theor. Phys. Suppl. **117**, 415 (1994)

20. J. Mareš, J. Žofka: Phys. Lett. **249B**, 181 (1990)

21. M.M. Sharma, M.A. Nagarajan, P. Ring: Phys. Lett. **B312**, 377 (1993)

22. C.J. Horowitz and B.D. Serot: Nucl. Phys. **A368**, 503 (1981)

23. P.-G. Reinhard et al.: Z. Phys. **A323**, 13 (1986); P.-G. Reinhard: Z. Phys. **A329**, 257 (1988)

24. C.B. Dover and A. Gal: Ann. Phys. **146**, 309 (1983)

25. X. Jin and M. Nielsen: Phys. Rev. **C51**, 347 (1995)

26. M.M. Nagels, T.A. Rijken, J.J. de Swart: Phys. Rev. **D20**, 1633 (1979)

27. M.M. Nagels, T.A. Rijken, J.J. de Swart: Phys. Rev. **D12**, 744 (1975); **D15**, 2547 (1977)

28. C.B. Dover and A. Gal: Progr. Part. Nucl. Phys. **12**, 171 (1984)

29. S.A. Chin and A.K. Kerman: Phys. Rev. Lett. **43**, 1292 (1979); E. Witten: Phys. Rev. **D30**, 272 (1984); E. Farhi, R.L. Jaffe: Phys. Rev. **D30**, 2379 (1984)

30. C.J. Batty, E. Friedman, A. Gal: Phys. Lett. **B335**, 273 (1994); Progr. Theor. Phys. Suppl. **117**, 227 (1994)

31. C.J. Batty: Phys. Lett. **87B**, 324 (1979)

32. A. Gal and C.B. Dover: Phys. Rev. Lett. **44**, 379 (1980); 962(E)

33. M. Barakat and E.V. Hungerford: Nucl. Phys. **A547**, 157c (1992)

34. R. Sawafta: Nucl. Phys. **A585**, 103c (1995)

35. R.S. Hayano et al.: Phys. Lett. **B231**, 355 (1989)

36. C.B. Dover, A. Gal, D.J.Millener: Phys. Lett. **138B**, 337 (1984)

37. T. Yamazaki, R.S. Hayano, O. Morimatsu, K. Yazaki: Phys. Lett. **B207**, 393 (1988)

38. C.E. Hyde-Wright (spokesperson): *Quasi-Free Strangeness Production in Nuclei*, CEBAF experiment 91-014

39. E.D. Cooper, S. Hama, B.C. Clark, R.L. Mercer: Phys. Rev. **C47**, 297 (1993)

Few-Body Systems Suppl. 9, 141–144 (1995)

Few-Body Systems

Σ Hypernuclear States

J. Dąbrowski, J. Rożynek

Sołtan Institute for Nuclear Studies, Hoża 60, PL-00-681, Warsaw, Poland

Abstract. Σ hypernuclear states produced in the (K^-, π^+) reaction are described as resonances of Σ in the s.p. potential with a repulsive bump near the nuclear surface, suggested in microscopic calculations. The coupled-channels impulse approximation for the inclusive (K^-, π^+) experiments is outlined.

The presently existing data on the Σ hypernuclear states produced in the strangeness exchange reactions (K^-, π) reveal two characteristic features: the states are narrow (their width $\Gamma_\Sigma \sim 5$ MeV), and their energy is positive ($E_\Sigma \sim 5 - 10$ MeV).

The present paper, based partly on the results obtained in [1,2], is an attempt at a theoretical description of both features of the Σ hypernuclear states. We apply the Σ single particle (s.p.) model in which the motion of Σ in the hypernucleus is described by the wave function $\psi_\Sigma(\boldsymbol{r})$ which is the solution of the s.p. Schrödinger equation with the s.p. potential $\mathcal{V}_\Sigma(r) = V_\Sigma(r) + \mathrm{i}W_\Sigma(r)$, where $W_\Sigma(r)$ represents the absorption due to the $\Sigma\Lambda$ conversion.

Let us consider the (K^-, π^+) reaction. To calculate the cross section for this reaction with the pion emerging in the direction $\hat{k}_\pi$ with the energy E_π and with Σ emerging in the direction $\hat{k}_\Sigma$, we apply the impulse approximation (with K^- and π^+ plane waves):

$$d^3\sigma/d\hat{k}_\Sigma d\hat{k}_\pi dE_\pi = \frac{E_K E_\pi M_\Sigma c^2 k_\pi k_\Sigma}{(2\pi)^5(\hbar c)^6 k_K} |t \int d\boldsymbol{r} \exp(-\mathrm{i}\boldsymbol{q}\boldsymbol{r})\psi(\Sigma, \boldsymbol{k}_\Sigma; \boldsymbol{r})^{(-)*}\psi_p(\boldsymbol{r})|^2, \tag{1}$$

where $\boldsymbol{q} = \boldsymbol{k}_\pi - \boldsymbol{k}_K$, ψ_p is the wave function of the target proton, and the Σ scattering state $\psi(\Sigma, \boldsymbol{k}_\Sigma; \boldsymbol{r})^{(-)}$ behaves asymptotically as $\exp(\mathrm{i}\boldsymbol{k}_\Sigma\boldsymbol{r})$ + incoming wave. Momenta of the respective particles (in units of $\hbar$) are denoted by $\boldsymbol{k}$, and their energies (including their rest masses) are denoted by E. We assume a zero-range spin-independent interaction for the elementary process $K^-p \to \pi^+\Sigma^-$ with a constant transition matrix t. Spins are suppressed in the notation. In the (inclusive) experiments performed so far, only the energy spectrum of pions at fixed $\hat{k}_\pi$ is measured. To calculate this spectrum, $[d^2\sigma/d\hat{k}_\pi dE_\pi]_\Sigma$, one has to integrate cross section (1) over $\hat{k}_\Sigma$.

The positions and widths of the peaks in the pion spectrum are approximately equal to the positions and widths of the resonances in the Σ scattering state. What we need to have narrow resonances at high energy is a sufficiently high barrier at the surface of the hypernucleus. Recently, Myint, Tadokoro and Akaishi [3,4], in their microscopic calculation, have suggested that indeed $V_\Sigma(r)$ has a repulsive bump $U_{\rm B}(r)$ near the nuclear surface. (A surface repulsion is also compatible with the latest analysis of Σ^- atoms [5].)

To see whether such repulsive bump may lead to the observed Σ hypernuclear states, let us consider the case of the (K^-, π^+) reaction on ^{16}O at $p_K = 450$ MeV/c (at $\theta \simeq 0°$) investigated at CERN by Bertini et al. [6]. (The analysis in [1] of cases with different kinematics leads to similar conclusions.) By analyzing the measured pion spectrum, they located two narrow peaks at $B_\Sigma = -5.9 \pm 1$ and -12.4 ± 1 MeV. Here B_Σ denotes the separation (binding) energy of Σ^- from the hypernucleus produced with the nuclear core left in its ground state.

The data in [6] as well as other existing data on Σ hypernuclei are not very accurate. In this situation, we restrict ourselves to a simplified calculation, and assume for $\mathcal{V}_\Sigma$ the form

$$\mathcal{V}_\Sigma(r) = -(V_0 + \mathrm{i}W_0)\theta(R - r) + V_1\delta(r - R), \tag{2}$$

with $R = 3$ fm, $V_0 = 20$ MeV (compatible with the model D of the Nijmegen baryon-baryon interaction), and $V_1 = \int dr U_{\rm B}(r) \simeq 20$ MeV, where $U_{\rm B}(r)$ is the repulsive bump suggested in [3,4]. Similarly, the s.p. proton potential (which we need to calculate ψ_p) is assumed in the form $-V_p\theta(R\text{-}r) - V_{\rm LS}\boldsymbol{l}\boldsymbol{s}\delta(r\text{-}R)$, with V_p and $V_{\rm LS}$ adjusted to the empirical proton s.p. energies in the $p_{1/2}$ and $p_{3/2}$ states, $\epsilon_p(p_{1/2})$ and $\epsilon_p(p_{3/2})$. For the depth of the absorptive potential we use the value $W_0 = 2.5$ MeV, calculated for Σ in nuclear matter in [7].

The calculated results together with the data of [6] are shown in Fig. 1. Since the data of [6] are only counting rates, the calculated results include a normalization to match the overall magnitude of the data. The contributions denoted by $p_{1/2}$ and $p_{3/2}$ result from the K^- interaction with $p_{1/2}$ and $p_{3/2}$ protons in ^{16}O. In the case of the $p_{1/2}$ contribution, the final state of the nuclear core is a $p_{1/2}$ hole in ^{16}O, i.e., the ground state of ^{15}N, and we have $-B_\Sigma = E_\Sigma = (\hbar k_\Sigma)^2/2M_\Sigma$. In the case of the $p_{3/2}$ contribution, the final nuclear configuration is a $p_{3/2}$ hole in ^{16}O, i.e., an excited state of ^{15}N with the excitation energy $E^* = \epsilon_p(p_{1/2}) - \epsilon_p(p_{3/2}) = 6.6$ MeV, and we have $-B_\Sigma = E_\Sigma + E^*$. In the continuum, the calculated curve has two peaks (whose width $\Gamma_\Sigma \sim 8$ MeV) at $B_\Sigma = -4.8$ and -11.3 MeV, i.e., very close to the location of the peaks found in [10]. The dominant contribution to the $p_{1/2}(p_{3/2})$ curves comes from the $p_{1/2}(p_{3/2})$ component of the final state of Σ, and the peaks correspond to the p state resonance of Σ in $\mathcal{V}_\Sigma$. Thus the two peaks are substitutional states $[p_{1/2}, p_{1/2}^{-1}]_{\Sigma p}$, $[p_{3/2}, p_{3/2}^{-1}]_{\Sigma p}$ in which Σ is in the continuum, with E_Σ in the vicinity of the resonance energy.

Our main conclusion is that it appears possible to explain the peaks in the pion spectrum observed in the (K^-, π) reactions as resonances in the Σ s.p.

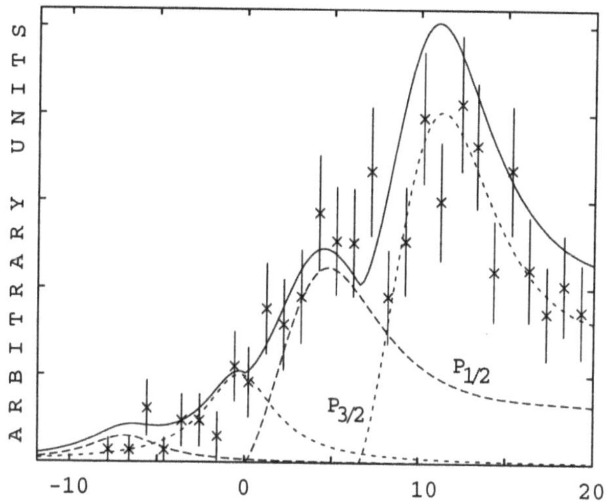

Figure 1. Pion spectrum from (K^-, π^+) reaction on ^{16}O at $p_K = 450$ MeV/c and $\theta = 0°$ calculated in [2]. The solid curve is the total spectrum, the dotted curves are contributions of the K^- interaction with $p_{1/2}$ and $p_{3/2}$ protons. Data are from [6].

potential with a repulsive bump near the nuclear surface, whose magnitude agrees with the microscopic estimate given in [3,4].

However, there is the following problem: the cross section $[d^2\sigma/d\hat{k}_\pi dE_\pi]_\Sigma$ calculated with $W_0 \neq 0$ should be compared to the results of an experiment in which the outgoing pion is observed in coincidence with an outgoing Σ, and not to the inclusive experiments like those described in [6], in which the detected pions may be accompanied by emitted Λ particles. A satisfactory solution of the problem requires introducing the Λ channel explicitly into the description of the strangeness exchange reaction. The simplest way of doing it is to apply the effective two-channel approach used in [8], and a coupled-channels impulse approximation (which corresponds to the coupled-channels Born approximation proposed by Penny and Satchler [9]).

In this approach, the state of the hyperon is represented by two components: the Σ component ψ and the Λ component χ. They satisfy the system of two coupled Schrödinger equations:

$$\{-\frac{\hbar^2}{2M_\Sigma}[\Delta + k_\Sigma^2] + V_\Sigma(r)\}\psi(\boldsymbol{r}) = V_\text{x}(r)\chi(\boldsymbol{r}), \tag{3}$$

$$\{-\frac{\hbar^2}{2M_\Lambda}[\Delta + k_\Lambda^2] + V_\Lambda(r)\}\chi(\boldsymbol{r}) = V_\text{x}(r)\psi(\boldsymbol{r}), \tag{4}$$

where V_Λ is the Λ s.p. potential, and the coupling potential V_x is the result of folding the two-body interaction for the $\Sigma N \to \Lambda N$ process into the density of the nuclear core.

To calculate $[d^2\sigma/d\hat{k}_\pi dE_\pi]_\Sigma$, we find the solutions $\psi(\Sigma, \boldsymbol{k}_\Sigma; \boldsymbol{r})^{(-)}$ and $\chi(\Sigma, \boldsymbol{k}_\Sigma; \boldsymbol{r})^{(-)}$ of Eqs. (5-6), which satisfy the asymptotic conditions:

$[\psi(\Sigma, \boldsymbol{k}_\Sigma; \boldsymbol{r})^{(-)} - \exp(i\boldsymbol{k}_\Sigma \boldsymbol{r})] \rightarrow$ incoming wave, $\chi(\Sigma, \boldsymbol{k}_\Sigma; \boldsymbol{r})^{(-)} \rightarrow$ incoming wave. The solution $\psi(\Sigma, \boldsymbol{k}_\Sigma; \boldsymbol{r})^{(-)}$ is then used in (1) and (2). To calculate the pion spectrum accompanied by Λ emission, $[d^2\sigma/d\hat{k}_\pi dE_\pi]_\Lambda$, we find the solutions $\psi(\Lambda, \boldsymbol{k}_\Lambda; \boldsymbol{r})^{(-)}$ and $\chi(\Lambda, \boldsymbol{k}_\Lambda; \boldsymbol{r})^{(-)}$ of Eqs. (5-6), which satisfy the asymptotic conditions: $[\chi(\Lambda, \boldsymbol{k}_\Lambda; \boldsymbol{r})^{(-)} - \exp(i\boldsymbol{k}_\Lambda \boldsymbol{r})] \rightarrow$ incoming wave, $\psi(\Lambda, \boldsymbol{k}_\Lambda; \boldsymbol{r})^{(-)} \rightarrow$ incoming wave. With the solution $\psi(\Lambda, \boldsymbol{k}_\Lambda; \boldsymbol{r})^{(-)}$ we calculate the triple differential cross section

$$d^3\sigma/d\hat{k}_\Lambda d\hat{k}_\pi dE_\pi = \frac{E_K E_\pi M_\Lambda c^2 k_\pi k_\Lambda}{(2\pi)^5(\hbar c)^6 k_K} |t \int d\boldsymbol{r} \exp(-i\boldsymbol{q}\boldsymbol{r})\psi(\Sigma, \boldsymbol{k}_\Sigma; \boldsymbol{r})^{(-)*}\psi_p(\boldsymbol{r})|^2, \tag{5}$$

and by integrating it over $\hat{k}_\Lambda$, we obtain $[d^2\sigma/d\hat{k}_\pi dE_\pi]_\Lambda$. Notice that in the case of (K^-, π^+) reaction we do not have direct transitions to the Λ component χ.

Preliminary results obtained for the (K^-, π^+) reaction on the ^{16}O target indicate that the cross section for the reaction accompanied by Λ emission, $[d^2\sigma/d\hat{k}_\pi dE_\pi]_\Lambda$, reveals similar resonances as $[d^2\sigma/d\hat{k}_\pi dE_\pi]_\Sigma$. Thus by adding the two cross sections one obtains the pion spectrum in an exclusive experiment, which is similar to the spectrum discussed in the present paper.

Finally, let us stress that the existing data on Σ hypernuclei are not very accurate and it would be highly desirable to perform the experiments with an improved statistics. It should be mentioned, that in the case of the strangeness exchange reaction on ^{9}Be, the recent experiments performed at BNL [10] seem not to reveal the peaks in the pion spectra reported earlier.

Acknowledgement. This research was partly supported by Komitet Badań Naukowych under Grant No. PB 2-0956-91-01.

References

1. J Dąbrowski and J. Rożynek: Phys. Lett. **B323**, 99 (1994)
2. J. Dąbrowski and J. Rożynek: Izv. V. U. Z. Fizika (in print)
3. K. S. Myint et al.: Progr. Theor. Phys. **82**, 112 (1989)
4. S. Tadokoro and Y. Akaishi: Phys. Rev. **C42**, 2591 (1990)
5. C. J. Batty: Nucl. Phys. **A585**, 229c (1995)
6. R. Bertini et al.: Phys. Lett. **B158**, 19 (1985)
7. J. Dąbrowski: Phys. Lett. **B139**, 7 (1984)
8. M. Kohno et al.: Nucl. Phys. **A470**, 609 (1987)
9. S. K. Penny and G. R. Satchler: Nucl. Phys. **53**, 145 (1964)
10. R. Sawafta: Nucl.Phys. **A585**, 103c (1995)

Few-Body Systems Suppl. 9, 145–149 (1995)

Few-Body Systems

$S = -2$ Nuclear Phenomena and ΞN Interaction

Y. Yamamoto*

Physics Section, Tsuru University, Tsuru, Yamanashi 402, Japan

Abstract. The ΞN G-matrix interaction is derived from the Nijmegen OBE model, and applied to the calculations of Ξ^--nucleus bound states with use of the DDHF formalism. The obtained Ξ^- binding energies in ^{12}C are consistent with the values indicated by the events of simultaneous emissions of two Λ hypernuclei observed in the KEK-E176 experiment. In spite of shallow binding of Ξ^- in light nuclei, our Ξ-nucleus potentials are shown to become deep with increase of mass number owing to the strong attraction in ΞN odd states.

It is very important to explore the properties of ΞN and $\Lambda\Lambda$ ($S = -2$) interactions in hypernuclear phenomena, because free-space scattering experiments will be extremely limited or impossible even in future. The useful guiding for ΞN and $\Lambda\Lambda$ interactions can be obtained from the SU(3) invariant OBE models, which are successful for single-Λ hypernuclei. In these models the SU(3) symmetric coupling constants between meson nonets and baryon octets are determined by utilizing rich NN scattering data as well as limited ΛN and ΣN ones, which automatically gives the coupling constants in the ΞN and $\Lambda\Lambda$. We adopt here the OBE models by the Nijmegen group, where the model D and F are abbreviated as ND and NF, respectively [1]. In the ΛN channel, the strange and non-strange meson contributions of ND and NF are very different from each other: The former contribution is small (large) and the latter one is large (small) in the ND (NF) case, though the summed ones are comparable. This difference appears explicitly in the $\Lambda\Lambda$ channel, because only the non-strange mesons contribute in this channel. It was shown that the $\Lambda\Lambda$ sector of ND (NF) is (not) attractive enough to reproduce the experimental $\Lambda\Lambda$ binding energies by taking the reasonable value of the hard-core radius (similarly to the one in the NN channel) [2]. Thus, the data of double-Λ nuclei play decisive roles to test the $\Lambda\Lambda$ interactions given by the OBE models. Similar situation can be seen in the ΞN channel. Considering qualitatively, the attractive parts

**E-mail address:* yamamoto@ins.u-tokyo.ac.jp

of Ξ-nucleus interactions are determined by the $T = 0$ scalar $(\epsilon + \epsilon')$ and vector $(\omega + \phi)$ mesons, because $T = 1/2$ strange mesons do not contribute and moreover the terms including $\boldsymbol{\sigma}_1 \cdot \boldsymbol{\sigma}_2$ and/or $\boldsymbol{\tau}_1 \cdot \boldsymbol{\tau}_2$ are canceled for the $Z = N$ nuclear core. These parts are very different from each other between ND and NF. The attraction of NF is quite insufficient to make a Ξ-nucleus bound state. Thus, it is very important for checking SU(3) OBE models to investigate the $\Xi N/\Lambda\Lambda$ interactions in hypernuclear phenomena.

We can get the important conjecture for the Ξ^--nucleus potential from the two events of simultaneous emissions of two Λ hypernuclei (KEK-E176). These data give the energy difference between the initial Ξ^- state and the final twin Λ state, considered as the binding energy B_Ξ between Ξ^- and the nucleus. The most probable assignment for the first event [3] is

$$\Xi^- + {}^{12}\mathrm{C} \rightarrow {}^4_\Lambda\mathrm{H} + {}^9_\Lambda\mathrm{Be} \quad (B_\Xi = 0.54 \pm 0.20 \text{ MeV}) .$$

On the other hand, there are three possibilities for the second event [4]:

$$\begin{aligned}
&(\mathrm{A}) \quad \Xi^- + {}^{12}\mathrm{C} \rightarrow {}^4_\Lambda\mathrm{H} + {}^9_\Lambda\mathrm{Be} \quad (B_\Xi = 3.70 \pm 0.19 \text{ MeV}) \ , \\
&(\mathrm{B}) \quad \Xi^- + {}^{12}\mathrm{C} \rightarrow {}^4_\Lambda\mathrm{H} + {}^9_\Lambda\mathrm{Be}^* \quad (B_\Xi = 0.62 \pm 0.19 \text{ MeV}) \ , \\
&(\mathrm{C}) \quad \Xi^- + {}^{12}\mathrm{C} \rightarrow {}^4_\Lambda\mathrm{H}^* + {}^9_\Lambda\mathrm{Be} \quad (B_\Xi = 2.66 \pm 0.19 \text{ MeV}) \ ,
\end{aligned}$$

where ${}^4_\Lambda\mathrm{H}^*$ and ${}^9_\Lambda\mathrm{Be}^*$ are the excited states. It is considered that the Ξ^-'s were captured from the 1s-, 2s- or 2p-bound states: These large binding energies of Ξ^- can not be obtained for 3s-, 3p-, 3d- and higher bound states with any Ξ-nucleus interaction of normal strength. Among three possibilities for the second event, (B) is quite consistent with the previous event: They can be interpreted as the 2p-Ξ^- captures ($B_{\Xi^-} = 0.4 \sim 0.7$ MeV), considering that the probability of s-state capture is very small in the cascade calculations.

Table 1. Partial-wave contributions to U_Ξ (Γ_Ξ) at $k_\mathrm{F} = 1.35 \text{ fm}^{-1}$

	${}^1S_0$	${}^3S_1$	${}^1P_1$	${}^3P$	Total
$T=0$	−3.4 (1.2)	1.3	−1.9	−2.7 (.03)	
$T=1$	−1.5	1.3	−2.8	−9.0	−18.8 (1.3)

Now, we derive the ΞN G-matrix interaction with ND [5], which has the imaginary part because of the ΞN-$\Lambda\Lambda$ coupling. The hard-core radii (r_c) are taken similarly to the NN channels: $r_\mathrm{c} = 0.52$ (0.48) fm in the $T = 0$ ${}^1S_0$ (the other) channels. It should be noted that the data of double-Λ hypernuclei (${}^{10}_{\Lambda\Lambda}\mathrm{Be}$ and ${}^{13}_{\Lambda\Lambda}\mathrm{B}$) can be reproduced consistently by taking $r_\mathrm{c} = 0.52$ fm in the $T = 0$ ${}^1S_0$ channels [2]. The calculated values of the well depth U_Ξ and the conversion widths Γ_Ξ for $k_\Xi = 0$ in nuclear matter at normal density are −18.8 and 1.3 MeV, respectively. In Table 1 we show the partial-wave contributions to U_Ξ and Γ_Ξ (in parentheses). The value of U_Ξ is fairly shallower than the experimentally-established one of U_Λ (∼ −30 MeV). The small values of obtained conversion widths are because of two reasons: The statistical weight of the main ΞN-$\Lambda\Lambda$ coupling channel ($T = 0$ ${}^1S_0$) is small, and the ΞN-$\Lambda\Lambda$ coupling interaction of ND is weak. In Fig.1 the even- and odd-contributions

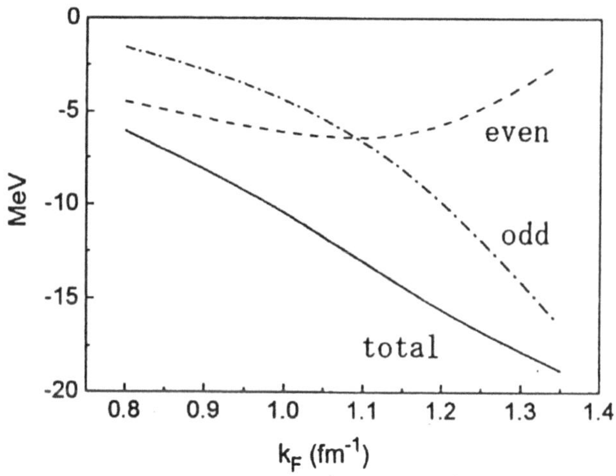

Figure 1. Even and odd contributions to U_{Ξ}

are plotted as a function of k_F by solid and dotted curves, respectively. It is notable that the odd-state contributions to U_{Ξ}, which are insensitive to r_c, are quite large compared to the cases of U_N and U_{Λ}. The OBE part of the ΞN interaction has no exchange force, because two units of strangeness cannot be carried by any mesons. This Wigner-type character of the ΞN interaction leads to the strong odd state attraction.

By using the obtained G-matrix interactions we perform the DDHF calculations for Ξ^--nucleus bound states, where the Skyrme III force is used for the core nucleus. In Table 2 the results for the $\Xi^- + ^{12}C$ system are shown, where the Coulomb interaction is included between Ξ^- and ^{12}C. It should be noted that the $2p$-state energy is consistent with the above experimental indication. The values of root mean square radii $\sqrt{\langle r^2 \rangle}$ show that the $1s$ state is hypernuclear, the $2s$, $3p$ and $3d$ ones are atomic, and the $2p$ one is *intermediate*. In the case of using the Woods-Saxon (WS) potential by Dover and Gal (the depth

Table 2. $\Xi^- + ^{12}C$ bound states with G-matrix interaction

orbit	B_{Ξ} (MeV)	$\sqrt{\langle r^2 \rangle}$ (fm)	Γ_{Ξ} (MeV)
$1s$	3.48	4.0	0.40
$2s$	0.36	20.	0.01
$2p$	0.61	8.8	0.08
$3s$	0.15	47.	.003
$3p$	0.19	33.	.001
$3d$	0.13	43.	.000

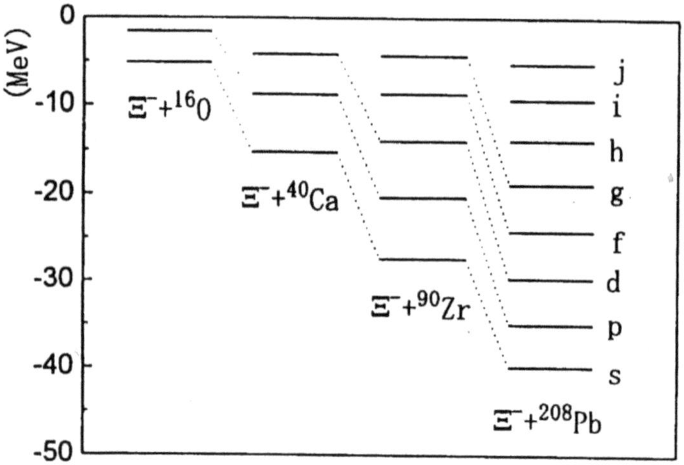

Figure 2. Ξ^- bound states in ^{16}O, ^{40}Ca, ^{90}Zr and ^{208}Pb

$V_0 = -24$ MeV, $r_0 = 1.1$ fm and $a = 0.65$ fm) [6], the larger binding energies 11.9 MeV and 2.36 MeV are obtained for the 1s- and 2p-state, respectively. Taking $V_0 = -16$ MeV, we get 7.00 MeV (1s) and 0.58 MeV (2p). Then, the 2p energy is similar to ours, but the 1s one is fairly larger. This means that our DDHF Ξ-nucleus potential is of longer range than the Dover-Gal's one. In this connection it is interesting that our result is also consistent with the following interpretation for the above data: The first event $\Rightarrow B_{\Xi^-}(2p) \sim 0.53$ MeV and the second one (A) $\Rightarrow B_{\Xi^-}(1s) \sim 3.70$ MeV. Anyway, our result means that the Ξ potential in ^{12}C is very shallow: If the Coulomb interaction is switched off, $B_{\Xi^-}(1s)$ is only 1.0 MeV in spite of the fairly large binding energy 18.8 MeV in nuclear matter. This is because the attractive contributions from ΞN odd states are relatively small in light systems.

Next, let us calculate energy spectra of Ξ^- bound states in various nuclei (^{16}O, ^{40}Ca, ^{90}Zr and ^{208}Pb). The result is shown in Fig.2, where Ξ^--atomic states are not given. In spite of the small values of B_{Ξ^-} in ^{12}C, there appear many high angular-momentum states in heavy nuclei owing to remarkable increase of the Coulomb attractions. If the Coulomb force is switched off in the case of ^{208}Pb, bound states are up to the f-orbit ($B_\Xi(1s) = 16.1$ MeV). In the case of $^{208}_{\Lambda}$Pb, known experimentally, bound states are up to the g-orbit ($B_\Lambda(1s) \sim 26.3$ MeV). In addition, it should be noted that the strong odd-state attraction originated from the above Wigner-type character of the ΞN interaction works more efficiently in heavy nuclei. In order to demonstrate the mass dependence of our DDHF potential compared with the WS one, we have to try to adjust the depth parameter V_0 so as to reproduce our calculated value of the 1s-state energy for each system. The obtained mass dependence of V_0 is shown in Fig.3. The possibility was pointed out in [7] to observe Ξ^- bound

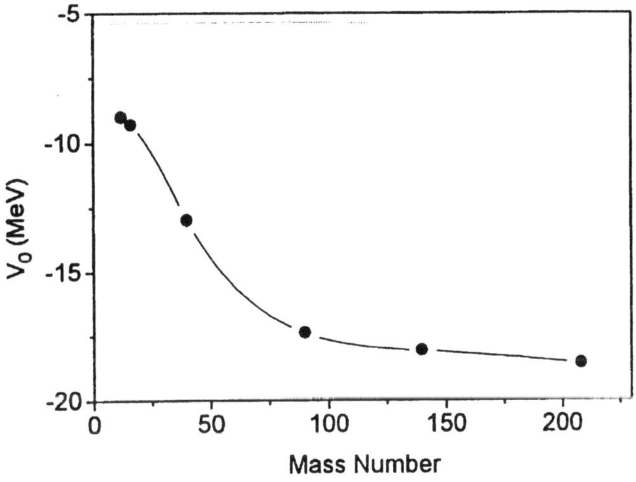

Figure 3. Mass dependence of V_0

states with high angular momenta in heavy systems by (K^-, K^+) reactions.

In conclusion, the present consideration based on the realistic ΞN interaction ND leads to fairly shallow potentials between Ξ and light nuclei consistently with the recent experimental indication. However, because of the strong odd-state attraction (Wigner-type character) of the ΞN interaction, our Ξ-nucleus potential becomes deep remarkably with increase of mass number.

References

1. M. M. Nagels, T. A. Rijken and J. J. deSwart: Phys. Rev. **D15**, 2547 (1977); **D20**, 1633 (1979)
2. Y. Yamamoto, H. Takaki and K. Ikeda: Prog. Theor. Phys. **86**, 867 (1991)
3. S. Aoki et al.: Prog. Theor. Phys. **89**, 493 (1993)
4. S. Aoki et al.: Phys. Lett. **355B**, 45 (1995)
5. Y. Yamamoto et al.: Prog. Theor. Suppl. No. **117**, 361 (1994)
6. C. B. Dover and A. Gal: Ann. Phys. **146**, 309 (1983)
7. T. Tadokoro, H. Kobayashi and Y. Akaishi: Phys. Rev. **C51**, 2656 (1995)

Few-Body Systems Suppl. 9, 150–154 (1995)

Few-Body Systems

Properties of the Hypertriton

K. Miyagawa[1], H. Kamada[2], W. Glöckle[3]

[1] Department of Applied Physics, Okayama University of Science 1-1 Ridai-cho, Okayama 700, Japan

[2] Paul Scherrer Institute, CH-532 Villigen PSI, Switzerland

[3] Institut für Theoretische Physik II, Ruhr Universität Bochum, 44780 Bochum, Germany

Abstract. The Faddeev equations for the hypertriton are solved precisely using the Nijmegen and Jülich hyperon-nucleon interactions with full inclusion of the $\Lambda - \Sigma$ conversion. For the Nijmegen interaction, the hypertriton turns out to be bound at the experimental value. Thereby the $\Lambda - \Sigma$ conversion is crucial.

1 Introduction

The hypertriton ${}^3_\Lambda$H is the bound nuclear system of lowest mass including one hyperon Y (Λ or Σ). Since the hyperon-nucleon scattering data are still extremely poor, it provides important information on the hyperon-nucleon forces, especially it can serve as a test for recently developed meson-theoretical forces. Furthermore, the study of the scattering states for this coupled $\Lambda NN - \Sigma NN$ system enables us to realize the energy dependence of the $\Lambda - \Sigma$ conversion at higher energies beyond the Λd threshold.

In addition, the hypertriton decays weakly to the $3N$ system with or without a pion. Since it is now possible to obtain exact 3-body bound and scattering wave functions using the Faddeev formalism, this hypernucleus provides a good testing ground for the study of weak decay processes [1].

In this paper, we solve the Faddeev equations in a coupled channel formalism, and present binding energies of the hypertriton, using the Nijmegen soft core YN potential [2] and the Jülich YN potential in a OBE potential parametrization [3]. We also calculate various expectation values of the hypertriton, analyze the property of this state and scrutinize the YN interactions.

2 Formalism

We number the two nucleons by 1 and 2 and the Λ or Σ by 3. Since we keep both states for the Λ and Σ explicitly, the wave function for the hypertriton is represented by the two-component vector

$$\underset{\sim}{\psi} = \begin{pmatrix} \psi_{N_1 N_2 \Lambda} \\ \psi_{N_1 N_2 \Sigma} \end{pmatrix} . \tag{1}$$

By introducing the Faddeev decomposition $\underset{\sim}{\psi} = \underset{\sim}{\psi}^{(12)} + \underset{\sim}{\psi}^{(13)} + \underset{\sim}{\psi}^{(23)}$, we obtain the matrix Faddeev equations

$$\underset{\sim}{\psi}^{(ij)} = \frac{1}{E - \underset{\sim}{H}_0} \underset{\sim}{T}_{ij} \sum_{kl \neq ij} \underset{\sim}{\psi}^{(kl)} , \tag{2}$$

where, the two-body T matrices obey the matrix Lippmann-Schwinger equations

$$\underset{\sim}{T}_{ij} = \underset{\sim}{V}_{ij} + \underset{\sim}{V}_{ij} \frac{1}{E - \underset{\sim}{H}_0} \underset{\sim}{T}_{ij} . \tag{3}$$

Here, $\underset{\sim}{V}_{12}$ is a diagonal 2×2 matrix, the diagonal elements of which are the nucleon-nucleon interaction $V_{N_1 N_2}$, while $\underset{\sim}{V}_{13}$ and $\underset{\sim}{V}_{23}$ have in addition off-diagonal components which connect the ΛNN and ΣNN systems:

$$\underset{\sim}{V}_{i3} = \begin{pmatrix} V_{N_i \Lambda, N_i \Lambda} & V_{N_i \Lambda, N_i \Sigma} \\ V_{N_i \Sigma, N_i \Lambda} & V_{N_i \Sigma, N_i \Sigma} \end{pmatrix} \quad i = 1, 2 . \tag{4}$$

Assuming that $\underset{\sim}{\psi}$ is antisymmetrized in the two nucleons 1 and 2, the coupled matrix equations (2) are simplified correspondingly. After partial wave decomposition, the Faddeev equations are solved in momentum space. For details, we refer the reader to ref. [4].

3 Results

We already presented [4] the calculations for the Jülich YN interaction in an OBE potential parametrization (model Ã) [3]. We solved Faddeev equations for the coupled ΛNN and ΣNN systems precisely. The hypertriton turned out to be unbound. Now we present calculations for another OBE model of the YN interaction, the Nijmegen soft core potential [2], for which the hypertriton is bound. We use the Bonn B, Paris and Nijmegen 93 potentials for the NN sector. The binding energies are −2.37 MeV for Bonn B, −2.36 MeV for Paris and −2.36 MeV for Nijmegen 93, all of which are within the experimental errors of -2.355 ± 0.05. In a previous reports [5], we pointed out that the Jülich $^1S_0$ force is quite different from the Nijmegen one. We shall also see below that the ΣN potential energy in the hypertriton as given by the Jülich potential is much bigger than the one due to the Nijmegen.

We analyze the structure of the hypertriton by calculating the total potential and kinetic energies and their individual contributions:

$$<V>=<V_{NN}>+<V_{YN}>,$$

$$<V_{NN}>=<V_{NN}>_\Lambda+<V_{NN}>_\Sigma,$$

$$<V_{YN}>=<V_{\Lambda N,\Lambda N}>+<V_{\Lambda N,\Sigma N}>+<V_{\Sigma N,\Lambda N}>+<V_{\Sigma N,\Sigma N}>. \quad (5)$$

$$<T>=<T_{NN}>+<T_{Y-NN}>,$$

$$<T_{NN}>=<T_{NN}>_\Lambda+<T_{NN}>_\Sigma,$$

$$<T_{Y-NN}>=<T_{\Lambda-NN}>+<T_{\Sigma-NN}>. \quad (6)$$

The results based on the Nijmegen YN interactions are displayed in Tab. 1. Note that the expectation values of the YN interactions include factor 2 resulting from the two nucleons. In Tab. 1, the sum $<V_{NN}>+<T_{NN}>$ amounts to -1.77 MeV, which is close to the deuteron binding energy as can be expected. The remaining binding energy of -0.57 MeV comes from the hyperon. By using the wave function obtained, we determine the probability of the (ΣNN) states in the hypertriton, which is only 0.005. However, as shown in Tab. 1, the Λ particle alone brings in more kinetic energy than the magnitude of its potential energy. It is only due to the $\Lambda-\Sigma$ conversion that the potential energy of the hyperon wins and the hyperon provides -0.57 binding energy. Note that the contributions of the $^1S_0$ and $^3S_1-{}^3D_1$ components of the YN potential energies are not in the ratio 3 : 1 as often quoted in the literature but are comparable.

In Tab. 2, we show the corresponding expectation values due to the Jülich YN interaction. In order to get the correct hypertriton binding energy, the Jülich interaction is multiplied by a factor 1.04. The results are strikingly different from the Nijmegen case. The contribution from the ΣN and $\Lambda-\Sigma$ transition potential are much bigger than in the Nijmegen case and it mainly comes from the $^1S_0$ force component. In Fig. 1, we also compare the ΛN total cross sections for the Nijmegen and Jülich interactions, separated into the contributions from the $^1S_0$ and $^3S_1$ states. The cross sections to the $^3S_1$ states are similar, but those to the $^1S_0$ states are quite different. The Jülich $^1S_0$ curve shows a small bump which comes from the relatively strong coupling to the ΣN states. This $^1S_0$ behavior of the Jülich interaction is presumably responsible for the underbinding of the hypertriton.

4 Summary and Outlook

We solved the (matrix) Faddeev equations for the hypertriton using the Nijmegen and Jülich hyperon-nucleon interactions with full inclusion of the $\Lambda-\Sigma$

Table 1. Various kinetic and potential energy contributions of Eqs. (5) and (6) in the hypertriton. The Nijmegen YN and Nijmegen93 NN interactions are used. The shorthand notation $< V_{\Lambda-\Sigma} >$ is used for $< V_{\Lambda N,\Sigma N} > + < V_{\Sigma N,\Lambda N} >$. All numbers are in units of MeV.

component	$< V_{\Lambda N,\Lambda N} >$	$< V_{\Lambda-\Sigma} >$	$< V_{\Sigma N,\Sigma N} >$	$< V_{YN} >$
$^1S_0$	-1.60	-0.38	0.02	-1.95
$^3S_1 - {}^3D_1$	0.02	-1.54	-0.06	-1.57
all	-1.58	-1.94	-0.02	-3.54
	$< V_{NN} >_\Lambda$		$< V_{NN} >_\Sigma$	$< V_{NN} >$
all	-22.22		-0.03	-22.25
	$< T_{NN} >_\Lambda$		$< T_{NN} >_\Sigma$	$< T_{NN} >$
all	20.25		0.23	20.48
	$< T_{\Lambda-NN} >$		$< T_{\Sigma-NN} >$	$< T_{Y-NN} >$
all	2.18		0.79	2.97

Table 2. Various kinetic and potential energy contributions in the hypertriton. The Jülich YN and the Bonn B NN interactions are used. The Jülich interaction is multiplied by a factor 1.04 to get the correct binding energy.

component	$< V_{\Lambda N,\Lambda N} >$	$< V_{\Lambda-\Sigma} >$	$< V_{\Sigma N,\Sigma N} >$	$< V_{YN} >$
$^1S_0$	2.66	-8.48	-5.63	-11.46
$^3S_1 - {}^3D_1$	-0.67	-0.66	-0.07	-1.41
all	1.99	-9.16	-5.76	-12.93
	$< V_{NN} >_\Lambda$		$< V_{NN} >_\Sigma$	$< V_{NN} >$
all	-17.38		-0.21	-17.59
	$< T_{NN} >_\Lambda$		$< T_{NN} >_\Sigma$	$< T_{NN} >$
all	15.51		2.20	17.71
	$< T_{\Lambda-NN} >$		$< T_{\Sigma-NN} >$	$< T_{Y-NN} >$
all	2.21		8.25	10.46

conversion. The Jülich potential (the energy-independent version $\tilde{\text{A}}$) does not lead to a bound hypertriton, whereas the Nijmegen one does bind the hypertriton at the experimental value. We found that the $\Lambda - \Sigma$ conversion is crucial for the binding. In the future, it will be possible to measure nonmesonic weak decays of the hypertriton and the $\Lambda NN - \Sigma NN$ scattering states. This will be a beautiful laboratory a) to test weak processes in an interacting few-baryon system and b) to study the energy dependence of the $\Lambda - \Sigma$ conversion from the Λd to the Σd threshold and beyond. Theoretical studies in that direction are planned.

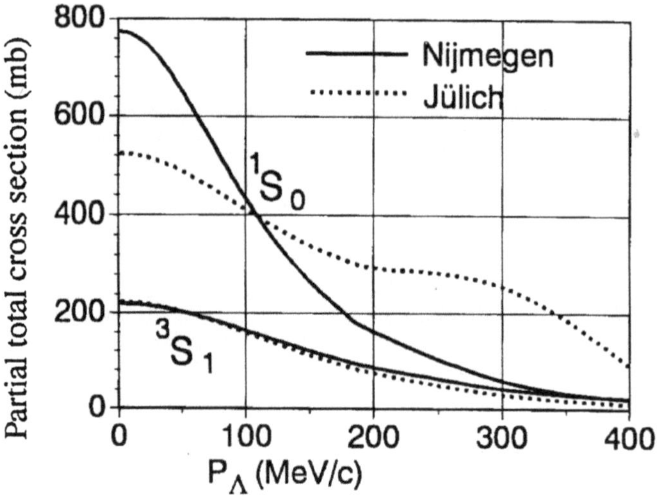

Figure 1. Partial total ΛN cross sections σ_s and σ_t to the $^1S_0$ and $^3S_1$ states, respectively, where the total cross section is $\sigma_{tot} = \frac{3}{4}\sigma_t + \frac{1}{4}\sigma_s$. Comparison of the predictions of the Nijmegen and Jülich interactions.

References

1. C. Bennhold, A. Ramos, D.A. Aruliah, U. Oelfke: Phys. Rev. **C45**, 947 (1992)

2. V. Stoks: Private Communication;
   P.M.M. Maessen, Th.A. Rijken, J.J. de Swart: Phys. Rev. **C40**, 2226 (1989)

3. A.G. Reuber, K. Holinde, J. Speth: Czech. J. Phys. **42**, 1115 (1992)

4. K. Miyagawa and W. Glöckle: Phys. Rev. **C48**, 2576 (1993)

5. K. Miyagawa and W. Glöckle: Nucl. Phys. **A585**, 169c (1995)

Few-Body Systems Suppl. 9, 155–160 (1995)

Few-Body Systems

Σ-Hypernuclear Production by K^- Capture in Flight

Toru Harada*

Department of Social Information, Sapporo Gakuin University, Ebetsu, Hokkaido 069, Japan and Institute of Socio-Information and Communication Studies (ISICS), University of Tokyo, Bunkyō-ku, Tokyo 113, Japan

Abstract. On the viewpoint of shell-model and cluster-model descriptions, Σ-hypernuclear productions by ^{9}Be(in-flight K^-,$\pi^\mp$) reactions are investigated theoretically within the DWIA framework. The resultant spectra are compared with the recent experimental data at BNL.

1 Introduction

The discovery of narrow Σ-hypernuclear states with $\Gamma \sim 8$ MeV on a ^{9}Be target at CERN [1] made us much excited as a "narrow width" puzzle, since the width of Σ in nuclei was believed to be $\sim$25 MeV due to the strong $\Sigma N \rightarrow \Lambda N$ conversion process. Despite theoretical efforts [2], this data was of such poor statistics that we could not get definitive conclusion. Recently, better statistical observations on the ^{9}Be target at BNL required that there are no narrow unbound Σ excitations [3]. It is important to confirm theoretically the spectra for solving the puzzle.

In this note, we investigate Σ-hypernuclear productions by ^{9}Be(in-flight K^-,$\pi^\mp$) reactions in order to explore the structures of A=9 hypernuclei ($^9_\Sigma$Be,$^9_\Sigma$He). We discuss the $\pi^\mp$ spectra in the DWIA framework by shell-model and cluster-model calculations.

2 Shell-model calculations

The spectral response by ^{9}Be(in-flight K^-,π^+) reactions leads us to clarify the nucleus-Σ potential for A=9 hypernuclei. First, we consider a shell model for Σ-hypernuclear states, which succeeds in describing the $^9_\Lambda$Be [4]. In order to construct microscopically the nucleus-Σ potential, we introduce the density-dependent effective ΣN interaction $\hat{g}_{\Sigma N}(r;\rho)$, which is obtained by applying

* *E-mail address:* harada@rikvax.riken.go.jp

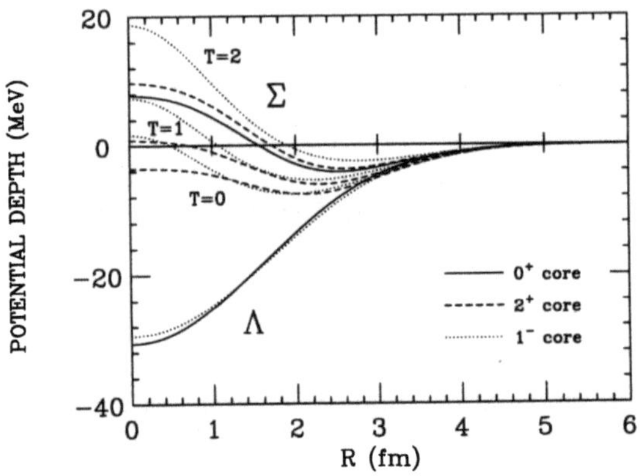

Figure 1. The real part of (^{8}Be,^{8}B,^{8}Li)-Σ potentials together with the Λ one.

the Brueckner g-matrix theory with the ΣN absorptive potential (SAP-3) [5] determined to fit available $\Sigma^{\pm}p$ scattering data at low energies (χ^2/F=1.1). Note that the effective ΣN interaction has a strong spin-isospin dependence.

2.1 Nucleus-Σ folding potential

We calculate the nucleus-Σ potential by folding the effective ΣN interaction in the local density approximation (LDA) [5]. It is given as

$$\begin{aligned}\hat{U}(R) &= \left\langle \left[\Phi_\alpha\{\boldsymbol{r}_i\} \otimes \mathcal{Y}^{\tau}_{(\ell\frac{1}{2})j}\right]_J \middle| \sum_{i=1}^{A} \hat{g}_{\Sigma N}(|\boldsymbol{r}_i - \boldsymbol{R}|;\rho_{\mathrm{G}}) \middle| \left[\Phi_{\alpha'}\{\boldsymbol{r}_i\} \otimes \mathcal{Y}^{\tau'}_{(\ell'\frac{1}{2})j'}\right]_J \right\rangle \\ &= U^0(R) + \frac{1}{A}U^{\tau}(R)(\boldsymbol{T}_{\mathrm{C}} \cdot \boldsymbol{t}_{\Sigma}), \qquad (1)\end{aligned}$$

where Φ_α and $\boldsymbol{R}$ denote the wave function of the A=8 core-nucleus and the radial coordinate between the nucleus-Σ system, respectively; ρ_{G} is the nuclear density at the center-of-mass of the interacting ΣN pair. Although the isospin dependence of Eq.(1) is induced by the Lane term, its contribution becomes small as the mass-number A increases.

Figure 1 displays the real parts of the potentials for the nucleus-Σ system. We find that there appears a repulsion near the nuclear center, and that the depth for Σ is quite shallower than that for Λ. If we employ nucleus-Σ folding potentials for light nuclei, we can explain the energy shifts and widths of the Σ^- atomic X-ray (χ^2/F=0.71) and the Σ-production spectra of the recent ^{12}C(K^-,π^+) data [5].

2.2 ^{9}Be(in-flight K^-,$\pi^\mp$) spectrum

The double-differential cross section in the DWIA framework is given by

$$\frac{d^2\sigma}{d\Omega dE_\pi} = \frac{1}{[J_i]}\sum_{M_i}\sum_{f} |\langle f|\hat{\mathcal{O}}|i\rangle|^2 \delta(E_\pi + E_f - E_K - E_i) \tag{2}$$

with

$$\hat{\mathcal{O}} = \int d\boldsymbol{r}\, \chi^{(-)*}_{\mathbf{k}_\pi}(\boldsymbol{r})\chi^{(+)}_{\mathbf{k}_K}(\boldsymbol{r}) \sum_{j=1}^{A} \overline{f}_{(Y\tau_Y)} \hat{U}_-(j)\delta(\boldsymbol{r}-\boldsymbol{r}_j), \tag{3}$$

where $\overline{f}_{(Y\tau_Y)}$ is the in-medium elementary amplitude of $K^-N\to\pi Y$ reactions [2]; $\chi^{(-)*}_{\mathbf{k}_\pi}$ and $\chi^{(+)}_{\mathbf{k}_K}$ are the distorted waves of the outgoing pion and of the incoming kaon, respectively. The kaon is assumed to be absorbed by a nucleon in p- and s-states on the ^{9}Be target. We obtain a spectroscopic factor of the nucleon pick-up by the OXBASH calculation, and take these energies from (p,2p) experiments. We calculate the $\pi^\mp$ spectrum by applying the Green's function method, which has an advantage of taking account of contributions from both Σ-escaping and Λ-conversion processes.

Figure 2 displays the calculated spectrum in the ^{9}Be(in-flight K^-,π^-) reaction at p_K= 586 MeV/c (θ=4°), together with recent experimental data at BNL [6]. The reaction can populate the $^9_\Sigma$Be states with T=0, 1 and 2. We

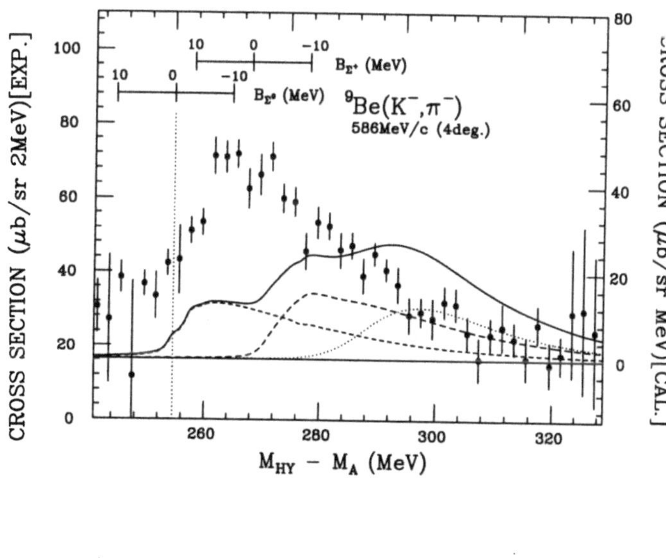

Figure 2. ^{9}Be(in-flight K^-,π^-)$^9_\Sigma$Be spectrum at 586 MeV/c (θ=4°) calculated in the shell model. The short-(long-) dashed and dotted curves denote contributions of the valence (the others) p^{-1} and s^{-1} nucleon states, respectively.

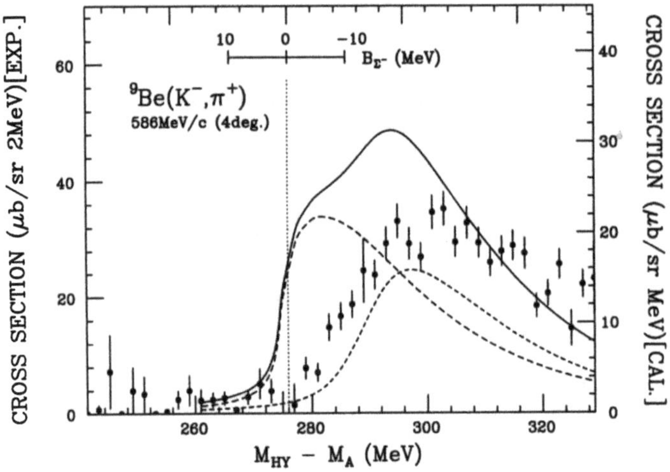

Figure 3. ^{9}Be(in-flight K^-,π^+)$^9_\Sigma$He spectrum at 586 MeV/c (θ=4°) calculated in the shell model. The dashed and dotted curves denote contributions of the p^{-1} and s^{-1} nucleons, respectively.

renormalize arbitrarily the calculated spectrum in order to compare the shapes with the data. We find that in shell-model calculation the Σ^0 population near $B_{\Sigma^0}\simeq-12$ MeV is not satisfied while its population near highly-excited region ($B_{\Sigma^0}\simeq-50$ MeV) is overestimated. On the other hand, Fig.3 displays the calculated spectrum in the ^{9}Be(in-flight K^-,π^+) reaction at p_K= 586 MeV/c (θ=4°) to be compared with the BNL data [6]. The $^9_\Sigma$He states with only T=2 can be produced by the reaction. We show that the Σ^- population near the ^{8}Li-Σ^- threshold is too large. Therefore, the shell-model calculation fails to reproduce the data. It would be caused by distribution of the spectroscopic factor in the shell model.

3 Cluster-model calculations

When employing the microscopic cluster model [7] of α+3N+Σ in $^8_\Sigma$Be ($^8_\Sigma$Li), we showed that there appears a cluster phenomenon of α+$^4_\Sigma$He (α+$^4_\Sigma$H) as the well-known di-molecular α+α structure of the ^{8}Be ground state. It is very different from the structure of $^8_\Lambda$Li ($^8_\Lambda$Be), where a Λ-hyperon plays a role of glue between α and 3N. In fact, the observation of the $^4_\Sigma$He bound state[1] in the ^{4}He(stopped K^-,π^-) reaction at KEK were reported [8], while the exclusive K^-^{4}He→$\pi^-\Lambda pd$ spectrum at ZGS might be reminiscent of a threshold cusp.

[1] $B^{\rm exp}_{\Sigma^+}$= 2.8±0.7 MeV, $\Gamma^{\rm exp}$= 12.1±1.2 MeV.

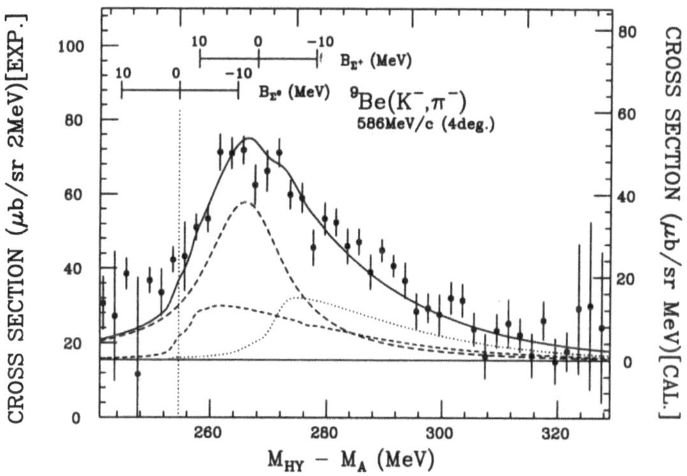

Figure 4. ^{9}Be(in-flight K^-,π^-)$^9_\Sigma$Be spectrum at 586 MeV/c (θ=4°) calculated in the cluster model. The short-dashed curve is of the valence p^{-1} neutron. The long-dashed and dotted curves are of bound and continuum states, respectively, where the kaon is absorbed by α in the ^{9}Be target.

These characteristic feature come from $(\boldsymbol{\tau}_N \cdot \boldsymbol{t}_\Sigma)(\boldsymbol{\sigma}_N \cdot \boldsymbol{\sigma}_\Sigma)$ term due to the $\pi+\rho$ meson exchange in the ΣN interaction.

In $^9_\Sigma$Be, therefore, the cluster configuration of

$$[\alpha + {}^4_\Sigma\mathrm{He} + n] + [\alpha + {}^4_\Sigma\mathrm{H} + p] \tag{4}$$

is expected to be dominant. We assume that the binding energy of $^9_\Sigma$Be measured from the α+$^4_\Sigma$He+n threshold is $\Delta B \simeq 2$ MeV. Figure 4 illustrates the ^{9}Be(in-flight K^-,π^-) spectrum at 586 MeV/c (θ=4°) in the simplest cluster description. Such $^4_\Sigma$He population makes the spectral shape changed drastically, where the kaon is absorbed by α in the ^{9}Be target, so that the obtained spectrum can reproduce the data. The enhancement near $B_{\Sigma^0} \simeq -12$ MeV in the spectrum might be identified as the cluster phenomenon of α+$^4_\Sigma$He($^4_\Sigma$H)+N. To underlie the structure of $^9_\Sigma$Be, we make progress in a sophisticated cluster calculation.

4 Conclusion

The Σ-hypernuclear studies by ^{9}Be(in-flight K^-, $\pi^\mp$) reactions give us valuable information to establish behavior of the Σ-hyperon in $^9_\Sigma$Be and $^9_\Sigma$He. The shell-model calculation can not reproduce the recent experimental data at BNL, but the cluster model is possible. The result brings us the cluster-model description of $\alpha+{}^4_\Sigma\mathrm{He}({}^4_\Sigma\mathrm{H})+N$ in ${}^9_\Sigma\mathrm{Be}$, and it implies the evidence of the ${}^4_\Sigma\mathrm{He}$ bound state.

This work was based on that performed in collaboration with H. Kitagawa and S. Okabe.

References

1. R. Bertini et al.: Phys. Lett. **B90**, 375 (1980)
2. C.B. Dover, D.J. Millener, A. Gal: Phys. Rep. **184**, 1 (1989)
3. R. Sawafta: Nucl. Phys. **A585**, 103c (1995)
4. E.H. Auerbach et al.: Ann. Phys. **148**, 381 (1983)
5. T. Harada, S. Shinmura, Y. Akaishi: (to be submitted)
6. Y. Shimizu: Thesis. Univ. of Tokyo 1995 (unpublished)
7. S. Okabe, T. Harada, Y. Akaishi: Nucl. Phys. **A514**, 613 (1990)
8. H. Outa et al.: Prog. Theor. Phys. Suppl. **117**, 171 (1994)

Few-Body Systems Suppl. 9, 161–164 (1995)

Few-Body Systems

Hyperfragment Production in Annihilation of Antiprotons Stopping in Nuclei

Yu.A. Batusov

Laboratory of Nuclear Problems, JINR, 141980 Dubna, Moscow Region, Russia

Abstract. Investigation of the production of light hyperfragments in the annihilation of antiprotons stopping in nuclei of nuclear photoemulsion and of their meson decays has revealed [1] that the lower limit of production probability of ${}^3_\Lambda$H and ${}^4_\Lambda$H per antiproton stopping in the nuclear photoemulsion is $(6.1 \pm 3.5)10^{-4}$. Mechanism schemes for the production and decay of such hyperfragments are presented. Arguments are presented in favor of one of the production channels of light hyperfragments probably being the sequential decay of hypernuclei produced in annihilation of antiprotons on heavy (Ag, Br) nuclei in photoemulsion. It may turn out to be that the production of such hypernuclei is due to fission of excited recoil nuclei.

The first results of an investigation of the formation and mesonic decay of light hyperfragments (H.F.) produced in the annihilation of stopping antiprotons on nuclei (C, N, O, S) in nuclear photoemulsion were presented in ref. [1]. The lower limit for the production probability of ${}^3_\Lambda$H and ${}^4_\Lambda$H hyperfragments per antiproton stopping in nuclear photoemulsion was found to be $(6.1 \pm 3.5) \cdot 10^{-4}$. The main characteristics of the three (H.F.) events detected in the experiment, the possible scheme of their formation and decay, and, also, the primary reactions of antiproton annihilation are presented in the table (ΣE_i is the total energy involved in the reaction, and ΣE_f is the total apparent final energy released in the reaction). The antiproton annihilation reaction involving formation of the hypernucleus ${}^4_\Lambda$H on the ^{32}S nucleus (event III) seems to be quite exotic, since the percentage of sulphur in nuclear photoemulsion only amounts to 0.06% of the total number of nuclei present in the emulsion. Therefore, the alternative possibility was examined for ${}^4_\Lambda$H being produced in annihilation of an antiproton on a bromine nucleus, since the content of bromine in the photoemulsion amounts to 33.7% and the capture of a stopping antiproton on a heavy nucleus is much more probable.

In this case the formation of a light hypernucleus may proceed in the following sequence: a heavy excited hypernucleus is produced at the first stage

Table 1.

NNo	The antiproton annihilation processes on nuclei	$\lvert M^{calc.}_{II} - M_{\Lambda H}\rvert$	$(\Sigma E_i - \Sigma E_f)$	Possible scheme for H.F. production
I	$\tilde{p} + {}^{12}C \rightarrow K^+ + \pi^- + \pi^+ + \pi^- + \pi^0 +$ $+ p + d + \alpha + {}^4_\Lambda H$ $\hookrightarrow \pi^- + \alpha$ $\tilde{p} + {}^{14}N \rightarrow K^+ + \pi^- + \pi^+ + \pi^- + \pi^0 +$ $+ p + 2\alpha + {}^4_\Lambda H$ $\hookrightarrow \pi^- + \alpha$	2.7 [MeV]	≤ 164.5 [MeV] ≤ 171.4 [MeV]	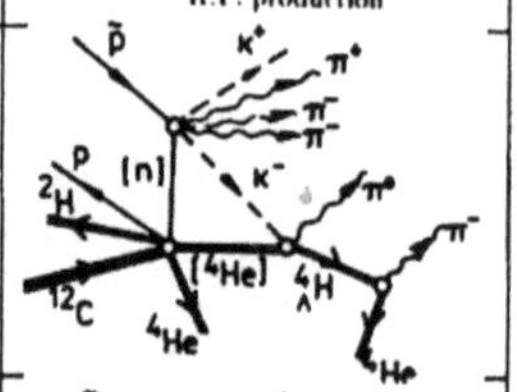
II	$\tilde{p} + {}^{14}N \rightarrow K^+ + \pi^- + \pi^+ + \pi^0 + n +$ $+ p + d + {}^6He + {}^3_\Lambda H$ $\hookrightarrow \pi^- + p + d$	7.2 [MeV]	1166.7 [MeV]	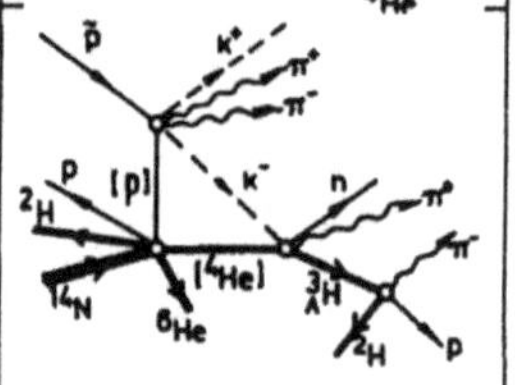
III	$\tilde{p} + {}^{31}S \rightarrow K^+ + \pi^- + \pi^0 + p + d +$ $+ 2t + 2\,{}^3He + 3\alpha + {}^4_\Lambda H$ $\pi^- + p + d + n \hookleftarrow$	0.1 [MeV]	≤ 281.6 [MeV]	

of the process, after the capture of the antiproton by the nucleus and prompt departure of the fast meson component from the nucleus; at the second stage, the high excitation energy results in such a hypernucleus undergoing quasifission, for instance, into two fragments, one of which captures the Λ-particle; at the third stage, the excited fragment-hypernucleus decays into light fragments, among which there is the ${}^4_\Lambda$H. A similar scheme for the production and decay of heavy hypernuclei was proposed and confirmed by the PS-177 CERN collaboration in studies of antiproton annihilation at rest on ^{238}U and ^{209}Bi [2]. It was stressed in ref. [2] that investigation of such events can provide a unique possibility for deriving new information on the dynamics of nuclear fission. Therefore, detailed studies of any possible formation channels are important.

Schematically, the annihilation of a stopping antiproton on a ^{79}Br nucleus, in the case of event III from the table, may be represented as follows:

$$\begin{array}{l} \tilde{p} + {}^{79}\mathrm{Br} \rightarrow K^+ + \pi^- + m\pi^0 + {}^{78}_{\Lambda}\mathrm{Se}^* \\ \qquad\qquad\qquad\qquad\qquad \swarrow \qquad \searrow \\ \qquad\qquad\qquad {}^{24}_{\Lambda}\mathrm{Na}^* \qquad\qquad {}^{54}\mathrm{V}^* \\ \qquad\qquad\quad \swarrow \qquad\qquad\qquad\qquad \searrow \\ 2t + 2\,{}^3\mathrm{He} + 2\,{}^4\mathrm{He} + {}^4_\Lambda\mathrm{H} \qquad p + d + {}^4\mathrm{He} + {}^{47}\mathrm{K}(\mathrm{K}^{47-mn} + mn) \\ \qquad\qquad\qquad \hookrightarrow \pi^- + p + d + n \end{array} \tag{1}$$

(here $m = 0, 1, 2, ...$ depending on the number of π^0-mesons and neutrons).

To make the energy balance for the initial energy and the energy released in reaction (1) be fully in accordance with the scheme presented and with available experimental data, it is sufficient to assume the annihilation of the antiproton to be accompanied by the production of a sole π^0-meson with an energy of about 100 MeV, and, maybe, by several neutrons escaping from the residual ^{47}K nucleus.

To check the possibility of fission processes occurring in antiproton annihilation on heavy nuclei (Ag, Br) in photoemulsion we have performed an additional analysis of 4880 annihilation stars registered in ref. [3]. For further consideration events were selected with more than eight strongly ionizing "h"-prongs, i.e. annihilation stars produced in antiproton capture by heavy nuclei. We have found 481 such events. One of the features that can be considered as indicating a fission process is the forward-backward asymmetry of the outgoing fission fragments. If the fragments produced are in a strongly excited state (which is to be expected in antiproton annihilation), a forward-backward asymmetry of the "h"-tracks should be observed in their decay products. To test this assumption, the opening angles were measured between the two tracks of all possible pairs of "h"-prongs in each chosen event. The procedure for determining the spatial asymmetry of the outgoing fragments of the excited residual nucleus considered is the following: first, the maximum opening angle $h_{\widehat{ij}}$ was determined among the "h"-prong pairs of the given event; then, a plane perpendicular to the prongs forming the angle $h_{\widehat{ij}}$ was made to pass through its bisector, and the opening angle $h_{\widehat{kl}}$ was determined for the pair of "h"-prongs closest to the indicated plane and pointing in the opposite direction; the $\cos\Psi_{\widehat{kl}}$ values for these space angles are presented in a two-dimensional plot (each event is represented by a point in the $\cos\Psi_{\widehat{kl}} = f(\cos\Psi_{\widehat{ij}})$ plane; $\cos\Psi_{\widehat{ij}}$ corresponds to the maximum angle, and $\cos\Psi_{\widehat{kl}}$ corresponds to the opposite angle).

The distribution of the maximum opening angles for the h-particles (h_i and h_j; ($\Psi_{\widehat{ij}}$)) is presented in the upper projection of the Fig. 1, on the right is the projection for the angles $\Psi_{\widehat{kl}}$ of the opposite pair of h-particles; the dotted line shows the results of calculations performed within the framework of the optical-cascade model [2].

From the figure it follows that the experimental distributions significantly contradict computations. The forward-backward asymmetry coefficient of the $\cos\Psi_{\widehat{ij}}$ distribution for the experimentally selected events is $K_E = -(0.76 \pm 0.05)$, while the value obtained from optical-cascade calculations equals $K_T = (0.190 \pm 0.005)$. The respective values for the angular distribution of the opposite pair ($\cos\Psi_{\widehat{kl}}$) are $K_E = (0.60 \pm 0.04)$ and $K_T = (0.025 \pm 0.002)$.

An analysis of the angular distributions (see Fig. 1) reveals that at least $(20 \pm 2)\%$ of the events with the pairs having opening angles $\Psi_{\widehat{kl}}$ exceeding 90° may be considered as quasifission of heavy nuclei in the photoemulsion into two excited fragments due to annihilation of stopping antiprotons in the photoemulsion.

To conclude, there exists a finite probability for annihilation of antiprotons

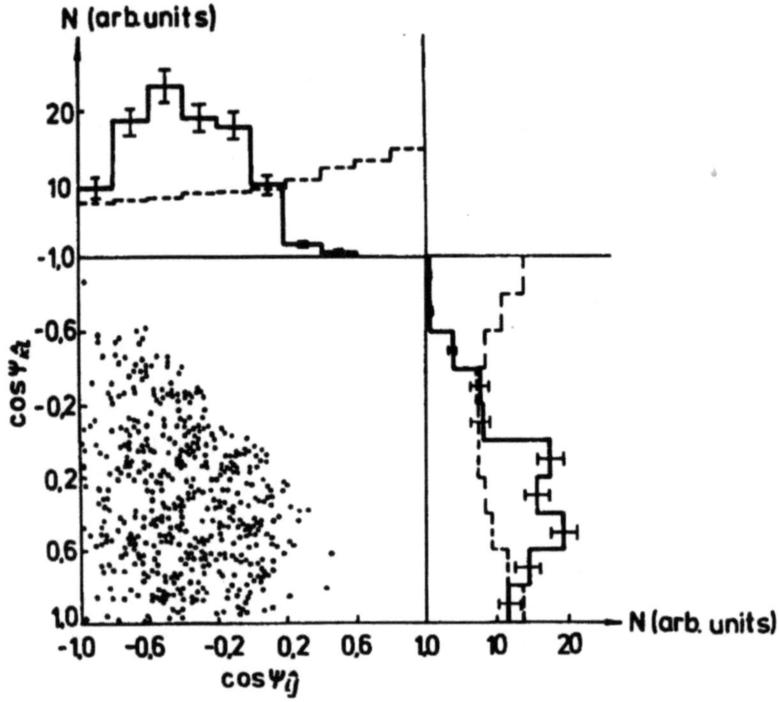

Figure 1. Dependence of the opposite angle $\cos\Psi_{\widehat{k\,l}}$ on the largest opening angle $\cos\Psi_{\widehat{i\,j}}$ of the pair of "h"-prongs for each event of antiproton annihilation on a heavy nucleus of the photoemulsion. The dotted line is the result of calculations by the optical-cascade model [3].

on heavy nuclei (Ag, Br) in photoemulsion, involving the production of hypernuclei leading to the quasifission of nuclei resulting in the production, in the final decay stage of the excited fission fragment, of an outgoing light hyperfragment.

References

1. F. Balestra et al.: Yad. Fiz. **56**, 6 (1993)
2. J.P. Bouquet et al.: Z. Phys. **A342**, 183 (1992)
3. Yu.A. Batusov et al.: JINR Comm. E1-90-486, Dubna 1990
4. A.S. Botvina et al.: Preprint INR II-0369 1984

Few-Body Systems Suppl. 9, 165–168 (1995)

Few-Body Systems

The Hypernuclei with Neutron Halo

L. Majling *

Bogoliubov Laboratory of Theoretical Physics, JINR, 141980 Dubna, Russia and Nuclear Physics Institute, CAS, Řež, Czech Rep.

Abstract. The hypernucleus ${}^{7}_{\Lambda}$He is used as an example for demonstrating possibilities of studying the structure of the neutron halo.

1 Introduction

In the last few years, a new branch of nuclear physics, namely, physics of light nuclei near neutron drip line have been constituted [1] and some interesting phenomena has been discovered.

In addition to the well-known neutron halo [2] it is noteworthy to mention a new β-delayed particle-decay modes [3] indicating substantial differences of the structure of these species.

Recently, we suggest to study Λ-hypernuclei with a large neutron excess and neutron halo [4].

There are many examples of the stabilizing influence of the Λ-hyperon: besides a large number of stable hypernuclei with unstable nuclear core [5] (${}^{6}_{\Lambda}$He (^{5}He), ${}^{7}_{\Lambda}$Be (^{6}Be), ${}^{9}_{\Lambda}$Be (^{8}Be), ${}^{10}_{\Lambda}$B (^{9}B)), also hypernuclear γ-quanta from the particle-unstable nuclear excited state have been registered [6]:

$$
\begin{array}{llll}
{}^{7}\mathrm{Li}\,(K^-,\pi^-\gamma_{2.0})\,{}^{7}_{\Lambda}\mathrm{Li} & (3^+ \to 1^+ & \mathrm{in} & {}^{6}\mathrm{Li}) \\
{}^{9}\mathrm{Be}\,(K^-,\pi^-\gamma_{3.1})\,{}^{9}_{\Lambda}\mathrm{Be} & (2^+ \to 0^+ & \mathrm{in} & {}^{8}\mathrm{Be}).
\end{array}
$$

Now is a proper time to revive interest in such 'exotic' hypernuclei, since in the nearest future a substantial breakthrough is expected in hypernuclear physics when new facilities such as CEBAF [7], and ϕ-factory DAΦNE [8] in Frascatti start to operate and neutral meson spectrometer will be installed at Brookhaven [9]; reactions $\gamma + p \to \Lambda + K^+$ and/or $K^- + p \to \Lambda + \pi^0$ produce hypernuclei with some excess of neutrons.

*This work has been supported in part from Grant Agency of CAS, Grant No. 148401.

2 Hypernucleus $^7_\Lambda$He and the Structure of the Neutron Halo in ^{6}He

Although the most spectacular nucleus of that class is ^{11}Li, some characteristics of the neutron halo, e.g., three particle (core+$n+n$) bound state, are studied in the simplest ^{6}He nucleus [10].

The wave functions of 0^+ and 2^+ ($T=1$) states are extremely simple in the $\alpha+n+n$ model:

$$|2^+1>= \cos\,\theta_2\;|^{31}D_2>+\;\sin\,\theta_2\;|^{33}P_2>,$$

$$|0^+1>= \cos\,\theta_0\;|^{31}S_0>+\;\sin\,\theta_0\;|^{33}P_0>,$$

therefore, the structure of the $^7_\Lambda$He states is also transparent:

$$|\tfrac{5}{2}^+>=|2^+>\times|s_\Lambda>,$$

$$|\tfrac{3}{2}^+>= \cos\,\beta\,|2^+>\times|s_\Lambda>+\sin\,\beta\,|1^+>\times|s_\Lambda>,$$

$$|\tfrac{1}{2}^+>= \cos\,\alpha\,|0^+>\times|s_\Lambda>+\sin\,\alpha\,|1^+>\times|s_\Lambda>.$$

The location of the thresholds and energy levels in ^{6}He nucleus [11] and $^7_\Lambda$He hypernucleus is displayed in Fig.1.

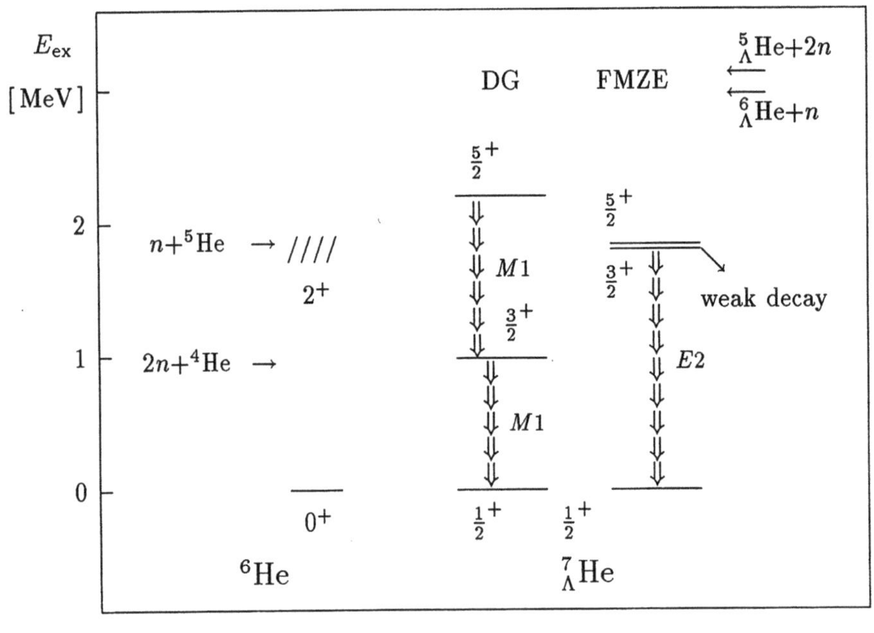

Figure 1. Spectrum of the ^{6}He (experiment) and $^7_\Lambda$He (calculations [12], [13])

The current calculations (see also [14]) locate the doublet of excited states *well below* the neutron threshold (2.92 MeV). Obviously, the electromagnetic decay rates for the excited states

$$\frac{5}{2}^+ \xrightarrow{M1} \frac{3}{2}^+ \xrightarrow{M1} \frac{1}{2}^+ \qquad \text{and/or} \qquad \frac{5}{2}^+ \xrightarrow{E2} \frac{1}{2}^+$$

depend not only on the ΛN interaction (sin α, sin β), but also on the ^{6}He nucleus structure (sin θ_J, location of the 1^+ state). The exhaustive analysis of the problem has been done by Dalitz and Gal almost 20 years ago [12] with conclusion that *these excited states are on the edge of being isomeric.* Both $\Gamma(M1)$ are very sensitive to the weights of the $|^{33}P_J>$ components and an $E2$ transition is possible only due to the distortion of the α-particle core.

Recently, the $^5D_0$-state of α-particle has been discovered in various experiments with polarized deuterons [15]. The most striking example is $^2\mathrm{H}(d,\gamma)^4\mathrm{He}$ process. Owing to the D-state in α-particle, the $E2$ transition $^5S_2 \rightarrow {}^5D_0$ occurs as seen for the astrophysical factor.

The $^5D_0$ component of α-particle is relevant mainly to the four-nucleon configuration with the Young partition $[f] = [22]$. In this case, a very loosely bound six-nucleon configuration with $[f] = [222]$ appears even in the ground state of ^{6}He nucleus. We mention here the failure in description of the quadrupole moment of ^{6}Li nucleus in the $\alpha + N + N$ approach [16] using very sophisticated interactions and model (antisymmetrized version of the multicluster dynamic model with Pauli projection) but *with structureless α-particle.*

A number of basis states in fully microscopic ($A = 6$) Translationally Invariant Shell Model (TISM) growths very quickly; even in the $(0{+}2)\hbar\omega$ space there are 31 states with $J^\pi T = 0^+1$ (and 73 states with 2^+1). Table 1 demonstrates the distribution of two protons and four neutrons on different shells for some TISM configurations. The number of 'active' protons is really large in the $s^2p^4[222]$ configuration, which may be crucial for $E2$ transition.

Table 1. Distribution of protons (N_p^l) and neutrons (N_n^l) for various TISM configurations

$[f](\lambda,\mu)$	s^4p^2 [42](20) N_p^l	N_n^l	$s^3p^2d^1$ [33](02) N_p^l	N_n^l	s^2p^4 [222](02) N_p^l	N_n^l
$2d-2s$			0.48	0.96		
$1p$	0.0	2.4	0.64	1.28	1.6	3.2
$0s$	2.0	1.6	0.88	1.76	0.4	0.8

So, the detection of γ-quanta in the reaction

$$^7\mathrm{Li}\,(\gamma, K^+\gamma_{1.8})\,{}^7_\Lambda\mathrm{He}\ \ (\text{ref. [7]})\quad \text{and/or}\quad {}^7\mathrm{Li}\,(K^-, \pi^0\gamma_{1.8})\,{}^7_\Lambda\mathrm{He}\ \ (\text{ref. [9]})$$

will be of great value and reveal tiny details of the neutron halo structure.

Another interesting feature of hypernuclei with the neutron halo is demonstrated in ref. [17].

3 Conclusion

There is an interesting program of the experiment and interpretation to be pursued in the next several years.

Acknowledgement. It is a pleasure to thank R. Eramzhyan, R.H. Dalitz, V. Fetisov, S. Frullani, D. Lanskoy, T. Motoba, Y. Penionzhkevich, M. Sotona and T. Tretyakova for illuminating discussions.

References

1. A.A. Ogloblin, Y.E. Penionzhkevich: In: *Treatise on Heavy Ion Science* (D.A.Bromley ed.) vol. **8**, p.261. Plenum Press 1989
2. P.G. Hansen: Nucl. Phys. **A553**, 89c (1993)
3. A.C. Mueller, B.M. Sherrill: Ann. Rev. Nucl. Part. Sci. **43**, 529 (1993)
4. L. Majling: Nucl. Phys. **A585**, 211c (1995)
5. R. Dalitz: In: *Interaction of High-Energy Particles with Nuclei* (Proc. Int. School 'Enrico Fermi', Course 38) p.89. New York, London: Ac. Press 1967
6. M. May et al.: Phys Rev. Lett. **51**, 2085 (1983)
7. R.A. Schumacher: Nucl. Phys. **A585**, 63c (1995); and these Proceedings; S. Frullani: these Proceedings
8. M. Agnello et al.: FINUDA, a Detector for Nuclear Physics at DAΦNE, INFN Report, LNF-93/021; Nucl. Phys. **A585**, 271c (1995)
9. H.A. Thiessen, Jen-Chieh Peng: AGS Proposal E-907
10. M.V. Zhukov et al.: Phys. Rep. **231**, 151 (1993)
11. F. Ajzenberg-Selove: Nucl. Phys. **A490**, 1 (1988)
12. R.H. Dalitz, A. Gal: Nucl. Phys. **B1**, 1 (1967); J. Phys. **G4**, 889 (1978)
13. V.N. Fetisov et al.: Z. Phys. **A339**, 399 (1991)
14. O. Richter, M. Sotona, J. Žofka: Phys. Rev. **C43**, 2753 (1991); D.J. Millener et al.: Phys. Rev. **C31**, 499 (1985)
15. H.R. Weller, D.R. Lehman: Ann. Rev. Nucl. Part. Sci. **38**, 563 (1988)
16. G.G. Ryzhikh et al.: Nucl. Phys. **A563**, 247 (1993); V.I. Kukulin: these Proceedings
17. T. Tretyakova: these Proceedings

Few-Body Systems Suppl. 9, 169–176 (1995)

Few-Body Systems

Pion-Nucleon Partial-Wave Analysis and the $NN\pi$ Coupling Constant

Rob G. E. Timmermans *

Kernfysisch Versneller Instituut, Zernikelaan 25, 9747 AA Groningen, the Netherlands

Abstract. After some general remarks about partial-wave analysis, the method and first conclusions of a new analysis of πN scattering data across the $\Delta(1232)$ are summarized. In particular, a mini-review of the state-of-the-art with respect to the pion-nucleon coupling constant is given.

1 Introduction

In recent years, the pion-nucleon system has been investigated extensively at meson factories like LAMPF, TRIUMF, and PSI, resulting in a wealth of new high-quality scattering data. At CEBAF, a large program is planned to study the nucleon-resonance spectrum.

Energy-dependent partial-wave analyses (PWA) of πN scattering data have been performed by only a few groups. The Karlsruhe-Helsinki group of Höhler and collaborators [1-3] employed strong dispersion-relation constraints from "Mandelstam analyticity" [4]. The famous "KH80 solution" dates from 1980, and was partially updated in 1985. The Carnegie-Mellon–LBL group of Cutkosky and collaborators [5, 6] also implemented constraints from Mandelstam analyticity and additional input from Regge-pole theory. This work also dates from 1980. The VPI&SU group of Arndt and collaborators [7, 8] in their latest analysis from 1994 use constraints from forward and fixed-t dispersion relations.

Especially the Karlsruhe-Helsinki PWA has had acclaim because of the sophisticated methods used. For instance, in a recent experimental paper, high-quality new data "are compared exclusively with the predictions of the Karlsruhe-Helsinki phase-shift analysis of 1979/80, because this analysis respects Mandelstam analyticity and unitarity" [9].

* *E-mail address:* timmermans@KVI.nl

Here we report on some first results from a new energy-dependent PWA of modern πN scattering data, where we focus on the method and the result extracted for the $NN\pi$ coupling constant. This PWA should be viewed as an *alternative* to the above PWA's in that it uses not the language of dispersion relations, but that of field theory. In view of the history of dispersion-relation techniques in the πN field, dating back to the 1960's when this formalism was especially fashionable, it should perhaps be said that the difference is indeed mainly one of language and not of physics, and that the limited number of proofs of dispersion relations all start from field theory. As witnessed by the chiral-perturbation-theory industry, field theory has made a come-back in the πN field.

The main goals of the present PWA are: (*i*) to build a modern consistent πN database; (*ii*) to test the predictions from chiral perturbation theory [10-14] for the threshold region; and (*iii*) at a later stage, to study the resonance spectrum. At present, a solution exists for both π^+p and π^-p, π^0n scattering from threshold across the $\Delta(1232)$ up to 725 MeV/c pion lab momentum (600 MeV kinetic energy), i.e. essentially the region of the new meson-factory data.

2 Pion-nucleon partial-wave analysis

The aim of an energy-dependent partial-wave analysis is to determine in a model-independent manner the phase-shift parameters as a function of energy, the accuracy being limited in principle only by the quality of the available database. PWA's are sometimes by non-experts [15] confused with amplitude analyses. A πN *amplitude analysis* at a certain momentum would require three independent experimental quantities (e.g. differential cross section, polarization, and one of the spin-rotation parameters) for every scattering angle and each isospin. These, of course, are not available in general. A *single-energy* PWA, on the other hand, can suffer from multiple solutions ("phase-shift ambiguities," e.g. the Fermi-Yang and Minami ambiguities in the early days of πN PWA's), is generally less stable to the addition of new data, and is also more sensitive to the fitting of noise. All these problems disappear in *energy-dependent* PWA's, especially when (model-independent) theoretical constraints are applied: *the best phase shifts are those obtained from a good energy-dependent PWA*.

The present πN PWA uses essentially the same methods as those used by the Nijmegen group in PWA's[1] of the low-energy pp and np data [16-18]. The same strategy has also led to the first PWA of the low-energy $\bar{p}p$, $\bar{n}n$ data [19]. The hallmarks of this method are: (*i*) good analyticity properties of the amplitudes; (*ii*) a careful treatment of charge-dependence effects, such as the electromagnetic interaction; and (*iii*) a consistent enforcement of the rules of statistics.

It is important that the amplitudes from a PWA satisfy general analyticity requirements. The present PWA implements a strategy that goes back to

[1]See on World Wide Web: `http://thef-nym.sci.kun.nl/NN/NN.html`

a 1944 paper by Landau [20] and that is similar to, but much more powerful than the so-called modified effective-range expansion. One could call this the implementation of *Cauchy-Riemann analyticity.* It starts from the simple observation that the low-energy region is the long-wavelength limit: long-range interactions give rise to rapid variations with energy of the amplitudes, while short-range interactions lead to smooth and slow variations. If one looks at the singularity structure of the S matrix, then the long-range forces correspond to near-by singularities, and the short-range forces to far-away singularities.

Once we take into account the relevant long-range forces exactly, the remainder can be parametrized smoothly and analytically. This is done by including the long-range forces as potentials in the relativistic Schrödinger equation which is solved with an energy-dependent boundary condition at $r = b = 1.2$ fm. This is essentially the Feshbach-Lomon approach [21]. Outside $r = b$, the spin-0–spin-1/2 "improved" Coulomb potential [22] supplemented with the vacuum-polarization potential [23] is used as electromagnetic interaction. The strong interaction for $r > b$ consists of the tail of $\rho(770)$, $\varepsilon(760)$, and pomeron exchange. The widths of $\rho(770)$ and $\varepsilon(760)$ are included, leading to a potential of two-pion-exchange range [24].

The boundary condition, the P matrix [25, 26], is parametrized smoothly as a function of energy and represents all short-range interactions. Right-hand cuts due to the coupling to inelastic channels can be treated with a complex boundary condition for the relevant partial waves. The πN resonances, such as the $\Delta(1232)$ are represented by *poles* in the P matrix [25]. The Born terms from nucleon exchange in the s- and u-channel are included.

Special care is taken to account properly for electromagnetic effects in order to arrive at hadronic phase shifts. This formalism [22] has been succesfully developed and used in the pp, np, and $\bar{p}p$ PWA's.

Statistics is an essential ingredient in an energy-dependent PWA. Data sets with too high or too low $\chi^2_{\rm min}$ are removed, using generalizations of the 3 s.d. criterion explained in detail in ref. [16]. It is *not* judged if experiments are right or wrong, only if data sets are statistically acceptable, yes or no. This procedure is well accepted in the NN field, but has not been applied consistently in πN PWA's. A consistent database is important for a reliable estimate of the errors on important quantities such as the πN σ–term. Well-documented systematic errors can be treated as well. A simple example is an overall normalization error on a group of observables.

The fit to the modern low-energy database leads to a $\chi^2_{\rm min}/{\rm df} = 1.18$, where all available scattering data from 1980 on are included, as well as some important data sets from the "old days." Special attention has been given to the inconsistencies in the $\pi^+ p$ scattering data. The solution to this problem is actually straightforward. The 1976 Bertin data below $T_{\rm L} = 100$ MeV, the cornerstone of the Karlsruhe-Helsinki PWA at low energies, have to be "floated": the normalizations predicted by the PWA vary from 0.77 at 20.8 MeV to 0.94 at 95.9 MeV. The Bertin data at 67.4 MeV have to be rejected. Also the 1983 Ritchie data are floated. The normalization of the 1983 Frank data varies between 0.94 and 1.12, within the quoted experimental errors. It is important that one does *not* include total cross sections, because of ambiguities in the definition in the presence of Coulomb forces; only ***partial total cross sections*** should be included. After this limited data pruning, one gets for the π^+p database $\chi^2_{\rm min}/{\rm df} = 1.11$, where expected is $\langle\chi^2_{\rm min}/{\rm df}\rangle = 1.00 \pm 0.06$.

3 The pion-nucleon coupling constant

An important quantity in nuclear and hadronic physics is the pion-nucleon coupling constant, defined here by the pseudovector interaction

$$L_{\rm int} = \frac{f_{NN\pi}}{m_\pi}\left(i\overline{\psi}\gamma_\mu\gamma_5\boldsymbol{\tau}\psi\right)\partial^\mu\boldsymbol{\pi} \ . \tag{1}$$

The scaling mass m_π is conventionally introduced to make the coupling constant $f_{NN\pi}$ dimensionless. Numerically it is equal to the charged-pion mass. We will distinguish two different coupling constants: the coupling of the neutral pion to protons $f_{pp\pi^0}$ and the charged-pion coupling to nucleons defined by $2f_{\rm c}^2 = f_{pn\pi^+} f_{np\pi^-}$. In case of isospin symmetry (charge independence) one has $f_{pp\pi^0}^2 = f_{\rm c}^2 \equiv f_{NN\pi}^2$.

In recent years, the precise value of $f_{NN\pi}$ has been the subject of many studies. Prior to 1987, the accepted value was

$$f_{\rm c}^2/4\pi = 0.079(1) \ , \tag{2}$$

from the 1980 Karlsruhe-Helsinki PWA. Determinations of the coupling constant from energy-dependent PWA's of the pp, np, and $\overline{p}p \to \overline{p}p$, $\overline{n}n$ data by the Nijmegen group consistently gave a value significantly lower [27]. The most recent numbers are:

$$f_{pp\pi^0}^2/4\pi = 0.0745(6) \ , \tag{3}$$

from the pp PWA below 350 MeV [16-18];

$$f_{\rm c}^2/4\pi = 0.0748(3) \ , \tag{4}$$

from the np PWA below 350 MeV [18]; and

$$f_{\rm c}^2/4\pi = 0.0732(11) \ , \tag{5}$$

from the $\overline{p}p$ PWA below 925 MeV/c [19]. All values are *at the pion pole* and all the quoted errors are statistical. The numbers for $f_{pp\pi^0}$ and $f_{\rm c}$ are consistent with charge independence. In the case of the pp PWA, many systematic checks could be made and the systematic error on $f_{pp\pi^0}$ was shown to be small [27]. For instance, a form factor was explicitly added to the one-pion-exchange potential to study the influence of the cut-off mass on $f_{NN\pi}$. The results clearly showed that the coupling was determined at the pole and not at $t = 0$. Also, the neutral-pion mass was determined in nice agreement with the experimental value. In view of all these studies, and because the low-energy pp database is of such high quality, the value $f_{pp\pi^0}^2/4\pi = 0.0745(6)$ is probably the most reliable number for the pion-nucleon coupling constant available. It corresponds to the pseudoscalar coupling constant

$$g^2_{NN\pi}/4\pi = 13.47(11)\ , \tag{6}$$

while the 1980 Karlsruhe-Helsinki value translates to

$$g^2_{NN\pi}/4\pi = 14.28(18)\ . \tag{7}$$

In addition to these Nijmegen numbers, Arndt and collaborators at VPI&SU have obtained "low" values for f_c from energy-dependent PWA's of low-energy πN scattering data [7, 8]. The most recent VPI&SU value is [8]

$$f^2_c/4\pi = 0.076(1)\ . \tag{8}$$

It is straightforward to extract the pion-nucleon coupling constant from the present πN PWA across the $\Delta(1232)$. The result is

$$f^2_c/4\pi = 0.0741(8)\ , \tag{9}$$

where the error is statistical only. No attempts have been made yet to study systematic errors, so this number is preliminary.

Despite the fact that all these modern determinations agree well, a "low" pion-nucleon coupling constant was considered controversial, no doubt partly because of the canonical status of the Karlsruhe-Helsinki analysis. So it is perhaps useful to point out the following.

(*i*) The meson-factory pion-nucleon scattering data were not yet available in 1980, when Koch and Pietarinen performed their PWA. The 1980 solution does a poor job on these new high-quality data, and the Karlsruhe-Helsinki PWA has not been updated with a comprehensive fit to the modern database.

(*ii*) The Karlsruhe-Helsinki method uses dispersion-relation constraints which require a value for the pion-nucleon coupling constant *as input*; therefore, the extraction of f_c is little more than a consistency check, and *not* really an unbiased determination.

The present πN PWA once more, *ad nauseam* perhaps, provides evidence for a "low" pion-nucleon coupling constant. We rest our case.

4 The Goldberger-Treiman discrepancy

The pion is the Goldstone boson of the (approximate) chiral symmetry of the strong interactions, spontaneously broken down in the vacuum to ordinary isospin symmetry. When the up- and down-quark masses go to zero (the "chiral limit") the pion becomes massless. As a result of this chiral symmetry, the pion couples derivatively to the nucleon (for massless pions, the coupling vanishes as the momentum goes to zero), so pseudovector coupling is preferred, since it gives *naturally* small scattering lengths. The magnitude of the coupling is fixed in terms of axial-vector weak-decay constants. This, according to Nambu [28], is the meaning of the Goldberger-Treiman relation [29], which reads

$$\frac{f_{NN\pi}}{m_\pi} = \frac{g_{\rm A}}{2f_\pi} , \tag{10}$$

where $g_{\rm A} = 1.2573(28)$ is minus the Gamow-Teller coupling in neutron β-decay [30] and $f_\pi = 92.4(2)$ MeV is the pion decay constant [31]. In the chiral limit this relation, a low-energy theorem, is *exact*. The discrepancy $\Delta_{\pi N}$ in the real world, defined by

$$\frac{f_{NN\pi}}{m_\pi}(1-\Delta_{\pi N}) = \frac{g_{\rm A}}{2f_\pi} , \tag{11}$$

is therefore a measure of explicit chiral-symmetry breaking due to nonzero quark masses. Like the πN σ−term, it is linear in the quark masses, so proportional to m_π^2.

If we use the recommended value [27] for the pion-nucleon coupling constant $f_{NN\pi}^2/4\pi = 0.0745(6)$ at the pion pole, we find

$$\Delta_{\pi N} = 1.9(5)\% , \tag{12}$$

whereas the old Karlsruhe-Helsinki value $f_{NN\pi}^2/4\pi = 0.079(1)$ would give $\Delta_{\pi N} = 4.7\%$. Chiral perturbation theory does not predict the value of $\Delta_{\pi N}$, except that it is of order $(m_u + m_d)/2\Lambda$, where Λ is a typical hadronic scale. Depending on what one takes for Λ, e.g. the nucleon or $\rho(770)$ mass, or the chiral-symmetry breaking scale $\Lambda_{\chi\rm SB} \approx 1$ GeV, or the QCD scale $\Lambda_{\rm QCD} \approx 250$ MeV, one gets a number of about 1 to 2% for $\Delta_{\pi N}$, using standard quark masses $m_u \approx 4$ MeV, $m_d \approx 7$ MeV. Other considerations also prefer a small $\Delta_{\pi N}$, less than about 2% [32]. For instance, if one naively blames the discrepancy on a simple form factor interpolating between the pion pole and $t = 0$, viz.

$$\frac{f_{NN\pi}}{m_\pi}\exp(-m_\pi^2/2\Lambda^2) = \frac{g_{\rm A}}{2f_\pi} , \tag{13}$$

one gets $\Lambda \approx 800$ MeV for $\Delta_{\pi N} \approx 2\%$, but an unrealistic $\Lambda \approx 450$ MeV for $\Delta_{\pi N} \approx 5\%$.

Acknowledgement. Part of this work was included in the research program of the Stichting voor Fundamenteel Onderzoek der Materie (FOM) with financial support from the Nederlandse Organisatie voor Wetenschappelijk Onderzoek (NWO).

References

1. R. Koch and E. Pietarinen: Nucl. Phys. **A336**, 331 (1980)
2. G. Höhler: In: *Pion-Nucleon Scattering*, edited by H. Schopper, Landolt-Börnstein (New Series, Group 1, Vol. 9b, Pt. 2). Berlin: Springer-Verlag 1983; Nucl. Phys. **A508**, 525c (1990)
3. R. Koch: Z. Phys. C **29**, 597 (1985); Nucl. Phys. **A448**, 707 (1986)
4. S. Mandelstam: Phys. Rev. **112**, 1344 (1958); Rep. Prog. Phys. **25**, 99 (1962)
5. R.L. Kelly and R.E. Cutkosky: Phys. Rev. **D20**, 2782 (1979); R.E. Cutkosky, R.E. Hendrick, J.W. Alcock, Y.A. Chao, R.G. Lipes, and J.C. Sandusky: *ibid.* **D20**, 2804 (1979); R.E. Cutkosky, C.P. Forsyth, R.E. Hendrick, and R.L. Kelly: *ibid.* **D20**, 2839 (1979)
6. R.E. Cutkosky: In: *Baryon 1980*, edited by N. Isgur (Proceedings of the IVth International Conference on Baryon Resonances), p. 19. Toronto: University of Toronto 1980
7. R.A. Arndt, Z. Li, L.D. Roper, and R.L. Workman: Phys. Rev. Lett. **65**, 157 (1990); R.A. Arndt, Z. Li, L.D. Roper, and R.L. Workman: Phys. Rev. **D43**, 2131 (1991)
8. R.A. Arndt, R.L. Workman, and M.M. Pavan: Phys. Rev. **C49**, 2729 (1994)
9. Ch. Joram et al.: Phys. Rev. **C51**, 2144 (1995)
10. S. Weinberg: Phys. Rev. Lett. **17**, 616 (1966); Phys. Rev. **166**, 1568 (1968)
11. Y. Tomozawa: Il Nuovo Cimento **46A**, 707 (1966)
12. J. Gasser, M.E. Sainio, A. Svarc: Nucl. Phys. **B307**, 779 (1988); J. Gasser: Nucl. Phys. **B279**, 65 (1987)
13. J. Gasser, H. Leutwyler, and M.E. Sainio: Phys. Lett. **B253**, 252 (1991); *ibid.* **B253**, 260 (1991)
14. V. Bernard, N. Kaiser, and U.-G. Meissner: Phys. Lett. **B309**, 421 (1993)
15. J.-M. Richard, Phys. Rev. **C52**, 1143 (1995)
16. J.R. Bergervoet, P.C. van Campen, W.A. van der Sanden, and J.J. de Swart: Phys. Rev. **C38**, 15 (1988)
17. J.R. Bergervoet, P.C. van Campen, R.A.M. Klomp, J.-L. de Kok, T.A. Rijken, V.G.J. Stoks, and J.J. de Swart: Phys. Rev. **C41**, 1435 (1990)
18. V.G.J. Stoks, R.A.M. Klomp, M.C.M. Rentmeester, and J.J. de Swart: Phys. Rev. **C48**, 792 (1993)

19. R. Timmermans, Th.A. Rijken, and J.J. de Swart: Phys. Rev. **C50**, 48 (1994); *ibid.* **C52**, 1145 (1995)

20. L. Landau and J. Smorodinsky: J. Phys. USSR **8**, 154 (1944)

21. H. Feshbach and E.L. Lomon: Ann. Phys. (NY) **29**, 19 (1964)

22. G.J.M. Austen and J.J. de Swart: Phys. Rev. Lett. **50**, 2039 (1983)

23. L. Durand III: Phys. Rev. **108**, 1597 (1957)

24. J.J. de Swart and M.M. Nagels: Fortschr. Phys. **26**, 215 (1978)

25. R.L. Jaffe and F.E. Low: Phys. Rev. **D19**, 2105 (1979); R.L. Jaffe: In: *Asymptotic Realms of Physics*, edited by A.H. Guth, K. Huang, and R.L. Jaffe, p. 100. The MIT Press 1983

26. B.L.G. Bakker and P.J. Mulders: Adv. Nucl. Phys. **17**, 1 (1986)

27. V. Stoks, R. Timmermans, and J.J. de Swart: Phys. Rev. **C47**, 512 (1993)

28. Y. Nambu: Phys. Rev. Lett. **4**, 380 (1960)

29. M.L. Goldberger and S.B. Treiman: Phys. Rev. **110**, 1178 (1958)

30. D. Dubbers, W. Mampe, and J. Döhner: Europhys. Lett. **11**, 195 (1990); Particle Data Group, Phys. Rev. **D45**, June 1994

31. B.R. Holstein: Phys. Lett. **B244**, 83 (1990)

32. C.A. Dominguez: Rivista del Nuovo Cimento **8**, 1 (1985)

Few-Body Systems Suppl. 9, 177–186 (1995)

Few-Body Systems

The πNN Dynamics on $NN - NN, pp - \pi d, \pi d - \pi d$, and $pp - \pi NN$ below 1 GeV

T. Ueda, Y. Ikegami, K. Tada, K. Kameyama

Faculty of Science, Ehime University, Matsuyama, Ehime, Japan, 790

Abstract.

$NN - NN, pp - \pi d, \pi d - \pi d$, and $pp - \pi NN$ processes at incident proton laboratory energies $T_L \leq 1$ GeV are studied in a unified and unitary framework of the πNN dynamics. The three-body calculation is made with the πN interactions in the P_{11}, P_{33}, S_{11}, and S_{31} states and the NN interactions in the ${}^3S_1 - {}^3D_1, {}^1S_0$, and ${}^3P_2$ states . Additionally, (A) the backward going pion contribution at $\pi N \Delta$ vertex, (B) the $\pi N - \rho N$ coupling in the P_{33} state, (C) the heavy meson exchange in $NN \rightarrow NN$ driving term, and (D) the effect of the off-shell structure in $\pi N - \rho N\, P_{33}$ interaction are taken into account.

1 Introduction

Modern calculations of the $NN \rightarrow NN$, $pp \rightarrow \pi d$, $\pi d \rightarrow \pi d$, and $pp \rightarrow \pi NN$ processes below 1 GeV take into account fully the higher-order contributions of nucleon and Δ degrees [1, 2]. Among all, advanced ones use $NN \rightarrow \pi NN$ equation which is based on Alt-Grassberger-Sandhas (AGS) type three-body formalism on the πNN system [1 – 4] and extended to couple with the two nucleon states. They take into account the three-body channels of both πN interacting pair with N spectator and NN interacting pair with π spectator and also the two nucleon channels in a unified and unitary framework.

An interesting subject is whether the system has any resonance or quasi-bound state. The experimental evidences have been reported by Argonne and other groups [5]. We ask whether they need explicitly subhadronic degree to be explained. So far we have studied this problem in the πNN dynamics [6 – 10].

In 1982, we investigated the analytic structure on the complex energy plane for the NN partial wave amplitudes in the three-body calculation [7]. We find a pole at the mass 2116 – i61 MeV for $I = 1, J^P = 2^+$ amplitude and a pole at the mass 2155 – i60 MeV for $I = 1, J^P = 3^-$ amplitude. The energy-dependent structures appearing in these partial wave amplitudes are generated by the pole structures [7]. Namely the genuine resonances are there.

Since 1984, we have used basically the $NN - \pi NN$ equation. However we have modified it by introducing additional elements [8, 9]. Without the modification, the standard $NN \to \pi NN$ equation is inadequate to describe the strong energy-dependent structures in the $I = 1$ $J^P = 2^+$ and 3^- amplitudes.

Our approach involves the merits in the input elements which are additional to the standard $NN - \pi NN$ equation. Its effects on the πNN system have been discussed in previous references as follows: (A) The contribution of the pion which goes backward in time at $\pi N \Delta$ vertex [9], (B) the $\pi N \to \rho N$ channel coupling in the Δ state [8], (C) the heavy-meson-exchange contributions to $NN \to NN$ driving term [8], and (D) a good choice of the off-shell structure of the $\pi N - P_{33}$ interaction [9]. Elements (A), (B), and (D) are proved essentially important to describe the anomalous energy dependence of the amplitudes indicating three-body resonances in $pp - {}^1D_2$ and $pp - {}^3F_3$ channels [1, 8]. Without element (A), the effect of the πN P_{33} interaction is too weak to reproduce the strong structure in the ${}^1D_2$ and ${}^3F_3$ amplitudes. A good choice of the off-shell structure of the $\pi N - P_{33}$ interaction makes the enhancement of the effect of (A). Element (B) suppresses the effect of (A) in the structure in the $I = 1$ $J^P = 2^+$ amplitude. Otherwise the structure is generated too strongly. Therefore, without (A), (B), and (D) one fails to reproduce the structures in both the $I = 1$ $J^P = 2^+$ and 3^- amplitudes. Element (C) is of course very important in all the NN amplitudes. Without (C), the general trend of the NN amplitudes which are consistent with phase shift analysis result is lost.

In the present paper, we make an improved version of this approach to $pp \to pp$, $pp \to \pi d$, $\pi d \to \pi d$, and $pp \to pn\pi^+$ processes by involving the new input elements: The nucleon non-pole part of the πN P_{11} interaction which is primarily given by the contribution of the Roper resonance, the πN S_{11} and S_{31} interactions; the NN ${}^1S_0$ interaction due to OBEP. These two-body interactions have been provided in one of our recent works [11] based on new πN phase shifts given by Arndt et al. [11].

Another subject concerns the πNN system near the π threshold, where the S wave configurations between N and N as well as π and N make a major contribution. There, the Δ contribution is not appreciable. The NN ${}^3P_0$ is a good place to detect the contributions from the S wave configurations, since near the π threshold the S wave configurations dominate those amplitudes. We study the NN ${}^1S_0$ contribution to the πNN system in this paper.

2 Three-body equation

The $NN - \pi NN$ equation is constructed by extending the AGS equation according to Blankleider and Afnan [4]. This is written for anti-symmetrized amplitudes $X_{\alpha\beta}$ as follows.

$$X_{\alpha\beta} = Z_{\alpha\beta} + \sum_{\gamma} Z_{\alpha\gamma} \tau_\gamma X_{\gamma\beta}, \tag{1}$$

where $Z_{\alpha\beta}$ are the particle rearrangement terms and τ_γ are the propagator terms. The additional elements (A), ..., (D) are embedded in the $Z_{\alpha\beta}$ and τ_γ.

In Eq.(1) we consider the two nucleon channel N and the three particle channels denoted by the interacting pair Δ and d. Δ means the πN interacting pair other than the nucleon pole state with remaining nucleon as spectator. Namely it involves not only actual Δ, but also other πN eigenstates such as the non-pole part of the P_{11} state, the S_{11} state, and so on. d means the NN interacting pair with remaining pion as spectator.

Involving the additional elements (A), ... , (D), this paper takes into account the two-body interactions in the following pair-particle states: The P_{11}, P_{33}, S_{11}, and S_{31} states for the πN pair and the ${}^3S_1 - {}^3D_1$, ${}^1S_0$, and ${}^3P_2$ states for the NN pair.

The $NN \leftarrow NN$ driving term consists of the sum of the π rearrangement contribution $Z_{NN}(\pi)$ and the one-boson exchange contributions due to ρ, ω, σ, S^*, δ, η, and Π, respectively:

$$Z_{NN} = Z_{NN}(\pi) + B(\rho) + B(\omega) + B(\sigma) + B(S^*) + B(\delta) + B(\eta) + B(\Pi), \quad (2)$$

where $Z_{NN}(\pi)$ is given by

$$Z_{NN}(\pi) = g^N_{\pi N}(\boldsymbol{p}_\alpha) G_{\pi NN}(W) g^N_{\pi N}(\boldsymbol{p}_\beta). \quad (3)$$

G_{ABC} is the free Green function of the ABC system with the energy W and $g^A_{iB}(\boldsymbol{p})$ represents the vertex form-factor for particle A decaying into particles i and B with relative momentum $\boldsymbol{p}$. $B(i)$ in Eq.(2) is given by the field theoretical calculation of the one-i-exchange diagram whose details are given in refs. [1, 12, 13]. The Π is the three-pion correlated state with the same quantum number with the pion [13].

The $NN \leftarrow \Delta N$ and $N\Delta \leftarrow \Delta N$ driving terms are put as follows:

$$Z_{N\Delta} = Z_{N\Delta}(\pi) + Z_{N\Delta}(\rho) + Z'_{N\Delta}(\pi), \quad (4)$$

$$Z_{\Delta\Delta} = Z_{\Delta\Delta}(\pi) + Z_{\Delta\Delta}(\rho) + Z'_{\Delta\Delta}(\pi), \quad (5)$$

where $Z_{N\Delta}(i)$ and $Z_{\Delta\Delta}(i)$ are similarly defined with Eq.(3); the $Z'_{N\Delta}(\pi)$ and $Z'_{\Delta\Delta}(\pi)$ represent the contributions from the pion going backward in time to Δ. The $\pi d \leftarrow NN$ driving term Z_{dN} is also similarly defined with Eq.(3).

3 Two-body interactions

For the $\pi N - P_{11}$ interaction let us consider the separable potential of a rank two form for three channels : πN, $\pi\Delta$, and σN for the $\pi N - P_{11}$ interaction. We impose the following condition on the πN amplitude: The form factor at the nucleon pole is the πNN vertex form factor in the one pion exchange potential derived from the nucleon-nucleon analysis. The amplitude generates the nucleon pole with the correct pole residue. By fitting the πN P_{11} phase shift [11] up to incident pion laboratory energy $T^\pi_{\mathrm{L}} = 500$ MeV, we determine the form factor. We use the value for the form-factor parameter $\Lambda_\pi = 898.50$ MeV which is reasonable in recent argument [13]. We divide the amplitude into the pole and non-pole terms which are assigned to the N and Δ channels respectively.

For the $\pi N - \rho N$ P_{33} interaction, let us assume the πN and ρN two channel potential in a separable form. This makes the πN and ρN coupling of the additional element (B). The interaction which is described in reference [14] is used for this purpose. Furthermore, the backward going pion contribution of the additional element (A) is involved for the $NN - N\Delta$ and $N\Delta - \Delta N$ driving terms as in ref.[14].

For the πN S_{11} and S_{31} interactions let us consider a rank one and one channel form of the separable potential which works enough well for the incident pion laboratory energies $T_{\rm L}^{\pi} \leq 400$ MeV. At these energies, the inelasticity is negligible.

For ${}^3S_1 - {}^3D_1$ interaction, we use the deuteron wave function due to Ueda-Riewe-Green (URG II) [15]. The separable potential is derived from the deuteron wave function in a rank one form [10].

The phenomenological ${}^1S_0$ and ${}^3P_2$ interactions are used as in ref.[8]. In the case of Fig.4, however, we use the ${}^1S_0$ interaction due to OBEP.

The short range part of the NN interaction for a large momentum transfer $\Delta q \gtrsim 1.1$ GeV/c is phenomenologically modified according to ref.[10].

4 Numerical results and summary

The $pp - pp$ theoretical results for the phase parameters fit qualitatively the phase shift analysis results by Arndt et al. and Nagata et al. [16], except for the ${}^3P_1$ and ${}^3P_0$ phase parameters at $T_{\rm L} \gtrsim 700$ MeV. The deviations at these parameters are possibly removed by the fine tuning of the OBE parameters. (See Fig.1.)

The energy-dependent structures in the $pp-{}^1D_2$, ${}^3F_3$, and ${}^3P_2$ phase parameters at 400-1000 MeV are reproduced. The amplitudes make looping behavior in qualitative consistence with the results by Arndt et al. and Nagata et al. [16]. Furthermore, the $pp - \pi d$ and $\pi d - \pi d$ amplitudes with $J^P = 2^+$, 3^-, and 2^- make looping in qualitative consistence with the empirical ones by Hiroshige et al. [17]. These results provide a necessary condition for the existence of the dibaryons with $J^P = 2^+$, 3^-, and 2^-. (See Figs.2 and 3.)

The πN $-$ P_{11} non-pole, S_{11}, and S_{31} contributions are as follows. For $pp-pp$, $pp-\pi d$, and $\pi d-\pi d$ processes, the theoretical results with and without the P_{11} non-pole, S_{11}, and S_{31} contributions are obtained. The characteristic points of those contributions are found as follows:

1. The P_{11} non-pole contribution is remarkable in the amplitudes with $J^P = 2^-$ in all the processes. It is particularly remarkable at $T_{\rm L} \gtrsim 800$ MeV: The Roper resonance region. (See Fig. 1.)

2. The S_{11} contribution is remarkable in the amplitudes with $J^P = 2^+$ in all the processes at $T_{\rm L} \gtrsim 500$ MeV.

3. The S_{31} contribution is remarkable in the amplitudes with both $J^P = 2^-$ and 2^+ in all the processes at $T_L \gtrsim 500$ MeV. In this case, the S_{31} contribution is in the opposite direction to those from the P_{11} non-pole as well as S_{11}. (See Fig. 1.)

4. In all the pp partial-wave-amplitudes, the P_{11} non-pole contributions work attractively while the S_{11} and S_{31} do repulsively.

5. In the pp $^1S_0$ amplitude, the P_{11} non-pole contribution makes a considerable absorption effect.

The NN $^1S_0$ contribution to the πNN system is remarkable in the absorption parameter ρ of the $^3P_0$ NN amplitude at $T_L = 300 - 400$ MeV. With that contribution, the theoretical result makes a fit to the empirical one. (See Fig.4.)

The contribution from the non-pole part of the πNP_{11} state is considerable in the $I = 0, {}^3D_1$ state. A mild structure is found in our theoretical phase shifts with the full ingredient at $T_L = 400-1050$ MeV. The structure disappears without the P_{11} non-pole part contribution, mostly due to the Roper resonance. (See Fig.3.)

The observables for $pp \rightarrow \pi^+ pn$ processes are simultaneously calculated with those of the other processes at T_L=800, 510, 465, and 430 MeV. The data are qualitatively or almost quantitatively reproduced as some examples are shown.(See Fig.5.)

To investigate the resonance structures in the $J^P = 2^+$ and 3^- system, we make the amplitudes for the systems which are the average over the relative momentum between the spectator and the interacting pair and just depend on the energy. We find that the Argand plots of the amplitudes make anti-clockwise looping, similarly to the cases of the two-body final states.

5 Conclusion

We have presented the theoretical results on $pp - pp$, $pp - \pi d$, and $\pi d - \pi d$ processes with the new input elements of the P_{11} non-pole, the S_{11}, and S_{31} contributions which are additional to the ingredients of the earlier works [1, 10]. We observe that the new input elements make considerable effects in the $pp - {}^3P_2$ amplitudes.

We find that the $I = 1, J^P = 2^+$, 3^-, and 2^- amplitudes make clearly anti-clockwise looping in all the amplitudes of $pp - pp$, $pp - \pi d$, $\pi d - \pi d$, and $pp \rightarrow \pi^+ pn$ processes and provide the necessary condition for the existence of the resonances. This is consistent with our earlier results [1, 10].

Another interesting result is found for the πNN system near the π threshold, where the NN and πN S wave configurations make a major contribution. There, the Δ contribution is not appreciable. The NN $^3P_0$ amplitude is suitable to detect the contributions from the S wave configurations. We find that the NN $^1S_0$ contribution due to the OBEP to the πNN system is remarkable in the absorption parameter ρ of the NN $^3P_0$ amplitude at $T_L = 300 - 400$ MeV.

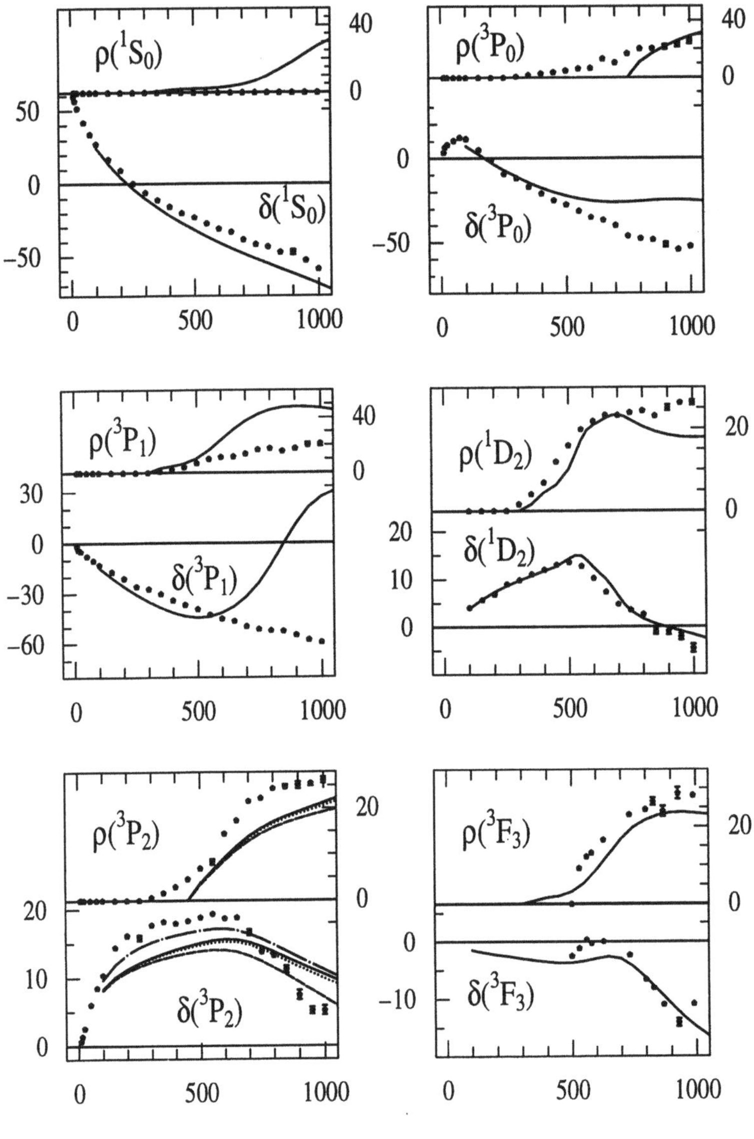

Figure 1. The theoretical *pp* phase parameters are compared with the phase shift analysis results [16]. The abscissa represents T_L in units of MeV, while the ordinate is given in units of degree. The solid curves indicate the theoretical results with full ingredients. The dashed, the dotted and the dot-dashed curves in $^3P_2$ indicate those without the P_{11} non-pole, without the S_{11} and without the S_{31} contributions, respectively.

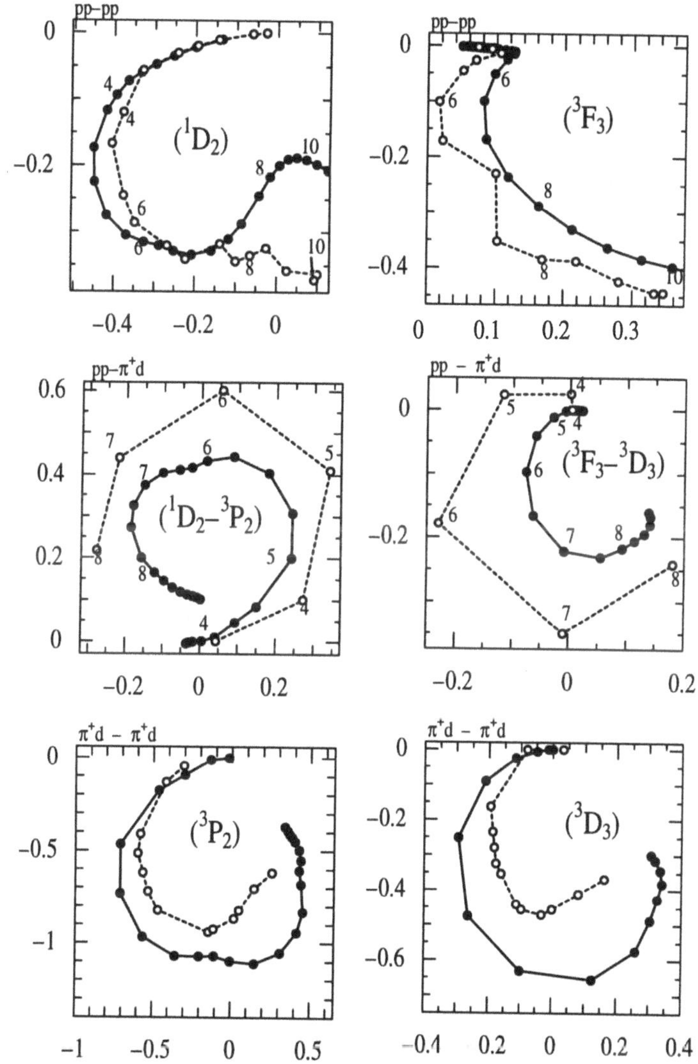

Figure 2. The Argand plots of the $J^P = 2^+$ and 3^- amplitudes. The curves for $pp - pp$ and $pp - \pi d$ indicate our theoretical results with the solid circles and the phase shift [16] and the amplitude [17] analysis results with the open circles. The numbers associated with the curves show the energy T_L in units of 100 MeV. The curves for the $\pi d - \pi d$ indicate our results at $T_\pi = 10 - 360$ MeV and the empirical results [17] at $T_\pi = 65 - 275$ MeV.

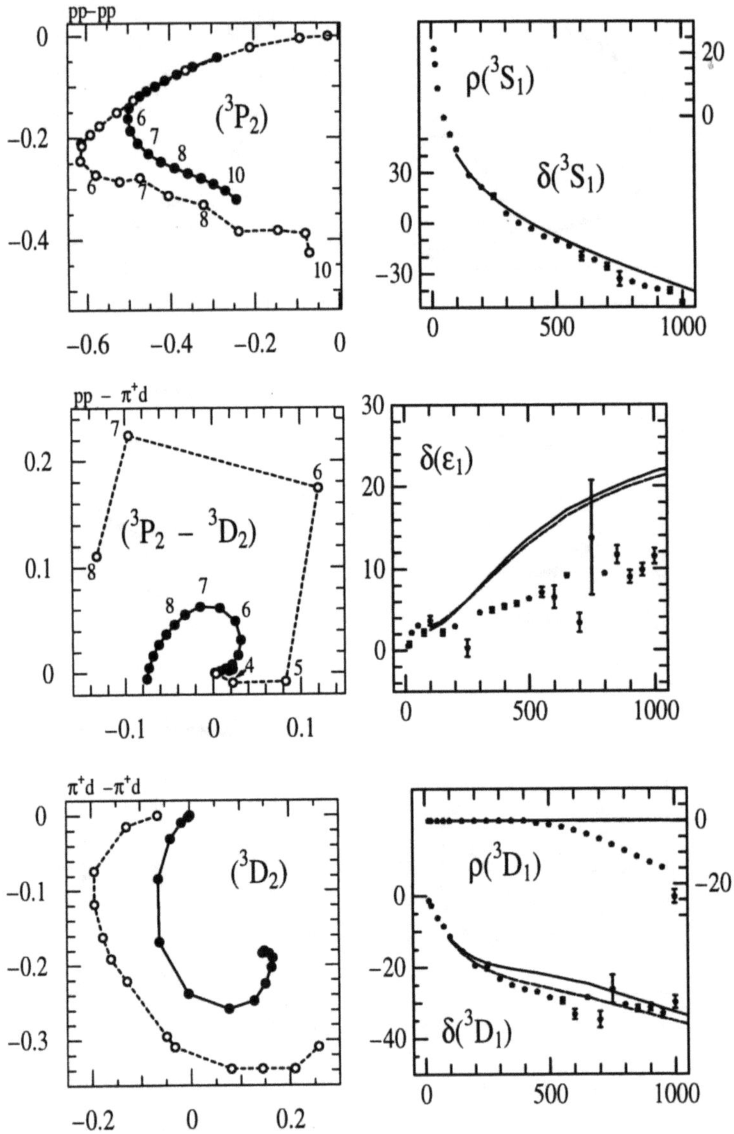

Figure 3. Left: The Argand plots of the $J^P = 2^-$ amplitude. See the figure caption of Fig.2. Right: The $I = 0, J^P = 1^+$ $np - np$ phase parameters. See the figure caption of Fig.1.

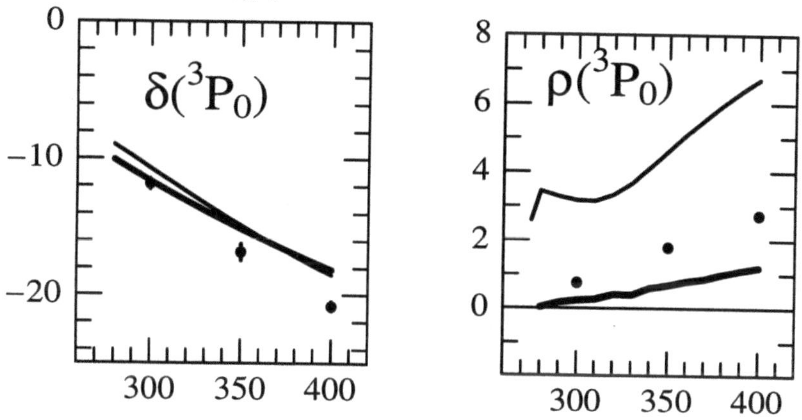

Figure 4. The $^3P_0$ phase parameters indicating the effects from the NN $^1S_0$ interaction due to the OBEP [15]. The cases with and without the NN $^1S_0$ interaction are represented by a solid and bold curves respectively. See also the caption of Fig.1.

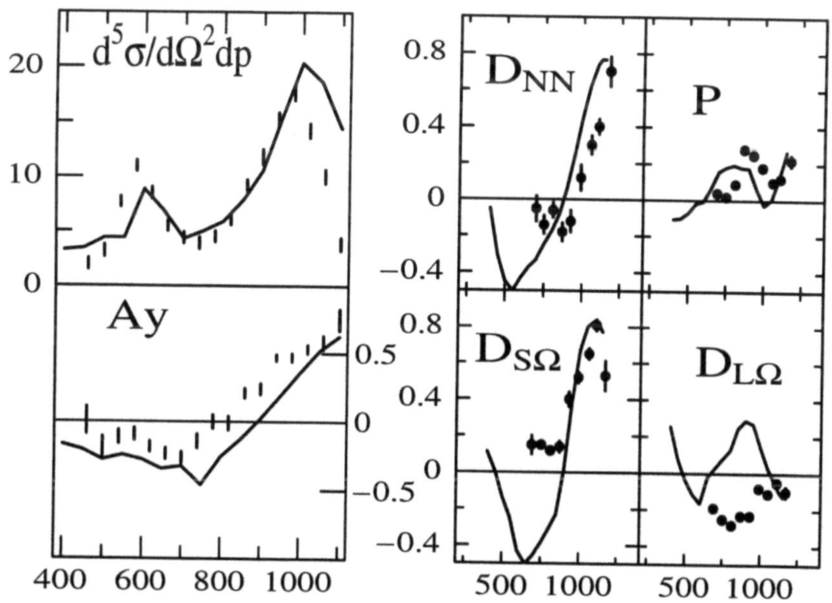

Figure 5. Left: The differential cross section [μb (sr^2 MeV$/c)^{-1}$] and A_y of $pp-\pi^+pn$ at 800 MeV with $\theta_p = 14.5°, \theta_\pi = 42°$. Right: The observables $D_{NN}, P, D_{S\Omega}$ and $D_{L\Omega}$ of $pp-\pi^+pn$ at 800 MeV with $\theta_p = 15°$ and $\theta_\pi = 95-135°$.

In conclusion we have made a qualitative description of $pp-pp$, $pp-\pi d$, $\pi d-\pi d$, and $pp\rightarrow\pi^{+}pn$ at $T_{\rm L}\leq 1$ GeV in terms of hadronic degree without introducing subhadronic degree explicitly. The dibaryon phenomena in the present processes are described in hadronic degree.

Acknowledgement. Numerical calculation has been supported by RCNP, Osaka University.

References

1. T. Ueda: Nucl. Phys. **A463**, 69c (1987)
2. H. Garcilazo and T. Mizutani: *πNN System*. Singapore: World Scientific 1990
3. E. O. Alt, P. Grassberger and W. Sandhas: Nucl. Phys. **A139**, 209 (1969)
4. B. Blankleider and I. R. Afnan: Phys. Rev. **C24**, 1572 (1981); Y. Avishai and T. Mizutani: Nucl. Phys. **A352**, 399 (1981)
5. A. Yokosawa: Phys. Rep. **64**, 47 (1980)
6. T. Ueda: Phys. Lett. **B79**, 487 (1978)
7. T. Ueda: Phys. Lett. **B119**, 281 (1982)
8. T. Ueda: Phys. Lett. **B141**, 157 (1984)
9. T. Ueda: Phys. Lett. **B175**, 10 (1986)
10. T. Ueda: Nucl. Phys. **A573**, 511 (1994)
11. T. Ueda and Y. Ikegami: Prog. Theor. Phys. **91**, 85 (1994); R. A. Arndt, J.M. Ford and L.D. Roper: Phys. Rev. **D32**, 1085 (1985)
12. T. Ueda and A. E. S. Green: Phys. Rev. **C18**, 337 (1978)
13. T. Ueda: Phys. Rev. Lett. **68**, 142 (1992)
14. T. Ueda: Prog. Theor. Phys. **76**, 729 (1986)
15. T. Ueda, F. Riewe and A. E. S. Green: Phys. Rev. **C17**, 1763 (1978)
16. R. A. Arndt, I.I. Strakovsky and R. L. Workman: Phys. Rev. **50**, 2731 (1994); J. Nagata, H. Yoshino and M. Matsuda: Prog. Theor. Phys. **93**, 559 (1995)
17. N. Hiroshige, W.Watari and M. Yonezaw: Prog. Theor. Phys. **72**, 1146 (1984); N. Hiroshige et al.: Prog. Theor. Phys. **84**, 941 (1990) and private communication

Few-Body Systems Suppl. 9, 187–191 (1995)

Few-Body Systems

Pion Double Charge Exchange in p-shell Nuclei

W. Fong[1*], J.L. Matthews[1], M.L. Dowell[1†], E.R. Kinney[1**], S.A. Wood[1††], P.A.M. Gram[2], G.A. Rebka, Jr.[3], D.A. Roberts[3***]

[1] Department of Physics and Laboratory for Nuclear Science, Massachusetts Institute of Technology, Cambridge, MA 02139 USA

[2] Los Alamos National Laboratory, Los Alamos, NM 87545 USA

[3] Department of Physics, University of Wyoming, Laramie, WY 82071 USA

Abstract. Doubly differential cross sections for inclusive pion double charge exchange (DCX) on ^{6,7}Li, ^{9}Be, and ^{12}C have been measured at the Los Alamos Meson Physics Facility with incident positive and negative pions of energies 120, 180, and 240 MeV. These new data, together with our previous studies of DCX on nuclei with A = 3, 4, 16, 40, 103, and 208, provide a detailed picture of the systematics of this reaction over the entire periodic table.

1 Introduction

The simplest mechanism for the pion-induced double charge exchange (DCX) process, e.g., (π^+, π^-), consists of two sequential single charge exchange (SCX) interactions, e.g., $\pi^+ n \rightarrow \pi^0 p$ followed by $\pi^0 n \rightarrow \pi^- p$, in which two (like-charge) nucleons are involved. A semi-classical calculation based on this mechanism, assuming the two nucleons to be in a degenerate Fermi gas and the SCX cross sections to be identical to those on the free proton, failed to reproduce the inclusive data for DCX on ^{16}O [1]. For this nucleus as well as for heavier nuclei [1], the measured outgoing pion spectra more closely resembled the distribution of events in four-body phase space.

A very different situation was observed in DCX on the lightest nuclei: ^{3}He [2] and ^{4}He [3], in which the spectra bore no resemblance to phase space but rather,

*Present address: Johns Hopkins Oncology Center, Baltimore, MD 21287 USA

†Present address: Joint Institute for Laboratory Astrophysics, Boulder, CO 80309 USA

**Present address: Department of Physics, University of Colorado, Boulder, CO 80309 USA

††Present address: Continuous Electron Beam Accelerator Facility, Newport News, VA 23606 USA

***Present address: Department of Physics, University of Michigan, Ann Arbor, MI 48109 USA at forward angles, exhibited a prominent double-peaked structure. A calculation by Kinney [4] based on the sequential SCX picture was found to reproduce the shapes of the forward-angle spectra in both of these nuclei. One might hypothesize that sequential SCX is the dominant DCX reaction mechanism and that its signature (a double-peaked spectrum at forward angles) is obscured in heavier nuclei by more complex processes which become more probable as the number of nucleons in the target increases.

It is thus of interest to obtain data on nuclei intermediate in mass between $A = 4$ and $A = 16$, to observe the evolution of the spectrum shape from double-peaked [2, 3] to phase-space-like [1].

2 Experiment

The measurements were performed in the high-energy pion channel ("P^3") at LAMPF, using a magnetic spectrometer to detect the outgoing pions at angles between 25° and 130°, over a kinetic energy range between 10 MeV and the kinematic limit. A set of multiwire proportional chambers determined the position of a particle at the focal plane, and thus its momentum, and a combination of plastic scintillator and threshold Čerenkov detector served to identify the pions. Absolute cross sections were obtained by observation of πp elastic scattering and comparison of the results to the known cross sections. Detailed discussions of the experimental apparatus, procedure, and data analysis may be found in refs.[1, 5].

The spectra [5] observed in the (π^+, π^-) reaction on ^{6,7}Li and ^{9}Be at 25° and incident energy 240 MeV are shown together with those for ^{4}He [4], and ^{16}O [1] in Fig. 1. One does indeed observe the transition between a pronounced double-peaked structure and a smooth spectrum whose shape closely resembles the distribution of events in four-body phase space [1].

3 Comparison with Theory

The sequential SCX model of Kinney [4] has been applied [5] to the data on ^{6}Li, taking account of the fact that the one of the two nucleons involved in DCX may be in the p-shell. The results are shown in Fig. 2. The calculation is seen to represent the measured spectra only qualitatively at best. The agreement between theory and data is significantly inferior to that found for ^{3,4}He.

An intranuclear cascade calculation, which allows the possibility of more than two interactions, has been developed by Oset and co-workers [6]. This more microscopic theory, which obtains the DCX cross sections simultaneously with those for other inclusive pion-nucleus reaction channels, might be expected to succeed best in heavy nuclei. It was found that this calculation reproduced some of the observed features of the DCX data for $A = 16 - 208$ [1], but the quantitative agreement was poor. This results of applying this theory to the ^{6}Li data [5] are also shown in Fig. 2. We note that a similar disagreement was found for ^{9}Be [7].

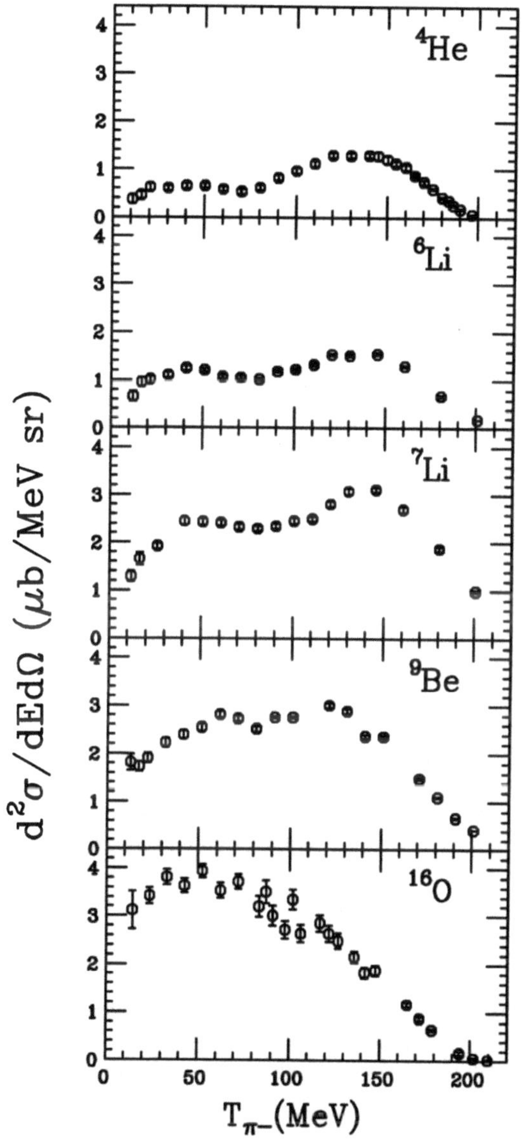

Figure 1. Doubly differential cross sections for (π^+, π^-) on a series of light nuclei from A=3 to A=16 at T_{π^+} = 240 MeV, θ_{π^-} = 25°

It is seen that neither calculation accurately represents the data. It might be concluded from Fig. 2 that the truth lies somewhere between the two models. That of Kinney [4] ignores the presence of the $A-2$ nucleons, except in that the interactions are required to take place in some average nuclear potential.

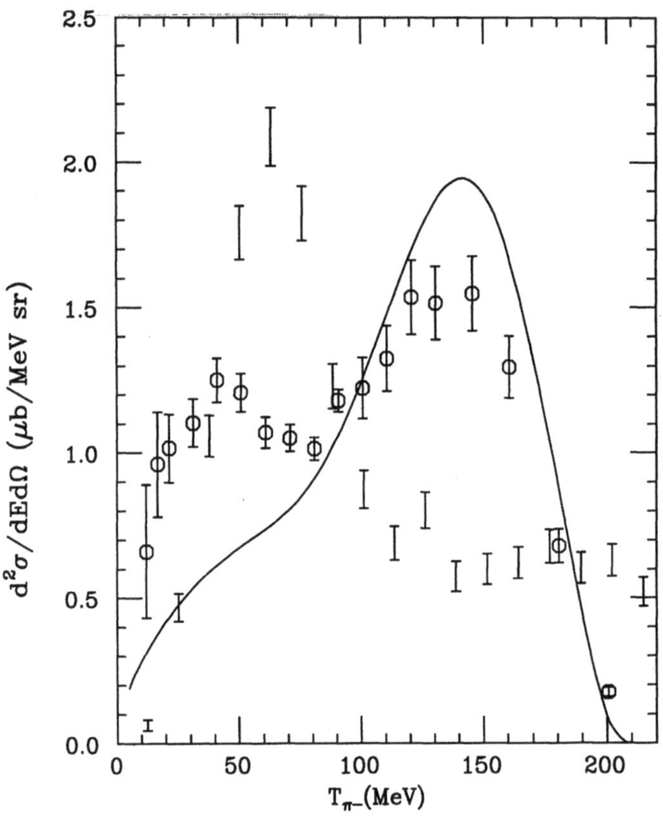

Figure 2. Comparison of the measured doubly differential cross section (open circles) for ^{6}Li(π^+, π^-) at $T_{\pi^+} = 240$ MeV and $\theta_{\pi^-} = 25°$ with theory. The solid curve represents the calculation of Kinney [4] and the bars that of Oset [8, 5]

That of Oset and co-workers [6, 7] perhaps overestimates the contribution of multiple interactions, leading to an overenhancement of the low-energy region of the spectra. Clearly, more theoretical work on the inclusive DCX process is required.

4 Phenomenological Discussion

By examining the systematics of the total reaction cross sections for DCX, we found that we could account for the Z- and A-dependence of the (π^+, π^-) and (π^-, π^+) reactions by means of a picture of sequential SCX occurring in competition with other processes [9]. One would expect the competition with quasi-free scattering (without charge exchange) to be important in the $\Delta(1232)$ resonance region, since the Clebsch-Gordan coefficients predict the cross section ratios $\sigma(\pi^- n \to \pi^- n)/\sigma(\pi^- p \to \pi^0 n) = \sigma(\pi^+ p \to \pi^+ p)/\sigma(\pi^+ n \to \pi^0 p) = 4.5$.

Comparing the DCX reactions on ^{6}Li and ^{7}Li, in the absence of competition one would expect the (π^-, π^+) cross sections to be identical, since only the protons are involved, and the (π^+, π^-) cross sections to differ by a factor of 2, since their probability should be proportional to $Z(Z-1)$. "Turning on" the competition should not affect the ^{7}Li/^{6}Li ratio for (π^+, π^-), since there are three protons in both cases. "Turning on" the competition in (π^-, π^+) should decrease the ratio from unity, since there are 33% more neutrons in ^{7}Li available for competing reactions. In fact, the observed ratios, averaged over the cross sections at 180 and 240 MeV, are 2.20 ± 0.14 and 0.75 ± 0.07. The data thus seem to bear out these simple arguments, of which a slightly less oversimplified version can be found in ref.[9].

5 Conclusion

The present results on DCX in *p*-shell nuclei are consistent with the hypothesis that sequential SCX is the dominant reaction mechanism in the $\Delta(1232)$ resonance region, although current theoretical efforts have not yet achieved quantitative agreement with the data. The effect of adding as few as two nucleons to a ^{4}He "core" is seen to be significant.

Acknowledgement. This work was supported in part by the United States Department of Energy.

References

1. S.A. Wood et al.: Phys. Rev. **C46**, 1903 (1992)
2. M.E. Yuly: Los Alamos National Laboratory Report LA-12559-T 1993
3. E.R. Kinney et al.: Phys. Rev. Lett. **57**, 3152 (1986)
4. E.R. Kinney: Los Alamos National Laboratory Report LA-11417-T 1988
5. W. Fong: Ph.D. Thesis, Massachusetts Institute of Technology 1994 (unpublished)
6. E. Oset et al.: Phys. Lett. **B165**, 13 (1985); L.L. Salcedo et al.: Nucl. Phys. **A484**, 557 (1988); M.J. Vicente et al.: Phys. Rev. **C39**, 209 (1989)
7. M. Vicente-Vacas and E. Oset: In: *Pion-Nucleus Double Charge Exchange*, p. 120. Singapore: World Scientific 1990
8. E. Oset: Private Communication
9. P.A.M. Gram et al.: Phys. Rev. Lett. **62**, 1837 (1989)

Few-Body Systems Suppl. 9, 193–202 (1995)

Few-Body Systems

What is so Special About Eta-Meson Physics?

B. M. K Nefkens

Department of Physics and Astronomy, UCLA, Box 951547, Los Angeles, CA 90095-1547, USA

Abstract. The η is a remarkable particle as regards its production, interaction, and its various suppressed, rare and forbidden decay modes. Eta production is always large; near threshold it is dominated by 3 special S-wave resonances. The ηN and $\eta \Lambda$ scattering lengths are attractive and large; this has generated speculations about the possible existence of a new type of nuclear matter, bound or quasi bound eta-mesic nuclei and eta-mesic hypernuclei. η is an isospin singlet, η production is therefore an isospin selective reaction which is particularly helpful in baryon spectroscopy. $\pi^0 - \eta$ mixing causes intrinsic (non Coulombic) violation of charge symmetry, which allows for new measurements of the mass difference between the up and down quarks at different four-momentum transfers. The η is a Goldstone boson; this provides opportunities for unique probes of chiral perturbation theory. Finally, the η is an eigenstate of the charge-conjunction and CP operators; this makes possible a variety of tests of C and CP invariance of flavor-non-changing strong/electromagnetic interactions.

1 Introduction

The eta-meson is associated with a wide variety of interesting nuclear and particle physics. Firstly, η production by π, K^-, γ, p and d near threshold is surprisingly large, it is associated with three unusual S-wave resonances. The ηN S-wave scattering length, $a_{\eta N}$, is attractive and large, unlike $a_{\pi N}$ which is repulsive and small. Secondly, η production is an important isospin selector e.g. $\pi^- p \to n\eta$ selects the pure $I = \frac{1}{2}$ states while the common pion scattering reaction involves a mixture of N^* and Δ^* states. Thirdly, the η which has G-parity +1 mixes with the π^0-meson which has G-parity -1. This $\pi^0 - \eta$ mixing induces potentially large charge-symmetry breaking. It provides a good way to measure the mass difference of the up and down quarks under different kinematical conditions. Fourthly, the η is a Goldstone boson so the not-so-rare η decays provide unique tests of chiral perturbation theory. Finally, the η is an eigenstate of the charge-conjugation operator, C, and of CP. This is special, it is a rarity in nature and it makes possible unique tests of C, CP and even CPT invariance in flavor-non-changing interactions in rare η decays.

2 Properties of the η

The η has the following vital statistics. Its rest mass is 547.4 MeV/c^2 which is 58% of the proton's mass. The quantum numbers of the η are zero, this includes the spin, isospin, electric charge, strangeness and other quark-flavor numbers, lepton number and baryon number. The η half life is 5×10^{-19} sec, which corresponds to a width of 1.2 keV. Charge conjugation and G-parity are +1. Thus the η has all the appearances of being massive vacuum, the only exception is that parity and CP are −1. The short η half life precludes making an η beam that travels a practical distance. Instead, precise η decay measurements can be made using tagged etas that are produced in a reaction such as $\pi^- p \rightarrow \eta n$ and $pd \rightarrow \eta\, ^3\mathrm{He}$.

The simplest allowed η decay mode is $\eta \rightarrow 2\gamma$. It is a second order electromagnetic transition, this explains the observed η half life. This process is actually forbidden in the limit of massless quarks, but takes place by the grace of the Wess-Zumino term. The strong decay $\eta \rightarrow 2\pi$ is forbidden by P and CP invariance while $\eta \rightarrow 3\pi$ is not allowed by charge symmetry conservation. The decay $\eta \rightarrow 3\pi$ occurs as a result of the quark-mass term in the QCD Lagrangian and thus provides a way for determining the up-down quark mass difference.

3 Eta-Meson Production

The η is produced prolifically in high energy reactions [1]. In heavy ion collisions it is the second most abundantly produced particle, after the pion, as much as 10%.

The large η threshold production is intimately connected with the existence of three special S-wave resonances, listed in Table 1. These resonances have a Q-value near zero and they decay predominantly under η emission, which is contrary to expectations based on the available phase space, favoring π emission. It is not difficult to speculate [2] on the possible existence of a similar type S-wave Ξ^* (1870) resonance that belongs to the same SU(3) octet as the Σ^*(1750). Truly spectacular would be an S-wave Ω^*(2220) resonance with a major η decay branch because it would have to be classified as a member of a decuplet.

The value of the ηN scattering length, $a_{\eta N}$, has been the subject of several recent analyses [3-6] and a new AGS experiment [7]. The various evaluations are not yet in complete agreement but it is safe to say that $a_{\eta N} > 0.6$ fm. This large value has raised the prospect of a new type of nuclear matter, an eta-mesic nucleus [8]. This is a nuclear system in which the η also provides strong binding between the nucleons. One could consider it a novel nuclear state with an excitation energy of $\sim$ 540 MeV. The width is subject to debate. The original prediction by Liu and Haider [8] was a narrow width of about 9 MeV. We expect that the pionic decay of the $S_{11}(1535)$, which accounts for about 1/3 to 1/2 of the decay rate, remains an open channel for a bound eta-mesic nucleus resulting in a width of 50-100 MeV. A direct search for a narrow bound eta-mesic nucleus via a narrow proton peak in the reaction $\pi^+ + {}^{16}\mathrm{O} \rightarrow p + {}^{15}_{\eta}\mathrm{O}$

Table 1. Three isobars with a well established large η decay branch and one proposed new such isobar. Sp= spectral notation, Q is the energy release, e.g., $Q = m(S_{11}) - m_n - m_\eta$, R is the ratio of branching fractions to η and π (or K) decay modes. + The pole position is around 1487 MeV which implies that $Q \sim 2$ MeV.

symbol	I, J^P	Sp	mass (MeV)	width (MeV)	η decay (%)	Q (MeV)	R
$N^*(1535)$	$\frac{1}{2}, \frac{1}{2}^-$	S_{11}	$\approx 1535^+$	≈ 150	45-55	$\sim 50^+$	$\eta N/\pi N \sim 1$
$\Lambda^*(1670)$	$0, \frac{1}{2}^-$	S_{01}	1670±5	≈ 35	15-35	7 ± 5	$\eta\Lambda/\bar{K}N \sim 1$
$\Sigma^*(1750)$	$1, \frac{1}{2}^-$	S_{21}	≈ 1760	≈ 90	15-55	20±20	$\eta\Sigma/\pi\Sigma > 2$
$\Xi^*(1870)$	$\frac{1}{2}, \frac{1}{2}^-$	S_{11}	?			?	

was unsuccessful [9] at an incident pion momentum of 800 MeV/c. It has been pointed out [4] that this momentum is too high as the ηN scattering will have turned from attractive near threshold to repulsive.

The original value of the ηN scattering length, $a_{\eta N} = 0.27$, used in ref. [8] was based on a poor choice of input data. It required an eta-mesic nucleus to have a mass number larger than $A \approx 10$. The currently favored value, $a_{\eta N} > 0.6$ fm, makes both $\eta\,^3$He and $\eta\,^4$He possible candidates for a quasi-bound or even a bound eta-mesic nucleus. Indicative is a large η-nucleus scattering length which has been obtained in recent calculations [5, 10, 11]. Besides having a large and attractive η-nucleus scattering length, an eta-mesic nucleus will manifest itself in production processes such as $pd \to \eta\,^3$He near threshold by a characteristic momentum dependence of the invariant production amplitude as discussed by Wilkin [5]. A recent measurement [12] appears to bear this out.

Since the expected width of 50-100 MeV of eta-mesic nuclei will make direct detection in a two-body production reaction very challenging it is of interest to search for other η-mesic systems. Recently we have considered the possible existence of an eta-mesic hypernucleus [4], this was prompted by the outcome of a determination of the $\eta\Lambda$ scattering length finding a large value, $a_{\eta\Lambda} \sim 0.6$ fm. Since the width of the relevant resonance, the $\Lambda^*(1670)$, is much narrower than the $N^*(1535)$, see Table 1, the η-mesic hypernucleus would have a width of only 10 MeV. An eta-mesic hypernucleus would more likely be a bound state than an eta-mesic nucleus since the Q-value of the $\Lambda^*(1670)$ is closer to zero than that of the $N^*(1535)$. This suggests a direct 2-body reaction search such as [4]

$$K^- + {}^{12}\mathrm{C} \to n + {}^{11}_{\eta\Lambda}\mathrm{B}\,, \tag{1}$$

as well as a careful threshold measurement of η production to measure the invariant amplitude e.g. in $K^- + {}^{12}\mathrm{C} \to \eta + {}^{12}_{\Lambda}\mathrm{B}$. Also of interest is

$$K^- + {}^{3}\mathrm{He} \to \eta + {}^{3}_{\Lambda}\mathrm{H}\,. \tag{2}$$

Elsewhere [1] we have discussed the hints for exotica found in a large η production in $pd \to \eta\,^3$He especially at 180° and also in $\pi^- \,^3\mathrm{He} \to \eta\,^3\mathrm{H}$. We want to emphasize the large difference in the threshold production of π^0 and η in the $pd \rightarrow {}^3\mathrm{He}X$ reaction. Both π^0 and η are Goldstone bosons and one might expect some similarities in their threshold-production characteristics. Experiment has found the opposite. The dependence of $\sigma_t(pd \rightarrow \pi^0\,{}^3\mathrm{He})$ is $\sigma_t \sim p_\pi$ just as expected for an S-wave production of a pseudoscalar meson and predicted long time ago by Gell-Mann and Watson [13]. The angular distribution has a near-exponential decrease with increasing four-momentum transfer, suggestive of a $(d - {}^3\mathrm{He})$ transition-form-factor dominated process; the forward-backward asymmetry in the angular distribution is a factor of 5 when $p_{\pi^0} = 14$ MeV/c or $\lambda_{\pi^0} \simeq 8$ fm; this makes the strict dependence of σ_t on p_{π^0} even more remarkable. The behavior of $\sigma_t(pd \rightarrow \eta\,{}^3\mathrm{He})$ is very different, increasing initially much stronger with p_η than an S-wave reaction, becoming nearly constant a few MeV above threshold. The angular distribution, however, is completely isotropic up to the maximum energy measured, 11 MeV above threshold, where $p_\eta = 75$ MeV/c and $\lambda_\eta = 2.7$ fm.

4 Baryon Spectroscopy with Eta-mesons

The η is an isospin singlet and therefore two-body η production reactions are isospin selective. For example $K^-p \rightarrow \eta\Lambda$ concerns only the pure $I = 0$ states and $K^-p \rightarrow \eta\Sigma$ only pure $I = 1$. The well known elastic scattering reaction $K^-p \rightarrow K^-p$ is a mixture of $I = 0$ and 1 states requiring an additional experiment of the difficult charge exchange reaction. η production has not been measured in much detail because of experimental limitations. The proposed move of the famous SLAC Crystal Ball [14] to a π^- and K^- beam at the AGS would improve the situation dramatically. The Crystal Ball is a multiphoton spectrometer with good energy and angular resolution and 94% solid angle coverage. It would be particularly useful in the search for heavy S- and P-wave N^* resonances which have been predicted to exist by various models especially the successful Isgur-Karl quark model [15].

Another unique aspect of η-baryon spectroscopy derives from the fact that low energy etas are a potential signature of the $N^*(1535)$, $\Lambda^*(1670)$ or $\Sigma^*(1750)$ resonances. This allows the identification of certain sequential decays of massive baryons. Consider the reaction $N^* \rightarrow \pi^0\eta N$, this is really the decay of the N^* under π^0 emission to the $N^*(1535)$. Such an identification has never been made; the existing studies of sequential decays are based on two-pion emission and in general are far from definitive.

5 $\pi^0 - \eta$ Mixing and Charge Symmetry Breaking

Free quarks do not exist according to QCD and thus it is not possible to measure directly the mass of a quark. However, the mass difference between the up and down quarks can be readily obtained from suitable measurements of charge symmetry breaking [16]. A clean way is from a measurement of $\pi^0 - \eta$ mixing. This works as follows. There is only one term in the QCD Hamiltonian density which depends explicitly on the quark mass namely,

$$H_m = m_d d\bar{d} + m_u u\bar{u} + \cdots \tag{3}$$

The matrix element between the π^0 and η meson states is therefore

$$\langle \pi^0 | H_m | \eta \rangle = \frac{1}{2}\sqrt{2}\langle u\bar{u} - d\bar{d} | m_d d\bar{d} + m_u u\bar{u} | u\bar{u} + d\bar{d} - s\bar{s}\rangle \frac{1}{3}\sqrt{3} = \frac{1}{3}\sqrt{6}(m_u - m_d). \tag{4}$$

A possible way to study this mixing is via the decay of the $S_{11}(1535)$ resonance because of its large branching ratio to the η channel. This decay has two contributions, one direct decay, $S_{11} \rightarrow N + \eta$, described by the amplitude A_η, and one indirect one via the π^0 which then turns into an η as a result of $\pi^0 - \eta$ mixing, this is represented by the amplitude $A_{\pi^0 \rightarrow \eta}$. As a result the decay amplitudes for the charged and neutral states are

$$A(S_{11}^+ \rightarrow p\eta) = A_\eta + A_{\pi^0 \rightarrow \eta},$$

$$A(S_{11}^0 \rightarrow n\eta) = A_\eta - A_{\pi^0 \rightarrow \eta}, \tag{5}$$

the + and - signs come from the Clebsch-Gordan coefficients that are associated with the isospin (1/2 $\rightarrow$ 1/2, 1) transition.

Experimentally this can be measured using the ratio

$$R_\eta = \frac{d\sigma(\pi^+ d \rightarrow pp\eta)}{d\sigma(\pi^- d \rightarrow nn\eta)} \simeq \left| \frac{A_\eta + A_{\pi^0 \rightarrow \eta}}{A_\eta - A_{\pi^0 \rightarrow \eta}} \right|^2 \tag{6}$$

in the region of the S_{11} resonance from $p_\pi = 610$ to 850 MeV/c. Charge symmetry requires this ratio to be 1.0. The experiment [17] has recently been performed at the AGS. Near $p_\pi = 750$ MeV/c the on-line results show R_η to be around 0.9 indicating a dramatic violation of charge symmetry due to $\pi^0 - \eta$ mixing.

Recently, it has been argued that the $\rho - \omega$ mixing matrix element which is quite similar to the one for $\pi^0 - \eta$ mixing and which is extracted from the on-mass- shell mesons, should change significantly off-shell [18,19]. The off-shell behavior affects the contribution of meson mixing to charge symmetry breaking in low energy nuclear physics. The off-shell behavior is hard to measure so we suggest instead to investigate charge symmetry breaking for different values of the four-momentum transfer. This may be done by measuring R_η at different regions of the (ηNN) Dalitz plot.

6 Chiral Symmetry

The challenging region for applying QCD is at low energies where perturbation theory is not applicable. This covers all of nuclear physics, threshold meson production, the strong decays of mesons and baryons, and so forth. One of the promising approaches to a variety of low energy strong interaction processes is based on chiral symmetry which is a property of massless QCD. The main idea is the following. Massless quarks come in two varieties, righthanded ones when the momentum and spin vectors are parallel and lefthanded ones with antiparallel alignment. The QCD Lagrangian has no helicity-flip term hence the handedness of the massless quarks is unaffected by strong interactions. This implies that there must be a new symmetry, which stems from the fact that the Lagrangian is invariant under the interchange of all left and right handed quarks, and is called chiral symmetry.

Consider a certain system of particles that has a Lagrangian which is invariant under some symmetry. Furthermore, this system has a continuum family of ground-state solutions which are related to each other by the symmetry but which individually are not invariant under the symmetry. This is misleadingly called a spontaneously broken symmetry. This rather subtle property is prominent in certain gauge theories of massless quanta. Such a spontaneously broken symmetry implies the existence of a new **massless** field particle [20]. This is Goldstone's theorem and conceptually has a simple origin. Define the ground state to have $E = 0$ then there must be other states with $E = 0$ and their corresponding particles must have zero mass! The general theorem states that whenever a continuous global symmetry is spontaneously broken the particle spectrum must have a massless, spin-zero boson called a Goldstone boson [21]. There are 3 quarks implying the existence of $3 \times 3 - 1 = 8$ Goldstone bosons which are the π, K and η. To take into account that the physical quarks are not massless, Weinberg [22] has introduced an expansion of the Lagrangian in powers of the pion mass and momentum. The algebra that governs this is called chiral perturbation theory, it has been developed by Gasser and Leutwyler [23]. Chiral perturbation theory has already given important relations that govern pion-pion scattering, threshold production of the Goldstone bosons, the decay of many systems under emission of a soft pion and a large variety of meson decays. A particularly interesting example is the decay $\eta \to \pi^0\gamma\gamma$ which allows probing the p^6 order term, because the lower order terms are forbidden or much suppressed by various symmetries.

7 Testing Charge Conjugation and CP Invariance

Matter is distinct from antimatter in most instances. For example, proton and antiproton have opposite electric charge. Since they have also opposite baryon number, they may annihilate each other. The neutron has its spin and magnetic dipole moment vectors antiparallel, while those of the antineutron are parallel. There is a small number of mesons which are their own antiparticles, they must be neutral and flavorless such as a π^0 and η. Their quark structure is $q\bar{q}$. These special mesons can be used to investigate if matter and antimatter interact identically which is difficult to test otherwise. We anticipate, as a consequence of the CPT theorem, that particles and antiparticles have identical lifetimes and rest masses but that does not imply that they have identical interactions, this is well known for the case of neutrino and antineutrino interactions.

C-invariance of the strong interactions has not been tested extensively [24].

Table 2. Testing C, CP, T and CPT in rare η decays.

Decay	Symmetry	Observable	Exp.Limit
$\eta \to \pi^0 \ell^+ \ell^{-(a)}$	C	rate	$< 5\times10^{-6}$
$\eta \to \pi^0 \ell^+ \ell^{-(a)}$	$C, CP^{(b)}$	charge asym.	
$\eta \to \pi^0 \mu^+ \mu^-$	T	μ^+ trans.pol.	
$\eta \to \pi^0 \ell^+ \ell^{-(a)}$	CPT	decay correl.	
$\eta \to 3\gamma$	C	rate	$< 5\times10^{-4}$
$\eta \to \mu^+ \mu^-$	P,CP	μ^+ long.pol.	
$\eta \to \pi^+ \pi^-$	P,CP	rate	$< 1.5\times10^{-3}$
$\eta \to \pi^0 \pi^0$	P,CP	rate	
$\eta \to 4\pi^0$	P,CP	rate	
$\pi \to \pi^+ \pi^- \ell^+ \ell^{-(a)}$	P,CP$^{(b)}$	decay correl.	
$\eta \to \pi^+ \pi^- \gamma$	P,CP	spectral shape$^{(c)}$	
$\eta \to \pi^+ \pi^- \pi^0$	C	charge asym.	$< 3\times10^{-3}$
$\eta \to \pi^+ \pi^- \gamma$	C	charge asym.	$< 1\times10^{-2}$
$\eta \to \pi^0 \pi^0 \gamma$	C	rate	

$^{(a)}\ell$ stands for μ or e
(b) in the limit of the one photon intermediate state
$^{(c)}$ bremsstrahlung component

The best limit is $A_{\not{C}} < 0.1 A_C$ where $A_{\not{C}}$ is the C-violating and A_C is the C-conserving strong amplitude. The review of Particle Properties lists the absence of the decay mode $\eta \to \pi^0 \mu^+ \mu^-$ as our best test of C-invariance. This decay is a second order electromagnetic transition and its decay rate must be less than that for $\eta \to \pi^0 \gamma\gamma$ modified by the $\gamma\gamma \to \mu^+\mu^-$ transition factor. Even if C were maximally violated we expect that $BR(\eta \to \pi^0 \mu^+ \mu^-) \lesssim 3\times10^{-5}$. In view of the fact that the experimental upper limit, see Table 1 first line, for this is $< 5 \times 10^{-6}$, the estimate $A_{\not{C}} < 0.1 A_C$ is a very generous one.

The η meson, whose quark composition is $|\eta\rangle = \frac{1}{3}\sqrt{3}|u\bar{u} + d\bar{d} - s\bar{s}\rangle$, is an eigenstate of the C-operator. Since $\eta \to 2\gamma$ is allowed the decay $\eta \to 3\gamma$ is forbidden if C is to be conserved. There are other interesting 3-body η decays involving one or more pions that are possible for testing C-invariance, they are listed in Table 2.

The preponderance of matter over antimatter in the universe is known as the cosmological matter problem. It is not explained by the small CP violation known from K^0-decay and requires another, much stronger, source of CP

Table 3. Other interesting rare η decays.

Decay Mode	New Physics	Needed Sensitivity
$\eta \to e^+e^-$	leptoquark	$BR \sim 10^{-9}$
$\eta \to \mu e$	lepton fam. viol.	$BR \sim 10^{-10}$
$\eta \to \pi \ell \nu$	2nd class current	$BR \sim 10^{-11}$
$\eta \to K e \nu$	ΔS weak int.	$BR \sim 10^{-13}$
$\eta \to \pi\pi\nu\bar{\nu}$	neutral weak current	$BR \sim 10^{-10}$
$\eta \to \ell^+\ell^-\gamma$	form factor	$BR \sim 10^{-4}$
$\eta \to \pi^0\gamma\gamma$	chiral perturb.th.	10% in spectr.
$\eta \to \pi^+\pi^-\gamma$	chiral perturb. th.	5% in shape
$\eta \to 3\pi^0$	chiral perturb. th.	1% in shape

violation. It may come as a surprise to many nuclear/particle physicists, that there is no direct test of CP outside the neutral K-meson system [25]. Since the η is also an eigenstate of the CP operator, η decays make possible new tests of CP invariance [24,25]

Consider $\eta \to \pi^0\pi^0$, this is forbidden by CP-invariance analogous to $K_L \to \pi^0\pi^0$. However, it is also forbidden by parity which is known to hold in strong interaction to the level of about 10^{-6}. The major η decays are electro-strong interactions; note that at the level of 10^{-7} η decay becomes a weak interaction which violates parity thereby making $\eta \to 2\pi^0$ an honest test of CP invariance in a flavor-non-changing transition. There are practical limitations to making a sensitive search for this decay mode because of serious background from $2\pi^0$ production that inevitably accompanies all η production. We suggest therefore to search for $\eta \to 4\pi^0$ instead. It is another test of P and CP and the decay rate compared to $\eta - 2\pi^0$ is suppressed by 3 orders of magnitude due to phase space. Thus, $\eta \to 4\pi^0$ should be searched for at the level 10^{-10} to have a true test of CP-invariance. This is a clean decay, practically without background. Some other tests of CP invariance that are possible in η decay are listed in Table 2.

η decay can be used to test other invariances as well, e.g. lepton-family conservation in $\eta \to \mu e$. Such tests have been discussed elsewhere [25]. The most interesting one are listed in Table 3.

8 Conclusions

The η is a fascinating particle. It is produced abundantly near threshold which involves three special S-wave resonances with a Q-value close to zero. The ηN scattering length is attractive and large. This has given rise to speculations about the possible existence of a new type of nuclear matter, the eta-mesic nucleus or even an eta-mesic hypernucleus. $\pi^0 - \eta$ mixing is a rich source of violations of charge symmetry that is largely unexplored. A new charge symmetry breaking of 10% has been observed recently in η production by pions on deuterium.

The η is a Goldstone boson, it allows new and unique investigations of the p^6 term of chiral perturbation theory in measurements of the rate and spectrum of the decay $\eta \to \pi^0\gamma\gamma$ which has a $BR \sim 10^{-4}$.

Finally, the eta-meson being an eigenstate of the C and CP operator, offers a variety of tests of these basic symmetries in the unexplored flavor-non-changing electro-strong interaction.

Acknowledgement. This work is supported in part by the U.S.A. DOE.

References

1. B.M.K. Nefkens: Proc. Int. Conf. on *Mesons and Nuclei at Intermediate Energies*, Dubna, Russia, 1994, p.667. Singapore: World Scientific 1994
2. S.F. Tuan: Phys. Rev. **D46**, 4095 (1992)
3. A. Švarc et al.: Talk presented at this Conference; Phys. Rev. **C** (submitted)
4. V.V. Abaev and B.M.K. Nefkens: preprint UCLA-10-P25-227; Phys. Rev. **C** (submitted)
5. C. Wilkin: Phys. Rev. **C47**, 938 (1993)
6. M. Arima et al.: Nucl. Phys. **A543**, 613 (1992)
7. AGS Exp. #909, W.J. Briscoe and W. B. Tippens, spokesmen
8. L.C. Liu and Q. Haider: Phys. Rev. **C34**, 1845 (1986)
9. R. E. Chrien et al.: Phys. Rev. Lett. **60**, 2595 (1995)
10. S. Wycech, A. Green, J. Niskanen: Preprint HU-TFT-95-8 University of Helsinki
11. S.A. Rakityansky et al.: Los Alamos e-print archive: Nucl - th/9504020 (1995); Phys. Lett. B (submitted)
12. R. Kessler: Thesis. UCLA 1992; the Saturne η collaboration (to be published)
13. M. Gell-Mann and K.M. Watson: Ann. Rev. Nucl. Sc. **4**, 219 (1954)

14. AGS proposal #913, Baryon Spectroscopy with the Chrystal Ball, M.E. Sadler and W. B. Tippens, spokesmen; AGS proposal #914 Neutral Hyperon Spectroscopy, B.M.K. Nefkens, A Efendiev, N. Kozlenko, spokesmen

15. R. Koniuk and N. Isgur: Phys. Rev. **D21**, 1968 (1980)

16. G.A. Miller, B.M.K. Nefkens and I. Slaus: Phys. Reports **194**, 1 (1990)

17. AGS Exp. #890, B.M.K. Nefkens, R. Chrien, J.C. Peng, spokesmen

18. T. Goldman et al.: Few Body Systems **12**, 193 (1992)

19. M. J. Iqbal et al.: Preprint TRI-PP-95-14 TRIUMF; Phys. Rev. Lett. (submitted)

20. J. Donoghue: Int. School of *Low-Energy Antiprotons*, Erice 1990; Preprint TH-56670/90 CERN

21. J. Kane: *Modern Elementary Particle Physics.* New York: Addison-Wesley 1987

22. S. Weinberg: Physica **96A**, 327 (1979)

23. J. Gasser and H. Leutwyler: Nucl. Phys. **B250**, 465; 517; 539 (1985)

24. B.M.K. Nefkens: *New Vistas in Physics with High-Energy Pion Beams* Santa Fe, N.M. Oct. 1992, ed. by B. Gibson and J. McClelland, p.91. Singapore: World Scientific 1993

25. B.M.K. Nefkens: Workshop on *Future Directions in Particle and Nuclear Physics at Multi-GeV Hadron Facilities*, ed. D. Geesaman, BNL-52389, March 1993, p. 491

Few-Body Systems Suppl. 9, 203–211 (1995)

ηN S-wave Scattering Length, Limitations of the Single Resonance Model, Predictions of the Three Coupled Channel, Multiresonance and Unitary Model

A. Švarc, M. Batinić, I. Šlaus

Rudjer Bošković Institute, POB. 1016, 41001 Zagreb Croatia

Abstract. It is shown from the first principles that only two physical quantities are needed to calculate the ηN S-wave scattering length if only one resonance is used per partial wave: the πN elastic T-matrix at the η production threshold, and the threshold slope of the total η production cross section. Taking πN elastic data from most of the πN elastic partial wave analyses, and the threshold slope of the total η production cross section of 21.2 ± 1.8 μb the absolute lower and upper limits of the real and imaginary part of the ηN S-wave scattering length within the single resonance model are given.

The value of the ηN S-wave scattering length is extracted in the three coupled channel, multiresonance and unitary model for two cases: when the number of resonances in the S-wave is restricted to one, and for the full model. It is demonstrated that the single resonance restriction of a multiresonance model obeys the aforementioned limitations, and the full calculation gives a significantly higher value for the scattering length real part. Our conclusion that the single resonance models are not sufficient to reliably predict the ηN S-wave scattering length is supported by other, non-single resonance models.

1 Introduction

In 1985, Bhalerao and Liu [1] have constructed a coupled channel isobar model for the $\pi N \to \pi N$, $\pi N \to \eta N$ and $\eta N \to \eta N$ T- matrices with πN, ηN and $\pi \Delta$ ($\pi\pi N$) as isobars. A single resonance separable interaction model for S_{11}, P_{11}, P_{33} and D_{13} partial waves has been used. They have used only πN elastic scattering data as a constraint while their prediction for the η production cross section has been compared with, at that time the most recent data [2]. Their conclusion has been that the S-wave ηN interaction is attractive, and they have extracted the two solutions for the S-wave scattering length: $a^1_{\eta N} = (0.27 + \mathrm{i}\,0.22)$ fm, and $a^2_{\eta N} = (0.28 + \mathrm{i}\,0.19)$ fm.

The recent single resonance model by Bennhold-Tanabe [3] is limited to T_π less than 700 MeV. It describes the dominant peak in the total η production cross section pretty well on the gross scale, but significantly fails in giving details like the exact peak position. The reason for the wrong peak position of the Bennhold-Tanabe analysis is that it relies on the data of ref. [2], which suffer from a serious beam momentum calibration error at lower energies [4]. Though the extracted ηN S-wave scattering length is not published in the quoted reference, we have found out that it is $a_{\eta N} = (0.25 + \mathrm{i}\,0.16)$ [5]. Notice that the small imaginary part completely corresponds to total η production cross section data of Brown et al. [2].

For both single resonance models, Bhalerao-Liu and Bennhold-Tanabe [1, 3], the value of the imaginary part of the $a_{\eta N}$ is far too low because they have been using much lower slope of the total η production cross section at η production threshold than it has been nowadays generally accepted [6]. In the single resonance model, we can renormalize the ηN S-wave scattering length values (see ref. [7]) so that the imaginary part reproduces the accepted η production total cross section slope at threshold $\sigma_{\pi^- p \to \eta n} = (21.2 \pm 1.8)$ μb. In that case we obtain somewhat higher values for all published S-wave scattering length values: $a_{\eta N}^{\mathrm{BL}^1} = (0.38 + \mathrm{i}\,0.31)$ fm; $a_{\eta N}^{\mathrm{BL}^2} = (0.44 + \mathrm{i}\,0.30)$ fm; $a_{\eta N}^{\mathrm{BT}} = (0.46 + \mathrm{i}\,0.29)$ fm.

Wilkin [8] based his calculation on an S-wave threshold enhancement calculation, used the η total cross section near threshold to fix the imaginary part of the T-matrix and obtained the real part by fitting the $\pi^- p \to \eta n$ production cross section up to the center of mass momentum in the ηn system of 1.2 (fm^{-1}). He quotes the value of $a_{\eta N} = (0.55 \pm 0.20 + \mathrm{i}\,0.30)$ fm. Notice the huge error in the real part of $a_{\eta N}$, which reflects the fact that the analysis of ref. [8] is not sufficiently constrained.

Abaev and Nefkens [9] have also used a form of an S-wave single resonance model, adjusted the resonance parameters to reproduce the $\pi^- p \to \eta n$ production channel to the best of their ability and extracted the S-wave scattering length as: $a_{\eta N} = (0.62 + \mathrm{i}\,0.30)$ fm.

Arima et al [10] have studied the nature of not one, but two S-wave resonances $\mathrm{S}_{11}(1535)$ and $S_{11}(1650)$ concerning their couplings with the ηN channel using the two quark-model wave functions with pure intrinsic spin states for the isobars. The dynamical coupling of the isobars to πN and ηN channels are described by the meson-quark coupling. They have obtained the S-wave scattering length $a_{\eta N} = (0.98 + \mathrm{i}\,0.37)$ fm what notably deviates from earlier extracted values.

Within the framework of the K-matrix approach with the two S_{11} resonances $S_{11}(1535)$ and $S_{11}(1610)$, Sauermann et al. [11] have fitted πN elastic PWA [12] and have obtained quite a good agreement with the η production cross section. However, the slope of the total η production cross section is significantly lower than the generally accepted value obtained in ref. [6] (17.0 vs. 21.2), see ref. [13]. They find yet the value for the ηN S-wave scattering length of $a_{\eta N} = 0.51 + \mathrm{i}\,0.21$ fm. The importance of that work is that it directly claims that introducing another resonance in the S-wave significantly influences the obtained values of the $\eta N N^*$ coupling constant of the first resonance. Very similar conclusions about the necessity to include more than one resonance have been reached by our group [14] within the framework of the different model.

In this article we show that the mutual agreement of all afore listed results, with the exception of Wilkin, Arima et al. and Sauermann et al. (which are not single resonance models) [8, 10, 11], is to be expected since the knowledge of πN elastic T-matrix, originating from any of a world collection of PWA [12, 15, 16, 17], and the threshold value of the $\pi^- p \rightarrow \eta n$ cross section are sufficient to calculate the ηN S-wave scattering length value in any, single resonance model. All cited models indeed do assume using only one resonance for the dominant S_{11} partial wave. Variations among the obtained ηN S-wave scattering length values can be completely attributed to the somewhat different assessment of the input data.

We show that the complete knowledge of the ηN S-wave T-matrix is essential in order to extract the real part of the S-wave scattering length. Therefore, attempts not founded on a multichannel, multiresonance, unitary representation of the S-wave T-matrix should be considered as a rough estimate only. The S-wave scattering length, coming out of our three coupled channel, multiresonance analyses [14] gives higher values for the scattering length magnitude than it has previously been reported in all publications with the exception of Arima et. al. [10] and Sauermann et.al. [11]. All values of the extracted $a_{\eta N}$ using the world collection of πN elastic η threshold T-matrix values are given in Table 1 and all other $a_{\eta N}$ values given on the literature are given in Table 2.

2 Formalism

2.1 *Limitations of the Single Resonance Models*

We have given the full formalism in [18], so we shall just outline the relevant result.

Using elementary formulae characteristic for the single resonance model, some isospin algebra, optical theorem and the detailed balance we get the lower bound for the imaginary part of the ηN S-wave scattering length:

$$\mathrm{Im}\ a_{\eta N,0} \geq \frac{3p_\pi^2}{8\pi} \frac{\sigma^{\mathrm{tot}}_{\pi^- p \rightarrow \eta n}(W_\eta)}{p_\eta}, \tag{1}$$

p_η is the η c.m. momentum and p_π is the c.m. momentum of the particles in the πN elastic channel at ηN threshold W_η. Using the experimental value of the η production total cross section near ηN threshold W_η:

$$\frac{\sigma^{\mathrm{tot}}_{\pi^- p \rightarrow \eta n}(W_\eta)}{p_\eta} = (21.2 \pm 1.8)\ \mu\mathrm{b/MeV} \tag{2}$$

taken from ref. [6], we obtain the appropriate constraint on the imaginary part of the ηN scattering length based exclusively on the optical theorem:

$$\mathrm{Im}\ a_{\eta N,0} \geq (0.24 \pm 0.02)\ \mathrm{fm}. \tag{3}$$

Knowing πN elastic T matrix and $\pi^- p \to \eta N$ near ηN threshold total cross section slope (see Eq. (2)) the isospin algebra gives:

$$a_{\eta N,0} \approx \frac{3}{2}\frac{p_\pi^2}{4\pi}\frac{T_{\pi\pi}(W_\eta)}{|T_{\pi\pi}(W_\eta)|^2}\frac{\sigma^{\text{tot}}_{\pi^- p\to\eta n}(W_\eta)}{p_\eta}\,. \tag{4}$$

The result is independent on the details of the model, e.g. on the T-matrix parametrization and the number of channels, which, because of unitarity, can not be lower than three if both, π elastic and η channel are included.

There are a lot of πN elastic phase shift analyses, let us mention just the few of them: [12, 15, 16, 17]. The $\pi^- p \to \eta n$ experimental total cross section is also quite well known [6]. Using the particular πN elastic phase shift analysis and $\pi^- p \to \eta n$ total cross section as input we get the distribution of values for the ηN scattering length for the single resonance model using only Eq. (2). We analyze the sensitivity of the distribution to the input.

2.2 Three Coupled Channel, Multiresonance and Unitary Model

Our model is based on the three coupled channel, multiresonance and unitary model developed by Cutkosky et. al. [15]. The starting point is setting up the three coupled channel T-matrix having all partial waves noncoupled because of spin zero of one of the interacting particles.

The $\pi N \to \eta N$ partial wave T matrices are matrix elements of the three channel partial wave T^{JL} matrix which is given as:

$$T^{JL} = \begin{pmatrix} T^{JL}_{\pi\pi} & T^{JL}_{\pi\eta} & T^{JL}_{\pi\pi^2} \\ T^{JL}_{\eta\pi} & T^{JL}_{\eta\eta} & T^{JL}_{\eta\pi^2} \\ T^{JL}_{\pi^2\pi} & T^{JL}_{\pi^2\eta} & T^{JL}_{\pi^2\pi^2} \end{pmatrix}$$

where various channels are denoted by the index π for πN, η for ηN and π^2 for all other channels ($\pi\Delta$, ρN, $\pi\pi N$, ...). The third channel is effectively described as a two body process $\pi^2 N$ with π^2 being a quasiparticle with a different mass chosen for each partial wave. We have fixed the channel masses, for each partial wave independently. Let us just mention that there is no mixing between different partial waves like e.g. in nucleon-nucleon scattering due to the fact that both, π and η are spin singlets.

Upon obtaining the full partial wave T-matrix, we reproduce the πN elastic partial waves by Höhler et.al. [12], and obtain a prediction for the $\pi N \to \eta N$ and $\eta N \to \eta N$ partial wave T-matrices. Using the prediction for the S-wave ηN elastic T-matrix, we extract the value for the ηN S-wave scattering length in a standard way.

3 Results and Conclusions

In Table 1 and Fig. 1 we show the possible spread in the ηN S-wave scattering length within the framework of any single resonance model. Numerical

Table 1. The ηN S-wave scattering length within the framework of any single resonance model, using πN elastic T matrix at ηN threshold from different phase shift analyses [12, 15, 16, 17] and η production total cross section [6] from Eq. (2) as the input. The πN T matrix is given in the first column, and resulting ηN S-wave scattering length in the second one.

	$T_{\pi N\,\pi N}(W_\eta)$	$a_{\eta N}$ [fm]
solution CMB [15]	$(0.376 + i0.439)$	$(0.269 + i0.315)$
solution KH80 [12]	$(0.332 + i0.393)$	$(0.301 + i0.356)$
solution KA84 [16]	$(0.390 + i0.374)$	$(0.320 + i0.307)$
solution FA84 [17]	$(0.400 + i0.339)$	$(0.348 + i0.296)$
solution CV90 [17]	$(0.412 + i0.328)$	$(0.356 + i0.283)$
solution KV90 [17]	$(0.418 + i0.334)$	$(0.350 + i0.279)$
solution SM90 [17]	$(0.407 + i0.303)$	$(0.379 + i0.282)$
solution FA93 [17]	$(0.403 + i0.344)$	$(0.344 + i0.293)$
solution WI94 [17]	$(0.408 + i0.339)$	$(0.348 + i0.289)$
solution SP95 [17]	$(0.399 + i0.338)$	$(0.350 + i0.296)$

values for the ηN scattering length in Table 1 and the open circles of Fig. 1. are based on the afore described model independent approach. The square at Fig. 1 represents the domain where the ηN S-wave scattering length can be if we use the present πN elastic T-matrix values at η production threshold (given in Table 1) and the slope of total η production cross section of ref. [6] as input. Error bars are predominantly a consequence of the error bar of the total η production cross section slope at the η production threshold and the contribution from the T-matrix uncertainty is ~ 5 %.

In Table 2 we give the results for the ηN scattering length given elsewhere together with the input: πN elastic T-matrices and the near threshold η production total cross section when possible. A compilation of all results is shown in Fig. 1.

Open circles represent the ηN S-wave scattering length values obtained on the basis of πN elastic T-matrices of refs. [12, 15, 16, 17] and the η production total cross section given in Eq. (2). Values of the ηN scattering length which are obtained using the questionable data of [2], in particular in refs. [1, 3] can be renormalized according to the procedure given in ref. [14]. Full normal and inverse triangles represent the original and modified scattering length values of ref. [1], cross and encircled cross represent the original and modified scattering length of ref. [3] and the full square gives the value given by ref. [9]. Their value is much higher then in other single resonance models because of unreasonably low value for the πN elastic T-matrix value at the η production threshold. The full circles represent the value of ref. [8]. The huge error bar for the real part of the ηN scattering length (≈ 50 %) indicates that the model is not

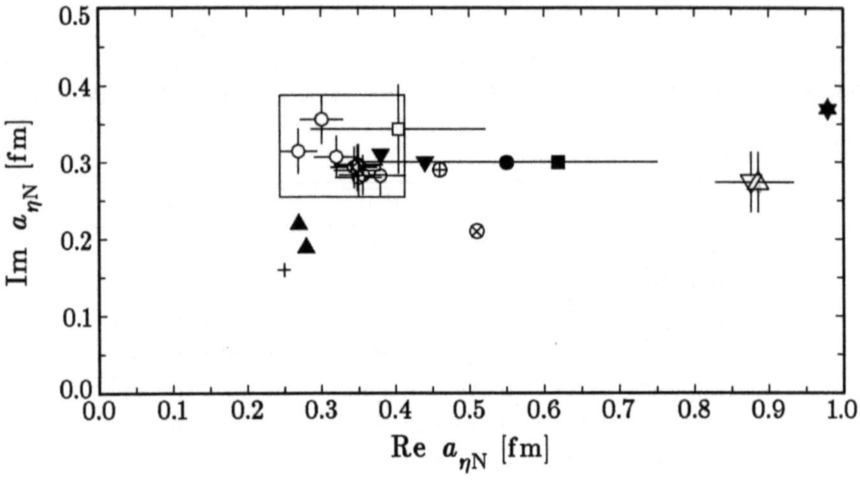

Figure 1. ηN S-wave scattering length. Open circles represent the ηN S-wave scattering length values obtained on the basis of πN elastic T-matrices of refs. [12, 15, 16, 17] and the η production total cross section given in Eq. (2). Full normal and inverse triangles represent the original and modified scattering length values of ref. [1], cross and encircled cross represent the original and renormalized value of ref. [3] and the full square gives the value given by ref. [9]. The full circles represent the value of ref. [8]. Open square represents the value obtained in our full model [14] restricted to the single resonance only and an open normal and inverse triangles represent the values of our full, multiresonance model with three and four resonances P_{11} partial wave, respectively. The full star represents the value of ref. [10], and the encircled x represent the value of ref. [11].

sufficiently constrained. Open square represents the value obtained in our full model [14] restricted to the single resonance only. Open normal and inverse triangles represent the values of our full, multiresonance model using three and four poles in the P_{11} partial wave, respectively. The full star represents the value of ref. [10], and the encircled x represents the value of Sauermann et al. [11]. The value of Arima et al. [10] is somewhat bigger than our value [7], but they have used bigger value for the total η production cross section slope. Exactly in the same way, the total η production cross section for the Sauermann et al. is somewhat lower than the recommended value used in our publication [14] giving the imaginary part of the ηN scattering length dangerously close to the generally accepted optical theorem value. If appropriate renormalization of both calculations is done [7] in order to reproduce the total cross section near η-threshold slope of Binnie et al. [6], we expect that their values will become much closer to our value.

The ηN S-wave scattering length obtained in the framework of the full model, when other resonances and the background term in particular are explicitly introduced in the unitary way is significantly more attractive than any

Table 2. Results of the ηN S-wave scattering length analyses. The scattering length is given in the first column, and the πN elastic T matrix at ηN threshold and corresponding η production total cross section near threshold are given in the last two columns.

	$a_{\eta N}$ [fm]	$T_{\pi\pi}$(1487 MeV)	$\sigma^{tot}_{\pi^- p \to \eta n}(W_\eta)/p_\eta$ [μb/MeV]
BL	(0.27 + i0.22)	(0.38 + i0.31)*	15.0
	(0.28 + i0.19)	(0.37 + i0.25)*	13.4
mBL	(0.38 + i0.31)	(0.38 + i0.31)*	21.2
	(0.44 + i0.30)	(0.37 + i0.25)*	21.2
BT	(0.25 + i0.16)	(0.37 + i0.24)*	11.5
mBT	(0.46 + i0.29)	(0.37 + i0.24)*	21.2
AN	(0.62 + i0.30)	(0.298 + i0.145)	20.2
SR	(0.404 + i0.343)	(0.345 + i0.293)	21.2
Wi	(0.55 + i0.30)	–	–
Ar	(0.98 + i0.37)	(0.47 + i0.37)	–
Sa	(0.51 + i0.21)	(0.43 + i0.40)*	17.0**
full	(0.886 + i0.274)	(0.373 + i0.331)	21.2
	(0.876 + i0.274)	(0.375 + i0.330)	21.2

BL	Bhalerao-Liu [1]	mBL	"modified" Bhalerao-Liu [14]
BT	Bennhold-Tanabe [3]	mBT	"modified" Bennhold-Tanabe [3]
AN	Abaev and Nefkens [9]	SR	our SR model [14]
Wi	Wilkin [8]	Ar	Arima et al. [10]
Sa	Sauermann et al. [11]	full	our full model [14]
*	data are read off the graphs	**	ref. [13]

of the predictions obtained within the limits of the single resonance models. All single resonance models, including the reduction of our model to the single resonance case, with good πN elastic T-matrix at threshold lie within the limits which are obtained using realistic ηN production cross section, and the πN elastic T-matrix. However, we consider the prediction of ref. [9] as non-reliable because the threshold value of the πN elastic T-matrix differs significantly from worldwide accepted values. The solution of Arima et al. [10] and Sauermann et al. [11] are much closer to our final value because their models are *not* of single resonance nature proving the consistency of the suggested analyses.

Conclusion:

For the single resonance model the simple, model independent mechanism for extracting the ηN S-wave scattering length from the near threshold values of the πN elastic T-matrices and the η production total cross section exists. All reported values, based on the single S-wave resonance assumption agree with our model independent results, bearing in mind differences in chosen input. The realistic ηN S-wave scattering length based on the full, three coupled channel, multiresonance and manifestly unitary model [14] is, and should be very different from any single resonance prediction. The coupled channel process can not be reliably described using a simplified, single resonance model. The analysis of [14] gives the new prediction for the ηN S-wave scattering length with much more attraction than previously reported. The findings for the ηN scattering length of all other attempts to extract the ηN S-wave scattering length , which include more than one resonance per partial wave [8, 10, 11] confirm the conclusion that the single resonance model is insufficient to give a confident value for the analyzed quantity. The results for the ηN S-wave scattering length of models which include more then one resonance in the S-wave [10, 11], or model which is not relying on the single resonance assumption [8] tend to be much closer to the value predicted by our three coupled channel, multiresonance and unitary model. Therefore, new, very precise measurements of η production total cross section [19] are needed to determine ηN scattering length much better, but the extraction has to be made in a multiresonance model. Otherwise, because of complexity of the problem, even very precise measurements can not give a good result if a correct extracting procedure for ηN scattering length is not applied.

References

1. R.S. Bhalerao, L.C. Liu: Phys. Rev. Lett. **54**, 865 (1985)

2. R.M. Brown et al.: Nucl. Phys. **B153**, 89 (1979)

3. C. Bennhold, H. Tanabe: Nucl. Phys. **A350**, 625 (1991)

4. M. Clajus, B.M.K. Nefkens: in πN Newsletter, **No 7**, ed. G. Höhler, W. Kluge, B.M.K. Nefkens, 76 (1992)

5. C. Bennhold: Private Communication

6. D.M. Binnie et al.: Phys. Rev. **D8**, 2793 (1973)

7. M. Batinić, I. Šlaus, A. Švarc: submitted for publication to Phys. Rev. C, available as paper nucl-th/9502017 at **xxx.lanl.gov.** or **babbage.sissa.it**

8. C. Wilkin: Phys. Rev. **C47**, R938 (1993)

9. V.V. Abaev, B.M.K. Nefkens: Private Communication

10. M. Arima, K. Shimizu, K. Yazaki: Nucl. Phys. **A543**, 613 (1992)

11. Ch. Sauermann, B.L. Friman, W. Nörenberg: Phys. Lett. **B341**, 261 (1995)

12. G. Höhler: in *Landolt-Börnstein Elastic and Charge Exchange Scattering of Elementary Particles*, Vol. **9**, Subvolume **b**: *Pion Nucleon Scattering*, **Part 2** (1983)

13. Ch. Sauermann: Private Communication

14. M. Batinić et al.: Phys. Rev. **C51**, No 5 (1995)

15. R.E. Cutkosky et al.: Phys. Rev. **D20**, 2804 (1979); R.E. Cutkosky et al.: Phys. Rev. **D20**, 2839 (1979)

16. R. Koch: Z. Physik **C29**, 597 (1985); R. Koch: Nucl. Phys. **A448**, 707 (1986);

17. Solutions FA84, CV90, KV90, SM90, FA93, WI94, and SP95 obtained from SAID via INTERNET in January 1995, from Department of Physics, Virginia Polytechnic Institute and State University, Blacksburg, VA 24061, USA, telnet: VTINTE.PHYS.VT.EDU, user: PHYSICS, Password: QUANTUM

18. M. Batinić, A. Švarc: submitted for publication to Few-Body Systems, available as paper nucl-th/9503020 at `xxx.lanl.gov.` or `babbage.sissa.it`

19. AGS Experiment 909, May-June 1995.

Few-Body Systems Suppl. 9, 213–218 (1995)

Few-Body Systems

Photo- and Electroproduction of Eta Mesons on Nucleons and Nuclei

L. Tiator[1], C. Bennhold[2], S.S. Kamalov[3], G. Knöchlein[1], F.X. Lee[4], L.E. Wright[5]

[1] Institut für Kernphysik, Johannes Gutenberg-Universität Mainz, 55099 Mainz, Germany

[2] Center of Nuclear Studies, Department of Physics, The George Washington University, Washington, D.C., 20052, USA

[3] Laboratory of Theoretical Physics, JINR Dubna, Head Post Office Box 79, SU-101000 Moscow, Russia

[4] TRIUMF Theory Group, 4004 Wesbrook Mall, Vancouver, B.C., V6T 2A3, Canada

[5] Institute of Nuclear and Particle Physics, Department of Physics, Ohio University, Athens, Ohio 45701, USA

Abstract. Eta photo- and electroproduction off the nucleon is investigated in an approach that contains Born terms, vector meson and nucleon resonance contributions. In a comparison with the new Mainz data we find a large sensitivity on the elementary ηNN coupling. Our analysis results in a pseudoscalar ηNN coupling with a coupling constant of $g^2_{\eta NN}/4\pi$=0.4. Furthermore, we also study coincidence cross sections for eta electroproduction and present calculations for structure functions and kinematical conditions that are most sensitive to the $S_{11}(1535)$ and the $D_{13}(1520)$ resonances. Finally, we show results on the inclusive eta photoproduction off complex nuclei with a very good agreement with recent data from Mainz.

1 Introduction

Over the last two years eta photoproduction from the nucleon has been measured at Mainz and at Bonn with an accuracy of more than an order of magnitude better than in older experiments up to 30 years ago. At Mainz, the TAPS collaboration has obtained high-quality data for angular distributions and total cross sections for photon energies between threshold and 790 MeV that may be considered to be a qualitative break-through in the experimental field [1]. Similarly, data at the higher energies up to 1150 MeV will be provided soon by the Phoenics collaboration at ELSA [2]. In contrary, the data basis for electroproduction of eta mesons is still very scarce. It is limited to a few older Bonn data [3] with large error bars and a more recent investigation of the ELAN collaboration at ELSA at very small momentum transfer, $Q^2 = 0.056\,\mathrm{GeV}^2$ [4]. Since eta production is strongly dominated by the $S_{11}(1535)$ resonance, its coupling to other resonances is difficult to extract even from very precise angular distributions. Instead, polarization observables will have to play a major role in order to constrain the small resonance couplings. Such experiments proposed at GRAAL in Grenoble will cover the energy range up to 2 GeV in the c.m. frame and provide also a first look at η' production. Moreover, a series of experiments has been planned at CEBAF to study both η and η' production by use of polarization degrees of freedom.

Most attempts to describe eta photoproduction on the nucleon have involved Breit–Wigner functions for the resonances and either phenomenology or a Lagrangian approach to model the background. These models which contain a large number of free parameters were then adjusted to reproduce the few available data [5]. In a very different approach, ref. [6] derived a dynamical model which employs $\pi N \rightarrow \pi N, \pi N \rightarrow \pi\pi N$ and $\pi^- p \rightarrow \eta n$ to fix the hadronic vertex as well as the propagators and the $\gamma N \rightarrow \pi N$ to construct the electromagnetic vertex. This calculation represents a prediction rather than a fit to the $\gamma N \rightarrow \eta N$ reaction.

Here we extend the model of ref. [6] by taking into account the background from s, u–channel nucleon Born terms and ρ, ω exchange in the t–channel. Since the resonance sector is fixed in our approach and also the vector meson couplings can be obtained from independent sources, we can use this model to extract information on the ηNN coupling [7]. Furthermore we will show the sensitivity of $(e, e'\eta)$ coincidence cross sections on different nucleon resonances and will apply our model to inclusive eta photoproduction on complex nuclei.

2 Eta Photoproduction on the Nucleon

In contrast to the πN-interaction, little is known about the ηN-interaction and, consequently, about the ηNN vertex. The uncertainty regarding the structure of the ηNN vertex extends to the magnitude of the coupling constant. This coupling constant $g^2_{\eta NN}/4\pi$ varies between 0 and 7 with the large couplings arising from fits of one boson exchange potentials. In SU(3) flavor symmetry the values of the coupling constant is between 0.8 and 1.9. Smaller values are supported by NN forward dispersion relations and analyses of the strangeness content of the proton [9] with $g^2_{\eta NN}/4\pi + g^2_{\eta' NN}/4\pi \leq 1.0$. Nevertheless, from the above discussion it seems clear that the ηNN coupling constant is much smaller compared to the corresponding πNN value of around 14, see ref. [7].

In Fig. 1, we show the sensitivity of the total cross section close to threshold when varying the coupling constant from 0 to 3 for the PS and from 0 to 10 for the PV form. There is a large variation of more than a factor of 2 at 750 MeV for the PS case while changing the value with PV structure modifies the

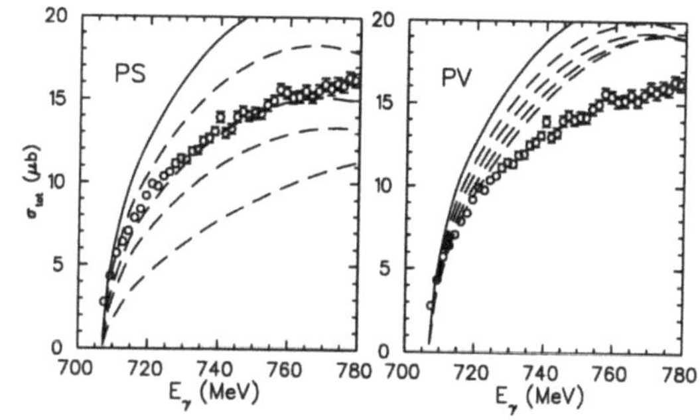

Figure 1. Total cross section for the process (γ, η) on the proton calculated with PS and PV Born terms. The full curve contains no Born terms, while the dashed lines are (from the top down) obtained with $g^2_{\eta NN}/4\pi$=0.1, 0.5, 1.0, and 3.0 for PS-coupling, and $g^2_{\eta NN}/4\pi$=1.0, 3.0, 6.0 and 10.0 for PV-coupling, respectively. The experimental data is from Mainz [1].

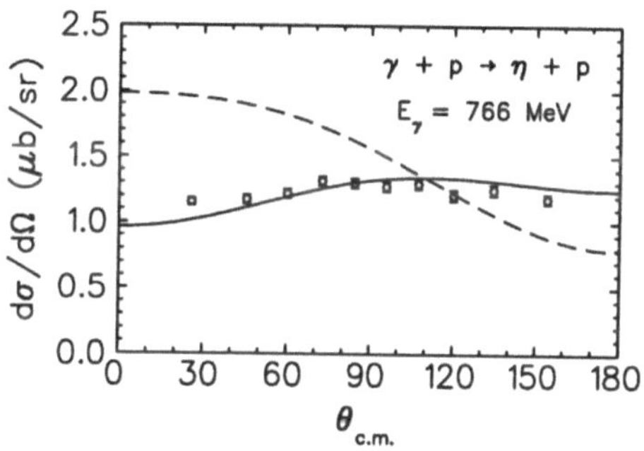

Figure 2. Differential cross section for eta photoproduction at 766 MeV photon lab. energy. The solid and dashed lines are calculations in PS coupling with coupling constant 0.4 and in PV coupling with 10, respectively. The data is from Mainz [1].

cross section only by a relatively small amount.

Here the very precise results of the new Mainz experiment can clearly distinguish between the different models preferring a pseudoscalar coupling scheme with a best-fit coupling constant of $g^2_{\eta NN}/4\pi$=0.4. Furthermore, in Fig. 2 the Mainz experiment gives another even stronger argument favouring PS coupling with a small coupling constant.

3 Eta Electroproduction on the Nucleon

Without additional polarization of electrons and nucleons the differential cross section for eta electroproduction can be split into 4 parts, the transverse (T), the longitudinal (L), the longitudinal-transverse (TL) and the transverse-transverse (TT) coincidence cross sections [8]. Figure 3 presents the longitudinal cross section σ_L for kinematics of ref. [3] in different models, which are more

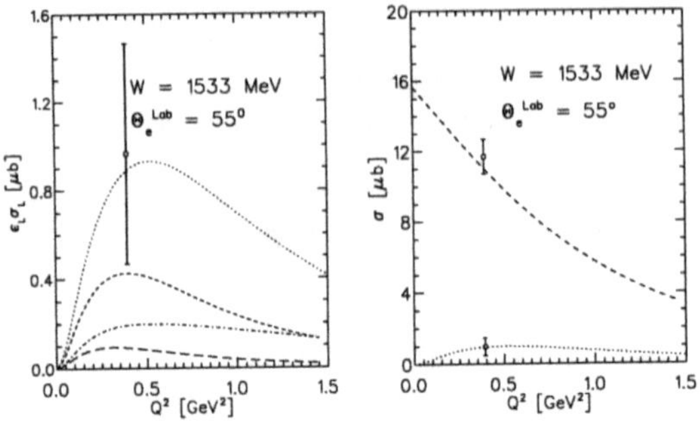

Figure 3. Longitudinal and transverse total cross sections for $p(e,e'\eta)p$. Left: Different models for the L_{0+} of the S_{11}-resonance. Non-relativistic constituent quark model [10], relativized constituent quark model [11], light cone model [12], from top to bottom. The long dashed line is the contribution of the non-resonant background. Right: Transverse (upper) and longitudinal (lower) total cross sections in our standard parametrization. The data points are from ref. [3].

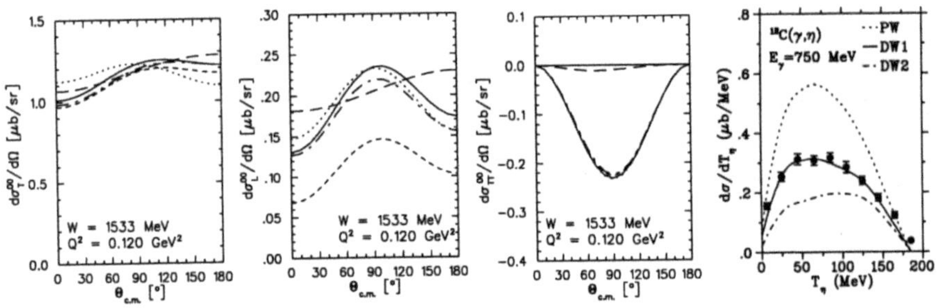

Figure 4. Unpolarized differential cross sections $d\sigma_{\mathrm{T}}/d\Omega$, $d\sigma_{\mathrm{L}}/d\Omega$, $d\sigma_{\mathrm{TL}}/d\Omega$ and $d\sigma_{\mathrm{TT}}/d\Omega$, at $W = 1533$ MeV and $Q^2 = 0.120$ GeV^2. The solid curve is our standard calculation (M1), for the short dashed line the resonant parts of the E_{0+} and L_{0+} multipoles were calculated in the light-cone model [12], the dotted line is a calculation without $P_{11}(1440)$, the long dashed line without $D_{13}(1520)$, and the dash-dotted line without non-resonant background.

or less consistent with the existing data point. In our standard calculations we will use the non-relativistic constituent quark model with the amplitude of the resonant part of the L_{0+} multipole normalized to the data point from [3]. The reason for the large experimental error bar is due to the fact that the longitudinal excitation of the $S_{11}(1535)$ is weaker than the transverse one. This can be clearly seen in Fig. 3. The transverse/longitudinal separation of the inclusive cross section, $\sigma_{\mathrm{tot}} = \sigma_{\mathrm{T}} + \varepsilon_{\mathrm{L}}\sigma_{\mathrm{L}}$, leads to a longitudinal cross section $\varepsilon_{\mathrm{L}}\sigma_{\mathrm{L}}$, which is smaller than the transverse σ_{T} by more than an order of magnitude. The data point for the transverse cross section, however, is in good agreement with our model prediction.

In Fig. 4 we show the 4 coincidence cross sections for $p(e,e'\eta)p$. The transverse cross section $d\sigma_T/d\Omega$ has a rather flat angular distribution as in the photoproduction case. The longitudinal cross section $d\sigma_L/d\Omega$ is smaller than the transverse by about one order of magnitude and has a maximum around 100° due to the $D_{13}(1520)$ contribution. Of course, the magnitude of this cross section depends strongly on the prediction of the different quark models. The cross sections $d\sigma_{TT}/d\Omega$ and $d\sigma_{TL}/d\Omega$ are very sensitive to the presence of the resonance $D_{13}(1520)$. The transverse-transverse interference cross section almost vanishes without this resonance, and the shape of the transverse-longitudinal interference cross section changes completely when the resonance is decoupled.

4 Inclusive Eta Photoproduction on Complex Nulei

In Fig. 5 we show our results for inclusive eta photoproduction on ^{12}C and ^{40}Ca. In the simple IA the nuclear cross section is A times the nucleon cross section. However, nucleon Fermi motion and eta absorption and rescattering leads to a drastic reduction that scales like $A^{0.68}$ [13] in the total cross section over a range of nuclei from C to Pb. While our plane wave calculation is about a factor of 2 above the data, the eta distortion with our first-order optical potential (DW1) describes the data very well. On the other hand, the potential DW2 of ref. [14] has a much stronger absorption and leads to cross sections well below the data.

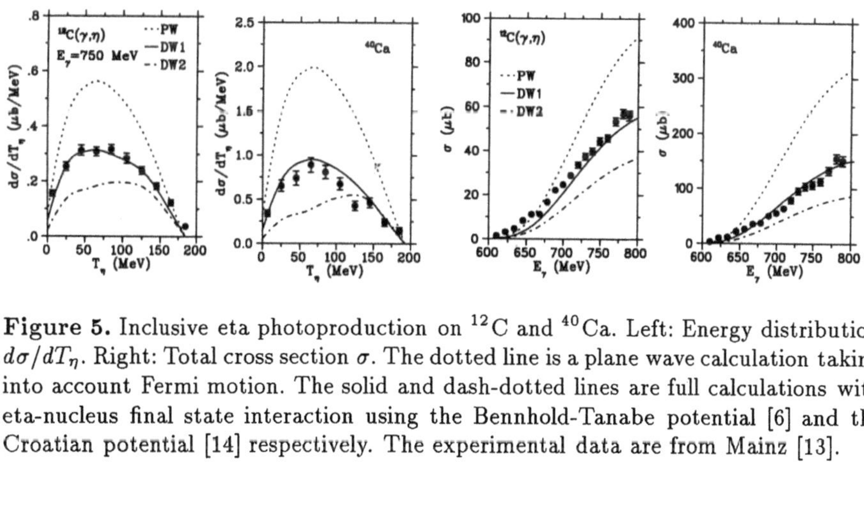

Figure 5. Inclusive eta photoproduction on ^{12}C and ^{40}Ca. Left: Energy distribution $d\sigma/dT_\eta$. Right: Total cross section σ. The dotted line is a plane wave calculation taking into account Fermi motion. The solid and dash-dotted lines are full calculations with eta-nucleus final state interaction using the Bennhold-Tanabe potential [6] and the Croatian potential [14] respectively. The experimental data are from Mainz [13].

5 Conclusion

We have presented a model for eta photoproduction on the nucleon that includes nucleon Born terms and t-channel vector meson exchanges and nucleon resonances $S_{11}(1535)$, $P_{11}(1440)$ and $D_{13}(1520)$. The resonance sector is fixed by using data from hadronic and electromagnetic reactions such as pion scattering and photoproduction and pion-induced eta production. Vector meson couplings are determined from their radiative decay widths and the NN-interaction. Using the new experimental data from Bonn and Mainz we are able to determine the ηNN coupling constant around 0.4 with a clear preference for the PS coupling scheme. Further experiments with polarized beam and target as well as coincidence cross sections for $(e, e'\eta)$ will give the possibility to study also smaller resonances as $D_{13}(1520)$ and $P_{11}(1440)$ as well as longitudinal resonance excitations, especially for the $S_{11}(1535)$. Inclusive eta photoproduction from complex nuclei shows a strong sensitivity to the eta-nucleus interaction. Such informations can not be obtained by direct scattering experiments as in the pion case. Future exclusive quasi-free experiments of etas from individual nuclear shells can shed more light on the eta-nucleus interaction as well as on possible medium effects of nucleon resonances in nuclei.

Acknowledgement. This work was supported in part by the Deutsche Forschungsgemeinschaft (SFB201), the U.S. DOE grant DE-FG02-95-ER40907, the Heisenberg-Landau program, and a NATO Collaborative Research Grant.

References

1. B. Krusche et al.: Phys. Rev. Lett. **74**, 3736 (1995)
2. M. Breuer: Dissertation. Bonn 1994
3. H. Breuker et al.: Phys. Lett. **74B**, 409 (1978); C. Nietzel: Dissertation. Bonn 1978
4. M. Wilhelm: Dissertation. Bonn 1993
5. M. Benmerrouche, N.C. Mukhopadhyay, J.F. Zhang: Phys. Rev. **D51**, 3237 (1995)
6. C. Bennhold and H. Tanabe: Nucl. Phys. **A530**, 625 (1991)
7. L. Tiator, C. Bennhold, S.S. Kamalov: Nucl. Phys. **A580**, 455 (1994)
8. G. Knöchlein, D. Drechsel, L. Tiator: Zeitschrift für Physik **A** (in print)
9. T. Hatsuda: Nucl. Phys. **B329**, 376 (1990)
10. L. A. Copley, G. Karl, E. Obryk: Nucl. Phys. **B14**, 302 (1969)
11. M. Warns et al.: Z. Phys. **C45**, 613; 627 (1990)
12. W. Konen, H.J. Weber: Phys. Rev. **D41**, 2201 (1990)
13. M. Röbig-Landau: Dissertation. Giessen 1995
14. M. Batinic et al.: Phys. Rev. **C51** (1995) (in print)

Few-Body Systems Suppl. 9, 219–222 (1995)

Few-Body Systems

Fully Relativistic Calculation of the $\pi d \rightarrow \eta NN$ Process, With the Final State Interaction Included

M. Batinić[1], I. Šlaus[1], A. Švarc[1], B.M.K. Nefkens[2]

[1] Rudjer Bošković Institute, Zagreb, Croatia
[2] University of California, Los Angeles, USA

Abstract. A fully relativistic calculation of the $\pi d \rightarrow \eta NN$ process, based on the leading diagram, is presented. The complete knowledge about the contributing vertices is taken into account. The role of the $\pi N \rightarrow \eta N$ vertex was studied by using 1) only S_{11}, P_{11}, and D_{13} partial waves single resonances, and 2) three coupled channel multiresonance unitary model for eight lowest partial waves. The final state NN interaction has been taken into account effectively using the Jost function recipe.

1 Introduction

Few body reactions are very often described by using elementary two body processes as the input.

As $\pi N \rightarrow \eta N$ partial wave analysis already exists [1], and data for $\pi^{\pm} d$ reaction testing charge symmetry breaking (CSB) near η production threshold are coming [2], the η production in pion deuteron collisions is a good candidate to test (extend) our knowledge of η production amplitudes in pion-nucleon collisions.

2 Formalism

2.1 *Invariant Matrix Element*

The invariant matrix element, describing the $\pi d \rightarrow \eta NN$ based on the dominant diagram shown in Fig. 1, using Bjorken-Drell [3] conventions, is:

$$\mathcal{M}_{\lambda_d,\lambda_1\lambda_2} = \frac{1}{2}\left\{ \overline{u}_{\lambda_1}(p_1)\left[A(W,\cos\theta) + \not{p}_\eta B(W,\cos\theta)\right] \frac{\not{p}_d - \not{p}_2 + m}{(p_d - p_2)^2 - m^2} \right.$$
$$\left. [\not{\epsilon}_{\lambda_d} G_a + (p_2 \cdot \epsilon_{\lambda_d}) G_b]\, v_{\lambda_2}(p_2) \;-\; (1 \leftrightarrow 2) \right\} \qquad (1)$$

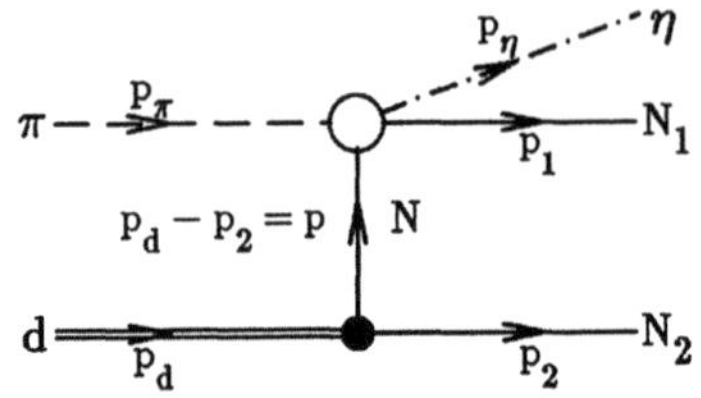

Figure 1. The dominant diagram for the $\pi d \to \eta NN$ process

with notation as in Fig. 1: p_π, p_d, p_η, p_1, and p_2 are particle's momenta and deuteron and nucleon's helicities are denoted by λ_d, λ_1, and λ_2. The u and v are nucleon spinors. The relevant vertices (πN-ηN and n-p-d) are taken into account.

2.1.1 πN-ηN vertex

The πN-ηN vertex has been represented by the invariant amplitude

$$A(W, \cos\theta) + \not{p}_\eta B(W, \cos\theta)\,, \tag{2}$$

with the standard on shell partial waves decomposition for A and B, similar to those for πN elastic case [4]. Only changes are to use partial wave T matrices $T_{l\pm}(\pi N \to \eta N)$ (no isospin index, ηN is truly $I = 1/2$ state) and replace

$$q^3 \longrightarrow \sqrt{q_\pi^3 q_\eta^3} \quad \text{and} \quad E \pm m \longrightarrow \sqrt{(E_i \pm m)(E_f \pm m)}$$

where q_π and q_η are initial pion and final eta momenta, E_i and E_f are initial and final nucleon energies, all in the local $\pi N(\eta N)$ center of mass system. Off shell effects are taken into account as in [5].

For $T_{l\pm}$, we always used on shell values of different models:

- Single resonance model (SR) using resonances in three partial waves S_{11}, P_{11}, and D_{13}, dominating eta production near threshold [1], with presently accepted resonance parameters.
- Three coupled channel multiresonance unitary model (CCMRU) in eight lowest partial waves [1], fitted to the πN elastic T matrices and $\pi^- p \to \eta n$ available experimental data.

2.1.2 The n-p-d vertex

The n-p-d vertex is described by

$$\not{\epsilon}_{\lambda_d} G_a + (p_i \cdot \epsilon_{\lambda_d}) G_b\,; \quad i = 1, 2 \tag{3}$$

where ϵ_{λ_d} is deuteron polarization vector and G_a and G_b are deuteron invariant functions depending on $(p_d - p_i)^2$. We used invariant expansion [6] fitted [7] to modern electron scattering data on the deuteron.

2.2 Cross Sections

The differential cross section in the c.m. for exclusive experiment is given by

$$\frac{d^5\sigma}{dp_\eta d\Omega_\eta d\Omega_1} = \frac{|\mathcal{M}_{\lambda_d,\lambda_1\lambda_2}|^2}{\sqrt{(p_\pi \cdot p_d)^2 - m_\pi^2 m_d^2}} \frac{p_\eta^2 p_1^2 m^2}{32(2\pi^5)E_\eta(p_1 E_2 - E_1 p_1 \hat{p}_2)}, \tag{4}$$

and if target is unpolarized and nucleon helicities are not detected we need to average over deuteron and sum over nucleon helicities. Inclusive and total cross sections are obtained by appropriate integrations.

2.3 The Final State Interaction

The final state interaction (FSI) is taken into account using the Jost function recipe. As the $^1S_0$ partial wave is dominant at low energies, we multiply the whole $\mathcal{M}_{\lambda_d,\lambda_1\lambda_2}$ by $^1S_0$ Jost function. Therefore, the FSI reduces the cross section by

$$F(k) \equiv |J_{^1S_0}(k)|^2 = \frac{[-1/a + rk^2/2]^2}{[-1/a + rk^2/2]^2 + k^2} \tag{5}$$

where k is NN relative c.m. momentum; a and r are $^1S_0$ NN scattering length and effective range respectively.

3 Results and Conclusion

We study the sensitivity to the variations in our standard model. As the standard we used CCMRU model with 4 resonances in P_{11} for πN-ηN vertex [1],

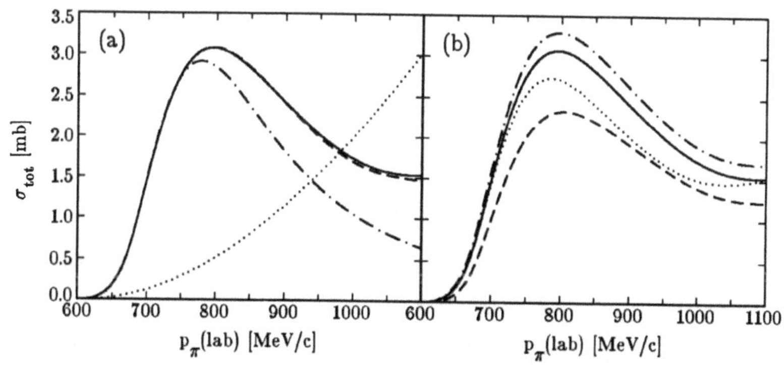

Figure 2. The $\pi^+ d \to \eta pp$ total cross sections. Sensitivities to the variations of the model. The full line represents our standard (see text). (a) Dashed line is CCMRU model [1] with 3 resonances in P_{11} partial wave; dash-doted line is SR model for πN-ηN vertex; dotted line represents phase space. (b) Dashed line is with NN FSI; dash-dotted line is with n-p-d parametrisation from ref. [6]; dotted line is with on shell πN-ηN vertex.

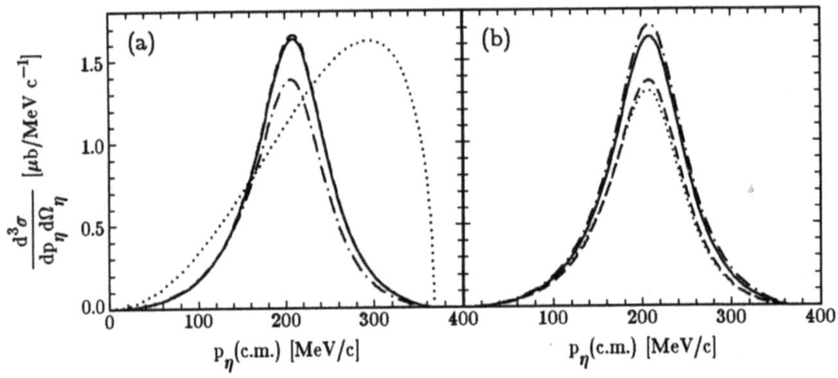

Figure 3. The $\pi^+d \rightarrow \eta pp$ inclusive cross sections for p_π(lab)=808 MeV/c, θ_η(c.m.)=90°. The labeling is the same as in Fig. 2.

off shell extrapolation as in ref. [5], n-p-d parametrisation from ref. [7], and with FSI switched off. For the illustration, we show $\pi^+d \rightarrow \eta pp$ total (Fig. 2) and inclusive (Fig. 3) cross sections. As can be expected, difference between SR and CCMRU models is small near η production threshold and it becomes large for $p_\pi(\text{lab}) > 750$ MeV/c. Difference between CCMRU model with 3 and 4 resonances in P_{11} partial wave is small. The final state interaction reduces cross sections, typicaly by 30 %, without significant change in shape. Off shell effects are important for both total and inclusive cross sections. Cross sections are sensitive to the n-p-d vertex parametrisation, larger differences are at higher energies. So, experimental data are badly nedeed to fix model details.

Our model predictions for the $\pi^+(\pi^-)d \rightarrow \eta pp(nn)$ processes are the same except for the kinematics and FSI.

References

1. M. Batinić et al.: Phys. Rev. **C51**, (1995)
2. AGS experiment 890, Jan-April 1995.
3. J.D. Bjorken and S.D. Drell: *Relativistic Quantum Mechanics.* New York: McGraw-Hill Book Company 1964
4. G. Höhler: In *Landolt-Börnstein: Elastic and Charge Exchange Scattering of Elementary Particles* (Vol. **9** Subvolume **b**: Pion Nucleon Scattering, Part 2) Berlin: Springer-Verlag 1983
5. A. König and P. Kroll: Nucl. Phys. **A356**, 345 (1981)
6. M. Gourdin et al: Nuovo Cim. **37**, 524 (1965)
7. M. P. Locher and A. Švarc: Z. Phys. **A338**, 89 (1991); Fizika **22**, 549 (1990)

Few-Body Systems Suppl. 9, 223–226 (1995)

Few-Body Systems

Photoproduction of Eta Mesons on Deuterium from Threshold to 1.2 GeV*

P.Hoffmann-Rothe[1], E.Hourany[1], M. Breuer[1], J.-P. Didelez[1†], M.Rigney[1], J. Ajaka[1], G.Anton[2], J.Arends[2], G.Berrier-Ronsin[1], W.Beulertz[2], G.Blanpied[4], A.Bock[2], G.v.Edel[3], R.Frascaria[1], K.Helbing[2], J.Hey[2], M.Krebeck[3], R.Maass[3], G.Nöldeke[2], B.Preedom[4], B.Ritchie[5], L.Rosier[1], B.Saghai[6], M.Schumacher[3], F.Smend[3], S.Whisnant[4], B.Zucht[1]

[1] IN2P3, Institut de Physique Nucléaire, 91406 Orsay, France

[2] Physikalisches Institut, Nussallee 12, 53117 Bonn, Germany

[3] Zweites Physikalisches Institut, Bunsenstr. 7-9, Göttingen, Germany

[4] University of South Carolina, Physics Dept, Columbia, SC 29208, USA

[5] Arizona State University, Physics Dept, Tempe, AZ 85281 USA

[6] Laboratoire DAPNIA/SPN, 91191 Gif-sur-Yvette, France

Abstract.

Measurements of the total and differential cross sections for η-meson photoproduction on ^{1}H, ^{2}D, and ^{14}N liquid targets from threshold to 1.2 GeV have been taken using the tagged Bremsstrahlung photon beam produced by the electrons extracted from the ELSA storage ring at Bonn. The reaction was identified by detecting the η decay products in the neutral meson spectrometer SPES0-2π, while the recoil baryons (proton, neutron, or deuteron) were detected by a variety of large angle scintillator detectors. We present here our experimental results on the deuterium target, both the comparison of the quasi-free proton and neutron as well as results on the coherent production. Our preliminary results for the counting rates integrated on the S_{11} excitation show significantly smaller figures for the neutrons than for the protons. This is consistent with an isoscalar part of the amplitude roughly ten times smaller than the isovector one and also explains the very low rate of coherent deuterons observed.

*Supported in part by BMFT, Germany, HC&M, Brussels, and the NSF, USA

† *E-mail address:* didelez@ipncls.in2p3.fr

1 Introduction

While there has been some recent activity, both experimental and theoretical on η photoproduction, most of the existing differential cross section data [1] and the 4 polarization data points [2] have been measured before 1970 typically using an untagged photon beam and detecting only one of the photons from the η. The recently published results [3] have concentrated on the study of the threshold behavior of η photoproduction on nucleons below the $S_{11}(1535)$ peak. The present Orsay-Bonn-Göttingen-Columbia collaboration has an extented program of η photoproduction both on nucleons and nuclei up to incident photon energies of 1.2 GeV, well above the $S_{11}(1535)$ resonance. Experiments have also been proposed at the Laser Compton Back-scattering facility GRAAL with photon energies up to 1.8 GeV with emphasis on polarization observables and at CEBAF until 2.2 GeV. All of these second generation experiments use tagged photon beams, high duty-cycle machines, and large angular acceptance detectors increasing greatly the potential to improve the experimental database.

2 Experiment

The η meson is a very selective probe for the study of the I=1/2 nucleon resonances (N^*), since it is an isospin I=0 particle. Furthermore, there are only a few of these resonances that have a significant decay mode to the η channel. For these reasons, the photoproduction of η meson is an ideal probe for the study of these resonances, in particular the S_{11} resonance.

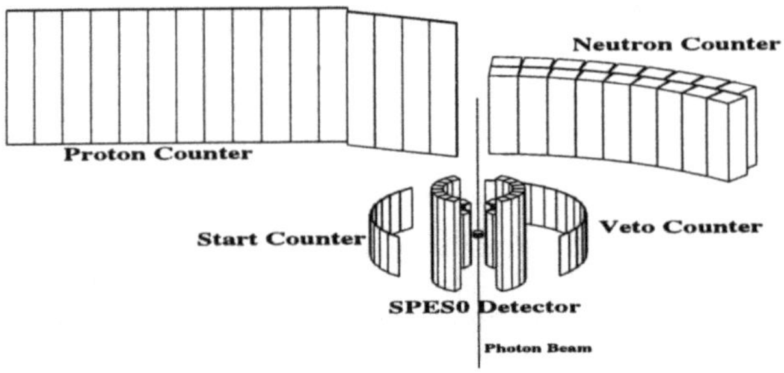

Figure 1. The experimental setup consisted of the neutral meson spectrometer SPES0-2π surrounding the target and the large scintillating hadron counters in the forward direction.

The collaboration presenting this paper has finished an extensive η photoproduction program measuring both total and differential cross sections on ^1H, ^2D, and ^{14}N 7 cm liquid targets as well as the target asymmetry using a dynamically polarized butanol target. All of these measurements have used the PHOENICS [4] tagged Bremsstrahlung photon beam up to 1.2 GeV at the electron stretcher ring ELSA in Bonn as well as the Orsay-built neutral meson spectrometer SPES0-2π [5] to identify the η by its decay products. The recoil baryons were detected using a variety of large scintillating detector systems po-

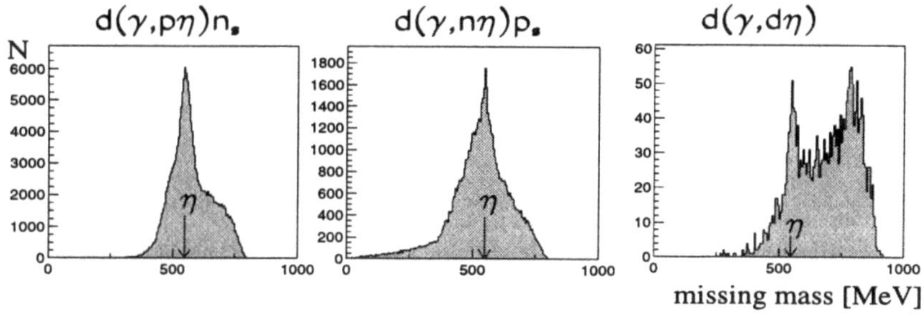

Figure 2. The η missing mass spectra corresponding to production on a QF proton, a QF neutron, and the coherent deuteron. All three spectra correspond to the same number of runs and represent about 30% of the data.

sitioned in the forward direction with a significant distance to permit a clean particle identification by time of flight. Figure 1 shows the main components of the detector system including the central SPES0-2π spectrometer and the large scintillating detectors (Proton Counters and Neutron Counters). Two other detection systems (not shown in Fig.1), the AMADEUS detector for increased recoiled deuteron acceptance and the SENECA detector array for increased neutron detection efficiency were placed at small angles during certain runs.

3 Discussion

The extraction of η photoproduction on the free neutron is included among the main goals of the present collaboration. This is obtained by measuring the production of the free proton on the Hydrogen target and the quasi-free production of the proton and neutron from the Deuterium target with the same experimental setup. Another important goal was a precise measurement of the coherent production from the deuteron through a clean identification of the final state. The one previous measurement [6] is now nearly 30 years old and still the large cross sections are not explainable by present theories. Lastly, we have made measurements also on a Nitrogen target in order to examine the nuclear medium effects on the excitation of the S_{11} resonance. A discussion of the first two points follows, while the interested reader is referred to the communication in the Varenna conference [7] for a detailed discussion of the last point.

The differential cross section on the free proton has been fitted on the whole energy range of the present experiment [8]. The formalism is based on an isobar model. In this approach, electric and magnetic multipole amplitudes are expressed in terms of various isospin-1/2 nucleonic resonances plus a smooth background including S and P waves. The resonances are described by energy dependent Breit-Wigner forms. The background arises mainly from the nucleon pole diagrams, and possibly at higher energies from t-channel vector meson exchange processes. The results confirm the major role played by the $S_{11}(1535)$ resonance, while the Roper resonance is not required to reproduce the data. Besides, among the phenomenological fits obtained, the one including only $S_{11}(1535)$, $D_{13}(1520)$, and $F_{15}(1680)$ resonances gives the most satisfactory agreement.

For the Deuterium target, we succeeded to identify completely the final states corresponding to the production of an η meson on a Quasi-Free (QF) proton, a QF neutron and the coherent deuteron. Figure 2 shows the corresponding peaks for the η missing mass. Although the analysis is still in a preliminary stage, it is already clear that neutrons are produced with a rate roughly 30% lower than the proton one. This generates an Isoscalar Amplitude about ten times smaller than the Isovector one. Therefore, the coherent deuteron should be produced with a rate hundred times smaller than the proton one as observed in Fig.2. It should be noted that we report here the first complete identification of the final state corresponding to the coherent production of a η meson on a Deuterium target. In ref.[6], only the recoil deuteron was detected, resulting in cross sections at least a factor five too high compared to our first estimations.

References

1. C.A. Heusch et al.: Phys. Rev. Lett. **17**, 573 (1966); C. Bacci et al.: Nuov. Cim. **45**, 983 (1966); C. Bacci et al.: Phys. Rev. Lett. **16**, 157 (1966); R. Prepost et al.: Phys. Rev. Lett. **18**, 82 (1967); E.D. Bloom et al.: Phys. Rev. Lett **21**, 1100 (1968); P.S.L. Booth et al.: Lett. Nuov. Cim. **2**, 66 (1969); B. Delcourt et al.: Phys. Lett. **B29**, 75 (1969); P.S.L. Booth et al.: Nucl. Phys. **B25**, 510 (1971); A. Christ et al.: Lett Nuov. Cim. **8**, 1039 (1973); P.S.L. Booth et al.: Nucl. Phys. **B71**, 211 (1974)

2. C.A. Heusch et al.: Phys. Rev. Lett. **25**, 1381 (1970)

3. S. Homma et al.: J. Phys. Soc. Japan. **57**, 828 (1988); S.A. Dytman et al.: Phys. Rev. **C51**, 2710 (1995); B. Krusche et al.: Phys. Rev. Lett. **74**, 3736 (1995); J.W. Price et al.: Phys. Rev. **C51**, R2283 (1995)

4. P. Detemple et al.: Nucl. Instr. Meth **A321**, 479 (1992)

5. M. Rigney: PhD Thesis. Univ. of S. Carolina 1992

6. R.L. Anderson, R. Prepost: Phys. Rev. Lett. **23**, 46 (1969)

7. M. Breuer et al.: 7th International Conference on *Nuclear Reaction Mechanisms*. Varenna, Italy, June 1994, Università degli Studi de Milano, Supplemento $n°$ 100, p. 584. ed. E. Gadioli, 1994

8. P. Hoffmann-Rothe, B. Saghai and F. Tabakin: XVth European Conference on Few-Body Problems, Peniscola, Spain, June 1995.

Few-Body Systems Suppl. 9, 227–230 (1995)

Few-Body Systems

On the possibility of an η–meson light nucleus bound state formation

S. A. Rakityansky [1,2], S. A. Sofianos[1], V. B. Belyaev [2], W. Sandhas [3]

[1] Physics Department, University of South Africa, P.O.Box 392, Pretoria 0001, South Africa

[2] Joint Institute for Nuclear Research, Dubna, 141980, Russia

[3] Physikalisches Institut, Universität Bonn, D-53115 Bonn, Germany

Abstract. The resonance and bound–state poles of the amplitude describing elastic scattering of η-meson off the light nuclei ^{2}H, ^{3}H, ^{3}He, and ^{4}He are calculated in the framework of a microscopic approach based on few–body equations. For each of the nuclei, the necessary two–body ηN–input allowing appearance of quasi–bound η–nucleus state, are also determined.

Estimations obtained in the framework of the first-order optical potential theory [1,2], put a lower bound on the nucleus atomic number A for which an η–nucleus bound state could exist, namely, $A \geq 12$. However, some speculations on the possibility of a formation of η–helium bound state still appear [3] despite the discouraging results of the first (and the sole) experiment [4] on direct search for bound states of η–meson with Lithium, Carbon, Oxygen, and Aluminium. All such speculations are based on the large negative values (~ -2 fm) of the real parts of η–nucleus scattering lengths calculated within a simplified optical–potential theory [3,5].

To the best of our knowledge, the only microscopic calculations of the scattering lengths was presented in our recent papers [6–8]. In these, it turned out that the η–helium scattering length could have even larger (negative) real part than earlier estimations. This raises more doubts on the validity of the above-mentioned constraint, $A \geq 12$, for the existence of η–nucleus bound states.

In the present work we examine the possibility of formation of a bound state in the η–meson d, t, ^{3}He, and ^{4}He systems, in the framework of a microscopic approach, namely, the Finite-Rank Approximation (FRA) of the nuclear Hamiltonian [9,10]. The approximate few–body equations of this approach (see ref.[7]) enable us to calculate the η–nucleus T–matrix $T(\boldsymbol{k}', \boldsymbol{k}; z)$ for any complex total energy z, i.e. for any point on the complex plane of the momentum $p = \sqrt{2\mu z}$. In this way we have numerically located the resonance state poles and (for each of the nuclei) we found a factor which enhances the ηN–attraction to generate quasi–bound state.

As an input information we used the ground-state wave functions of the nuclei involved and the two-body t–matrix $t_{\eta N}$. Firstly, we employed simple Gaussian–type functions. Secondly, we used the following separable form $t_{\eta N}(k', k; z) = \lambda/[(k'^2 + \alpha^2)(z - E_0 + i\Gamma/2)(k^2 + \alpha^2)]$ with $E_0 = 1535$ MeV $- (m_N + m_\eta)$ and $\Gamma = 150$ MeV [11]. In order to fix the parameter α, we make use of the results of refs.[12,13], where the same $\eta N \rightarrow N^*$ vertex function $(k^2 + \alpha^2)^{-1}$ was employed with α being determined via a two-channel fit to the $\pi N \rightarrow \pi N$ and $\pi N \rightarrow \eta N$ experimental data.

Due to experimental uncertainties and differences between the models for the physical processes, one can use three different values for the range parameter α, namely, $\alpha = 2.357\ \text{fm}^{-1}$ [12], $\alpha = 3.316\ \text{fm}^{-1}$ [13], and $\alpha = 7.617\ \text{fm}^{-1}$ [12]. Since there is no criterium for singling out one of them, we use all three in our calculation.

The remaining parameter λ is chosen to provide the correct zero-energy on-shell limit, i.e., to reproduce the known ηN scattering length. Like the range parameter α, the scattering length $a_{\eta N}$ is not well known, the estimated values being within the range $\text{Re}a_{\eta N} \in [0.27, 0.98]$ fm and $\text{Im}a_{\eta N} \in [0.19, 0.37]$ fm [14]. Our intention is to vary the $\text{Re}a_{\eta N}$ until the corresponding ηN attraction generates a bound state in the η–nucleus system. As the starting value we chose $(0.55 + i0.30)$ fm proposed by Wilkin [5]. Thus, we take $a_{\eta N} = (g0.55 + i0.30)$ fm, where g is an enhancing parameter.

Since $a_{\eta N}$ is complex, the η–nucleus Hamiltonian is non–Hermitian and its eigenenergies are generally complex. Hence, we do not expect to find a pole of $T(z)$ on the positive imaginary axis of the complex k–plane with any choice of the enchancing factor g. As was shown in ref.[15], when the interaction becomes complex the bound-state poles move into the second quadrant of the complex k–plane. Therefore we search in this quadrant in order to locate possible poles of $T(z)$. However, not all poles in the second quadrant stem from bound states. Indeed, the energy $E_0 = p_0^2/2\mu$ corresponding to a pole, has a negative real part only if p_0 is above the diagonal of this quadrant. Below the diagonal, where $\text{Re}E_0 > 0$, the pole is attributed to a resonance. Therefore this diagonal is the critical border, and when crossing from below, a pole becomes a quasi–bound state.

Fixing the enhancing factor g to the value $g = 1$ and making variations of the complex parameter $p = \sqrt{2\mu z}$ within the second quadrant, we located the poles close to the origin, $p = 0$, which are given in the Table 1.

It is seen, that for the ηd, ηt, and $\eta\,^3$He systems, these poles lie below the diagonal, i.e. in the resonance region, while the $\eta\,^4$He system has a quasi–bound state. On the other hand, all such poles are not far from the border separating the resonance and bound state domains. Hence, one can expect that small changes of the factor g could place the poles on this border. Following this idea, we varied g untill we found the factors which generates poles on the diagonal. They are given in the Table 2. These factors correspond to an ηN attraction, which just generates an η – nucleus binding with $\text{Re}E_0 = 0$. Further

Table 1. Positions $p_0 = \sqrt{2\mu E_0}$ of the poles of the η–nucleus amplitudes with $g = 1$. For each of the nuclei the calculations were done with three values of the range parameter α.

	p_0 (fm^{-1})	E_0 (MeV)	α (fm^{-1})
$\eta\, d$	−0.90254 + i0.35880	31.448 − i29.698	2.357
	−0.84562 + i0.32422	27.969 − i25.143	3.316
	−0.82460 + i0.30855	26.813 − i23.333	7.617
$\eta\, t$	−0.56125 + i0.24475	10.818 − i11.650	2.357
	−0.55747 + i0.27050	10.076 − i12.789	3.316
	−0.52717 + i0.28349	8.3770 − i12.675	7.617
$\eta\, ^3$He	−0.54791 + i0.25111	10.056 − i11.669	2.357
	−0.51111 + i0.30709	7.0788 − i13.312	3.316
	−0.47578 + i0.34354	4.5944 − i13.863	7.617
$\eta\, ^4$He	−0.15056 + i0.18278	−0.43713 − i2.2399	2.357
	−0.17940 + i0.24300	−1.0933 − i3.5484	3.316
	− 0.23100 + i0.30850	− 1.7016 − i5.8006	7.617

Table 2. The enhancing factors g moving the η–nucleus amplitude poles to the points $p_0 = \sqrt{2\mu E_0}$ on the diagonal. For each of the nuclei the calculations were done with three values of the range parameter α.

	g	p_0 (fm^{-1})	E_0 (MeV)	α (fm^{-1})
$\eta\, d$	1.654	−0.32545 + i0.32545	−i9.7134	2.357
	1.566	−0.33741 + i0.33741	−i10.440	3.316
	1.535	−0.33938 + i0.33938	−i10.566	7.617
$\eta\, t$	1.361	−0.33900 + i0.33900	−i9.7467	2.357
	1.310	−0.35424 + i0.35424	−i10.643	3.316
	1.260	−0.35378 + i0.35378	−i10.615	7.617
$\eta\, ^3$He	1.330	−0.34375 + i0.34375	−i10.022	2.357
	1.221	−0.36640 + i0.36640	−i11.386	3.316
	1.144	−0.38004 + i0.38004	−i12.247	7.617
$\eta\, ^4$He	0.955	−0.16164 + i0.16164	−i2.1267	2.357
	0.911	−0.19940 + i0.19940	−i3.2363	3.316
	0.899	− 0.25130 + i0.25130	− i5.1403	7.617

increase of g moves the poles up and to the right, enhancing the binding and reducing the widths of the states.

It is seen, that the real part of the $a_{\eta N}$, providing the critical binding of η–meson to a light nucleus, lies within the existing uncertainties, [0.27, 0.98] [14], for this value. Therefore, in reality the η–nucleus quasi–bound states can exist with $A \geq 2$. If this is not the case, then at least the near–threshold resonances (poles just below the diagonal) must exist. However, as one sees from both tables, the width of such quasi–bound and resonance states are small only for the $\eta\,^4$He system while for the other systems considered, they are rather large ~ 20 MeV, which means that it is difficult to detect such states in experiments.

References

1. Q. Haider and L.C. Liu: Phys. Lett. **B172**, 257 (1986)
2. L.C. Liu and Q. Haider: Phys. Rev. **C34**, 1845 (1986)
3. S. Wycech, A.M. Green, J.A. Niskanen: Los-Alamos e-print archive: nucl-th/9502022, (1995)
4. R.E. Chrien et al.: Phys. Rev. Lett. **60**, 2595 (1988)
5. C. Wilkin: Phys. Rev. **C47**, R938 (1993)
6. S.A. Rakityansky, S.A. Sofianos, V.B. Belyaev: in *Symposium on Effective Interactions in Quantum Systems*, Ed. S.A. Sofianos. Pretoria: UNISA December 1994
7. S.A. Rakityansky et al.: Los-Alamos e-print archive: nucl-th/9504020, (1995) and Phys. Lett. B (submitted)
8. V.B. Belyaev et al.: in: *European conference on few–body problems in physics* Peniscola, Spain, 5-9 June 1995. Few Body Ssystems Suppl. (to appear)
9. V.B. Belyaev and J. Wrzecionko: Sov. Journal of Nucl. Phys. **28**, 78 (1978)
10. V. B. Belyaev: in *Lectures on the theory of few-body systems*. Heidelberg: Springer-Verlag 1990
11. Particle Data Group: Phys. Rev. **D50**, 1319 (1994)
12. R.S. Bhalerao and L.C. Liu: Phys. Rev. Lett.: **54**, 865 (1985)
13. C. Bennhold and H. Tanabe: Nucl. Phys. **A530**, 625 (1991)
14. M. Batinic, A. Svarc: Los-Alamos e-print archive: nucl-th/9503020, (1995)
15. W. Cassing, M. Stingl: Phys. Rev. **C26**, 22 (1982)

Few-Body Systems Suppl. 9, 231–235 (1995)

Few-Body Systems

Description of Low-Energy Pion Double Charge Exchange Reactions

F. Šimkovic *, A. Faessler

Institut für Theoretische Physik der Universität Tübingen, Auf der Morgenstelle 14, 72076 Tübingen, Germany

Abstract. A new sequential model for low energy pion double charge exchange on nuclei is presented. We consider pion double charge exchange to be the second-order perturbation process in the s- and p- wave pion nucleon interaction Hamiltonian considering exactly the intermediate nuclear states. To our knowledge, this is the first sequential model showing that intermediate nucleus spectra could be the source of resonant-like structures in the double charge exchange amplitudes. A possible connection with the experimentally observed sharp peaks in the energy dependence of the double charge exchange cross section at $T_\pi \approx 50$ MeV is discussed.

1 Introduction

The recent much attention paid to low energy pion double charge exchange reaction (DCX),

$$\pi^+ + (A,Z) \rightarrow (A,Z+2) + \pi^-$$

is connected with the very narrow peak seen near $T_\pi = 50$ MeV in the energy dependence of the cross section, nearly independent of the nuclear target [1].

Many mechanisms for pion DCX have been suggested and studied. The simplest- sequential single charge exchange on two nucleons is believed to dominate this process [4, 5, 6]. However, none of the present sequential calculations of the DCX cross section is able to describe a peak in the region 30-50 MeV. The apparent failure of the conventional theory to explain this low-energy behaviour has renewed recently the speculation about a possible dibaryon-resonance with quantum numbers $J^p = 0^-, T =$ even and a mass of 2065 MeV, which may be the cause of the observed peak [1]. At present this idea is being explored [2, 3].

* *On leave from:* Bogoliubov Theoretical Laboratory, Joint Institute for Nuclear Research, 141980 Dubna, Moscow Region, Russia and Department of Nuclear Physics, Comenius University, Mlynská dolina F1, Bratislava, Slovakia

In most of the DCX studies, the role of the intermediate nuclear spectrum has been neglected by introducing a closure approximation [5]. The first calculation with explicit consideration of the intermediate nucleus has been done by Kaminski and Faessler in the framework of Quasiparticle Random Phase Approximation (QRPA) [7, 8]. They have described DCX reactions in the non-relativistic approximation using the formalism of the pion s- and p- wave charge changing operators. This model does not predict any narrow peaks in the energy dependence of the DCX cross section.

In this contribution we shall describe the DCX reaction as a second order process in s- and p-wave nucleon interaction by using the field theory approach without invoking closure approximation. This sequential model differs from that of ref. [7] especially by the treatment of the pion s-wave contribution to the DCX amplitude. We shall demonstrate that the discrete spectrum of the intermediate nucleus is a source of resonant-like structures in the energy dependence of the differential cross section.

2 Double Charge Exchange Amplitude

We shall assume the effective pion nucleon interaction Hamiltonian of the σ model in its non-linear form [9],

$$\mathcal{H}_I(x) = \mathcal{H}_{\pi NN}(x) + \mathcal{H}_{\pi\pi NN}(x), \tag{1}$$

where,

$$\mathcal{H}_{\pi NN}(x) = \frac{1}{2f_\pi}\bar{\psi}(x)\gamma_\mu\gamma_5\boldsymbol{\tau}\psi(x)\cdot\partial_\mu\boldsymbol{\phi}(x), \tag{2}$$

$$\mathcal{H}_{\pi\pi NN}(x) = \frac{1}{4f_\pi^2}\bar{\psi}(x)\gamma_\mu\boldsymbol{\tau}\psi(x)\cdot(\boldsymbol{\phi}(x)\times\partial_\mu\boldsymbol{\phi}(x)). \tag{3}$$

Here, $\mathcal{H}_{\pi NN}(x)$ is the pseudoscalar coupling interaction Hamiltonian of the nucleon $\psi(x)$ and pion $\boldsymbol{\phi}(x)$ fields and $\mathcal{H}_{\pi\pi NN}(x)$ is an effective interaction between nucleon and pion isovector currents. The non-relativistic reduction of $\mathcal{H}_{\pi NN}(x)$ and $\mathcal{H}_{\pi\pi NN}(x)$ leads to the effective πN p-wave and s-wave Hamiltonian, respectively [9]. The coupling strengths are completely determined by the pion decay constant $f_\pi \simeq 93$ MeV.

Clearly, the DCX process occurs in the second order perturbation theory of πN interaction. For the matrix element of the DCX process we have

$$< f|S^{(2)}|i >= \frac{(-\mathrm{i})^2}{2!}\int < f|T(\mathcal{H}_I(x)\mathcal{H}_I(y))|i > dxdy. \tag{4}$$

By inserting Eq.(1) into Eq.(4), assuming the non-relativistic impulse approximation for nuclear currents, adopting the plane wave approximation for pion wave functions and by using the closure relation for intermediate nuclear states $1 \equiv \sum_n |n >< n|$, we obtain

$$< f|S^{(2)}|i >= 2\pi\delta(E_f + k_{f0} - E_i - k_{i0})\frac{[M^p(\boldsymbol{k}_i,\boldsymbol{k}_f) + M^s(\boldsymbol{k}_i,\boldsymbol{k}_f)]}{(2\pi)^3\sqrt{k_{i0}k_{f0}}}, \tag{5}$$

where

$$M^p(\boldsymbol{k}_i,\boldsymbol{k}_f) = \frac{\mathrm{i}g_A^2}{(2f_\pi)^2}\sum_n\Big[\frac{< \acute{A}|O^{(p)}(-\boldsymbol{k}_f)|n >< n|O^{(p)}(\boldsymbol{k}_i)|A >}{E_n - E_i - k_{i0} - \mathrm{i}\varepsilon} + \frac{< \acute{A}|O^{(p)}(\boldsymbol{k}_i)|n >< n|O^{(p)}(-\boldsymbol{k}_f)|A >}{E_n - E_f + k_{i0} - \mathrm{i}\varepsilon}\Big], \quad (6)$$

$$M^s(\boldsymbol{k}_i,\boldsymbol{k}_f) = \frac{\mathrm{i}g_V^2}{(2f_\pi)^4}\sum_n \frac{1}{(2\pi)^3}\int\frac{d\boldsymbol{p}}{2p_0}\times \Big\{\frac{(p_0+k_{f0})(p_0+k_{i0})}{E_n - E_i + p_0 - k_{i0} - \mathrm{i}\varepsilon} < \acute{A}|O^{(s)}(-\boldsymbol{k}_f - \boldsymbol{p})|n >< n|O^{(s)}(\boldsymbol{k}_i + \boldsymbol{p})|A > + \frac{(p_0-k_{f0})(p_0-k_{i0})}{E_n - E_f + p_0 + k_{i0} - \mathrm{i}\varepsilon} < \acute{A}|O^{(s)}(\boldsymbol{k}_i + \boldsymbol{p})|n >< n|O^{(s)}(-\boldsymbol{k}_f - \boldsymbol{p})|A >\Big\}, \quad (7)$$

with

$$O^{(p)}(\boldsymbol{k}) = \sum_i \tau_i^+ \boldsymbol{\sigma}_i\cdot\boldsymbol{k}\ \exp(\mathrm{i}\boldsymbol{k}\cdot\boldsymbol{r}_i), \quad O^{(s)}(\boldsymbol{k}+\boldsymbol{p}) = \sum_i \tau_i^+ \exp(\mathrm{i}(\boldsymbol{k}+\boldsymbol{p})\cdot\boldsymbol{r}_i), \quad (8)$$

where $g_V = 1$ and $g_A = 1.25$. Here, $k_i \equiv (\boldsymbol{k}_i, \mathrm{i}k_{i0})$, $p \equiv (\boldsymbol{p}, \mathrm{i}p_0)$ and $k_f \equiv (\boldsymbol{k}_f, \mathrm{i}k_{f0})$ are respectively the four momenta of the incoming, intermediate and outgoing pion. $|A>$, $|\acute{A}>$ and $|n>$ are respectively the wave functions of the initial, final and intermediate nuclei with corresponding energies E_i, E_f and E_n. $M^p(\boldsymbol{k}_i,\boldsymbol{k}_f)$ ($M^s(\boldsymbol{k}_i,\boldsymbol{k}_f)$) is the p-wave (s-wave) scattering amplitude.

The intermediate nuclear spectrum can be found e.g. by shell model diagonalization for light nuclei or by using QRPA for heavy nuclei. The calculations of the strength distribution for the transitions to the intermediate nucleus show that practically the full strength is concentrated up to the excitation energy $E_{\text{exc}} \approx 20$ MeV. This value is much below the pion threshold. For $E_n - E_i \leq E_{\text{exc}} \ll k_{i0}$ the p-wave amplitude in Eq.(6) possesses no poles. Therefore, the sequential mechanism based on the p-wave πN interaction could not explain a resonant-like behaviour of the DCX cross section. We note that the p-wave amplitude $M^p(\boldsymbol{k}_i,\boldsymbol{k}_f)$ is strongly suppressed due to a mutual cancellation of the direct and crossed terms in the r.h.s. of Eq.(6). If we assume $E_n - E_i + k_{i0} \approx E_f - E_n + k_{i0} \approx k_{i0}$, the p-wave amplitude for the most interesting nuclear transitions $0^+{}_i \to 0^+_f$ is even equal to zero. Therefore, the p-wave amplitude is expected to be sensitive to the details of the intermediate nucleus.

For the s-wave amplitude analyses it is useful to rewrite the first denominator in the r.h.s. of Eq.(7) as follows:

$$\frac{1}{E_n - E_i + p_0 - k_{i0} - \mathrm{i}\varepsilon} = \mathcal{P}\frac{1}{E_n - E_i + p_0 - k_{i0}} + \mathrm{i}\pi\delta(E_n - E_i + p_0 - k_{i0}). \quad (9)$$

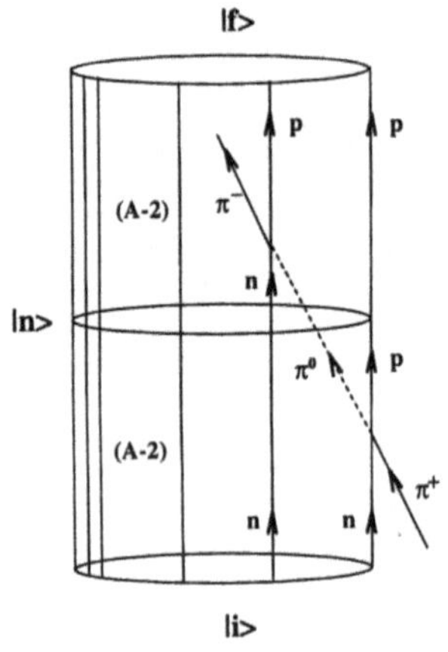

Figure 1. The sequential mechanism based on the s-wave πN interaction with real intermediate pion π^0.

We see that the argument of the delta function in Eq.(9) is zero, if the kinetic energy T_{kin} of the incoming pion obeys the relation

$$T_{\text{kin}} \geq E_n - E_i + m^*_{\pi^0} - m_{\pi^+}. \tag{10}$$

Here, $m^*_{\pi^0}$ is the effective mass of the intermediate pion π^0 in a dense nuclear medium and m_{π^+} is the mass of incoming pion π^+. If the kinetic energy T_{kin} fulfil the above relation, the following scenario would be possible. The incoming pion π^+ is absorbed by the initial nucleus (A, Z), which undergoes the transition to the real intermediate nuclear state $(A, Z+1)$ with emission of a real pion π^0. This pion is absorbed by the intermediate nucleus, which undergoes the transition to the final nucleus $(A, Z+2)$ with emission of the pion π^-. This mechanism is illustrated in Fig.1. The threshold of the intermediate real pion production would produce a resonant-like shape in the energy dependence of the differential cross section, directly connected with p- and s- wave DCX amplitudes as follows:

$$\frac{d\sigma}{d\Omega}(\theta_{k_i k_f}, q) = 2\pi \frac{|\boldsymbol{k}_f|}{|\boldsymbol{k}_i|} |M^p(\boldsymbol{k}_i, \boldsymbol{k}_f) + M^s(\boldsymbol{k}_i, \boldsymbol{k}_f)|^2. \tag{11}$$

Here, $\boldsymbol{q} = \boldsymbol{k}_i - \boldsymbol{k}_f$ is the momentum transfer between the incoming and outgoing pions. We note that the threshold and the corresponding position of the resonant-like shape depend on the eigenenergy E_n of the intermediate state. The each intermediate nuclear state could response to a peak in the DCX amplitude. However they are weighted by different strenghts in the DCX cross section and therefore most of them are suppressed. Without performing an adequate calculation of the differential cross section, we can do some estimations. Let us suppose that there is at least one favoured nuclear transition to intermediate nuclear state with energy $E_{\text{res}} \approx 10$ MeV. Knowing the position of the corresponding peak, we can deduce the effective mass of the intermediate pion. If the change of the mass of the intermediate pion π^0 in the nuclear medium is negligible, the position of the corresponding resonant-like shape is about $T_{\rm kin} \approx 5.5$ MeV. Some theories predict $m^*_{\pi^0}/m_{\pi^0} \approx 0.89$ [10] (m_{π^0} is mass of a free pion π^0). In this case the DCX cross section is smooth without peaks. However, if we accept the value $m^*_{\pi^0}/m_{\pi^0} \approx 1.3$, the theory predicts a peak at $T_{\rm kin} \approx 50$ MeV. Therefore, we see that there is a chance to explain the resonant-like behaviour of the energy dependence of the DCX differential cross section without invoking the exotic dibaryon mechanism.

3 Conclusion

We have proposed a new sequential model to study the DCX reactions. The pion–nucleon p-wave and s-wave interactions have been considered. We have shown that the p-wave amplitude is expected to be strongly suppressed due to a cancellation of the direct and crossed terms. The s-wave amplitude has been found to be sensitive to the effective mass of the intermediate pion and to the structure of the intermediate nucleus. To our knowledge, we are the first showing that in the framework of the sequential model the resonant-like shape of the DCX cross section is possible and that a useful information about the intermediate pion mass could be obtained. By assuming $m^*_{\pi^0}/m_{\pi^0} \approx 1.3$ the model predicts a peak at $T_{\rm kin} \approx 50$ MeV. It remains to be seen in future, whether the sequential model or the idea of a dibaryon-resonance will prevail in the explanation of the peculiar narrow peak. The QRPA calculation of the DCX cross section within the proposed model is in progress.

The authors are grateful to H. Clement, L. Glozman, W.A. Kaminski and A. Valcarce for interesting discussions.

References

1. R. Bilger, H.A. Clement, M.G. Schepkin: Phys. Rev. Lett. **71**, 42, (1993)
2. L.Ya. Glozman, A. Buchmann, A. Faessler: J. Phys. **G20**, L49 (1994)
3. A. Valcarce, H. Garcilazo, F. Fernandez: Phys. Rev. **C52**, (in print)
4. H. Sarafian, M.B. Johnson, E.R. Siciliano: Phys. Rev. **C48**, 1988 (1993)
5. E. Bleszynski, M. Bleszynski: Phys. Rev. Lett. **60**, 1483 (1988)
6. J.D. Vergados: Phys. Rev. **C44**, 276 (1991)
7. W.A. Kaminski, A.Faessler: Nucl. Phys. **A529**, 605 (1991)
8. W.A. Kaminski: Fiz. Elem. Chastits At. Yadra **26**, 362 (1995)
9. T. Ericson, W. Weise: *Pions and Nuclei*. Oxford: Clarendon Press 1988
10. G.E. Brown, M. Rho: Phys. Rev. Lett. **66**, 2720 (1991)

Few-Body Systems Suppl. 9, 237–240 (1995)

Few-Body Systems

Inclusive Pion Double Charge Exchange on Light Nuclei above 0.5 GeV

B.M. Abramov[1], S.A. Bulychjov[1], I.A. Dukhovskoi[1], A.I. Khanov[1], Y.S. Krestnikov[1], A.P. Krutenkova[1], V.V. Kulikov[1], M.A. Matsuk[1], I.A. Radkevich[1], A.N. Starodumov[1], A.I. Sutormin[1], E.N. Turdakina[1], M.J. Vicente Vacas[2]

[1] Institute of Theoretical and Experimental Physics, 117259 Moscow, Russia

[2] Departamento de Física Teórica and IFIC, Centro mixto Universidad de Valencia - CSIC, 46100 Burjassot, Spain

Abstract. Inclusive double charge exchange (DCX) reaction $A(\pi^-, \pi^+)X$ on ^{6}Li and ^{16}O was for the first time measured at incident kinetic energy $T_0 = 1.1$ GeV ($< \theta > \approx 5^\circ$) for outgoing particle momentum region where additional pion production is kinematically forbidden. The experiment was performed at secondary π^- beam of the ITEP proton synchrotron using the 3m magnet spectrometer. The measured double differential cross section at 1.1 GeV together with our previous results at 0.6 and 0.75 GeV [1] and with the lower energy data [2] and [3] shows rather slow decrease with energy. This result is in contradiction with the fast fall of small angle DCX cross section calculated [4] in the standard mechanism of two sequential single charge exchanges of pion on two like nucleons. Therefore, other approaches to DCX (for example, short range *NN* correlations, meson exchange currents, inelastic Glauber corrections *etc.*) are needed to explain the observed energy dependence.

1 Introduction

The goal of our experiment is to shed light on the mechanisms of pion DCX reaction as the process that occurs on at least two like nucleons only. We studied energy region above 0.5 GeV where in the framework of the standard DCX mechanism, i.e. two sequential single charge exchanges of pion (SSCX), the strong decrease of small angle DCX cross section is expected [5]. It was emphasized in [5] that predicted low value of DCX cross section near 1.3 GeV offers unique possibility to search for nonconventional mechanisms. Our recent measurements [1] definitely show that at 0.75 GeV we observed DCX signal at higher level than predicted in the framework of SSCX model. So, some nonconventional mechanisms are needed to explain our data in this energy region.

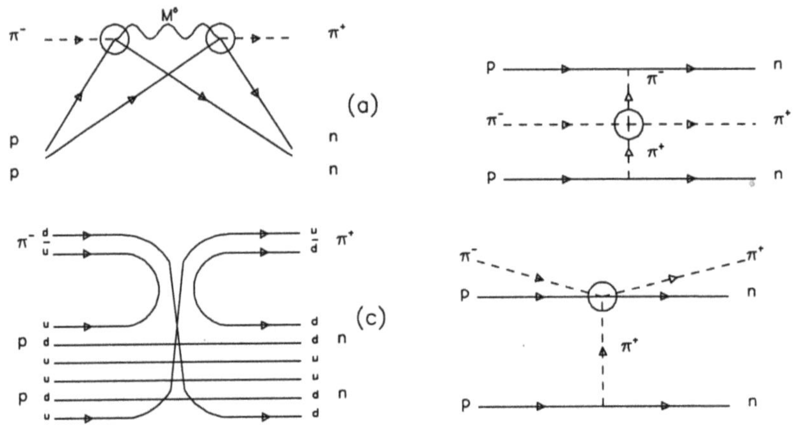

Figure 1. Mechanisms for DCX reaction $A(\pi^-, \pi^+)X$.
(a) Sequential single charge exchanges on two separate protons :
$M^0 = \pi^0$ is standard SSCX (elastic Glauber correction),
$M^0 = \eta^0$ is quasielastic Glauber correction,
$M^0 = \pi^+\pi^-, \pi^0\pi^0$ are inelastic Glauber corrections.
(b) Meson exchange currents (model dependent microscopic approaches).
(c) Double charge exchange on two correlated protons (model dependent six quark approaches).

Three DCX pictures could be considered. The first one consists of two sequential charge exchanges on two separate nucleons with different intermediate states (Fig. 1(a)) of which two pion state is of interest. The second one is so-called meson exchange current mechanism (Fig. 1(b)) while the third one considers DCX on a correlated pair of like nucleons at the quark level (Fig. 1(c)).

2 Experiment

We measured energy spectra for outgoing pions in the reactions $A(\pi^-, \pi^+)X$ on ^{6}Li, ^{7}Li, ^{12}C, and ^{16}O at $\theta = 0$ - $10°$ and incident pion kinetic energy $T_0 =$ 600, 750, and 1120 MeV in the region $\Delta T = T_0 - T \leq m_\pi = 140$ MeV (T is kinetic energy of positive charged particle) where $(\pi, 2\pi)$ reaction is kinematically forbidden.

The experiment was carried out with the help of 3m magnet spectrometer ITEP (see, *e.g.*, [6]) with spark chambers. The targets were placed in the middle of the magnet and the first part of it was used as a beam spectrometer and the second one as outgoing particle spectrometer. TOF measurement permitted to separate $e^+ + \pi^+$, p, and d. The Cherenkov counter tag was used off line for positron rejection.

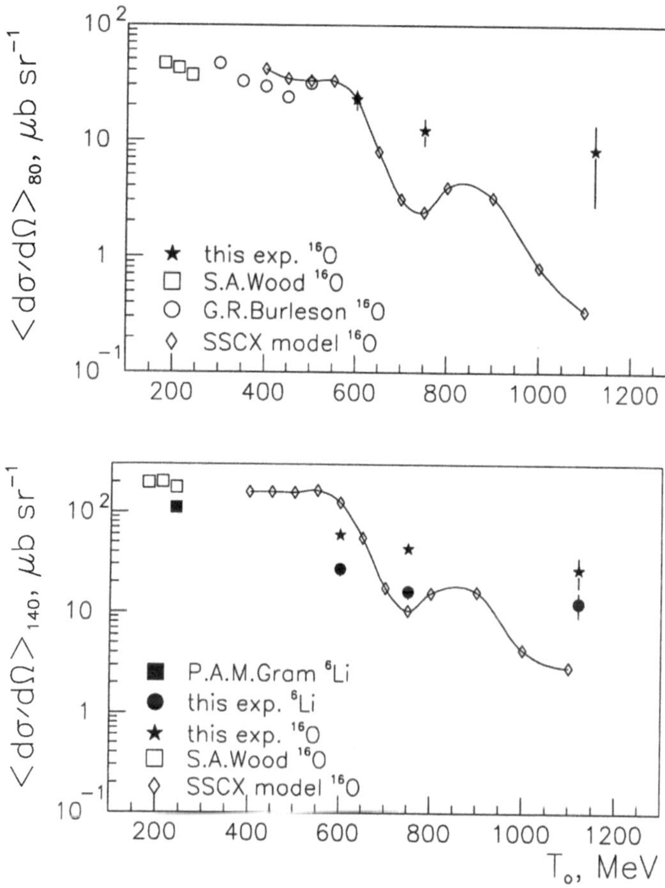

Figure 2. Energy dependence of double differential cross section for inclusive pion DCX on
(a) ^{16}O integrated over the ΔT region from 0 to 80 MeV,
(b) ^{6}Li and ^{16}O integrated over the ΔT region from 0 to 140 MeV.

3 Results

To study energy dependence, the measured double differential cross section was integrated over two regions: $0 \leq \Delta T \leq 80$ MeV and $0 \leq \Delta T \leq 140$ MeV. This partially integrated cross sections are shown in Fig.2 together with the data [2] and [7] and with the results [3] on the reaction $^{16}O(\pi^+, \pi^-)X$. These cross sections slowly fall with energy from 240 to 1120 MeV. The decrease of inclusive cross section that is now seen in direct way has been noticed earlier [3,8] as a tendency of background near the peaks of the reactions $^{16}O(\pi^+, \pi^-)^{16}Ne$ and $^{48}Ca(\pi^+, \pi^-)^{48}Ti$ to drop with increasing energy. As it is seen from Fig.2, our results confirm this observation and permit to state that inclusive pion DCX cross section continues to decrease in wider energy range (up to 1.1 GeV).

The results of theoretical calculations for SSCX mechanism of inclusive DCX on ^{16}O in the framework of cascade style Monte Carlo model [4], which was intended to describe all the reactions $A(\pi,\pi)X$ simultaneously, are also shown in the Fig.2 (SSCX model). The elementary πN cross sections were used and such medium effects as Fermi motion, binding energy, Pauli-blocking, and absorption were taken into consideration. The calculated cross section is in a reasonable agreement with experiment [3] for T_0 = 400 - 500 MeV. It should be pointed out that calculated cross section decreases substantially faster in the region 0.6 - 1.1 GeV. Within sequential single charge exchange mechanism, this decrease reflects fast fall of πN single charge exchange amplitude in this energy region.

4 Conclusions

We observed the inclusive pion DCX reactions on ^6Li and ^{16}O at incident kinetic energy 0.6, 0.75, and 1.1 GeV in kinematical region where no additional pions are produced. The measured double differential cross sections decrease rather slowly with incident energy. At 0.75 and 1.1 GeV, the cross sections are significantly larger than expected in the framework of usual sequential single charge exchange mechanism. This observation implies the necessity of using non-standard approach to explain our results. Apart from meson exchange current and six-quark mechanism, the contribution of two pion intermediate states to standard SSCX should be taken into account.

Acknowledgement. This work was partly supported by the grant INTAS 93-3455 and CICYT Contract No. AEN 93-1205.

References

1. B.M. Abramov et al.: Yad. Fiz. (to appear)
2. S.A. Wood et al.: Phys. Rev. **C46**, 1903 (1992)
3. G.R. Burleson: In:*Pion-Nucleus Double Charge Exchange*, Proceedings of the Second LAMPF Workshop on Pion-Nucleus Double Charge Exchange, Los Alamos, New Mexico, USA, p.79. Singapore: World Scientific 1990
4. M.J. Vicente Vacas, M.Kh. Khanhkasaev, S.G. Mashnik: Preprint FTUV/94-73 University of Valencia 1994, submitted to Nucl. Phys. **A**
5. E. Oset and D. Strottman: Phys. Rev. Lett. **70**, 146 (1993)
6. B.M. Abramov et al.: Nucl. Phys. **A372**, 301 (1981)
7. P.A.M. Gram: In:*Pion-Nucleus Physics: Future Directions and New Facilities at LAMPF*, Los Alamos 1987, (AIP Conference Proceedings 163), p.75. New York: American Institute of Physics 1988.
8. A.L. Williams et al.: Phys. Rev. **C43**, 766 (1991)

Few-Body Systems Suppl. 9, 241–244 (1995)

Few-Body Systems

Study of Reactions $\pi^- p \rightarrow \eta n$ and $\pi^- p \rightarrow \pi^0 n$ on Pion Beams of the PNPI Synchrocyclotron

I.V. Lopatin[1], V.V. Abaev[1], V.S. Bekrenev[1], E.A. Filimonov[1], A.B. Gridnev[1], M.R. Kan[1], N.G. Kozlenko[1], S.P. Kruglov[1], L.V. Lapochkina[1], A.Yu. Majorov[1], D.V. Novinsky[1], A.B. Starostin[1], V.V. Sumachev[1], B.M.K. Nefkens[2], J.W. Price[2], D.B. White[2], R.M. Clajus[2], M.E. Sadler[3], L.D. Isenhower[3], S.E. Garner[3], J.R. Phillips[3], J.A. Redmon[3]

[1] Petersburg Nuclear Physics Institute, Gatchina, Leningrad district, 188350 Russia

[2] University of California at Los Angeles, Los Angeles, California, 90024 USA

[3] Abilene Christian University, Abilene, Texas, 79699 USA

Abstract. Yields of the reaction $\pi^- p \rightarrow \eta n$ have been measured in the near-threshold region. The experiment was carried out by means of detecting neutron produced in this reaction and measuring their energies by TOF-technique. Measured momentum dependences of the reaction yield are compared with calculated ones.

Preliminary measurements of differential cross sections of $\pi^- p$ charge exchange scattering were also performed using coincidences between neutrons and gammas from the decay $\pi^0 \rightarrow 2\gamma$.

1 Introduction

A general program of studying pion-proton binary reactions with both neutral particles in the final state is now in progress at the Petersburg Nuclear Physics Institute. In the first stage of these investigations, we studied the reaction $\pi^- p \rightarrow \eta n$ in the near-threshold region. Obtaining accurate experimental data is very important for testing various theoretical models suggested for description pion-nucleon scattering and η production. Such data can be used also for extracting the ηN scattering length and for more accurate determining the η mass. Until now experimental information about cross section of the η production process is very scarce and contradictory, especially in the near-threshold region.

2 Experimental Study of η Production Process

The kinematics of the near–threshold η production is such that in the lab frame all neutrons are emitted in a relatively small forward cone. The maximum neutron lab. angle increases as the incident momentum increases. Most of events occur very close to the maximum possible angle - the so called Jacobian effect. Another specific feature of this process - two different CM angles (one in the forward hemisphere and the other in the backward one) correspond to each neutron lab. angle Θ_n . It means that two groups of neutrons with different kinetic energies are detected at a given Θ_n .

In the near–threshold region the η production occurs only in the S–state. In this assumption the number of detected neutrons is defined by

$$N_n = N_\pi \ N_p \ \varepsilon_n \ \sigma_{\rm t} \ A_n. \tag{1}$$

where N_π is the pion flux through a target, N_p is the number of protons in the target (cm^{-2}), ε_n is the detector's efficiency, $\sigma_{\rm t}$ is the total cross section, and A_n is the angular acceptance (i.e. fraction of total solid angle in the CM frame covered by the detector). Evidently, the product of two functions of the incident momentum one of which, $\sigma_{\rm t}$, steeply rises and the other, A_n , falls with the momentum gives a dependence characterized by a peak. For a real experiment, this dependence should be averaged over the momentum spread of a pion beam.

Main parts of experimental setup are a liquid hydrogen target and four neutron detectors. The liquid hydrogen target consists of a vertical cylinder 10 cm diameter and 12 cm high; the flask is made of aluminium 0.1 cm thick.

The neutron detectors were designed, constructed and calibrated by scientists of the University of California at Los Angeles. Each detector with size 26.7 cm $\times$ 25.4 cm $\times$ 25.4 cm consists of three independent subcounters viewed each by two photomultiplier tubes. The thickness was sufficient to provide a detection efficiency of more than 20%. To prevent irradiation of the neutron detectors by the incident pion beam, a special bending magnet was placed between the hydrogen target and the neutron detectors.

In order to provide simultaneous measurements in several neighbouring momentum bins of the incident pions, a special hodoscope of narrow vertical scintillation counters was installed in a dispersive part of a pion channel.

Measurements were made for nine neutron lab. angles (covering range 1 to 12 degrees) at central momenta of the pion beam 670, 680, 690, 695, 700, and 710 MeV/c for both hydrogen-filled and empty target. At each central momentum data taking was performed simultaneously for two momentum bins corresponding to the beam counters H4 and H5. The number of "good" events (i. e. events attributed to the $\pi^- p \rightarrow \eta n$ reaction) can be extracted from TOF–spectra. An example of such spectrum is shown in Fig. 1. One can see two peaks η_1 and η_2 corresponding to neutrons emitted in this reaction to the forward and backward hemispheres in the CM frame and a rather big continuous background under these peaks.

To determine a momentum dependence of the reaction yield we used a

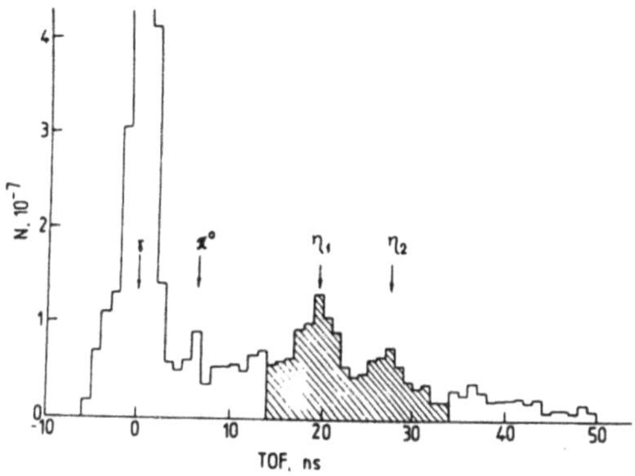

Figure 1. TOF-spectrum obtained for one of neutron subcounters; number of neutrons is normalized to one incident pion.

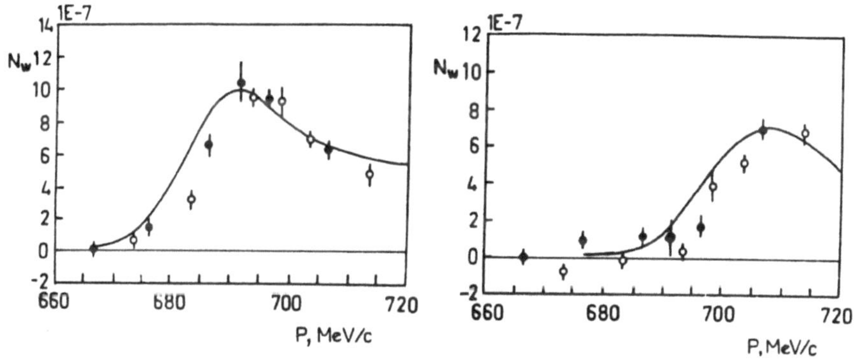

Figure 2. Momentum dependence of N_w obtained after subtracting background pedestal for neutron lab. angles 0.9° (left) and 10.5° (right). Two sets of data correspond to tagging incident pions by beam counters H4 (•) and H5 (o).

simplified version of data analysis. Instead of determining the number of "good" events (area under the η peaks) we counted the total number of events N_w in the "TOF-window" from 14 to 34 ns (shaded area in Fig.1). The momentum dependence of N_w is a superposition of an η production bump and a rather flat pedestal (due to two-pion production) which can be approximated by a straight line in the small momentum range investigated in the experiment. After subtracting this pedestal from the measured momentum dependence of N_w , one can obtain "pure" η production yield vs momentum of the incident pions. Examples of such dependences for two neutron lab. angles are presented in Fig.2. Two sets of points obtained for the beam counters H4 and H5 are shown in one and the same plot; they are shifted by ±0.5% (± 3.5 MeV/c) relative to the central beam momentum. Shown by curves in Fig.2 are results of calculations using equation (1); curves are normalized in maximums to the experimental data.

One can see that calculations reproduce shapes of experimental curves rather well; a possible explanation of some observing difference is that the real momentum dependence of σ_t does not coincide with the fit of all existing experimental data which was used in calculations. The results of calculations are very sensitive to the shape of the dependence $\sigma_t(p)$ in the near–threshold region, so in further processing we hope to determine the behaviour of $\sigma_t(p)$ near the threshold more accurately (using experimentally measured momentum dependences of the yield for all nine neutron lab. angles).

3 Measurement of Cross Sections of Reaction $\pi^- p \to \pi^0 n$

Values of differential cross section for $\pi^- p$ charge exchange scattering can be extracted, in principle, from π^0 peak in TOF-spectra similar Fig.1. But an ambiguous procedure of subtracting physical background leads to rather large systematic error. That is why we decided to use for such measurements a two–arms setup which detected neutron in coincidence with one gamma from the decay $\pi^0 \to 2\gamma$. Two kinds of total absorption electromagnetic calorimeters were built for detecting gammas. One was made of 12 Cherenkov lead glass blocks having each size 15 cm $\times$ 15 cm $\times$ 35 cm. Another consisted of 16 crystals CsI(Na) with individual size 6 cm $\times$ 6 cm $\times$ 30 cm.

The first run of measurements was performed at the momentum of the incident pions 710 MeV/c for both hydrogen-filled and empty target. We used the above described disposition of the neutron detectors which allowed to measure differential cross sections of $\pi^- p$ charge exchange scattering to the backward hemisphere (from 150° to 180° in the CM frame). Total statistics was about 10^3 π^0-mesons for each neutron angle providing the statistical accuracy on the level of $\pm 3\%$. We expect that additional systematic error will not exceed the same value. Data processing and analysis are under way now.

New accurate measurements of differential cross section of the reaction $\pi^- p \to \pi^0 n$ in the region of low-lying pion–nucleon resonances will allow to shed more light on a problem of charge symmetry breaking in pion–nucleon interaction [1].

This work is supported by Russian State Scientific-Technical Program "Fundamental Nuclear Physics" and by Russian Foundation for Fundamental Research.

References

1. V.V. Abaev and S.P. Kruglov: Preprint PNPI-1794 1992

Few-Body Systems Suppl. 9, 245–248 (1995)

Few-Body Systems

Theoretical Differential Cross Sections of πd Scattering for Pion Momenta up to 2 GeV/c

H. Garcilazo[1], L. Mathelitsch[2]

[1] Escuela Superior de Física y Matemáticas, Instituto Politécnico Nacional, Edificio 9, 07738 México D. F., Mexico

[2] Institut für Theoretische Physik, Universität Graz, Universitätsplatz 5 A-8010 Graz, Austria

Abstract. We have calculated the differential cross sections of elastic πd scattering for incident pion momenta between 500 and 2000 MeV/c using a relativistic three-body code. We got good agreement with experimental data below 700 MeV/c and fair agreement between 1 and 2 GeV/c. There are arguments that the poor reproduction of the data between 700 and 1000 MeV/c can be cured by the inclusion of intermediate states in the form of ρ, ω or η mesons.

Our relativistic calculation of elastic πd scattering takes into account the nucleon-nucleon ${}^1S_0$ and ${}^3S_1$- ${}^3D_1$ channels and all pion-nucleon S and P channels in a full three-body code and the pion-nucleon D-, F- and G-wave channels within an impulse approximation [1,2]. We applied the deuteron wave functions of the Paris [3] and Bonn potentials - here we used the so-called full version (BONNF) and the energy independent version which has been parametrized in momentum space (BONNQ) [4]. The πN onshell t-matrices are given by the Virginia Politechnic group [5]; an off-shell extension was provided by a form factor in exponential form with a cutoff parameter Λ which was taken to be $\Lambda = 600$ MeV/c.

Experimental data below 900 MeV/c were taken by several groups between scattering angles of 40° and 180°; between 1 and 2 GeV/c there exist data only above 120° [6].

Figures 1 and 2 show examples of the differential cross sections at pion incident momenta of 540 MeV/c and 2025 MeV/c (a detailed comparison of our results and experimental data at other pion momenta is given in ref. [7]). The reproduction of the experimental data is good at 540 MeV/c and reasonable at 2025 MeV/c; it should be noted that the later data are more than three orders of magnitude smaller than those at low momentum.

Figure 3 gives a different view of our results and the experimental data: The differential cross section at a scattering angle of 136° is plotted relative to

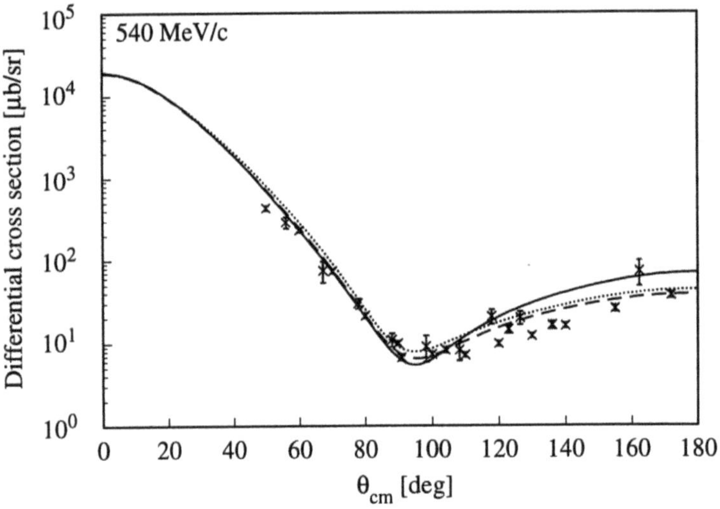

Figure 1. Differential cross section of elastic πd scattering at an incident pion momentum of $p_\pi = 540$ MeV/c. The solid line gives the result of the Paris potential, the dashed line of BONNF and the dotted line of BONNQ. The experimental data are from refs. [10,11,12].

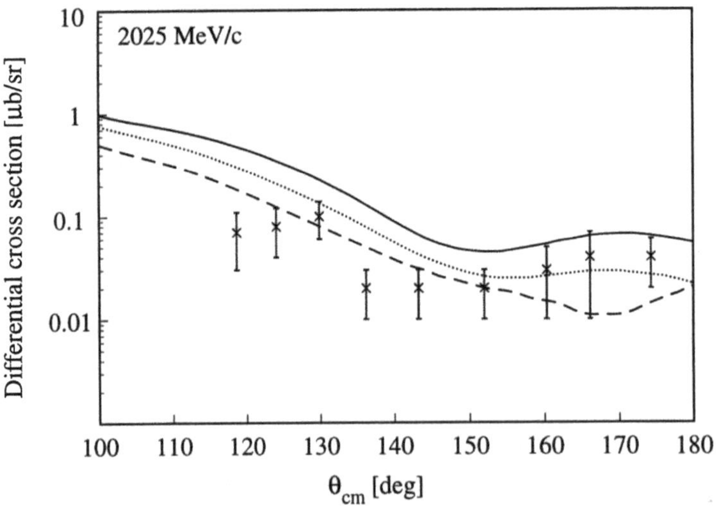

Figure 2. Differential cross section of elastic πd scattering at an incident pion momentum of $p_\pi = 2025$ MeV/c. Description as in Fig. 1. The experimental data are from ref. [6].

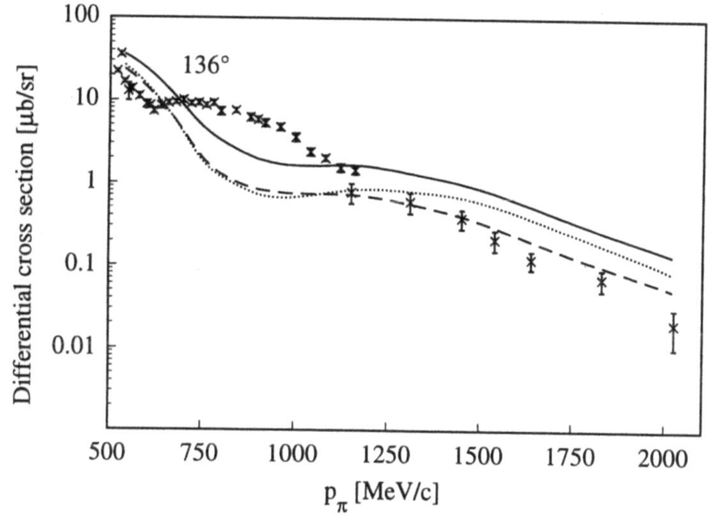

Figure 3. Differential cross section at a scattering angle of $\Theta = 136°$. Description as in Fig. 1. The experimental data are from refs. [6,10,11,13,14].

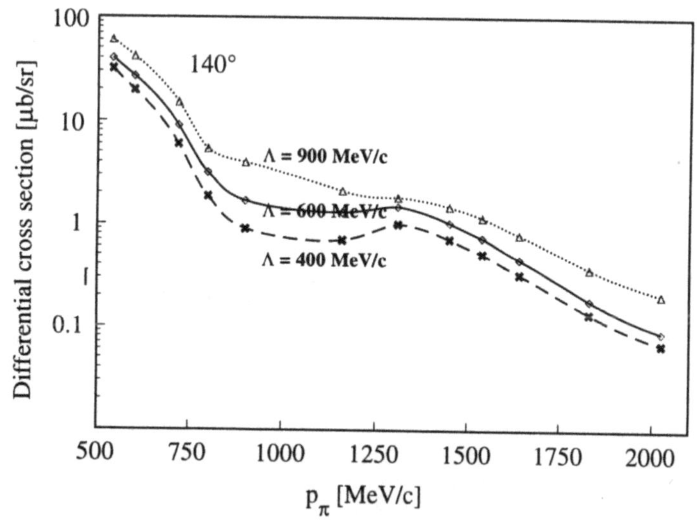

Figure 4. Differential cross section at a scattering angle of $\Theta = 140°$ at a scattering angle for different cutoff parameters.

the incident pion momentum. The graph shows fair agreement of our results with the experimental differential cross section below 600 MeV/c and above 1 GeV/c. From this graph one could read off that the Bonn potentials, especially the full version, are in slightly better agreement with the experimental cross sections than those of the Paris potential.

Figure 3 shows clearly that our model fails to reproduce the peak around 900 MeV/c. Abramov et al. [6] raised the question concerning an uncertainty in the three-body calculation due to the unknown off-shell extension. We therefore varied the cutoff parameter Λ (from 400 to 900 MeV/c) in the calculation with the Paris potential - the result is given in Fig. 4. A change of Λ produces a constant shift of the differential cross section and, to our opinion, the peak around 900 MeV/c cannot be explained by off-shell effects. We think that additional physical input into our theory, for example pion resonances in the intermediate state like ρ, ω and η (as was done partly by Kondratyuk and Lev [8]) are necessary for an improvement of the model.

References

1. H. Garcilazo: Phys. Rev. **C35**, 180 (1987)
2. H. Garcilazo: Phys. Rev. **C47**, 957 (1993)
3. M. Lacombe et al. : Phys. Rev. **C21**, 861 (1980);
   M. Lacombe et al. : Phys. Lett. **B101**, 139 (1981)
4. R. Machleidt, K. Holinde, and C. Elster: Phys. Rep. **149**, 1 (1987)
5. R.A. Arndt, R.L. Workman, Z. Li, and L.D. Roper: Phys. Rev. **C42**, 1853 (1990)
6. B.M. Abramov et al. : Nucl. Phys. **A542**, 579 (1992)
7. H. Garcilazo, L. Mathelitsch : Nucl. Phys. **A** (to be published)
8. M. Akemoto et al. : Phys. Rev. Lett. **50**, 400 (1983);
   L.A. Kondratyuk and F.M. Lev: Sov. J. Phys. **23**, 556 (1976)
9. M. Akemoto et al. : Phys. Rev. Lett. **51**, 1838 (1983)
10. R.C. Minehart et al. : Phys. Rev. Lett. **46**, 1185 (1981)
11. R.H. Cole et al. : Phys. Rev. **C17**, 681 (1978)
12. L.G. Dakhno et al. : Sov. J. Nucl. Phys. **31**, 326 (1980)
13. A.V. Kravtsov et al. : Nucl. Phys. **A322**, 439 (1978)
14. M. Akemoto et al. : Phys. Rev. Lett. **51**, 1838 (1983)

Few-Body Systems Suppl. 9, 249–252 (1995)

Few-Body Systems

Experimental Results on Pontecorvo Reactions with Strangeness

V.G. Ableev[1], M. Agnello[4], A. Andrighetto[3], F. Balestra[5], G. Belli[6], G. Bendiscioli[7], A. Bertin[8], G.C. Bonazzola[5], E. Botta[5], T. Bressani[5], M. Bruschi[8], M.P. Bussa[5], L. Busso[5], D. Calvo[5], M. Capponi[8], C. Cavion[9], B. Cereda[8], P. Cerello[5], C. Cicalò[2], M. Corradini[6], S. Costa[5], S. De Castro[8], O.Yu. Denisov[1], F. D'Isep[5], A. Donzella[6], L. Fava[5], A. Feliciello[5], L. Ferrero[5], A. Ferretti[8]l A. Filippi[5], V. Filippini[7], A. Fontana[7], D. Galli[8], R. Garfagnini[5], U. Gastaldi[9], B. Giacobbe[8], P. Gianotti[10], O. Gortchakov[1], A. Grasso[5], C. Guaraldo[10], F. Iazzi[5], A. Lanaro[10], E. Lodi Rizzini[6], M. Lombardi[9], V. Lucherini[10], A. Maggiora[5], S. Marcello[5], U. Marconi[8], G.V. Margagliotti[11], G. Maron[9], A. Masoni[2], I. Massa[8], B. Minetti[4], P. Montagna[7], M. Morando[3], F. Nichitiu[10], D. Panzieri[5], D. Parena[5], G. Pauli[11], C. Petrascu[10], M. Piccinini[8], G. Piragino[5], M. Poli[12], S.N. Prakhov[1], G. Puddu[2], R.A. Ricci[9,3], A. Rosca[10], E. Rossetto[5], A. Rotondi[7], A.M. Rozhdestvensky[1], A. Saino[7], P. Salvini[7], L. Santi[13], M.G. Sapozhnikov[1], N. Semprini Cesari[8], S. Serci[2], R. Spighi[8], P. Temnikov[2], S. Tessaro[11], F. Tosello[5], V.I. Tretyak[7], G. Usai[2], L. Vannucci[9], S. Vecchi[8], G. Vedovato[9], L. Venturelli[6], A. Vezzani[8], M. Villa[8], A. Vitale[8], A. Zenoni[6], A. Zoccoli[8], G. Zosi[5]

presented by P. Cerello

[1] Joint Institute for Nuclear Research, Dubna, SU-101000 Moscow, Russia

[2] Dip. di Scienze Fisiche, Università di Cagliari and INFN, I-09100 Cagliari, Italy

[3] Dip. di Fisica, Università di Padova and INFN, I-35100 Padova, Italy

[4] Politecnico di Torino and INFN, I-10125 Torino, Italy

[5] Istituto di Fisica, Università di Torino and INFN, I-10125 Torino, Italy

[6] Dip. di Elettronica per l'Autom. Industriale, Università di Brescia and INFN, I-25060 Brescia, Italy

[7] Dip. di Fisica Nucl. e Teor., Università di Pavia and INFN, I-27100 Pavia, Italy

[8] Dip. di Fisica, Università di Bologna and INFN, I-40100 Bologna, Italy

[9] Laboratori Nazionali di Legnaro dell' INFN, I-35100 Padova, Italy

[10] Laboratori Nazionali di Frascati dell' INFN, I-00044 Frascati, Italy

[11] Istituto di Fisica, Università di Trieste and INFN, I-34127 Trieste, Italy

[12] Dip. di Energetica, Università di Firenze and INFN, I-40100 Bologna, Italy

[13] Dip. di Fisica, Università di Udine and INFN, I-33100 Udine, Italy

Abstract. New results concerning the detection of Pontecorvo reactions with strangeness (hidden or open) are presented. With a very selective 2-prongs dedicated trigger, we could show evidence of the $\bar{p}d \rightarrow \Phi n$ reaction and an indication of the $\bar{p}d \rightarrow \Lambda^0 K^0$ reaction, reconstructed through the $\Lambda^0 \rightarrow p\pi^-$ decay.

1 Introduction

Pontecorvo reactions [1] are unusual $\bar{p}$ annihilations on light nuclei, involving more than one target nucleon.

From the theoretical point of view, their interest lies in the correlation between an eventual 6-quark component of the deuteron wave function and the yield of these reactions in annihilations, particularly at rest.

On the experimental side, the very small yield of Pontecorvo reactions, (from $\sim 10^{-5}$ to $\sim 10^{-7}$ depending on the final state) prevented a systematic study for many years. The advent of the CERN Low Energy Antiproton Ring (LEAR), which can provide excellent $\bar{p}$ beams, and the installation of the CRYSTAL BARREL [2] and OBELIX [3] spectrometers, with large geometrical acceptance and very selective trigger options, allowed to start a program dedicated to the identification of several different channels.

In this report we present for the first time results on the $\bar{p}d \rightarrow \Phi n$ and $\bar{p}d \rightarrow \Lambda^0 K^0$ reactions in gaseous deuterium at NTP.

2 Experimental Layout, Data Sample and Trigger

The OBELIX spectrometer, installed at the LEAR M2 beam line, is described in detail elsewhere [3]. For this measurement only the time of flight system (Tof), for particle identification and triggering, and the Jet Drift Chamber (JDC), for momentum measurement and particle identification via dE/dx, have been used. The first level trigger required a multiplicity 2 into both the inner (tofino) and outer (TOFONE) scintillator barrels and a topological correlation between the inner slabs (2 within a set of 3 adjacent must give signal, Fig. 1a). The second level fixed the correlation between the inner and outer hits: one of nine outer slabs facing each inner one should be hitted (Fig. 1b). The overall enhancement factor is about 100.

3 Data Analysis

The triggered events are submitted to some preliminary cuts, requiring two tracks reconstructed by the JDC with opposite charge and length greater than 50 cm: the sample is thus reduced from 8.5 to about 3.5 M events. The contamination from annihilations outside the target is smaller than 0.1%. The strategy for the final state selection is based on information coming both from the Tof and the JDC: particles are identified by dE/dx and then the residual contamination is removed by making use of the time of flight measurement. The requirements are very selective: the final contamination is at the level of 1% both for the K^+K^-X and for the $p\pi^-X$ final states. Once the final state

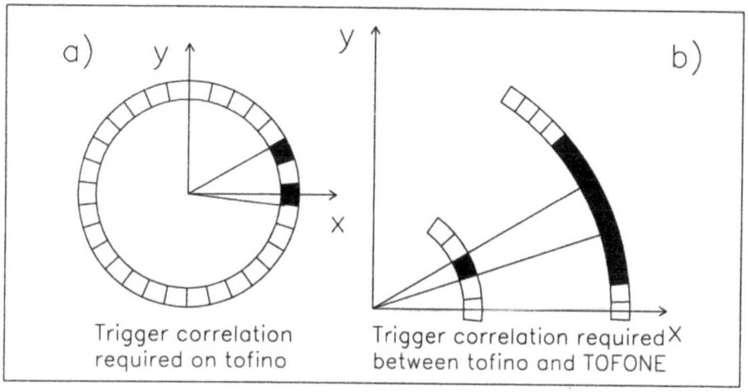

Figure 1. The Φ trigger: first (a) and second (b) level.

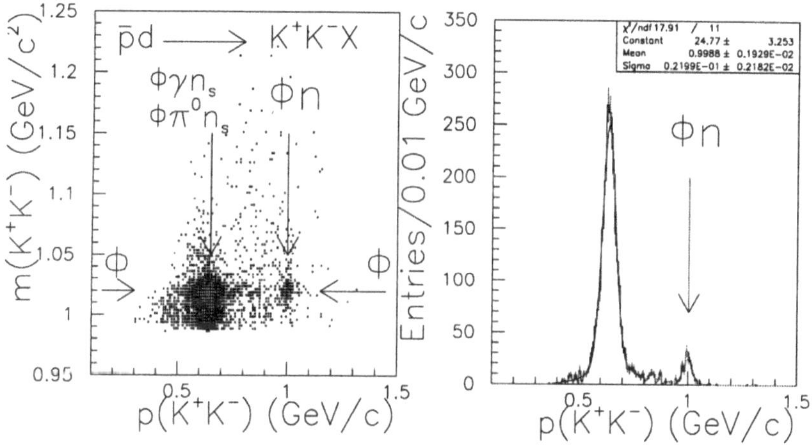

Figure 2. The correlation between the K^+K^- mass and momentum (a) and the momentum distribution for K^+K^- pairs in the Φ mass region: $|m(K^+K^-) - m(\Phi)| < 0.01$ GeV/c (b).

has been identified, the correlation between the momentum and the invariant mass of the detected pair gives information on the annihilation channel.

3.1 The K^+K^-X Final State

The result for the K^+K^- pair is shown in Fig.2a. The majority of the events clusters around the Φ mass, in two specific momentum regions: the first one, around 640 MeV/c, corresponds to annihilations on a single nucleon into $\Phi\pi^0$, $\Phi\gamma$, with a spectator neutron; the second one, around 1 GeV/c is the experimental evidence of the $\bar{p}d \rightarrow \Phi n$ Pontecorvo reaction. The momentum distribution for events selected in the Φ region (Fig. 2b), shows the Φn peak at $< p(\Phi) >= (999 \pm 2)$ MeV/c, compatible with the expected value of

1002 MeV/c, with a momentum resolution of 2.2%. The number of events in the peak is $N(\bar{p}d \to \Phi n) = (137 \pm 12)$.

3.2 The $p\pi^- X$ Final State

An analogous approach to the $p\pi^- X$ final state, shows a clean Λ^0 peak in the invariant mass distribution of the $p\pi^-$ system (Fig.3a). A fit to a gaussian function plus a linear background gives for the Λ^0 mass the value $m(\Lambda^0) = (1116 \pm 1)$ MeV/c^2. The same distribution, after the selection of the $p\pi^-$ momentum in the region expected for the binary reaction $\bar{p}d \to \Lambda^0 K^0$ ($|p(p\pi^-)$ (GeV/c) - 1.126$| < 0.050$), still shows a peak in the Λ^0 region, with $N(\bar{p}d \to \Lambda^0 K^0) = (29 \pm 5)$ events (Fig. 3b).

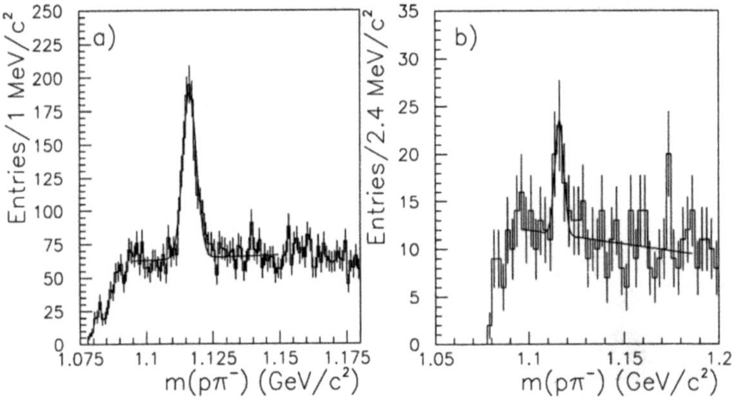

Figure 3. The $p\pi^-$ mass distribution for a) the whole sample and b) the events with $|p(p\pi^-)$ (GeV/c) - 1.126$| < 0.050$.

4 Results

The absolute branching ratios are under evaluation. A preliminary result concerning the $\bar{p}d \to \Phi n$, obtained with a relative normalization to the $\bar{p}pn_s \to \Phi\pi^0 n_s$ channel [4], is:

$$BR(\bar{p}d \to \Phi n) = (5.6 \pm 1.1) \cdot 10^{-6}$$

Moreover, the branching ratio for the $\bar{p}d \to \Lambda^0 K^0$ channel must be greater than 10^{-7}, which is the expected sensibility of the measurement.

References

1. B.M. Pontecorvo: Sov. Phys. JETP **3**, 966 (1956)
2. E. Aker et al.:, Nucl. Instr. Meth. **A321**, 69 (1992)
3. A. Adamo et al.: Sov. J. Nucl. Phys. **55**, 1732 (1992)
4. V.G. Ableev et al.: Nucl. Phys. (submitted)

Few-Body Systems Suppl. 9, 253–258 (1995)

Few-Body Systems

Some Multiple Scattering Effects in Antinucleon Annihilation on Nuclei

A.V. Stepanov, V.P. Zavarzina

Institute for Nuclear Research, Russian Academy of Sciences, Moscow 117312, Russia

Abstract. Antiproton annihilation cross sections on nuclei from ^{20}Ne to ^{208}Pb are calculated in the framework of a microscopic optical model in the kinetic energy range $E_L = 50$–2000 MeV both directly in the initial channel and after single quasi-free collision. The part of annihilation yield after one quasi-free nucleon knock-out from the nucleus is 20–25% of the conribution to the annihilation cross section in the initial channel. The sensitivity of this cross section to the parameters of antinucleon-nucleon scattering amplitude is investigated. Calculated ratios of volume to surface annihilation cross section are compared with experimental ones.

1. At low and intermediate energies the antiproton-nucleus interaction is dominated by strong annihilation. The numerous attempts to create the quark-gluon models of $\bar{p}p$ annihilation, studies of nucleon-nucleon correlation contributions to annihilation cross sections of antiprotons and lightest antinuclei on nuclei, probable observation of glueball signatures, the search for signatures of quark-gluon plasma droplets generation in nuclear volume–this is far from the total list of physical problems the solution of which is related with progress in antinucleon annihilation investigation.

An adequate approach to the above problems demands the knowledge of the extent of contributions of "usual" interaction mechanisms of antiprotons with nucleons of a target nucleus, for example antiproton rescattering. The analysis of data on secondary charged particle multiplicity [1,2] indicates that from 80% to 87% (at antiproton energies 180 MeV and 20 MeV, respectively) of the annihilation events on ^{20}Ne occur on the nuclear surface. The rest 20% and 13% of the cases, respectively, occur in the central region of the target nucleus. Due to the strong attenuation of the incident wave, most of the annihilations inside the nucleus occur after the preliminary quasi-free antiproton scattering on nucleons of the target.

In this paper we calculate various contributions to antiproton annihilation cross sections on nuclei at intermediate energies resulting from annihilation processes in the entrance channel, and after single quasi-free antiproton scattering, as well as total annihilation cross sections in the framework of the optical model with coupled channels. An approach similar to refs. [3, 4] is applied to the analysis of this problem.

2. The wave function $\Psi^{(+)}$, which describes the interaction of the incident antinucleon with the nucleus, obeys a many-body Schrödinger equation

$$(E - \widehat{H})\Psi^{(+)} = 0. \tag{1}$$

The projection operators P_0, P_1 and Q operate on the state space defined by this equation. These operators project onto the different regions of Hilbert space: P_0 projects $\Psi^{(+)}$ onto the nuclear ground state and P_1 onto the excited states of the nucleus. The operator Q projects $\Psi^{(+)}$ onto the state subspace, which contains a nucleus with mass number $A - 1$ in an excited state but not the antiproton.

The antiproton elimination from the initial channel is described by the optical potential (OP)

$$V_{\text{opt}}^{P_0 \to P_1 + Q} = P_0 \widehat{H}(P_1 + Q) \frac{1}{E - (Q + P_1)\,\widehat{H}(Q + P_1) + \mathrm{i}\eta} (P_1 + Q)\widehat{H} P_0. \tag{2}$$

Let us assume that the propagator $[E - (Q + P_1)\,\widehat{H}(P_1 + Q) + \mathrm{i}\eta]^{-1}$ is diagonal with respect to $P_1\Psi$ and $Q\Psi$. Then, we can decompose this OP in two parts:

$$V_{\mathrm{a}}^{P_0 \to Q} \simeq P_0 \widehat{H} Q \frac{1}{E - Q\widehat{H}Q + \mathrm{i}\eta} Q \widehat{H} P_0, \tag{3}$$

which corresponds to antiproton annihilation as resulting from the first collision, and

$$V_{\text{in}}^{P_0 \to P_1} \simeq P_0 \widehat{H} P_1 \frac{1}{E - P_1 \widehat{H} P_1 + \mathrm{i}\eta} P_1 \widehat{H} P_0, \tag{4}$$

describing processes following the first inelastic scattering of the antiproton.

3. The cross sections of interest, which describe the annihilations resulting from the first interaction of the antiproton with the nucleus, $\sigma_{\mathrm{a}}^{(1)}$, and after one antiproton inelastic scattering, $\sigma_{\mathrm{a}}^{(2)}$, may be written in the form

$$\sigma_{\mathrm{a}}^{(1)} = -\frac{2}{\hbar v_0} \mathrm{Im}\,\langle \psi_1 \,|\, H_{13} \frac{1}{E + \mathrm{i}\eta - H_{33}} H_{31} \,|\, \psi_1 \rangle \tag{5}$$

and

$$\sigma_{\mathrm{a}}^{(2)} = -\frac{2}{\hbar v_0} \mathrm{Im}\,\langle \psi_2 \,|\, H_{23} \frac{1}{E + \mathrm{i}\eta - H_{33}} H_{32} \,|\, \psi_2 \rangle, \tag{6}$$

where v_0 is the projectile velocity in the laboratory frame. Here, for the sake of simplicity, we use the notations

$$P_0 \,|\, \Psi^{(+)} \rangle = \psi_1, \quad Q \,|\, \Psi^{(+)} \rangle = \psi_3, \quad P_{11} \,|\, \Psi^{(+)} \rangle = \psi_2,$$

$$P_0 \widehat{H} P_0 = H_{11}, \quad P_1 \widehat{H} P_0 = H_{21}, \quad Q \widehat{H} P_0 = H_{31} \tag{7}$$

and so on.

Taking into account only direct coupling between the states $P_0\,|\Psi^{(+)}\rangle$ and $P_{11}\,|\Psi^{(+)}\rangle$ we can express this wave function in the form

$$\psi_2 \approx \left[E + \mathrm{i}\eta - H_{22} - H_{23}\frac{1}{E + \mathrm{i}\eta - H_{33}} H_{32}\right]^{-1} H_{21}\psi_1. \tag{8}$$

We have in the eikonal approximation (EA) the cross section of annihilation as a result of the first collision

$$\sigma^{(1)}_{\mathrm{aEA}} = \frac{\sigma^{\overline{N}N}_{\mathrm{a}}}{\sigma^{\overline{N}N}_{\mathrm{a}} + \sigma^{\overline{N}N}_{\mathrm{el}}}\sigma_{\mathrm{REA}}, \tag{9}$$

where σ_{REA} is the total reaction cross section in EA.

The expression (6) is still too complicated for further operations. To simplify (6) we use the closure approximation ($\langle\Delta\varepsilon\rangle$ - the average excitation energy, $\mathbf{k_1}$ and $\mathbf{k_2}$ are wave vectors of incident and single scattered antiproton, respectively, $k_2 = k_2(E - \langle\,\Delta\varepsilon\,\rangle\,)$ and an often applied approximate form of distorted wave $\widetilde{\psi}_1$

$$\widetilde{\psi}_1(\mathbf{r}) = \exp(\mathrm{i}\mathbf{kr})\,\exp\left[-\left(\frac{\mathrm{i}k_1R}{2}\right)U^{(1)}_{\mathrm{a}}(0)\right]. \tag{10}$$

Here $U^{(1)}_{\mathrm{a}}(0)$ is the value of the annihilation component of dimensionless OP in the center of the nucleus at the energy of an incident antiproton [5,6]. R is the depth of penetration of the incident wave into the nuclear matter. We consider R as an effective parameter to be determined.

Writing, in a manner similar to (10) the Green's function of single inelastic scattered antiproton

$$G^{(2)}_{\mathrm{opt}}(\mathbf{r} - \mathbf{r}') = \frac{1}{4\pi\,|\,\mathbf{r} - \mathbf{r}'\,|}\,\exp(\mathrm{i}k_2\,|\,\mathbf{r} - \mathbf{r}'\,| - \gamma_2|\,\mathbf{r} - \mathbf{r}'\,|), \tag{11}$$

where

$$\gamma_2 = |\,k_2\mathrm{Im}U^{(2)}_{\mathrm{a}}(0)\,|\,/2 \tag{12}$$

and using a standard parameterization of $\overline{N}N$ t-matrix $\tau^{\overline{N}N}(\mathbf{q})$ dependence on momentum transfer

$$\tau^{\overline{N}N}(\mathbf{q}) = \tau_0 e^{-\beta q^2/2} \tag{13}$$

after some integration, we obtain immediately the final expression for $\sigma^{(2)}_{\mathrm{a}}(k_1)$

$$\sigma^{(2)}_{\mathrm{a}}(k_1) = -\frac{k_2}{k_1}\,k_2 A\frac{\mathrm{Im}\,U^{(2)}_{\mathrm{a}}(0)}{(2\pi)^2}\,\exp\left[k_1 R\,\mathrm{Im}\,U^{(1)}_{\mathrm{a}}(0)\right]\,|\,\tau_0\,|^2\frac{e^{-\beta k_1^2}}{2\beta k_1}$$

$$\times\int\limits_0^\infty \frac{q\,\mathrm{d}q e^{-\beta q^2}\left[e^{2\beta q k_1} - e^{-2\beta q k_1}\right]}{[q^2 - k_2^2 + \gamma_2^2]^2 + 4\gamma_2^2 k_2^2}. \tag{14}$$

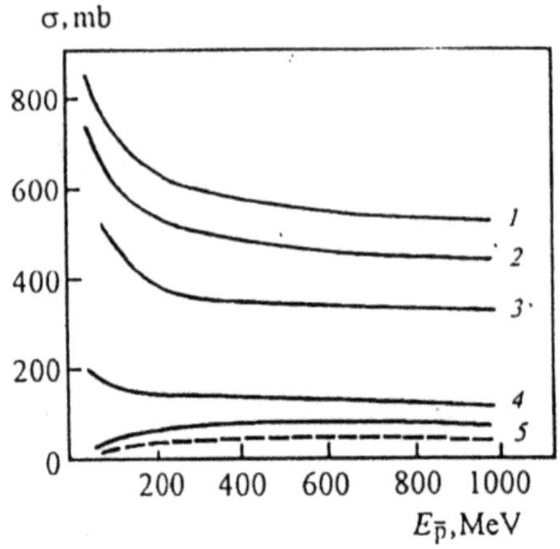

Figure 1. Total reaction cross section, σ_R (*1*), annihilation cross section, σ_a (*2*), $\sigma_a^{(1)}$ (*3*), $\Delta\sigma_a$ (*4*) and $\sigma_a^{(2)}$ (*5*) as functions of laboratory kinetic antiproton energy for $\bar{p}\,^{20}$Ne interactions. Dashed curve: $\sigma_a^{(2)}$ calculated from (14), where γ_2-parameter values are defined by the total optical potential, $U^{(2)}$.

4. Let us consider the values of the input parameters for σ_a, $\sigma_a^{(1)}$ and $\sigma_a^{(2)}$ calculations. We will use, as before [5,6], the standard equations relating $\tau^{\overline{N}N}$ in the projectile-target (PT) frame with the free $\overline{N}N$ scattering amplitude $M^{\overline{N}N}$ in this center-of-mass system. The values of $\sigma_T^{\overline{N}N}$, $\sigma_a^{\overline{N}N}$, β, $\rho = \mathrm{Re}M^{\overline{N}N}(0)/\mathrm{Im}M^{\overline{N}N}(0)$ are taken according to the papers [7–10]. The parameters of the point nucleon density were taken from works [9,11]. The free parameter R was found from the condition of equality of σ_a calculated in approximation for $\tilde{\psi}_1$ (10) to σ_a calculated in EA without approximation (10):

$$R = \frac{1}{|\,k_1 \mathrm{Im}\, U_a^{(1)}(0)\,|} \left| \ln \frac{\sigma_a(k_1)}{|\,k_1 \mathrm{Im}\, \tilde{U}_a^{(1)}(0)\,|} \right|. \tag{15}$$

Results of calculation of $\sigma_a^{(1)}$, $\sigma_a^{(2)}$, σ_a and σ_R for antiproton interaction with ^{20}Ne are shown in Fig. 1. The cross section of annihilation in all excited states of the target, $\Delta\sigma_a = \sigma_a - \sigma_a^{(1)}$, is also represented in the same figure. The energy dependence of σ_R, σ_a, $\sigma_a^{(1)}$ and $\Delta\sigma_a$ calculated by EA formulae follow in general the energy trend of the cross section $\sigma_T^{\overline{N}N}$ and $\sigma_a^{\overline{N}N}$ for antinucleon-nucleon collisions: these cross sections gradually decrease with increasing incident antiproton kinetic energy. At the same time $\sigma_a^{(2)}$ are practically constant in the E interval 0.3–1.0 GeV, then slowly decrease at $E > 1$ GeV and falls sharply when $E < 0.3$ GeV: $\sigma_a^{(2)}/\sigma_a^{(1)} = 0.25$–$0.3$ at $E = (0.3$–$1.0)$ GeV and $\sigma_a^{(2)}/\sigma_a^{(1)} = 0.05$ at $E = 0.05$ GeV. The cross section $\sigma_a^{(2)}$ is very sensitive to values of the slope parameter β. When the usually applied values of β are diminished 1.5 and 2.0

Table 1. Evaluated ratios $\Delta\sigma_a/\sigma_a^{(1)}$ and $\sigma_a^{(2)}/\sigma_a^{(1)}$ and experimental values of ζ = ratio of volume annihilation events to surface annihilation events (taken from Balestra et al. [1,2]*.

		$\bar{p}$ momentum, MeV/c					
		200	300	400	490	600	1400
		$\bar{p}$ kinetic energy, MeV					
		20	50	80	120	180	800
$\Delta\sigma_a/\sigma_a^{(1)}$	Ne		0.34	0.33	0.33	0.36	0.38
	Cu		0.38	0.39	0.40	0.41	0.44
	Pb		0.43	0.45	0.46	0.50	0.52
$\sigma_a^{(2)}/\sigma_a^{(1)}$	Ne		0.05	0.07	0.11	0.15	0.25
			(0.07)	(0.09)	(0.14)	(0.20)	(0.30)
	Cu		0.06	0.09	0.12	0.16	0.26
			(0.10)	(0.14)	(0.18)	(0.24)	(0.36)
	Pb		0.07	0.10	0.14	0.18	0.27
			(0.14)	(0.18)	(0.24)	(0.32)	(0.43)
	ζ_{Ne}	0.15				0.25	
	$\zeta_{Ag/Br}$	0.33	0.35	0.37	0.39		0.59

*/ The values of ratio $\sigma_a^{(2)}/\sigma_a^{(1)}$ in parentheses are calculated using (R) (15), where the substitutions $\sigma_a \to \sigma_R$ and Im $U_a^{(1)}(0) \to \mathrm{Im}U_{a+in}^{(1)}(0)$ are used.

times, the $\sigma_a^{(2)}$ arises at low energies. These factors, 2/3 and 1/2, are sufficient to eliminate of the decrease of $\sigma_a^{(2)}$ caused by the β sharp growth at low energies.

The cross section $\sigma_a^{(2)}$ is the inclusive annihilation cross section for all antiprotons, which undergoes at most one inelastic scattering. Let us consider with the help of formulae for $\sigma_a^{(2)}$ the exclusive annihilation cross section for antiprotons undergoing one (and only one!) inelastic scattering $\sigma_a^{(21)}$. In these formulae γ_2 is proportional to the imaginary part of the total optical potential (not only to the annihilation component OP as in (13)). The energy dependence of $\sigma_a^{(21)}$ repeats the energy trend of $\sigma_a^{(2)}$ and $\sigma_a^{(21)}/\sigma_a^{(2)} \approx 0.5$ for ^{20}Ne, ^{64}Cu and ^{208}Pb in the energy interval under consideration.

Table 1 contains the results of our calculations of $\Delta\sigma_a/\sigma_a^{(1)}$ and $\sigma_a^{(2)}/\sigma_a^{(1)}$ for $\bar{p}\,^{20}$Ne, $\bar{p}\,^{64}$Cu and $\bar{p}\,^{208}$Pb interactions and ζ (the ratio of the volume to surface annihilation events for $\bar{p}\,^{20}$Ne interactions and $\bar{p}$Ag and $\bar{p}$Br interactions in nuclear emulsion [1,2]). Evidently, $\Delta\sigma_a/\sigma_a^{(1)}$ and $\sigma_a^{(2)}/\sigma_a^{(1)}$ representing the same physical function calculated in the framework of different $\bar{p}$-nucleon interaction models, are the upper and lower bounds of ζ, respectively. When calculating $\sigma_a^{(2)}$, we use in Eq.(11) for Green's function $G_{opt}^{(2)}$ the decay constant

$\gamma_2(\sim \mathrm{Im}\, U_{\mathrm{opt}}^{(2)}(0))$ (12), which is proportional to the central nuclear density. Since the antiproton free path length even near the nuclear surface is small, after single collision the antiproton moves into a region of lower nuclear density. Therefore, more adequate would be a substitution in (11) by smaller values, γ_2. This results in a growth of $\sigma_{\mathrm{a}}^{(2)}$. The values of $\sigma_{\mathrm{a}}^{(1)}$ are best described by a linear dependence of $A^{2/3}$ while $\sigma_{\mathrm{a}}^{(2)} \sim A$, which corresponds to the predominantly peripheral character of interaction in the first case and a mainly volume one in the second.

In summary, the earlier studies of antiproton annihilation on nuclei are analyzed. It is shown that the main contribution to the annihilation cross section, $\sigma_{\mathrm{a}}(\approx 2/3)$, is due to the entrance channel annihilation cross section, $\sigma_{\mathrm{a}}^{(1)}$. The entrance channel annihilation takes place mainly at the nuclear surface. The results of our calculations of $\sigma_{\mathrm{a}}^{(1)}$ and $\sigma_{\mathrm{a}}^{(2)}$ (and $\Delta\sigma_{\mathrm{a}}$) are close to the estimations obtained by analysis of data [1,2] in spite of the approximate character of the calculations and uncertainties of analysis of data in [1,2].

This work was supported in part by the Fundamental Research Foundation of Russia (Grant N 95-02-05659a).

References

1. F. Balestra et al.: Nucl. Phys. **A452**, 573 (1986)
2. G. Piragino: in *Hadronic Physics at Intermediate Energy.* Winter School held at Folgaria, Italy, 1986 (Edited by Bressani T. and Ricci R.A.) p.293. Amsterdam, Oxford, N.Y., Tokyo: North Holland 1986; Yu.A. Batusov et al.: Europhys.Lett. **2**, 115 (1986)
3. H.C. Chiang and J. Hüfner: Nucl.Phys. **A349**, 466 (1980); W.Q. Chao, F. Hachenberg, J. Hüfner: Nucl.Phys. **A384**, 24 (1982)
4. K. Masutani and K. Yazaki: Phys. Lett. **B104**, 1 (1981); Nucl. Phys. **A407**, 309 (1983)
5. V.P. Zavarzina, A.V. Stepanov: Sov. J. Nucl. Phys. **43**, 543 (1986)
6. V.P. Zavarzina, A.V.Stepanov: Sov. J. Nucl. Phys. **54**, 27 (1991)
7. C.B. Dover and J.M. Richard: Phys. Rev. **C21**, 1466 (1980)
8. L.A. Kondratyuk and M.G. Sapozhnikov: Sov. J. Nucl. Phys. **46**, 56 (1987)
9. W.W. Buck, J.W. Norbury and J.W. Wilson: Phys.Rev. **C33**, 234 (1986)
10. H. Iwasaki et al.: Nucl.Phys. **A433**, 580 (1985)
11. H. Sakaguchi et al.: Phys. Rev. **C26**, 944 (1982)

Few-Body Systems Suppl. 9, 259–262 (1995)

Few-Body Systems

Latest achievements in dynamic calculations for light nuclei

V. I. Kukulin *

Institute of Nuclear Physics, Moscow State University, Moscow 119899, Russia

Abstract. A brief review of fully microscopic A-nucleon calculations of some light nuclei is presented. The results are compared to those of a multicluster dynamic model developed by the author with coworkers in Moscow State University during last decade. The comparison and an appropriate discussion show unambiguously that a construction of the consistent A-nucleon theory of light nuclei has to be started preferably with a trial multicluster wavefunction ansatz by completing it with minor components describing a (virtual or real) cluster breakup process.

1 Introduction

Today, quite accurate calculational methods toward quantitative treatment of three- and four-body systems with full complexity of realistic nuclear force have already been developed and widely used for calculations of particular systems.

Hence, a very interesting question arises on the modern possibility of similarly accurate treatment for five-, six-, etc. nucleon systems within the framework of a dynamic A-body approach. Solution for the last problem generally seems to be even more important than the precise three- and four-nucleon calculations. In contrast to 0s-shell nuclei, there is a very rich and abundant spectroscopic information on excited states of the nuclei in the region $A = 5 \div 12$ and up to now a huge volume of data for transition probabilities has been accumulated.

It should be emphasized, however, that all the wealthy database can be employed with maximal effectiveness only if one applies quantitative dynamic models with no fit or free parameters to the data interpretation. In opposite case, we attain often only *an appearance* of the explanation for processes, substituting lack of our knowledge (or roughness of it) by an arbitrary model parameter renormalization.

*This work has been supported by Russian Foundation for Fundamental Research (Grant No. 93-02-03376).

2 Dynamic many-body models

Main development in the $A > 4$ case has been made within a scope of variational and variational Monte-Carlo approaches [1-3]. Moreover, due to the high dimension of the problem, the variational approaches are also based on an idea of statistical trials [1,2,4-6].

Argonne group [1-3] traditionally uses the Green Function Monte-Carlo method (GFMC) for many-body calculations. Due to a tremendous dimensionality of the problem both in the configuration and spin-isospin spaces, the process of solution represents an extremely complicated numerical procedure, so that an employment of even large supercomputers requires many hours of runtime to reach some acceptable convergence [3].

Nevertheless, these calculations, together with the results attained, look to be quite attractive as an experience of direct solution for many-body quantum-mechanical problem with extraordinarily complicated particle interaction. From short summary of the results (see Table 1, 2nd row), it is evident that the overall agreement with the experiment (at least, for the state energies and rms charge radius) is quite good though the statistical error of the calculation is still a little bit large. Unfortunately, this approach also suffers from significant drawbacks and inconveniencies. First of all, it is an extreme labor- and time - consuming approach. In particular, in passing from ^{4}He to the $A = 6$ calculations, supercomputer time rises so drastically that one can, at the present level, foresee only very limited potential to extend the approach over $A > 6$ case (keeping the same statistical error).

The main difficulty in the approach is, however, met in attempts to take into account adequately the long range cluster-type correlations in light nuclei (and in many-body systems at all). Just because of the lack of those, the authors [3] have got a quite visible underbinding in $A = 6$ nuclei and especially for $A = 16$ case [2]. The underbinding in $A = 6$ case is especially evident if one takes into account the fact that the ^{4}He binding energy has been adjusted almost ideally (by choosing appropriately three-body force constants). Thus total error in six-nucleon binding energy must be attributed to the binding energy of two valence outer nucleons to the α-core, i.e. actually just to inadequate description of the cluster correlation in the approach. The same problem in the variational Monte-Carlo method as applied to ^{16}O-calculation looks to be even more serious [2].

Summarizing the results attained in the GFMC and variational Monte-Carlo approaches, one can say that the methods are quite accurate in the treatment of short-range correlations and inadequate in description of long-range and cluster-type correlations in many-nucleon systems.

In very recent years, there appeared also other fully microscopic dynamic calculations of light stable and neutron-deficient nuclei within the framework of a statistical trials method [5,6] proposed by the present author two decades ago [4]. This approach (referred to as stochastic variation methods (SVM)) combines idea of the statistical trials and big advantages of variational scheme on very flexible variational basis. Here, the statistical trials occur in a "parametric" multi-dimensional space (of nonlinear scale parameters) rather than in configuration and spin-isospin spaces as in GFMC-approach. This leads to a drastic optimization of the variational basis (based on multi-dimensional gaussians) and adaptation of it just to a particular task. In the passed years, the approach has been applied with a great success to numerous few- and many-body problems in the field of multiquark-, few-nucleon, mesomolecular-, hypernuclear-, atomic, etc. systems. Varga, Suzuki et al. were the first to show that the method works excellently for five-, six-etc. body systems [5,6]. In particular, their test calculations for $A = 4$ case with central MT-V force (including a strong repulsive core) have given the same results (with a high accuracy) as other methods (Faddeev-Yakubovsky eqs., etc.). The ground state energies for ^{6}Li and ^{6}He found with a high-quality effective $N - N$ force via the SVM [5] (see the 3rd row in Table 1) are in a reasonable accordance with experiment.

It can thus be concluded that the energies and (at least) main static observables of many nucleon systems with $A = 6, 7...$ can presently be found in the really *ab initio* microscopic calculations with the stochastic variational method.

3 Predictions of dynamic multicluster model and construction of a fully microscopic nuclear model

Now, we are prepared to compare at the example of six-nucleon system *ab initio* results of fully microscopic methods with those of macroscopic dynamic multicluster models (MDMP). Table 1 shows quite clearly that the MDMP-predictions (which do not include any free parameters) occur to be the most accurate among existing dynamic models. It is very important to check not only the accuracy for the energy predictions but also the quality of resulting nuclear wavefunctions (wf). The wf's can conveniently be tested through wide comparison with experiment for e.-m. form factors: elastic and transitional, longitudinal and transversal. The detailed comparison for the $A = 6$ and $A = 9$ e.m. form factors has been made in our recent works [7-10].

Thus this model can be a very appropriate starting point to construction of the consistent microscopic theory for light nuclei. The construction itself can be performed in a following way. We take as an ansatz for fully microscopic wf in a form of two *orthogonal* components:

$$\Psi_{\text{total}}(1, ...A) = \Psi_{\text{AMDMP}}(1, ...A) + \Delta\Psi_{\text{SVM}}(1, ...A)$$

where $\Psi_{\text{AMDMP}} = \hat{A}(\Phi_{a_1 a_2 ... a_n}\chi(\boldsymbol{R}_1, \boldsymbol{R}_2, ...\boldsymbol{R}_{n-1}))$ is antisymmetrised MDMP wf for n - cluster system, and $\Delta\Psi_{\text{SVM}} = det \mid \varphi_1(\boldsymbol{r}_1)\varphi_2(\boldsymbol{r}_2)...\varphi_A(\boldsymbol{r}_A) \mid$ is a Hartree-Fock type wf, which must be orthogonalised to Ψ_{AMDMP}, and hence corresponds to a highly excited mean field state. Here Ψ_{AMDMP} is a fixed component while Ψ_{SVM} is searched for starting from a realistic nuclear force. By this way, we supplement accurate *intercluster motion* emerged from the AMDMP by correct *one-particle motion* described through the modified Hartree-Fock part of the total wf. We elaborated a general computational technique for such a type of calculations which reduces all the integrals to simple analytical construction.

Table 1. Comparison of the state- and the Coulomb displacement energies and rms charge radii calculated by the fully microscopic GFMC-[3], SVM- [5] with the effective Minnesota NN-potential, and by $\alpha + 2N$ three-cluster dynamic MDMP-approaches [7]

Value	^{6}He(g.s.)	^{6}Li(g.s.)	^{6}Li(3^+)	$< R_p^2 >^{\frac{1}{2}}$	$\Delta(1)$	$\Delta(2)$
GFMC	27.3(4)	31.1(4)	28.2(3)	2.41(5)	2.5(1)	0.6(1)
SVM	30.07	34.59	-	2.22	-	-
MDMP*)	28.7	31.5	28.8	2.40	2.30	0.24
exp	29.3	32.0	29.3	2.43	2.35	0.34

$\Delta(1) = E(^6\text{Be}) - E(^6\text{He})$; $\Delta(2) = \frac{1}{2}\{E(^6\text{Be}) + E(^6\text{He})\} - E(^6\text{Li})$

*)For the MDMP-approach data, the α-particle binding energy is added in the entries of columns two to four to get comparison with the GFMC- and SVM-results.

References

1. S.C. Pieper, R.B. Wiringa, V.R. Pandaripande: Phys.Rev.Lett. **64**, 364 (1990)

2. R.B. Wiringa: Phys.Rev. **C43**, 1585 (1991)

3. B.S. Pudliner et al.: *Quantum Monte-Carlo calculations of* A $\leq$ 6 *nuclei.* Preprint, Argonne 1995

4. V.I. Kukulin: Izvestia Acad. Nauk SSSR (Bulletin of Soviet Ac. Sci.) **39**, 535 (1975); V.I. Kukulin and V.M. Krasnopolsky: J.Phys. G: Nucl.Phys. **3**, 785 (1977)

5. K. Varga and Y. Suzuki: *Precise solution of few-body problems with the stochastic variational method.* Preprint, RIKEN 1995

6. Y. Suzuki: Inv. Talk at 6th Int. Conf. on *Clusters in Nuclear Structure and Dynamics*, Strasbourg 1994

7. V.I. Kukulin et al.: Nucl.Phys. **A586**, 151 (1995)

8. G.G. Ryzhikh et al.: Nucl.Phys. **A517**, 221 (1990)

9. G.G. Ryzhikh et al., Proc. Int. Conf. *Weak and Electromagnetic Interactions*, Dubna 1992, p. 46

10. V.T. Voronchev et al.: Few Body Systems (to appear)

Few-Body Systems Suppl. 9, 263–266 (1995)

Few-Body Systems

Deuteron Compton Scattering*

T. Wilbois, P. Wilhelm, H. Arenhövel

Institut für Kernphysik, J. Gutenberg-Universität, D-55099 Mainz, Germany

Abstract. We have calculated deuteron Compton scattering below pion production threshold using different realistic *NN* potential models and explicit meson exchange current contributions. The gauge conditions for the current and two photon operators have been exploited extensively, so that our model fulfills the low energy theorems. Rescattering, meson exchange and nucleon substructure contributions are studied and compared with a previous calculation and recent experimental data.

As a pure electromagnetic process, photon scattering provides a powerful tool to investigate the internal structure of composite hadronic systems like nucleons and nuclei. Since in general the reaction is a two step process, the whole internal dynamics of the investigated system contributes to the observables via the intermediate propagation of the system. At very low energies, i.e., at energies where the wavelength of the incoming photon is small compared to the dimensions of the system, the photon scattering amplitude is governed by the well-known low energy theorem (LET) [1]. It states, that linear in the photon energy the amplitude is completely determined by global properties of the system like the charge, mass and magnetic moment and is independent from the internal structure. The latter contributes first in the next order in terms of the electric and magnetic polarizabilities.

In the last years there is a considerable effort in the experimental determination of the polarizabilities of the neutron. Because there is no free neutron target available one is forced to study to what extent these fundamental quantities can be extracted from other reactions like photon scattering off light nuclei or neutron scattering off heavy nuclei. Of course, due to its simple structure, photon scattering from the deuteron is best suited for such investigations. However, we want to emphasize that before one can extract any information about the bound neutron one first has to understand the internal dynamics of the nucleus itself. Otherwise the extracted information would be extremely model dependent.

*Supported by the Deutsche Forschungsgemeinschaft (SFB 201)

In this work we shall consider elastic photon scattering off the deuteron at intermediate energies below pion production threshold. This is the favorite energy regime in order to extract the neutron polarizabilities. In the spirit of the arguments mentioned above, the basic topic of this work is the investigation of the role of the NN interaction and meson exchange currents (MEC). To this end we have calculated the NN scattering off-shell T-matrix using a realistic modern interaction model with consistent exchange currents. The theoretical framework of our calculation is similar to a previous work of Weyrauch [2] in which a simple separable interaction model had been considered. In addition, some important contributions like explicit MEC beyond the Siegert operators, the center-of-mass (c.m.) part of the convection current and relativistic corrections to the two-photon operator as well as nucleon substructure contributions have been neglected there.

According to the explicit form of linear and quadratic terms in the e.m. interaction, the photon scattering amplitude is separated into a resonance amplitude $R_{\lambda'\lambda}(-\boldsymbol{k}',\boldsymbol{k})$ and a two-photon or "seagull" amplitude $B_{\lambda'\lambda}(-\boldsymbol{k}',\boldsymbol{k})$. One should keep in mind that $R^{fi}_{\lambda'\lambda}(-\boldsymbol{k}',\boldsymbol{k})$ and $B^{fi}_{\lambda'\lambda}(-\boldsymbol{k}',\boldsymbol{k})$ are not separately gauge invariant or in other words, the current and two-photon operators are related by corresponding gauge conditions [1, 3]. The resonance amplitude is given by the well-known expression

$$\begin{aligned} R^{fi}_{\lambda'\lambda}(\boldsymbol{k}',\boldsymbol{k}) &= \langle f|\boldsymbol{\epsilon}'^{\star}_{\lambda'}\cdot\boldsymbol{j}(-\boldsymbol{k}',2\boldsymbol{P}_f+\boldsymbol{k}')G(E_i+\omega)\boldsymbol{\epsilon}_{\lambda}\cdot\boldsymbol{j}(\boldsymbol{k},2\boldsymbol{P}_i+\boldsymbol{k})|i\rangle \\ &\quad +(\boldsymbol{\epsilon}\leftrightarrow\boldsymbol{\epsilon}'^{\star},\boldsymbol{k}\leftrightarrow-\boldsymbol{k}',\omega\leftrightarrow-\omega'). \end{aligned} \tag{1}$$

It contains the electromagnetic current operator $\boldsymbol{j}(\boldsymbol{k},\boldsymbol{P})$, describing the absorption and emission of the photons, and the full propagator $G(E_i+\omega)$, where $E_i+\omega$ is the energy available for the excitation of the system. Using the Lippmann-Schwinger equation, $G(E_i+\omega)$ can be separated into 'Born' terms describing the free propagation and rescattering contributions containing the full off-shell T matrix of the hadronic system.

In this work we shall restrict ourselves to intermediate NN-states only and the hadronic T matrix is calculated using the Bonn OBEPQ-B potential [4]. We would like to remark already here that the model dependence of the differential cross section due to the choice of different realistic NN potentials turns out to be very small (below 1%) in the energy region considered here. This encourages the use of elastic γd scattering as a tool to study the polarizability of the neutron. A detailed discussion of the model dependence will be reported elsewhere [5].

With the help of the multipole decomposition of the electromagnetic field one can formulate the photon scattering amplitude in terms of so-called generalized polarizabilities corresponding to an expansion in terms of the total angular momentum tranferred to the nucleus [3]. This expansion allows a clean separation between the dynamical and geometrical aspects of the reaction and enables a proper treatment of the singularity in the NN propagator. The latter is essential for the description of the imaginary part of the amplitude which describes the total photoabsorption cross section. Since the explicit expressions for the polarizabilities are quite lengthy the reader is referred to [3, 5].

With respect to the electromagnetic operators we have considered for the nuclear current operator the usual nucleonic one-body currents including the next to leading order relativistic corrections. Furthermore, explicit π and ρ MEC have been taken into account. However, it is well-known that due to the use of Siegert's theorem, a large body of the exchange current is already incorporated in the one-body operators. The same argument holds even better for the two photon contributions. It has been shown [6] that explicit mesonic seagull contributions are only important to produce the correct low energy limit. Thus these contributions can safely be neglected when Siegert operators are used in the calculation of the kinetic part.

So far, our model is based on structureless nucleons as constituents of the nucleus. In order to study the influence of nucleon substructure, we simply consider an effective description in terms of the nucleon (Compton) polarizabilities [7]. We would like to stress, that due to the unknown off-shell-behaviour, it is a priori not clear how much the free nucleon polarizabilities are changed for a bound nucleon. However, off-shell effects in the nucleon polarizabilities are expected to play a minor role due to the small binding energy compared to the large nucleon mass. Therefore, it is a legitimate starting point to use the values for the free nucleon polarizabilities [8] in this calculation in order to investigate the general importance of these contributions.

We start the discussion with our results for the differential cross section in the c.m. frame at three different energies as shown in Fig.1. In accordance with Weyrauch's results we find a rather small influence of rescattering contributions at 50 and 70 MeV. At forward angles the cross section is reduced by roughly 10%. In addition, it is seen that at these energies the MEC contributions are essentially contained in the Siegert operators. At 100 MeV we find a much stronger rescattering effect at forward and at backward angles, where the cross section becomes larger. This indicates that Weyrauch's use of a separable interaction model is not appropriate for describing the off-shell behavior of the NN interaction at medium energies. Note that Weyrauch did not find any rescattering effect at backward angles at all. As expected, the influence

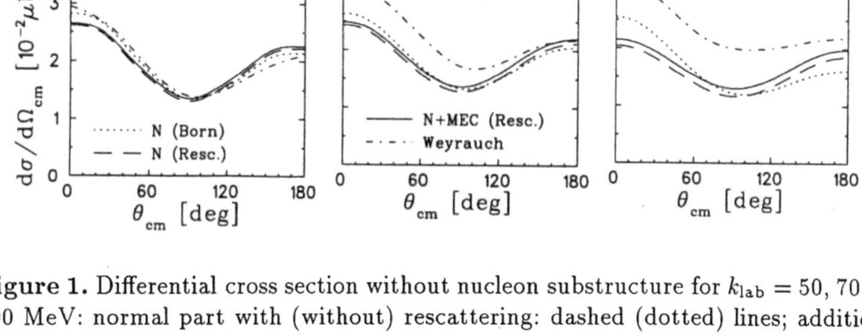

Figure 1. Differential cross section without nucleon substructure for $k_{\rm lab} = 50$, 70 and 100 MeV: normal part with (without) rescattering: dashed (dotted) lines; additional explicit MEC contributions: solid line; Weyrauch's results: dash-dotted curve

of explicit MEC is increasing with photon energy. Another important point is that our results are in general smaller than those of Weyrauch. This can be traced back to the additional ingredients of our calculation.

Finally, we shall consider the influence of the nucleon polarizabilities and compare with recent experimental data. In Fig.2 the lab frame cross section is shown for photon energies of 49 and 69 MeV. Apparently, the neglect of nucleon substructure leads to a cross section which clearly overestimates the experimental data for both energies. The influence of the nucleon polarizabilities leads to a considerable reduction of the cross section, but the calculation still overestimates the data at 49 MeV, where the size of the substructure contribution is comparable to the rescattering correction. However, before concluding that a larger electric polarizability is needed, which would decrease the cross section, further experimental data of higher accuracy are needed.

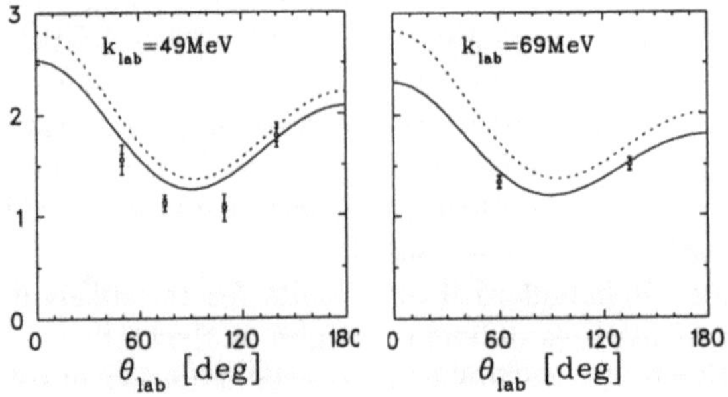

Figure 2. Differential cross section for $k_{\mathrm{lab}} = 49$ and 69 MeV compared to experiment [9]. Complete calculations with (solid) and without (dotted) nucleon polarizabilities

References

1. J.L. Friar: Ann. Phys. **95**, 170 (1975), and references therein
2. M. Weyrauch: Phys. Rev. **C41**, 880 (1990)
3. H. Arenhövel and M. Weyrauch: Nucl. Phys. **A457**, 573 (1986)
4. R. Machleidt: Adv. Nucl. Phys. **19**, 189 (1989)
5. T. Wilbois, P. Wilhelm, H. Arenhövel: to be published
6. M. Weyrauch and H. Arenhövel: Nucl. Phys. **A408**, 425 (1983)
7. A. L'vov and M. Schumacher: Nucl. Phys. **A548**, 613 (1992)
8. B.R. Holstein, A.M. Nathan: Phys. Rev. **D49**, 6101 (1994)
9. M.A. Lucas: Ph.D. thesis, University of Illinois at Urbana-Champaign, 1994.

Few-Body Systems Suppl. 9, 267–271 (1995)

Few-Body Systems

Electromagnetic Structure of Mesons in a Light-Front Constituent Quark Model

F. Cardarelli [1], I.L. Grach [2], I.M. Narodetskii [2], G. Salmè [3], S. Simula [3]

[1] Istituto Nazionale di Fisica Nucleare, Sezione Tor Vergata, Via della Ricerca Scientifica 1, I-00133 Roma, Italy

[2] Institute for Theoretical and Experimental Physics, 117259 Moscow, Russia

[3] Istituto Nazionale di Fisica Nucleare, Sezione Sanità, Viale Regina Elena 299, I-00161 Roma, Italy

Abstract. The elastic and transition electromagnetic form factors of mesons are investigated within a light-front constituent quark model. The general formulae for the space-like matrix elements of the one-body component of the electromagnetic current, including both Dirac and Pauli form factors of the constituent quarks, are presented. Our results for the pion charge form factor and the $\pi\rho$ transition form factor are reported and compared with those obtained within various relativistic approaches in a range of values of the four-momentum transfer accessible to CEBAF.

The investigation of elastic and transition electromagnetic (e.m.) form factors of mesons has recently received a renewed interest, because measurements of the pion and kaon charge form factors, as well as of the electroproduction cross section of vector mesons off the nucleon, are planned at CEBAF [1]. The aim of this contribution is to address the calculation of meson e.m. form factors, adopting the constituent quark (CQ) model for the description of the meson structure and the Hamiltonian light-front formalism [2] for a Poincaré-covariant treatment of the CQ degrees of freedom. For space-like values of the four-momentum transfer, all the relevant formulae needed for the calculation of the matrix elements of the one-body component of the e.m. current, including both Dirac and Pauli form factors for the CQ's, will be presented. Moreover, the results of our calculations of the pion charge form factor and of the $\pi\rho$ transition form factor will be reported and compared with those obtained from various relativistic approaches at momentum transfers accessible to CEBAF.

As is well known (cf. [2]), the light-front formalism allows an exact separation in momentum space between the center of mass and intrisic wave functions; therefore, in what follows, we will focus on the intrisic part of the meson wave function. Let us remind that the intrinsic light-front kinematical variables are $\boldsymbol{k}_\perp = \boldsymbol{p}_{q\perp} - \xi \boldsymbol{P}_\perp$ and $\xi = p_q^+/P^+$, where the subscript $\perp$ indicates the projection perpendicular to the spin quantization axis, defined by the vector $\hat{\boldsymbol{n}} = (0,0,1)$, and the *plus* component of a 4-vector $p \equiv (p^0, \boldsymbol{p})$ is given by $p^+ = p^0 + \hat{\boldsymbol{n}} \cdot \boldsymbol{p}$; eventually, $\tilde{P} \equiv (P^+, \boldsymbol{P}_\perp) = \tilde{p}_q + \tilde{p}_{\bar{q}}$ is the total light-front momentum of the meson. Omitting for simplicity the colour degrees of freedom, the requirement of Poincaré covariance for the intrinsic wave function $\langle \nu\bar{\nu}|JM\rangle$ of a meson with angular momentum J and projection M implies

$$\langle \nu\bar{\nu}|JM\rangle = \sqrt{A(\boldsymbol{k}_\perp,\xi)} \sum_{SM_S} R^{SM_S}(\xi, \boldsymbol{k}_\perp; \nu\bar{\nu})\ \tilde{w}^{JM}_{SM_S}(\boldsymbol{k}) \tag{1}$$

where $\nu, \bar{\nu}$ are the CQ spin variables; $k^2 \equiv k_\perp^2 + k_n^2$; $k_n \equiv (\xi - 1/2)M_0 + (m_{\bar{q}}^2 - m_q^2)/2M_0$; $M_0^2 = (m_q^2 + k_\perp^2)/\xi + (m_{\bar{q}}^2 + k_\perp^2)/(1-\xi)$; $A(\boldsymbol{k}_\perp,\xi) = M_0[1 - (m_q^2 - m_{\bar{q}}^2)^2/M_0^4]/4\xi(1-\xi)$. In Eq. (1) $\tilde{w}^{JM}_{SM_S}(\boldsymbol{k}) = \sum_{LM_L} w^{q\bar{q}}_{LSJ}(k^2)\, Y_{LM_L}(\hat{\boldsymbol{k}})\, \langle LM_L SM_S|JM\rangle$, where L is the orbital angular momentum, Y_{LM_L} the usual spherical harmonics and $w^{q\bar{q}}_{LSJ}(k^2)$ the radial wave function of the given (LSJ) channell. Moreover, the momentum-dependent spin factor R^{SM_S} is defined as: $R^{SM_S}(\boldsymbol{k}_\perp, \xi; \nu\bar{\nu}) = \sum_{\nu'\bar{\nu}'} \langle \nu|R_M^\dagger(\boldsymbol{k}_\perp, \xi, m_q)|\nu'\rangle$ $\langle \bar{\nu}|R_M^\dagger(-\boldsymbol{k}_\perp, 1-\xi, m_{\bar{q}})|\bar{\nu}'\rangle$ $\langle \frac{1}{2}\nu' \frac{1}{2}\bar{\nu}'|SM_S\rangle$, where the 2×2 irreducible representation of the generalized Melosh rotation [3] reads as follows

$$\langle \nu'|R_M(\boldsymbol{k}_\perp,\xi,m)|\nu\rangle = \chi^\dagger_{\nu'} \frac{m + \xi M_0 - \mathrm{i}\boldsymbol{\sigma}\cdot(\hat{\boldsymbol{n}} \times \boldsymbol{k})}{\sqrt{(m+\xi M_0)^2 + k_\perp^2}} \chi_\nu \tag{2}$$

with χ_ν being the two-component Pauli spinor and $\boldsymbol{\sigma} \equiv (\sigma_1, \sigma_2, \sigma_3)$ the usual Pauli matrices. Defining σ_0 as the identity 2×2 matrix, the spin factor R^{SM_S} can be cast in the following form

$$R^{SM_S}(\boldsymbol{k}_\perp, \xi; \nu\bar{\nu}) = \frac{1}{\sqrt{2}} \sum_{\alpha=0}^{3} a_\alpha^{SM_S}(\boldsymbol{k}_\perp,\xi)\ (\sigma_\alpha)_{\nu\bar{\nu}} \tag{3}$$

where the quantities $a_\alpha^{SM_S}(\boldsymbol{k}_\perp,\xi)$ are explicitly given by

$$\begin{aligned}
a_0^{00} &= -k_1(\lambda_q + \lambda_{\bar{q}})/D \quad , & a_0^{10} &= \mathrm{i}k_2(\lambda_q - \lambda_{\bar{q}})/D \\
a_1^{00} &= 0 \quad , & a_1^{10} &= (\lambda_q\lambda_{\bar{q}} + k_\perp^2)/D \\
a_2^{00} &= \mathrm{i}(\lambda_q\lambda_{\bar{q}} - k_\perp^2)/D \quad , & a_2^{10} &= 0 \\
a_3^{00} &= \mathrm{i}k_2(\lambda_q + \lambda_{\bar{q}})/D \quad , & a_3^{10} &= -k_1(\lambda_q - \lambda_{\bar{q}})/D \\
a_0^{1-1} &= (\lambda_q\lambda_{\bar{q}} - k_-^2)/\sqrt{2}D \quad , & a_0^{11} &= (\lambda_q\lambda_{\bar{q}} - k_+^2)/\sqrt{2}D \\
a_1^{1-1} &= -k_-(\lambda_q - \lambda_{\bar{q}})/\sqrt{2}D \quad , & a_1^{11} &= k_+(\lambda_q - \lambda_{\bar{q}})/\sqrt{2}D \\
a_2^{1-1} &= \mathrm{i}k_-(\lambda_q + \lambda_{\bar{q}})/\sqrt{2}D \quad , & a_2^{11} &= \mathrm{i}k_+(\lambda_q + \lambda_{\bar{q}})/\sqrt{2}D \\
a_3^{1-1} &= -(\lambda_q\lambda_{\bar{q}} + k_-^2)/\sqrt{2}D \quad , & a_3^{11} &= (\lambda_q\lambda_{\bar{q}} + k_+^2)/\sqrt{2}D
\end{aligned} \tag{4}$$

where $D = \sqrt{(\lambda_q^2 + k_\perp^2)\ (\lambda_{\bar{q}}^2 + k_\perp^2)}$; $\lambda_q = m_q + \xi M_0$; $\lambda_{\bar{q}} = m_{\bar{q}} + (1-\xi)M_0$ and $k_\pm = k_1 \pm \mathrm{i}k_2$.

The one-body component of the e.m. current operator is given by

$$\hat{I}_\mu = \sum_{j=q,\bar{q}} \left[F_1^{(j)}(Q^2)\ \gamma_\mu + F_2^{(j)}(Q^2)\ \mathrm{i}\sigma_{\mu\nu} \frac{q^\nu}{2m_j} \right] \tag{5}$$

where $Q^2 = -q \cdot q$ is the squared four-momentum transfer; $F_1^{(j)}(Q^2)$ and $F_2^{(j)}(Q^2)$ are the Dirac and Pauli CQ form factors, respectively (normalized as $F_1^{(q)}(0) = e_q, F_2^{(q)}(0) = \kappa_q$, with e_q and κ_q being the charge and anomalous magnetic moment of the CQ). Within the light-front formalism, all the space-like invariant form factors of a hadron can be determined using only the matrix elements of the *plus* component I^+ of the e.m. current operator, evaluated in a frame where $q^+ = 0$; the latter choice allows to suppress the contribution of the pair creation from the vacuum [4]. Thus, using Eqs. (1-5) and performing simple traces involving Pauli matrices, one gets

$$\langle J'M'|\hat{I}^+|JM\rangle = \sum_{j=q,\bar{q}} \sum_{\beta=1,2} F_\beta^{(j)}(Q^2)\ H_\beta^{(j)}(J'M', JM; Q^2) \tag{6}$$

with

$$H_1^{(j)}(J'M', JM; Q^2) = \int d\boldsymbol{k}_\perp d\xi\ \sqrt{A(\boldsymbol{k}_\perp, \xi) A(\boldsymbol{k}'_\perp, \xi)} \tag{7}$$

$$\sum_{SM_S} \tilde{w}_{SM_S}^{JM}(\boldsymbol{k}) \sum_{S'M_{S'}} \tilde{w}_{S'M_{S'}}^{J'M'}(\boldsymbol{k}') \sum_{\alpha=0}^{3} [a_\alpha^{(S'M_{S'})}(\boldsymbol{k}'_\perp, \xi)]^*\ a_\alpha^{(SM_S)}(\boldsymbol{k}_\perp, \xi)$$

$$H_2^{(j)}(J'M', JM; Q^2) = -\frac{Q}{2m_j} \int d\boldsymbol{k}_\perp d\xi\ \sqrt{A(\boldsymbol{k}_\perp, \xi) A(\boldsymbol{k}'_\perp, \xi)} \tag{8}$$

$$\sum_{SM_S} \tilde{w}_{SM_S}^{JM}(\boldsymbol{k}) \sum_{S'M_{S'}} \tilde{w}_{S'M_{S'}}^{J'M'}(\boldsymbol{k}')\ \{\delta_j [a_0^{(S'M_{S'})}(\boldsymbol{k}'_\perp, \xi)]^*\ a_2^{(SM_S)}(\boldsymbol{k}_\perp, \xi) +$$
$$\delta_j [a_2^{(S'M_{S'})}(\boldsymbol{k}'_\perp, \xi)]^*\ a_0^{(SM_S)}(\boldsymbol{k}_\perp, \xi) - [a_1^{(S'M_{S'})}(\boldsymbol{k}'_\perp, \xi)]^*\ a_3^{(SM_S)}(\boldsymbol{k}_\perp, \xi) +$$
$$[a_3^{(S'M_{S'})}(\boldsymbol{k}'_\perp, \xi)]^*\ a_1^{(SM_S)}(\boldsymbol{k}_\perp, \xi)\}$$

where $\delta_q = \mathrm{i}$; $\delta_{\bar{q}} = -\mathrm{i}$; $\boldsymbol{k}'_\perp = \boldsymbol{k}_\perp + (1-\xi)\boldsymbol{q}_\perp$ for $j = q$ and $\boldsymbol{k}'_\perp = \boldsymbol{k}_\perp - \xi \boldsymbol{q}_\perp$ for $j = \bar{q}$. Equations (6-8) represent our final result, since in terms of them any elastic or transition form factor involving mesons can be calculated. As a matter of fact, we have applied Eqs. (6-8) to the calculation of the elastic form factors of pseudoscalar [5] and vector [6] mesons, as well as of the $\pi\rho$ and $\pi\omega$ [7] transition form factors. In what follows we will focus on the charge form factor of the pion, $F_\pi(Q^2)$, and on the form factor corresponding to the radiative transition $\pi^+\gamma^* \to \rho^+$, $F_{\pi\rho}(Q^2)$. One has

$$F_\pi(Q^2) = F_1^{(\mathrm{V})}(Q^2) H_1(00,00;Q^2) + F_2^{(\mathrm{V})}(Q^2) H_2(00,00;Q^2) \tag{9}$$

$$F_{\pi\rho}(Q^2) = \frac{2\sqrt{2}}{3Q} [F_1^{(\mathrm{S})}(Q^2) H_1(11,00;Q^2) + F_2^{(\mathrm{S})}(Q^2) H_2(11,00;Q^2)] \tag{10}$$

where $F^{(S,V)}_{1(2)}$ are the isoscalar and isovector parts of the u and d CQ form factors, given by $F^{(S)}_{\alpha}(Q^2) \equiv 3[F^{(u)}_{\alpha}(Q^2) + F^{(d)}_{\alpha}(Q^2)]$ and $F^{(V)}_{\alpha}(Q^2) \equiv F^{(u)}_{\alpha}(Q^2) - F^{(d)}_{\alpha}(Q^2)$. Equations (9-10) have been calculated adopting the eigenfunctions of a light-front mass operator, constructed from the effective $q\bar{q}$ Hamiltonian of ref. [8] and using a simple parametrization for the CQ form factors (see, for more details, [7]). In Fig. 1 the results of our calculations are reported and compared with the predictions of a simple Vector Meson Dominance (VMD) model (i.e., $F^{VMD}(Q^2) = 1/(1 + Q^2/M^2_{\rho})$, with M_{ρ} being the ρ-meson mass), the results of the Bethe-Salpeter (BS) approach of refs. [9, 10] and those obtained in refs. [11, 12] using QCD sum rule techniques. It can be seen that the differences among the theoretical calculations of the pion form factor are quite small, so that the existing pion data do not discriminate among various models of the pion structure. On the contrary, the differences among various relativistic calculations of the $\pi\rho$ transition form factor are quite sizeable at $Q^2 > 1$ $(\mathrm{GeV}/c)^2$; therefore, the measurement of $F_{\pi\rho}(Q^2)$ could help in discriminating among various models of the meson structure.

In conclusion, the elastic and transition electromagnetic form factors of mesons have been investigated within a light-front constituent quark model. All the relevant formulae needed for the calculation of the space-like matrix

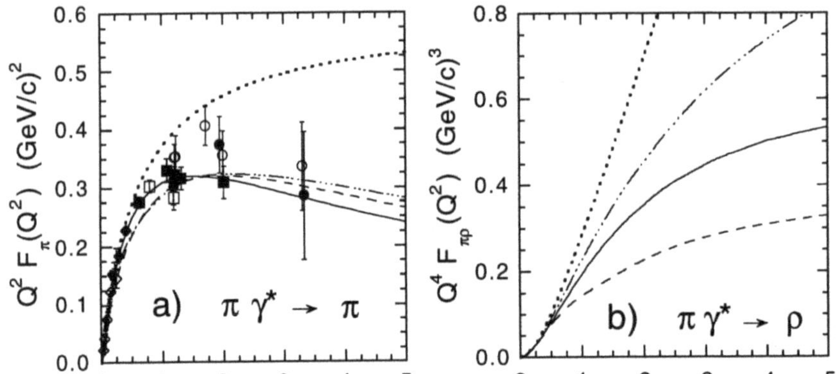

Figure 1. a) The elastic form factor of the pion (Eq. (9)), times Q^2, vs. Q^2. Our results, obtained using for w^{π}_{000} the wave function of ref. [8] and considering the CQ form factors of ref. [7], are represented by the solid line. The dotted, dot-dashed and dashed lines correspond to the predictions of a simple VMD model (ρ-meson pole only), the BS approach of ref. [9] and the QCD sum rule technique of ref. [11], respectively. b) The form factor of the radiative transition $\pi^+\gamma^* -> \rho^+$ (Eq. (10)), times Q^4, vs. Q^2. The solid line correspond to our results, obtained using for w^{π}_{000} and w^{ρ}_{011} the wave functions of ref. [8] and considering the same CQ form factors as in (a). The dotted, dot-dashed and dashed lines correspond to the predictions of a simple VMD- model (ρ-meson pole only), the BS approach of ref. [10] and the QCD sum rule technique of ref. [12], respectively. (After ref. [7]).

elements of the one-body component of the electromagnetic current, including both Dirac and Pauli form factors of the constituent quarks, have been presented. Our results obtained for the pion charge form factor and the $\pi\rho$ transition form factor have been compared with those resulting from various relativistic approaches, showing that the measurements of meson form factors planned at CEBAF could help in discriminating among different models of the meson structure.

Acknowledgements.
Two of the authors (I.L.G. and I.M.N.) acknowledge the financial support of the INTAS grant, Ref. No 93-0079.

References

1. CEBAF Proposal E-93-012: *Electroproduction of Light Quark Mesons* (M. Kossov, spokeman); CEBAF Proposal E-93-018: *Separation of Longitudinal and Transverse Amplitudes in Kaon Electroproduction* (O.K. Baker, spokeman); CEBAF Proposal E-93-021: *The Charged Pion Form Factor* (D. Mack, spokeman)
2. B.D. Keister and W.N. Polyzou: Adv. Nucl. Phys. **20**, 225 (1991); F. Coester: Progress in Part. and Nucl. Phys. **29**, 1 (1992)
3. H.J. Melosh: Phys.Rev. **D9**, 1095 (1974)
4. G.P. Lepage and S.J. Brodsky: Phys. Rev. **D22**, 2157 (1980); L.L. Frankfurt and M.I. Strikman: Nucl. Phys. **B148**, 107 (1979); M. Sawicki: Phys. Rev. **D46**, 474 (1992)
5. F. Cardarelli, I.L. Grach, I.M. Narodetskii, E. Pace, G. Salmé and S. Simula: Phys. Lett. **332B**, 1 (1994); and to appear in Phys. Rev. **D**, brief report
6. F. Cardarelli, I.L. Grach, I.M. Narodetskii, G. Salmé and S. Simula: Phys. Lett. **349B**, 393 (1995)
7. F. Cardarelli, I.L. Grach, I.M. Narodetskii, G. Salmé and S. Simula: submitted to Phys. Lett. **B**
8. S. Godfrey and N. Isgur: Phys. Rev. **D32**, 185 (1985)
9. H. Ito, W.W. Buck and F. Gross: Phys. Rev. **C45**, 1918 (1992); Phys. Lett. **287B**, 23 (1992)
10. H. Ito and F. Gross: Phys. Rev. Lett. **71**, 2555 (1993)
11. V.A. Nesterenko and A.V. Radyushkin: Phys. Lett. **115B**, 410 (1982)
12. V. Braun and I. Halperin: Phys. Lett. **328B**, 457 (1994)

Few-Body Systems Suppl. 9, 272–276 (1995)

Few-Body Systems

Structure of Λ Hypernuclei with Neutron Halo

T. Yu. Tretyakova[1]*, D. E. Lanskoy[2]†

[1] Frank Laboratory of Neutron Physics, Joint Institute for Nuclear Research, 141980 Dubna, Russia

[2] Institute of Nuclear Physics, Moscow State University, 119899 Moscow, Russia

Abstract. Properties of light hypernuclei with neutron halo are considered in the framework of the Skyrme-Hartree-Fock approach. Neutron halo states can become more pronounced as well as less bound on hyperon addition. Halo properties are sensitive to hyperonic interaction features, particularly, to the ability of hyperon to distort the core.

1 Introduction

In the last decade, nuclei near the neutron drip-line, particularly, light nuclei with neutron halo, became a new exciting field of nuclear physics [1]. Recently, Majling [2] discussed some prospects for production and properties of Λ hypernuclei with large neutron excess. Such systems may be of interest from various viewpoints:

•Embedding of Λ hyperon into loosely bound halo system can provide unique information on its response to a small perturbation.

•Hypernuclear interactions at low nuclear densities can be examined, specifically, role of 3-body ΛNN force or density-dependent ΛN force can be revealed.

•New hypernuclear species including ones with unstable nuclear cores can be observed.

•Charge symmetry breaking effects in ΛN interaction can be studied.

Neutron-rich hypernuclei can be produced in (K^-, π^+) [2, 13] or (π^-, K^+) reactions. Evaluation of the related cross sections and search for optimal kinematical conditions should be the subject of further studies. Some estimation for production cross sections of light hypernuclei including neutron-rich ones in heavy-ion collisions has been presented in [3]. The cross sections appear to be not extremely small.

**E-mail address:* ttretyak@nfsun1.jinr.dubna.su

†*E-mail address:* lanskoy@nsrd.npi.msu.su

2 The Model

We use the Skyrme-Hartree-Fock approach. The main advantage of this method for our purposes is its capability to take into account influence of hyperon not only on a halo but also on a compact core. It also enables one to deal with ΛN interaction of sufficiently complicated structure.

In this paper, the $^{12}_{\Lambda}$Be hypernucleus is mainly considered as an example of a halo hypernucleus. We treat the ground state $(1/2^+)$ of ^{11}Be as a pure neutron $2s_{1/2}$ halo state and the excited bound state $(1/2^-)$ as a $1p_{1/2}$ halo state near ^{10}Be(0^+). Possible configuration mixing is ignored, but distortion of the core (^{10}Be) by neutron (and also by hyperon in $^{12}_{\Lambda}$Be) is taken into account.

The majority of current approaches are known to fail in reproducing of anomalous level ordering in ^{11}Be and neighbouring nuclei. To put neutron separation energies to their proper values, we adopt the procedure by Bertsch et al. [4]. This prescription involves a renormalization of single-particle potential for halo neutron(s) and provides reasonable values for halo radii. Unfortunately, it violates self-consistency of the Hartree-Fock scheme. In this view we identify with the halo neutron separation energy the single-particle energy calculated (with opposite sign) and neglect rearrangement effects.

Poorly known spin-spin ΛN interaction (leading to doublet splitting) as well as charge symmetry breaking ΛN interaction are not taken into account in our consideration.

3 Properties of Halo Hypernuclei

The famous Sk3 parameter set is mainly used for the NN potential. For matter radii of beryllium isotopes, it gives 2.35, 2.99, and 2.58 fm at A=10, 11, and 12, correspondingly, which should be compared with 2.28±0.02, 2.71±0.05, and 2.57±0.05 fm [5].

We present calculations for the $^{12}_{\Lambda}$Be hypernucleus with three different ΛN parameter sets: SkSH1 [6], the 5th set from [8] (hereafter YBZ5), and the 3rd set from [7] (hereafter R3). Providing similar results for Λ binding energies, these sets possess rather different ability to distort the nucleonic core. Implications of 3-body ΛNN force for polarizing property of hyperonic interaction has been analyzed (e.g., ref. [9] and refs. therein). The SkSH1 set represents purely two-body potential and, therefore, contracts the core, the R3 set incorporates great 3-body force and dilutes the core, and the YBZ5 set exemplifies an intermediate case.

We calculate properties of the $^{12}_{\Lambda}$Be hypernucleus in the states built on the ground and the first excited states of ^{11}Be. In Fig.1, energy spectrum and rms radii of the hyperonic orbits, of the compact 10-nucleonic core, and of the halo neutron orbit are presented. It is seen that the halo $1p$ state is very sensitive to ΛN interaction properties. For the SkSH1 set, a striking result is obtained: hyperon addition pushes halo state upward to the threshold. That is due to core contraction by hyperon. The halo neutron single-particle potential becomes deeper in the central region but slightly narrower (Fig.2). Since the

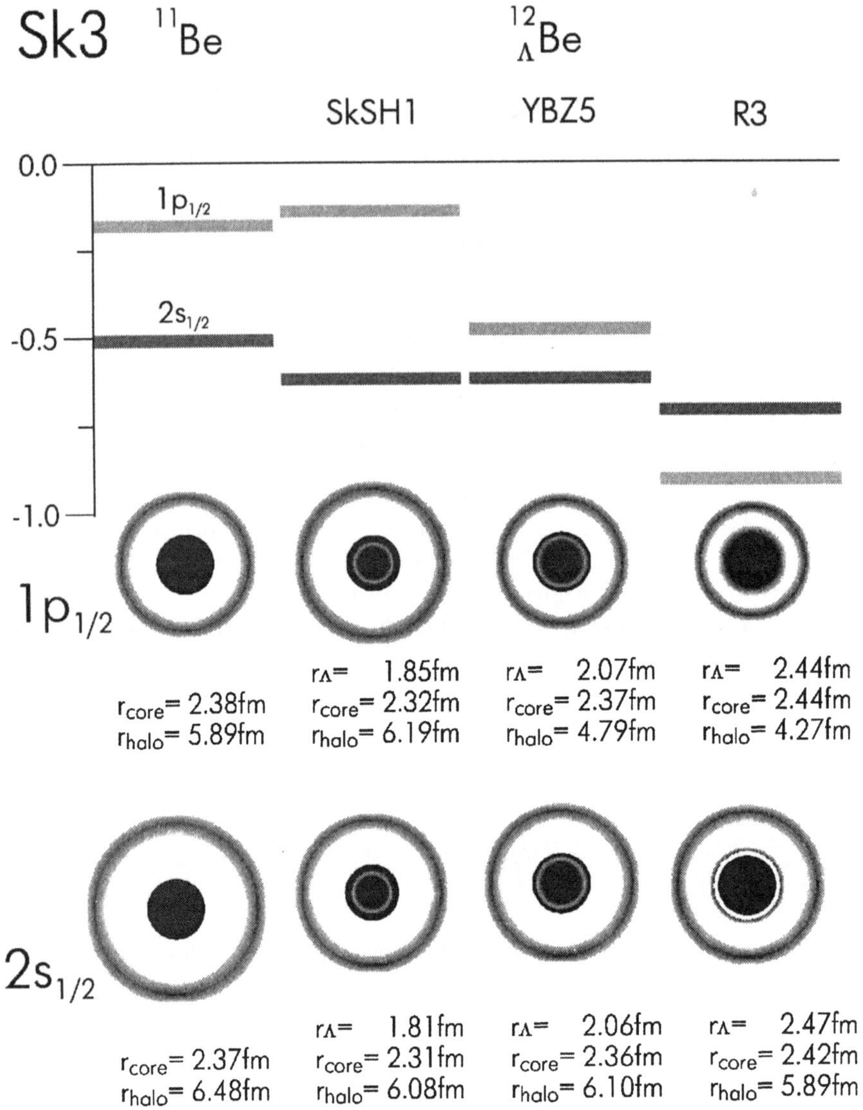

Figure 1. Spectrum of the halo neutron levels (MeV) and rms radii of Λ orbits (r_Λ), of neutron halo orbits (r_{halo}), and of the core (r_{core}) in ^{11}Be and $^{12}_{\Lambda}$Be for various ΛN potentials. The NN potential is Sk3.

halo neutron moves outside of the central region, it responds stronger to a very little reduction of the potential range than to additional attraction inside the core. The R3 set, otherwise, dilutes the core and therefore, the potential widens. The halo becomes more compact and tightly bound. In the case of the YBZ5 set, the core remains almost undistorted so the neutron feels only the hyperon attraction.

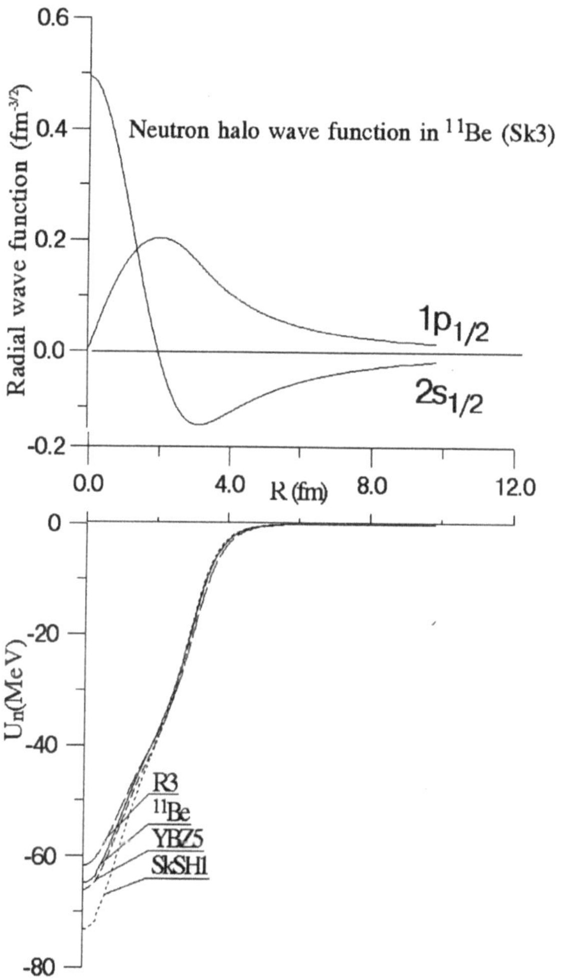

Figure 2. Neutron halo wave functions in ^{11}Be (upper part) and single-particle potential U_n acting on halo neutron in ^{11}Be and $^{12}_{\Lambda}$Be for various ΛN potentials (lower part).

The Sk3 potential is quite stiff. So the core polarization and related effects may be even greater. For instance, the soft Z_σ set [10] together with the SkSH1 potential predict even unbound $1p$ halo state.

The halo neutron in the $2s$ state moves for substantial time inside the core, and the core distortion does not affect the halo. This state is not sensitive to ΛN potential properties.

Recently, the ^{10}He nucleus was found to be unbound with separation energy of two neutrons S_{2n}=-1.07±0.07 MeV [11, 12]. This quantity is well reproduced with the Sk3 potential without any renormalization. As for contracting ΛN potentials, the $^{11}_{\Lambda}$He is, of course, also unbound. The R3 set predicts the bound state with S_{2n}=1.2 MeV. For undistorting potentials with moderate ΛNN contribution, like YBZ5, the separation energy appears to be just near zero. Therefore, existence of bound ${}^{11}_{\Lambda}$He is very sensitive to details of hyperonic interaction.

Halo response to core polarization, discussed above, is inherent mostly in $1p$ states. For instance, the ${}^{16}_{\Lambda}$C hypernucleus properties with the host nucleus ^{15}C both in the ground ($1/2^+$, S_n=1.2 MeV) and in the excited ($5/2^+$, S_n=0.5 MeV) states are slightly sensitive to ΛN potential properties.

Acknowledgement. We are thankful to V.N.Fetisov and L.Majling for useful discussions. This work was supported in part by Russian Foundation for Fundamental Investigations (RFFI), grant 94-02-04112.

References

1. I. Tanihata: Nucl. Phys. **A522**, 275c (1991)
2. L. Majling: Nucl. Phys. **A585**, 211c (1995)
3. H. Bando: Nuovo Cim. **102A**, 627 (1989)
4. G.F. Bertsch, B.A. Brown, H. Sagawa: Phys. Rev. **C39**, 1154 (1989)
5. I. Tanihata et al.: Phys. Lett. **B206**, 592 (1988)
6. F. Fernandez, T. Lopez-Arias, C. Prieto: Z. Phys. **A334**, 349 (1989)
7. M. Rayet: Nucl. Phys. **A367**, 381 (1981)
8. Y. Yamamoto, H. Bando, J. Žofka: Prog. Theor. Phys. **80**, 757 (1988)
9. D.E. Lanskoy, T.Yu. Tretyakova: Sov. J. Nucl. Phys. **49**, 248 (1989)
10. J. Friedrich, P.-G. Reinhard: Phys. Rev. **C33**, 335 (1986)
11. A.A. Korsheninnikov et al.: Phys. Lett. **B326**, 31 (1994)
12. A.N. Ostrowski et al.: Phys. Lett. **B338**, 13 (1994)
13. V.N. Fetisov: Nuovo Cim. **102A**, 307 (1989)

Few-Body Systems Suppl. 9, 277–280 (1995)

Few-Body Systems

Double-Strangeness Hypernuclei in the Skyrme-Hartree-Fock Approach

D. E. Lanskoy

Institute of Nuclear Physics, Moscow State University, 119899 Moscow, Russia

Abstract. The Skyrme-Hartree-Fock approach for $\Lambda\Lambda$ and Ξ hypernuclei is suggested. Parameters of $\Lambda\Lambda$ and ΞN potentials are found and A-dependences of binding energies of $\Lambda\Lambda$, Ξ^-, and Ξ^0 hypernuclei are discussed.

1 $\Lambda\Lambda$ Hypernuclei

Current interest to $S = -2$ hypernuclei is induced by growing experimental activity in this field [1-3]. In this view we extend the famous Skyrme-Hartree-Fock approach to $S = -2$ hypernuclei.

We use the simplified Skyrme $\Lambda\Lambda$ potential as in [4] (in usual notation):

$$V = \lambda_0\delta(\boldsymbol{r}_1 - \boldsymbol{r}_2) + \frac{1}{2}\lambda_1[k'^2\delta(\boldsymbol{r}_1 - \boldsymbol{r}_2) + \delta(\boldsymbol{r}_1 - \boldsymbol{r}_2)k^2] + \lambda_2\boldsymbol{k}'\delta(\boldsymbol{r}_1 - \boldsymbol{r}_2)\boldsymbol{k}. \quad (1)$$

Straightforward extension of the standard formalism [5, 6] to $\Lambda\Lambda$ hypernuclei was described elsewhere [7]. The only quantity used for parameter fit is the bond energy of two Λ's, $\Delta B_{\Lambda\Lambda}$ in ${}^{13}_{\Lambda\Lambda}$B [1, 8]. The last term in (1) relates to p-wave interaction. As we address here to the ground states only, the fit is restricted by λ_0 and λ_1. It is known that the first value is defined by the volume integral and the λ_1/λ_0 is connected with the potential range. We examine the following sets of parameters numbered in order of increasing range:

- S$\Lambda\Lambda$1: λ_0 =-312.6 MeV·fm^3, λ_1 =57.5 MeV·fm^5;
- S$\Lambda\Lambda$2: λ_0 =-437.7 MeV·fm^3, λ_1 =240.7 MeV·fm^5;
- S$\Lambda\Lambda$3: λ_0 =-831.8 MeV·fm^3, λ_1 =922.9 MeV·fm^5.

All of them give the proper $\Delta B_{\Lambda\Lambda}$ value in boron with the SkM* set for NN potential and the 5th set from [9] (hereafter YBZ5) for ΛN one.

In Fig. 1, the bond energies $\Delta B_{\Lambda\Lambda}$ in various $\Lambda\Lambda$ hypernuclei are presented. Apart from the YBZ5 set, we also employ the SkSH1 set [10] and the 3rd set from [6] (hereafter R3) for ΛN interaction. The first one is purely two-body and the second one incorporates great ΛNN force. The $\Lambda\Lambda$ interaction extracted from the experimental quantity is very sensitive to choice of ΛN interaction so

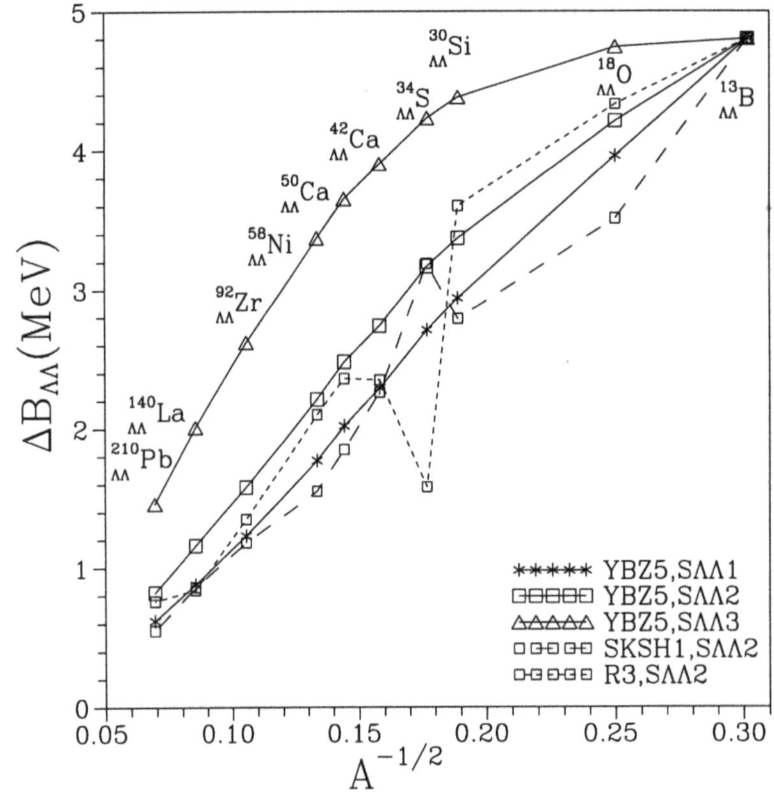

Figure 1. The bond energies of two Λ's in double-Λ hypernuclei for various ΛN and $\Lambda\Lambda$ interactions

a renormalization of a $\Lambda\Lambda$ potential in each case is needed [7]. This is mainly due to polarization of the nuclear core of $^{13}_{\Lambda\Lambda}$B with the SkSH1 and R3 sets. Note that the initial YBZ5 potential is almost core-undistorting.

For the YBZ5 set, the $\Delta B_{\Lambda\Lambda}$ falls monotonically with increasing A due to rising of the interhyperon distance. The greater is the range of the $\Lambda\Lambda$ potential, the slower is the $\Delta B_{\Lambda\Lambda}$ decrease. This behaviour is similar to those obtained in a 3-body $\Lambda+\Lambda+{}^A Z$ model [11]. Some irregularities appear with the SkSH1 and R3 sets. For $^{34}_{\Lambda\Lambda}$S, when four s-nucleons are added with respect to $^{30}_{\Lambda\Lambda}$Si, the radii of Λ orbits reduce with the two-body SkSH1 potential and $\Lambda\Lambda$ attraction increases. Otherwise, for the R3 set with great ΛNN repulsion, the radii of Λ orbits grow. Reverse situation takes place on addition of f nucleons in $^{50}_{\Lambda\Lambda}$Ca and $^{58}_{\Lambda\Lambda}$Ni. However, actual ^{32}S and ^{56}Ni are not magic nuclei, so A-dependence of occupation numbers and, hence, of $\Delta B_{\Lambda\Lambda}$ may be in reality more smooth.

For strongly polarizing ΛN potentials, the $\Delta B_{\Lambda\Lambda}$ differs from the magnitude of $\Lambda\Lambda$ interaction energy, $V_{\Lambda\Lambda}$, even in heavy hypernuclei (for example, $\Delta B_{\Lambda\Lambda}$ =0.8 MeV and $V_{\Lambda\Lambda}$ =0.4 MeV in $^{210}_{\Lambda\Lambda}$Pb with R3 and $S\Lambda\Lambda 2$).

2 Ξ Hypernuclei

The 6 known events are attributed, while ambiguously, to ground states of specific Ξ^- hypernuclei for which binding energies were measured ([12] and refs. therein). These data may be used for rough evaluation of ΞN interaction.

We adopt the same expression (1) for ΞN interaction with an additional density-dependent force $V_{\Xi NN} = (3/8)\lambda_3\delta(\boldsymbol{r}_N - \boldsymbol{r}_\Xi)\rho_N((\boldsymbol{r}_N + \boldsymbol{r}_\Xi)/2)$ in the usual form [13]. The imaginary part of ΞN interaction as well as its isospin dependence are neglected. The only difference in the Hartree-Fock scheme with respect to single-Λ hypernuclei [6] is the Coulomb $\Xi^- p$ interaction.

The following two parameter sets for $\Xi^- N$ Skyrme potential were found treating all the 6 events as Ξ^- hypernuclei (Fig. 2):
SΞN1: $\lambda_0 = -195.5$ MeV·fm^3, $\lambda_1 = 76$ MeV·fm^5, $\lambda_2 = 10.6$ MeV·fm^5, $\lambda_3 = 0$;
SΞN2: $\lambda_0 = -310$ MeV·fm^3, $\lambda_1 = -\lambda_2 = 133$ MeV·fm^5, $\lambda_3 = 2200$ MeV·fm^6.
The Sk3 set was used for NN interaction (the SkM* one gives similar results).

The main differences between these sets are lower binding in heavy hypernuclei and greater level spacing for SΞN1. For Ξ binding energy in nuclear matter, values $D_\Xi \approx 25 \div 28$ MeV (SΞN1) and $D_\Xi \approx 27 \div 29$ MeV (SΞN2), depending on saturation density, are obtained which are compatible with $D_\Xi = 24 \pm 4$ MeV [12].

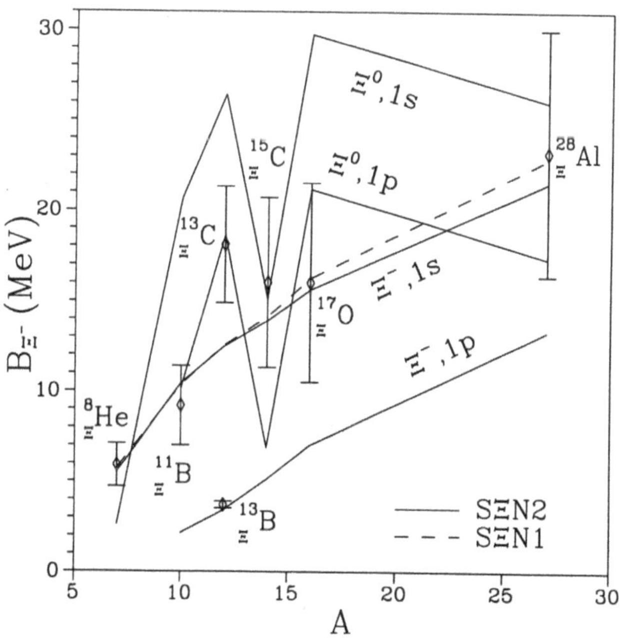

Figure 2. A-dependence of binding energies of Ξ^- hyperon $B_{\Xi^-} = M(\Xi^-) + M(^A(Z+1)) - M(^{A+1}_{\Xi}Z)$ in Ξ^- and Ξ^0 hypernuclei. The point for $^{13}_{\Xi}$B is from [2], the others are from [12]. Broken lines connect results of our calculation for Ξ^- and Ξ^0 in $1s$ and $1p$ states

The experimental point for ${}^{13}_{\Xi}$C violates the smooth A-dependence. It is seen from Fig. 2 that this quantity agrees well with binding energy of the ${}^{13}_{\Xi^0}$C hypernucleus with the hyperon in the $1p$ state, so this event most probably should be attributed to the Ξ^0 hypernucleus. Also other observed hypernuclei may be alternatively interpreted as Ξ^0 hypernuclei. Strong mixing of Ξ^- and Ξ^0 states may be suggested in ${}^{15}_{\Xi}$C and possibly in ${}^{28}_{\Xi}$Al. Note that in ${}^{11}_{\Xi}$B, closely spaced Ξ^- and Ξ^0 states are of opposite parity.

In all the Ξ hypernuclei considered (excepting ${}^{8}_{\Xi}$He), Ξ^0 states lie lower than Ξ^- ones because the nucleonic cores of the Ξ^- states are mainly proton-rich. For neutron-rich cores (e.g., ^{14}C, [14]) the picture is opposite. In heavy Ξ hypernuclei, Ξ^- states lie much lower due to Coulomb attraction.

Acknowledgement. I am indebted to T.Yu. Tretyakova for participation on the earlier stage of this work and also to A.M. Shirokov and Yu.A.Lurie for useful discussions of $\Lambda\Lambda$ bond energies. This work was supported in part by Russian Foundation for Fundamental Investigations (RFFI), grant 94-02-04112.

References

1. S. Aoki et al.: Prog. Theor. Phys. **85**, 1287 (1991)
2. K. Imai: Nucl. Phys. **A547**, 199c (1992)
3. G.B. Franklin: Nucl. Phys. **A585**, 83c (1995)
4. D.E. Lanskoy, T.Yu. Tretyakova: Z. Phys. **A343**, 355 (1992)
5. D. Vautherin, D.M. Brink: Phys. Rev. **C5**, 626 (1972)
6. M. Rayet: Nucl. Phys. **A367**, 381 (1981)
7. D.E. Lanskoy: In: *Proc. 23 INS Intern. Symp. "Nucl. and Part. Phys. with Meson Beams in the 1 GeV/c Region"*, ed. by S. Sugimoto, O. Hashimoto. Tokyo: Univ. Acad. Press 1995 (in print)
8. C.B. Dover et al.: Phys. Rev. **C44**, 1905 (1991)
9. Y. Yamamoto, H. Bando, J. Žofka: Prog. Theor. Phys. **80**, 757 (1988)
10. F. Fernandez, T. Lopez-Arias, C. Prieto: Z. Phys. **A334**, 349 (1989)
11. D.E. Lanskoy, Yu.A. Lurie, A.M. Shirokov: In: *Abstr. and Contrib. Pap. 23rd INS Intern. Symp. "Nucl. and Part. Phys. with Meson Beams in the 1 GeV/c Region"*, p.132. Tokyo: INS Univ. Tokyo 1995
12. C.B. Dover, A. Gal: Ann. Phys. **146**, 309 (1983)
13. D.E. Lanskoy, T.Yu. Tretyakova: Sov. J. Nucl. Phys. **49**, 987 (1989)
14. C.B. Dover, A. Gal, D.J. Millener: Nucl. Phys. **A572**, 85 (1994)

Few-Body Systems Suppl. 9, 281–284 (1995)

Few-Body Systems

Production of Double-Λ and Twin-Λ Hypernuclei in the Ξ^--Atomic Capture Reaction

T. Yamada and K. Ikeda†

Lab. of Physics, Kanto Gakuin University, Yokohama 236, Japan
†Dep. of Physics, Niigata University, Niigata 950-21, Japan

Abstract. The production rates of double-Λ and twin-Λ hypernuclei are discussed within the frame of the direct reaction picture.

1 Introduction

Recently the Ξ^--atomic capture reaction experiment [1, 2] has been performed at KEK to investigate the strangeness $S = -2$ world. Two kinds of interesting findings have been reported; one is the observation of a new double-Λ hypernucleus, ${}^{10}_{\Lambda\Lambda}$Be or ${}^{13}_{\Lambda\Lambda}$B, (one event), and another is that of twin-Λ hypernuclear productions, ^{12}C(Ξ^--stopped, ${}^{4}_{\Lambda}$H)${}^{9}_{\Lambda}$Be, (two events). It is important to clarify the mechanism of the double-Λ and twin-Λ hypernuclear productions in the Ξ^--atomic capture in light nuclei.

A ^{12}C-Ξ^- atomic state appears at about 40 MeV excitation energy from the ${}^{13}_{\Lambda\Lambda}$B ground state. Therefore, many production channels, double-Λ, twin-Λ and single-Λ hypernuclear channels are open in the Ξ^--atomic capture reaction. For double-Λ hypernuclear channels, the following channels are open; ${}^{12}_{\Lambda\Lambda}$Be+p (Q = 26.3 MeV), ${}^{12}_{\Lambda\Lambda}$B+n (25.9), ${}^{11}_{\Lambda\Lambda}$Be+d (19.4), ${}^{10}_{\Lambda\Lambda}$Be+t (19.2), ${}^{10}_{\Lambda\Lambda}$Be+d+n (12.0), ${}^{9}_{\Lambda\Lambda}$Li+α (22.8), ${}^{6}_{\Lambda\Lambda}$He+^{7}Li (15.0) and ${}^{5}_{\Lambda\Lambda}$H+^{8}Be (8.7) etc. (Q denotes the Q value measured from the ^{12}C+Ξ^- threshold.) For the two-body twin-Λ hypernuclear channels, only the three channels are open; ${}^{10}_{\Lambda}$Be+${}^{3}_{\Lambda}$H (Q=6.0 MeV), ${}^{9}_{\Lambda}$Be+${}^{4}_{\Lambda}$H (10.0) and ${}^{8}_{\Lambda}$Li+${}^{5}_{\Lambda}$He (13.6). It is interesting to study how large production rates their channels have, especially, double-Λ hypernuclear channels.

In order to calculate the production rates, three pictures have been proposed so far. First picture [3] is the *quasi-deuteron model*, in which a Ξ^- particle in an atomic orbit is absorbed into a *deuteron* in a nucleus (Ξ^- +"d" → "$\Lambda\Lambda$" +n+28 MeV) and then, a double-Λ hypernucleus is produced with a high-energy neutron emission. This model can explain the production of a double-Λ hypernucleus, while it is difficult to explain the production of a twin-Λ hypernucleus, because this event does not accompany with a high-energy neutron emission. Second picture [4] is the *compound double-Λ hypernuclear picture.* According to the picture, the production rate of double-Λ and twin-Λ hypernuclei etc. depends mainly on the Q value, namely, a hypernuclear channel with the larger (smaller) Q value has the larger (smaller) production rate. Therefore, the production rate of ${}^{8}_{\Lambda}$Li+${}^{5}_{\Lambda}$He should be larger than that of ${}^{9}_{\Lambda}$Be+${}^{4}_{\Lambda}$H, because the Q value for the former channel is larger than that of the latter. Moreover, an abundant production of double-Λ hypernuclei, in particular, ${}^{6}_{\Lambda\Lambda}$He, was predicted [3]. However, the experimental result shows that only one double-Λ hypernuclear event (${}^{10}_{\Lambda\Lambda}$Be or ${}^{13}_{\Lambda\Lambda}$B) and two twin-Λ hypernuclear events (${}^{9}_{\Lambda}$Be+${}^{4}_{\Lambda}$H) have been observed so far [2]. This may be due to the poor statistic (one double-Λ and two twin-Λ events of only about 30 Ξ^--capture events at rest). However, it may suggest another mechanism for the Ξ^--atomic capture reaction.

The third picture [5] is the *direct reaction picture*, which has been proposed by the present authors. According to this picture, 1) the Ξ^- particle in an atomic orbit interacts with a proton in ^{12}C and then, a highly-excited double-Λ hypernuclear state, ${}^{13}_{\Lambda\Lambda}$B, with a nuclear-core proton hole state (s- or p-hole states) is produced as an intermediate state ($E_{\mathrm{x}} \simeq 40$ MeV); 2) reflecting the structure and decay mode of the nuclear-core hole state, the intermediate state is fragmented to twin-Λ hypernucleus or double-Λ hypernucleus with emission of a nucleon, deuteron, triton etc. It should be noted that in our picture the characteristic of the fragmentation of the nuclear-core hole states play an important role in the Ξ^--atomic capture reaction.

The purpose of this paper is to discuss the double-Λ, twin-Λ and single-Λ hypernuclear production rates and show their characteristics for the following channels within the frame of the *direct reaction picture*: ${}^{9}_{\Lambda}$Be+${}^{4}_{\Lambda}$H, ${}^{8}_{\Lambda}$Li+${}^{5}_{\Lambda}$He, ${}^{10}_{\Lambda\Lambda}$Be+t, ${}^{9}_{\Lambda\Lambda}$Li+α, ${}^{6}_{\Lambda\Lambda}$He+^{7}Li, ${}^{5}_{\Lambda\Lambda}$H+^{8}Be and ${}^{12}_{\Lambda}$B+Λ together with their excited hypernuclear production channels.

2 Formulation

The production rate of a double-Λ hypernucleus, ${}^{A_1+2}_{\Lambda\Lambda}Z_1 +{}^{A_2} Z_2$, via the Ξ^--atomic capture into ^{12}C is given as

$$R_{fi} = \frac{\Gamma_{fi}}{\Gamma} = \frac{1}{\Gamma}\frac{2\pi}{\hbar}\int d\boldsymbol{k}\delta(E_f - E_i) \mid \langle \Phi^f_{\ell LJ}(\boldsymbol{k}) \mid V_{\Xi^- p\to\Lambda\Lambda} \mid \Phi(^{12}\mathrm{C})\varphi_J(\Xi)\rangle \mid^2, \tag{1}$$

$$\Phi^f_{\ell LJ}(\boldsymbol{k}) = \sqrt{\frac{A_1!A_2!}{12!}}\mathcal{A}'\{\phi_\ell({}^{A_1+2}_{\Lambda\Lambda}Z_1)\phi({}^{A_2}Z_2)\varphi^f_L(\boldsymbol{r};\boldsymbol{k})\}_J, \tag{2}$$

where $\varphi_J(\Xi)$ is the Ξ^--atomic wave function obtained by solving the Schrödinger equation with the Ξ^--nucleus potential and Coulomb potential. The total production rate Γ in Eq. (1) is defined as $\Gamma = \sum_f \Gamma_{fi}$, which is given by applying the closure approximation for all final hypernuclear states. The continuum wave function $\Phi^f_{\ell LJ}(\boldsymbol{k})$ is evaluated with use of the ${}^{A_1+2}_{\Lambda\Lambda}Z_1 -{}^{A_2} Z_2$ folding potential. The wave function of the double-Λ hypernucleus is obtained by the nuclear core + Λ + Λ folding potential model, where we use the YNG-D

Table 3.1. Calculated double-Λ, twin-Λ and single-Λ hypernuclear production rates $R(\%)$.

channel	Q(MeV)	R(3d-atomic capture)	R(2p-atomic capture)
${}^{9}_{\Lambda}$Be+${}^{4}_{\Lambda}$H	10.0	0.10	0.30
${}^{8}_{\Lambda}$Li+${}^{5}_{\Lambda}$He	13.6	0.06	0.31
${}^{10}_{\Lambda\Lambda}$Be+t	19.2	0.03	0.21
${}^{9}_{\Lambda\Lambda}$Li+α	22.8	0.04	0.09
${}^{6}_{\Lambda\Lambda}$He+^{7}Li	15.0	0.37	0.18
${}^{5}_{\Lambda\Lambda}$H+^{8}Be	8.7	0.45	0.06
${}^{12}_{\Lambda}$B+Λ	23.9	2.40	4.87

Λ-Λ interaction which reproduces the experimental binding energy of ${}^{13}_{\Lambda\Lambda}$B. For simplicity, the ^{12}C wave function is given in terms of the SU(3) $(\lambda\mu)$=(04)= [^{11}B(s-hole)$\otimes$(0s)]+[^{11}B(p-hole)$\otimes$(0p)] wave function, in which the ^{11}B(s-hole) and ^{11}B(p-hole) states are given as the SU(3) $(\lambda\mu)$=(04) and (31) states, respectively. The conversion potential $V_{\Xi^- p\to\Lambda\Lambda}$ is assumed to the δ-type function which works only for the ${}^{1}S$ state. On the other hand, the production rates of twin-Λ and single-Λ hypernuclear channels are also evaluated with use of the similar equations to Eqs. (1) and (2).

3 Results and discussion

According to the Ξ^- cascade calculation, the Ξ^- particle is absorbed mainly from the 3d-atomic orbit into ^{12}C but the 2p-atomic capture is not negligible. (Note that the absorbed atomic orbit is unknown experimentally.) Therefore, we give the calculated production rates in both the cases, which are given in Table 3.1. It should be noted that the calculated production rate R in Table 3.1 is the total sum for the ground-state and excited-state production channels, for example, $R({}^{9}_{\Lambda}\mathrm{Be}+{}^{4}_{\Lambda}\mathrm{H})= R({}^{9}_{\Lambda}\mathrm{Be}(0^+)+{}^{4}_{\Lambda}\mathrm{H})+ R({}^{9}_{\Lambda}\mathrm{Be}(2^+)+{}^{4}_{\Lambda}\mathrm{H})+ R({}^{9}_{\Lambda}\mathrm{Be}(4^+)+{}^{4}_{\Lambda}\mathrm{H})$.

In the case of the 3d-atomic capture, the characteristics of our results are given as follows: Concerning the twin-Λ hypernuclear production, the production rate for ${}^{9}_{\Lambda}$Be+${}^{4}_{\Lambda}$H, $R({}^{9}_{\Lambda}\mathrm{Be}+{}^{4}_{\Lambda}\mathrm{H})$, is about 1.7 times larger than $R({}^{8}_{\Lambda}\mathrm{Li}+{}^{5}_{\Lambda}\mathrm{He})$, and the production rate for ${}^{9}_{\Lambda}$Be(0$^+$)+${}^{4}_{\Lambda}$H is nearly equal to that for ${}^{9}_{\Lambda}$Be(2$^+$)+${}^{4}_{\Lambda}$H. These results are consistent with the experimental results that one of the two events in the ^{12}C(Ξ^--stopped, ${}^{4}_{\Lambda}$H)${}^{9}_{\Lambda}$Be reaction is ${}^{9}_{\Lambda}$Be(0$^+$)+${}^{4}_{\Lambda}$H and another is ${}^{9}_{\Lambda}$Be(2$^+$)+${}^{4}_{\Lambda}$H, and other twin-Λ channels have not been observed at the present stage [2]. Concerning the double-Λ hypernuclear production, the production rates for ${}^{9}_{\Lambda\Lambda}$Li+α and ${}^{10}_{\Lambda\Lambda}$Be+t are less than that for ${}^{9}_{\Lambda}$Be+${}^{4}_{\Lambda}$H, while those for ${}^{6}_{\Lambda\Lambda}$He+^{7}Li and ${}^{5}_{\Lambda\Lambda}$H+^{8}Be are about 4 times larger than that for ${}^{9}_{\Lambda}$Be+${}^{4}_{\Lambda}$H. This is due to the fact that the resonant states ($J^{\pi} = 2^+$) appear around the Ξ^-+^{12}C threshold for both the ${}^{6}_{\Lambda\Lambda}$He+^{7}Li and ${}^{5}_{\Lambda\Lambda}$H+^{8}Be channels with excited ^{7}Li and ^{8}Be states in our calculation. The sum of the double-Λ production rates is about five times larger than that of the twin-Λ production rates. On the other hand, the single-Λ hypernuclear production rate $R({}^{12}_{\Lambda}\text{B}+\Lambda)$ is much larger than the twin-Λ and double-Λ ones. The reason is that the overlap between the initial atomic state and final single-Λ hypernuclear state is considerably large in comparison with the twin-Λ and double-Λ hypernuclear cases.

In the case of the $2p$-atomic capture, the feature of hypernuclear production rates is a little different from that in the case of the $3d$-atomic capture. For the twin-Λ hypernuclear production, the production rates are about 3 times larger than those in the $3d$ case, and $R({}^{9}_{\Lambda}\text{Be}+{}^{4}_{\Lambda}\text{H})$ is nearly equal to $R({}^{8}_{\Lambda}\text{Li}+{}^{5}_{\Lambda}\text{He})$. The double-$\Lambda$ production rates in the $2p$-atomic case are smaller than those in the $3d$-atomic case. This is due to the fact that the resonant states ($J^{\pi} = 1^{-}$) do not appear around the $\Xi^{-}+{}^{12}\text{C}$ threshold for the double-Λ production channels. The sum of the double-Λ production rates is nearly equal to that of the twin-Λ production rates. This is in contrast to the case of $3d$-atomic capture.

The trapping probability for two strangenesses $P(\Lambda\Lambda)$ is defined as the sum of the double-Λ and twin-Λ production rates. In our calculation, $P(\Lambda\Lambda)$ is about 1.1~1.2 % in both the $3d$- and $2p$-atomic capture cases (see Table 3.1). This value is smaller than the experimental value $P(\Lambda\Lambda)$=2.6% [2]. It should be noted that only the six channels, ${}^{9}_{\Lambda}\text{Be}+{}^{4}_{\Lambda}\text{H}$, ${}^{8}_{\Lambda}\text{Li}+{}^{5}_{\Lambda}\text{He}$, ${}^{10}_{\Lambda\Lambda}\text{Be}+t$, ${}^{9}_{\Lambda\Lambda}\text{Li}+\alpha$, ${}^{6}_{\Lambda\Lambda}\text{He}+{}^{7}\text{Li}$ and ${}^{5}_{\Lambda\Lambda}\text{H}+{}^{8}\text{Be}$, are taken into account in the present calculation. In the actual situation, the channels other than the above six channels, for example, ${}^{12}_{\Lambda\Lambda}\text{Be}+p$ (Q=26.3 MeV) etc. should contribute to $P(\Lambda\Lambda)$. Therefore, it is important to estimate the production rates for other double-Λ and twin-Λ hypernuclear production channels together with single-Λ hypernuclear channels.

4 Summary

We have studied the fragmentation of a ${}^{12}\text{C}$-Ξ^{-} atomic system within the frame of the direct reaction picture and showed the calculated production rates of the seven channels with their excited channels. In the $3d$-atomic capture, which is most probable in the cascade calculation, $R^{\text{cal}}({}^{9}_{\Lambda}\text{Be}+{}^{4}_{\Lambda}\text{H}) > R^{\text{cal}}({}^{8}_{\Lambda}\text{Li}+{}^{5}_{\Lambda}\text{He})$ and $R^{\text{cal}}({}^{9}_{\Lambda}\text{Be}(0^{+})+{}^{4}_{\Lambda}\text{H}) \simeq R^{\text{cal}}({}^{9}_{\Lambda}\text{Be}(2^{+})+{}^{4}_{\Lambda}\text{H})$. These results are consistent with the present experimental data.

References

1. S. Aoki et al.: Prog. Theor. Phys. **85**, 951, 1287 (1991); **89**, 493 (1993)
2. K. Nakazawa: Nucl. Phys. **A585**, 75 (1995)
3. C. B. Dover et al.: Phys. Rev. **C44**, 1905 (1991)
4. Y. Yamamoto, M. Wakai and M. Sano: Prog. Theor. Phys. Suppl. No. **117**, 265 (1994)
5. T. Yamada and K. Ikeda: Prog. Theor. Phys. Suppl. No. **117**, 445 (1994)

Few-Body Systems Suppl. 9, 285–292 (1995)

Few-Body Systems

Subthreshold K^+ Production on Nuclei by π^+ Mesons

S.V. Efremov[1], E.Ya. Paryev[2]

[1] Bonner Nuclear Laboratory, Rice University, P.O. Box 1892, Houston, TX 77251-1892, USA

[2] Institute for Nuclear Research, Russian Academy of Sciences, Moscow 117312, Russia

Abstract. The inclusive K^+ mesons production in π^+–nucleus reactions in the subthreshold energy regime is analyzed with respect to the one–step ($\pi^+ n \to K^+ \Lambda$) and the two–step ($\pi^+ n \to \eta p_1$, $\eta p_2 \to K^+ \Lambda$) incoherent production processes on the basis of an appropriate folding model, which allows one to take into account the various forms of an internal nucleon momentum distribution as well as on– and off–shell propagation of the struck target nucleon. Contrary to proton–nucleus reactions primary reaction channel is found to be significant practically at all considered energies. Detailed predictions for the K^+ total and invariant differential cross sections from $\pi^+ C^{12}$–collisions at subthreshold energies are provided.

Introduction

An extensive investigations of the production of K^+ mesons in proton–nucleus [1–5] and nucleus-nucleus [6–10] reactions at incident energies lower than the free nucleon-nucleon threshold have been carried out in the past years. Because of the high K^+ production threshold (1.58 GeV) in the nucleon-nucleon collision and the rather weak K^+ rescattering in the surrounding medium compared to the pions, etas, antiprotons and antikaons, from these studies one hopes to extract some additional information about the properties of nuclear matter, reaction dynamics, in–medium properties of hadrons at both normal and high nuclear densities. However, because of the complexity of collision dynamics and uncertainties in elementary kaon production cross sections close to the production thresholds [10, 11], in spite of large efforts, subthreshold kaon production is still far from being fully understood. To better understand the phenomenon of the subthreshold kaon production in pA– and AA–interactions it is necessary to undertake experimental and theoretical investigation of the subthreshold kaon production in πA–collisions, because in such collisions one may hope to get a clearer insight into the nuclear structure and the production mechanism [1, 12]. It should be noted that at present there are no experimental data on the kaon and antikaon subthreshold production in pion–induced reactions. Theoretical study of the subthreshold K^- production in πA–collisions in the framework of the first–collision model has been performed elsewhere [12]. The aim of the present work is to explore the influence of an internal nucleon momentum distribution on the description of K^+ subthreshold production in $\pi^+ A$–interactions as well as to evaluate the contributions from primary and secondary channels to the K^+ production process in the subthreshold regime, using the appropriate folding model [12, 13].

1 Direct K^+ Production Process

Apart from participation in the elastic scattering an incident pion can produce a K^+ directly in the first inelastic πN–collision due to nucleon Fermi motion. Since we are interested in the far subthreshold region, we have taken into account the following elementary process which requires the least amount of energy and, hence, has the lowest free production threshold (0.76 GeV):

$$\pi^+ + n \to K^+ + \Lambda. \tag{1}$$

Neglecting the kaon rescatterings in the nuclear medium [8], we can represent the invariant inclusive cross section of K^+ production on nuclei by the initial π^+ meson with momentum $\boldsymbol{p}_0$ as follows [12, 13]:

$$E_{K^+}\frac{d\sigma^{(\mathrm{prim})}_{\pi^+ A\to K^+ X}(\boldsymbol{p}_0)}{d\boldsymbol{p}_{K^+}} = I_{\mathrm{V}}[A, \sigma^{\mathrm{tot}}_{\pi^+ N}(p_0)] \left\langle E_{K^+}\frac{d\sigma_{\pi^+ n\to K^+\Lambda}(\boldsymbol{p}_0, \boldsymbol{p}_{K^+})}{d\boldsymbol{p}_{K^+}}\right\rangle, \tag{2}$$

where

$$I_{\mathrm{V}}[A, \sigma^{\mathrm{tot}}_{\pi^+ N}(p_0)] = N\int \rho(\boldsymbol{r})d\boldsymbol{r}\exp\left[-\mu(p_0)\int\limits_{-\infty}^{0}\rho(\boldsymbol{r}+x\boldsymbol{\Omega}_0)dx\right], \tag{3}$$

$$\mu(p_0) = \sigma^{\mathrm{tot}}_{\pi^+ p}(p_0)Z + \sigma^{\mathrm{tot}}_{\pi^+ n}(p_0)N; \tag{4}$$

$$\left\langle E_{K^+}\frac{d\sigma_{\pi^+ n\to K^+\Lambda}(\boldsymbol{p}_0, \boldsymbol{p}_{K^+})}{d\boldsymbol{p}_{K^+}}\right\rangle = \int n(\boldsymbol{p}_{\mathrm{t}})d\boldsymbol{p}_{\mathrm{t}}\left[E_{K^+}\frac{d\sigma_{\pi^+ n\to K^+\Lambda}(\sqrt{s}, \boldsymbol{p}_{K^+})}{d\boldsymbol{p}_{K^+}}\right]. \tag{5}$$

Here, $E_{K^+}d\sigma_{\pi^+ n\to K^+\Lambda}(\sqrt{s}, \boldsymbol{p}_{K^+})/d\boldsymbol{p}_{K^+}$ is the free invariant inclusive cross section for the K^+ production in reaction (1); $\rho(\boldsymbol{r})$ and $n(\boldsymbol{p}_{\mathrm{t}})$ are the density and ground–state momentum distribution normalized to unity; $\boldsymbol{p}_{\mathrm{t}}$ is the internal momentum of the struck target nucleon just before the collision; $\sigma^{\mathrm{tot}}_{\pi^+ N}(p_0)$ is the total cross section of free $\pi^+ N$–interaction; Z and N are the numbers of protons and neutrons in the target nucleus ($A = N + Z$); $\boldsymbol{\Omega}_0 = \boldsymbol{p}_0/p_0$; $\boldsymbol{p}_{K^+}$ and E_{K^+} are the momentum and total energy of a K^+ meson, respectively; s is the $\pi^+ n$ center–of–mass energy squared. The expression for s is:

$$s = (E_0 + E_{\mathrm{t}})^2 - (\boldsymbol{p}_0 + \boldsymbol{p}_{\mathrm{t}})^2, \tag{6}$$

where E_0 and E_t are the projectile's total energy, given by $E_0 = \sqrt{p_0^2 + m_\pi^2}$, and the struck target nucleon total energy, respectively. In our calculations we will use two formulas for E_t. In the first case we take into account the recoil of the residual nucleus. Then the energy that the struck target nucleon brings into the collision is equal to [2]:

$$E_t = M_A - \sqrt{(-\boldsymbol{p}_t)^2 + M_{A-1}^2}, \tag{7}$$

where M_A and M_{A-1} are the masses of the initial target nucleus and the recoiling residual nucleus, respectively. It is easily seen that in this case the struck target nucleon is off–shell and for a large target nucleus E_t is approximately equal to the rest mass of nucleon m_N. In the opposite case we assume that the struck target nucleon is on–shell and E_t is given simply by:

$$E_t = \sqrt{\boldsymbol{p}_t^2 + m_N^2}. \tag{8}$$

The invariant inclusive cross section for K^+ production in the elementary process (1) has been described by the two–body phase space calculations normalized to the corresponding total cross section $\sigma_{\pi^+ n \to K^+ \Lambda}$ [14].

The internal nucleon momentum distribution, $n(\boldsymbol{p}_t)$, is a crucial point in the evaluation of the subthreshold production of any particles on a nuclear target. Therefore, we calculated the cross sections for the K^+ production in $\pi^+ C^{12}$–collisions using various types of $n(\boldsymbol{p}_t)$.

The standard shell–model momentum distribution [12]:

$$n(\boldsymbol{p}_t) = \frac{(b_0/\pi)^{3/2}}{A/4}\left\{1 + \left[\frac{A-4}{6}\right] b_0 \boldsymbol{p}_t^2\right\} \exp(-b_0 \boldsymbol{p}_t^2), \tag{9}$$

where $b_0 = 68.5(\mathrm{GeV}/c)^{-2}$.

The momentum distribution in which the part corresponding to the $1p_{3/2}$ shell and having the exponential fall–off at high momentum p_t, was inferred by Million [15] from the $(e, e'p)$ and (γ, p) experiments:

$$n(\boldsymbol{p}_t) = \frac{1}{A/4}\left\{n_{1/2}(\boldsymbol{p}_t) + \left[\frac{A-4}{4}\right] n_{3/2}(\boldsymbol{p}_t)\right\}, \tag{10}$$

where

$$n_{1/2}(\boldsymbol{p}_t) = (\pi\nu)^{-3/2} \exp(-\boldsymbol{p}_t^2/\nu), \tag{11}$$

$$n_{3/2}(\boldsymbol{p}_t) = C_1 n_{\mathrm{HO}}(\boldsymbol{p}_t) + C_2 n_{\exp}(\boldsymbol{p}_t), \tag{12}$$

$$n_{\mathrm{HO}}(\boldsymbol{p}_t) = \frac{2}{3}(\pi\nu)^{-3/2}(\boldsymbol{p}_t^2/\nu) \exp(-\boldsymbol{p}_t^2/\nu),$$

$$n_{\exp}(\boldsymbol{p}_t) = \left[\frac{1}{24\pi(p_t^0)^3}\right](p_t/p_t^0) \exp(-p_t/p_t^0) \tag{13}$$

and $\sqrt{\nu} = 127.0$ MeV/c, $p_t^0 = 55.0$ MeV/c, $C_1 = 0.997$, $C_2 = 0.003$.

The double Gaussian distribution with a large high–momentum tail extracted by Geaga et al. [16] from high–energy proton backward scattering:

$$n(\boldsymbol{p}_\mathrm{t}) = \frac{1}{(2\pi)^{3/2}(1+\alpha)}\left[\frac{1}{\sigma_1^3}\exp\left(-p_\mathrm{t}^2/2\sigma_1^2\right) + \frac{\alpha}{\sigma_2^3}\exp\left(-p_\mathrm{t}^2/2\sigma_2^2\right)\right], \tag{14}$$

where $\sigma_1 = 0.119$ GeV/c, $\sigma_2 = 0.230$ GeV/c. The parameter α which defines the high–momentum part in $n(\boldsymbol{p}_\mathrm{t})$ is 0.06 for C^{12} and is proportional to $A^{1/3}$ for other target nuclei.

Let us focus now on the two–step K^+ production mechanism.

2 Two–Step K^+ Production Process

Kinematical considerations show that in the bombarding energy range of our interest ($\epsilon_0 \leq 0.76$ GeV) the following two–step production process may contribute to the K^+ production in π^+A–interactions. An incident pion can produce in the first inelastic collision with an neutron also an η meson through the elementary reaction:

$$\pi^+ + n \to \eta + p. \tag{15}$$

We remind that the free threshold energy for this reaction is 0.56 GeV . Then the intermediate η meson, which is assumed to be on–shell, produces the kaon on a proton of the target nucleus via the elementary subprocess with the lowest free production threshold (0.20 GeV):

$$\eta + p \to K^+ + \Lambda, \tag{16}$$

provided that this subprocess is energetically possible. For instance, the maximum kinetic energy of η meson produced by a pion with the energy $\epsilon_0 = 0.76$ GeV on a target neutron at rest is about 0.33 GeV. Therefore, for the beam energies considered here, there is a region of eta's energy where the K^+ production process (16) occurs even if the proton is at rest. Due to the Fermi motion of the protons the production will be promoted by the target nucleus. It is thus desirable to evaluate the respective K^+ yield. In order to calculate the K^+ production cross section for π^+A–reactions from the secondary eta induced reaction channel (16) we fold the Fermi–averaged differential cross section for the η production in the reaction (15) (denoted by $< d\sigma_{\pi^+n\to\eta p}(\boldsymbol{p}_0,\boldsymbol{p}_\eta)/d\boldsymbol{p}_\eta >$) with the Fermi–averaged invariant differential cross section for K^+ production in this channel (denoted by $< E_{K^+}d\sigma_{\eta p\to K^+\Lambda}(\boldsymbol{p}_\eta,\boldsymbol{p}_{K^+})/d\boldsymbol{p}_{K^+} >$) and the effective number of np pairs per unit of square (denoted by $I_V[A,\sigma^{\mathrm{tot}}_{\pi^+N}(p_0),\sigma^{\mathrm{tot}}_{\eta N}(p_\eta),\vartheta_\eta]$), i.e.:

$$E_{K^+}\frac{d\sigma^{(\mathrm{sec})}_{\pi^+A\to K^+X}(\boldsymbol{p}_0)}{d\boldsymbol{p}_{K^+}} = \int d\boldsymbol{p}_\eta I_V[A,\sigma^{\mathrm{tot}}_{\pi^+N}(p_0),\sigma^{\mathrm{tot}}_{\eta N}(p_\eta),\vartheta_\eta]\times \tag{17}$$

$$\times\left\langle\frac{d\sigma_{\pi^+n\to\eta p}(\boldsymbol{p}_0,\boldsymbol{p}_\eta)}{d\boldsymbol{p}_\eta}\right\rangle\left\langle E_{K^+}\frac{d\sigma_{\eta p\to K^+\Lambda}(\boldsymbol{p}_\eta,\boldsymbol{p}_{K^+})}{d\boldsymbol{p}_{K^+}}\right\rangle,$$

where

$$I_V[A, \sigma^{tot}_{\pi^+N}(p_0), \sigma^{tot}_{\eta N}(p_\eta), \vartheta_\eta] = NZ \int\int d\boldsymbol{r} d\boldsymbol{r}_1 \theta(x_\parallel)\delta^{(2)}(\boldsymbol{x}_\perp)\rho(\boldsymbol{r})\rho(\boldsymbol{r}_1)\times$$
$$\times \exp\left[-\mu(p_0)\int\limits_{-\infty}^{0} \rho(\boldsymbol{r}_1 + x'\boldsymbol{\Omega}_0)dx' - \mu(p_\eta)\int\limits_{0}^{x_\parallel} \rho(\boldsymbol{r}_1 + x'\boldsymbol{\Omega}_\eta)dx'\right], \quad (18)$$

$$\boldsymbol{r} - \boldsymbol{r}_1 = x_\parallel \boldsymbol{\Omega}_\eta + \boldsymbol{x}_\perp, \boldsymbol{\Omega}_\eta = \boldsymbol{p}_\eta/p_\eta, \mu(p_\eta) = A\sigma^{tot}_{\eta N}(p_\eta). \quad (19)$$

Here, $\sigma^{tot}_{\eta N}(p_\eta)$ is the total cross section of the free ηN–interaction; $\boldsymbol{p}_\eta$, is the momentum of η meson; $\cos\vartheta_\eta = \boldsymbol{\Omega}_0\boldsymbol{\Omega}_\eta$. In our calculations the angular distributions $d\sigma_{\pi^+n\to\eta p}/d\Omega^*$ and $d\sigma_{\eta p\to K^+\Lambda}/d\Omega^*$ of the subprocesses (15) and (16) in the corresponding center–of–mass systems are assumed to be isotropic. For the total cross section $\sigma_{\pi^+n\to\eta p}$ of the reaction (15) we used the parametrization suggested in [17]. Because of the lack of knowledge about the total cross section $\sigma_{\eta p\to K^+\Lambda}$ of the elementary process (16), to estimate it we choose in this work the following natural way, which was used also in [18] for the evaluation of lambda production cross section in the reaction $\omega N \to K\Lambda$. The probability for producing a kaon in the reaction under consideration is given by the ratio of the $\sigma_{\eta p\to K^+\Lambda}$ to the ηp–inelastic cross section $\sigma^{in}_{\eta p}$. We assume that this ratio is equal to that of the $\sigma_{\pi^+n\to K^+\Lambda}$ to the π^+n–inelastic cross section $\sigma^{in}_{\pi^+n}$ at the same invariant energy $\sqrt{s}$. Taking into account that $\sigma^{in}_{\eta p} \approx \sigma^{in}_{\pi^+n} \approx 20$ mb [13, 18] in the eta and pion energy ranges of interest, we get:

$$\sigma_{\eta p\to K^+\Lambda}(\sqrt{s}) = \sigma_{\pi^+n\to K^+\Lambda}(\sqrt{s}). \quad (20)$$

Now let us discuss the results of our calculations in the framework of the approach outlined above.

3 Results and Discussion

The expected total cross sections for K^+ production in $\pi^+ + ^{12}$C–reactions from the primary $\pi^+n \to K^+\Lambda$ and secondary $\eta p \to K^+\Lambda$ channels calculated according to (2)–(14), (17)–(20) are shown in the left part of Fig.1 as functions of the laboratory energy ϵ_0 of the pion.

It is seen that the η induced production channel becomes comparable to the π^+n channel only at very low energies ($\epsilon_0 < 600$ MeV) if we adopt the off–shell assumption about the struck target nucleon as well as use the nucleon momentum distributions (9), (10) without a large high–momentum tails. The cross section of the K^+ production from the primary π^+n production channel in the energy region of $600 \leq \epsilon_0 \leq 650$ MeV, where this channel dominates, still strongly depends both on the choice of the nucleon momentum distribution and on the bombarding energy ϵ_0 (contrary to the secondary production channel). The cross section calculated with the momentum distribution (14) is larger by a factor of 2–10 than that calculated with the shell–model momentum distribution (9) in this region. The momentum distribution (10), deduced

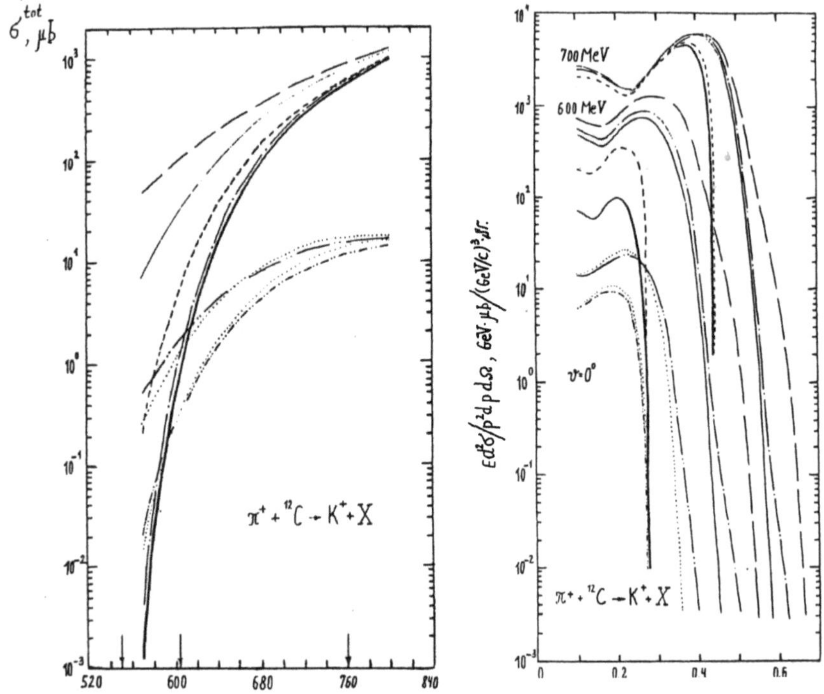

Figure 1. Left: The total cross sections for K^+ production in $\pi^+ + ^{12}$ C–reactions from primary $\pi^+ n \to K^+\Lambda$ and secondary $\eta p \to K^+\Lambda$ channels as functions of the laboratory energy of the pion. The cross sections from primary $\pi^+ n$–collisions: the heavy and the light solid lines are calculations with the shell–model momentum distribution (9) with off– and on–shell assumptions about the struck target neutron, respectively; the dot–dashed line is the calculation with the momentum distribution (10) and off–shell struck target neutron; the short– and the long–dashed lines are calculations with distribution (14) with large high–momentum tail with off– and on–shell assumptions about the struck target neutron, respectively. The cross sections from secondary ηp–collisions: the lower and upper dotted lines are calculations with the shell–model momentum distribution (9) with off– and on–shell assumptions about the struck target nucleons, respectively; the short– and the long–dashed lines with two dots are calculations with distribution (14) with off– and on–shell assumptions about the struck target nucleons, respectively. The arrows show the production thresholds on a free neutron and on off–shell neutron with the Fermi momentum of 250 MeV/c and the absolute production threshold.
Right: The inclusive invariant cross sections for the production of K^+ mesons in primary $\pi^+ n$– and secondary ηp–collisions at an angle of 0^0 as functions of the kaon momentum in the interaction of pions with the energies of 600 and 700 MeV with ^{12}C nuclei. The dot–dashed line denotes the same as above, but it is supposed that the struck target neutron is on–shell. The rest of the notation is the same as above.

from the $(e, e'p)$ and (γ, p) experiments, gives practically the same results as the shell–model momentum distribution (9). The values of the total production cross section in the energy region under consideration lie in the range of 0.5–30 μb. Such rapid energy dependence of the (π^+, K^+) total cross section in the energy region considered here is a characteristic signature of the $\pi^+ n \rightarrow K^+ \Lambda$ one–step production mechanism. Therefore, the measurement of the total K^+ production cross section at incident energies between 600 and 650 MeV seems to be quite promising to study both the kaon production mechanism and the high–momentum components within nucleus. However, it is important to emphasize that in order to get a more clear insight into the relative role of the primary and secondary reaction channels, further theoretical efforts are needed for a better understanding about the η induced elementary K^+ production process (16).

In the right part of Fig. 1 we show the results of our calculations of the inclusive invariant cross sections for the kaon production from the primary $\pi^+ n$– and secondary ηp–reaction channels at an angle of 0° in the interaction of pions with the energies of 600 (lower lines) and 700 MeV (upper lines) with ^{12}C nuclei. It is clearly seen that the secondary production channel practically does not contribute to the spectrum of emitted kaons at incident energies between 600 and 700 MeV. Also one can see that the high–momentum part of the spectrum of kaons from the one–step production process (1) is essentially determined by the nucleon momentum distribution, whereas its low–momentum part is not sensitively affected by the choice of the internal momentum distribution in the case of on–shell assumption about the struck target neutron.

Taking into account the considered above, we conclude that the measurement of the total and differential K^+ production cross sections at incident energies between 600 and 700 MeV offers the possibility to check the dominant rôle of the one–step production process in the subthreshold K^+ production as well as to study the high–momentum components within target nucleus.

4 Summary

We have calculated the total and differential cross sections for K^+ production from $\pi^+ +^{12}$ C–reactions in the subthreshold regime by considering incoherent primary pion–neutron and secondary eta–proton production processes within the framework of an appropriate folding model. It was shown that the one–step K^+ production mechanism clearly dominates at all subthreshold energies if the struck target nucleon is assumed to be on–shell, whereas for the off–shell assumption about the struck target nucleon the one–step and the two–step reaction mechanisms are of equal importance only at very low energies ($\epsilon_0 < 600$ MeV) in the case of use of the nucleon momentum distributions without a large high–momentum tails. The measurements of the inclusive K^+ production cross sections at incident pion energies between 600 and 700 MeV, as one may hope, allow one to obtain a clear information both on the production mechanism and on the high–momentum components within target nucleus.

Acknowledgement. We would like to thank V.Koptev for stimulating discussions on the initial stage of this study. We are also grateful to K.Oganesyan for interest in the work. One of us (E.Ya.P.) was supported in part by Grant No.N6K000 from the International Science Foundation as well as by Grant No.N6K300 from the International Science Foundation and Russian Government.

References

1. V.P.Koptev et al.: ZhETF **94**, 1 (1988)
2. A.Shor, V.P.Mendez, K.Ganezer: Nucl. Phys. **A514**, 717 (1990)
3. W.Cassing et al.: Phys. Lett. **238B**, 25 (1990)
4. H.Müller, K.Sistemich: Z. Phys. **A344**, 197 (1992)
5. A.Sibirtsev, M.Büscher: Z. Phys. **A347**, 191 (1994)
6. W.Cassing et al.: Nucl. Phys. **A545**, 123c (1992)
7. U.Mosel: Ann. Rev. Nucl. Part. Sci. **41**, 29 (1991)
8. X.S.Fang et al.: Nucl. Phys. **A575**, 766 (1994)
9. D.Miskowiec et al.: Phys. Rev. Lett. **72**, 3650 (1994)
10. C.Hartnack et al.: Nucl. Phys. **A580**, 643 (1994)
11. S.V.Efremov, M.V.Kazarnovsky, E.Ya.Paryev: Z. Phys. **A344**, 181 (1992)
12. S.V.Efremov, E.Ya.Paryev: *Subthreshold K^- production in pion–nucleus reactions.* Preprint INR-857/94, Moscow, 1994; Z. Phys. **A** (in press)
13. A.S.Iljinov, M.V.Kazarnovsky, E.Ya.Paryev: *Intermediate Energy Nuclear Physics.* Boca Raton: CRC Press, Inc. 1994
14. J.Cugnon, R.M.Lombard: Nucl.Phys. **A422**, 635 (1984)
15. B.Million: Phys. Rev. **C40**, 2924 (1989)
16. J.V.Geaga et al.: Phys. Rev Lett. **45**, 1993 (1980)
17. W.Cassing et al.: Z. Phys. **A340**, 51 (1991)
18. C.M.Ko, R.Yuan: Phys. Lett. **192B**, 31 (1987)

Few-Body Systems Suppl. 9, 293–296 (1995)

Few-Body Systems

Relativistic Versus Nonrelativistic ΛN Correlations in the Weak Decay of Hypernuclei

A. Parreño[1], E. Oset[2], A. Ramos[1]

[1] Dept. ECM, Facultat de Física, Diagonal 647, 08028 Barcelona, Spain
[2] Depto. de Física Teórica and *IFIC*. Centro Mixto Universidad de Valencia-CSIC, 46100 Burjassot (Valencia), Spain

Abstract. We establish the reasons for the different effect of short range correlations in the nonmesonic decay of Λ hypernuclei found by relativistic and nonrelativistic approaches. By means of a schematic microscopic model for the origin of correlations, the appropriate method to include them in nuclear processes, via a correlation function, is derived and is found to be the one used in the nonrelativistic approach.

1 Introduction

The short range nuclear forces between the interacting ΛN pair in the nonmesonic $\Lambda N \rightarrow NN$ decay generate short range correlations (SRC) which must be taken into account. The nonrelativistic calculations performed in nuclear matter [1-3] showed that the nonmesonic rate was reduced by a factor of up to two when SRC were introduced. Surprisingly, the relativistic calculation of ref. [4] found twice as much reduction with a similar correlation function. Our purpose is to understand the origin of such discrepancies.

2 Relativistic versus nonrelativistic treatment of SRC

The starting point for our derivation is the uncorrelated Feynman diagram for the $\Lambda N \rightarrow NN$ transition

$$\mathcal{M}_\pi = \int d^4x d^4y \overline{\psi}_{p_3}(\boldsymbol{x}) \Gamma^{\mathrm{W}} \psi^{\Lambda}_{p_1}(\boldsymbol{x}) \Delta_\pi(\boldsymbol{x}-\boldsymbol{y}) \overline{\psi}_{p_4}(\boldsymbol{y}) \Gamma^{\mathrm{S}} \psi_{p_2}(\boldsymbol{y}) , \qquad (1)$$

where $\psi_{p_i}(\boldsymbol{x})$ is the free baryon field of positive energy and $\Delta_\pi(\boldsymbol{x}-\boldsymbol{y})$ the pion propagator. For the weak vertex we take the parametrization $\Gamma^{\mathrm{W}} = G_{\mathrm{F}}\mu^2(A - B\gamma_5)$, where the empirical constants $A = 1.05$ and $B = 7.15$ have been adjusted to the free Λ-decay and determine the strength of the PV and

PC rates respectively. For the strong vertex we use the pseudoscalar coupling $\Gamma^{\mathrm{S}} = g_{\pi NN}\gamma_5$. The relativistic treatment [4] incorporates SRC by substituting:

$$\Delta_\pi(\boldsymbol{x}-\boldsymbol{y}) \rightarrow \Delta_\pi(\boldsymbol{x}-\boldsymbol{y}) f(|\boldsymbol{x}-\boldsymbol{y}|) \tag{2}$$

where $f(r)$ is an appropriate correlation function with $r = |\boldsymbol{x}-\boldsymbol{y}|$. Conversely, the nonrelativistic approach performs a nonrelativistic reduction of Eq. (1) to determine the transition potential $V(\boldsymbol{r})$ and defines the correlated potential as $\tilde{V}(\boldsymbol{r}) = f(r)V(\boldsymbol{r})$.

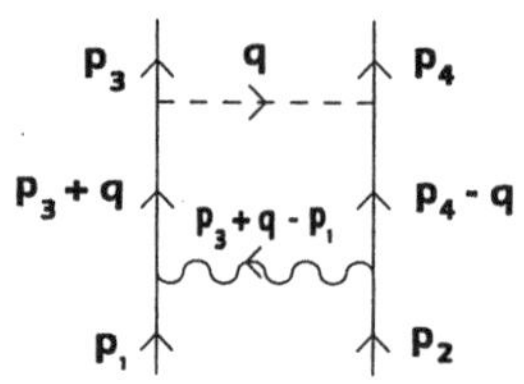

Figure 1. Interpretation of SRC via the simultaneous exchange of ω and π mesons

The formalism in coordinate space makes the differences between both approaches difficult to identify, but it becomes quite clear if a translation to momentum space is done. Taking into account the simultaneous exchange of other mesons (Fig.1), requires the evaluation of the loop integral

$$\int \frac{d^3q}{(2\pi)^3} D(\boldsymbol{q}) \overline{u}(\boldsymbol{p}_3)\gamma_5 u(\boldsymbol{p}_3-\boldsymbol{q}) \overline{u}(\boldsymbol{p}_4)\gamma_5 u(\boldsymbol{p}_4+\boldsymbol{q}) \ldots \tag{3}$$

where we have chosen the parity conserving part of the pion exchange amplitude. The dots symbolize additional elements of the amplitude which are not relevant for the present discussion. One can see that the matrix elements of the γ_5 operators are evaluated between spinors which depend upon the loop variable $\boldsymbol{q}$. However, in ref. [4], the matrix elements were evaluated between spinors of the external particles and were factorized out of the integral

$$\overline{u}(\boldsymbol{p}_3)\gamma_5 u(\boldsymbol{p}_1) \overline{u}(\boldsymbol{p}_4)\gamma_5 u(\boldsymbol{p}_2) \int \frac{d^3q}{(2\pi)^3} D(\boldsymbol{q}) \ldots \tag{4}$$

This is the reason of the discrepancies between the relativistic approach of ref. [4] and the nonrelativistic ones [1, 2, 3]. The origin is thus an incorrect treatment of the correlations in ref. [4] and not relativistic effects. In ref. [5] the momentum space nonrelativistic reductions of Eqs. (3) and (4), incorporating SRC, are derived. A subsequent Fourier transform allows to find the correlated potentials in coordinate space which can be split into central and tensor terms given by:

$$\begin{aligned} V^{\mathrm{C}}_{\mathrm{nr}}(\boldsymbol{r}) &= -\frac{g_{\pi NN}}{2M}\frac{G_{\mathrm{F}}\mu^2 B}{2\overline{M}}\frac{1}{3}\left(\mu^2\frac{\mathrm{e}^{-\mu r}}{4\pi r} - \delta(\boldsymbol{r})\right) f(r)\boldsymbol{\sigma}_1\boldsymbol{\sigma}_2 \\ V^{\mathrm{T}}_{\mathrm{nr}}(\boldsymbol{r}) &= -\frac{g_{\pi NN}}{2M}\frac{G_{\mathrm{F}}\mu^2 B}{2\overline{M}}\frac{1}{3}\mu^2\frac{\mathrm{e}^{-\mu r}}{4\pi r}\left(1 + \frac{3}{\mu r} + \frac{3}{(\mu r)^2}\right) f(r) S_{12}(\hat{\boldsymbol{r}}) \end{aligned} \tag{5}$$

for the nonrelativistic approach, and by

$$\begin{aligned}
\tilde{V}^{\mathrm{C}}_{\mathrm{rel}}(\boldsymbol{r}) &= -\frac{g_{\pi NN}}{2M}\frac{G_{\mathrm{F}}\mu^2 B}{2\overline{M}}\frac{1}{3}\left\{\left(\mu^2\frac{\mathrm{e}^{-\mu r}}{4\pi r}-\delta(\boldsymbol{r})\right)f(r)\right. \\
&\quad \left. -\mu^2\frac{\mathrm{e}^{-\mu r}}{4\pi r}\left(\frac{2}{\mu}\frac{df}{dr}-\frac{1}{\mu^2}\frac{d^2f}{dr^2}\right)\right\}\boldsymbol{\sigma}_1\boldsymbol{\sigma}_2 \\
\tilde{V}^{\mathrm{T}}_{\mathrm{rel}}(\boldsymbol{r}) &= -\frac{g_{\pi NN}}{2M}\frac{G_{\mathrm{F}}\mu^2 B}{2\overline{M}}\frac{1}{3}\mu^2\frac{\mathrm{e}^{-\mu r}}{4\pi r}\left\{\left(1+\frac{3}{\mu r}+\frac{3}{(\mu r)^2}\right)f(r)\right. \\
&\quad \left. -\left(\frac{3}{\mu r}+2\right)\frac{1}{\mu}\frac{df}{dr}+\frac{1}{\mu^2}\frac{d^2f}{dr^2}\right\}S_{12}(\hat{\boldsymbol{r}}) \qquad (6)
\end{aligned}$$

for the "*relativistic*" scheme of ref. [4]. Comparing both potentials, we observe some extra terms in the "*relativistic*" expression involving first and second derivatives of $f(r)$, which, as we will see, are responsible for the additional reduction of the rates.

3 Results and discussion

We present results for the nonmesonic decay width of $^5_\Lambda$He taking, for the correlation function, the parametrization:

$$f(r) = \left(1-\mathrm{e}^{-r^2/a^2}\right)^n + br^2\mathrm{e}^{-r^2/c^2}\ , \qquad (7)$$

with $a = 0.5$ fm, $b = 0.25$ fm^{-2}, $c = 1.28$ fm and $n = 2$, which provides a good description of realistic ΛN correlation functions obtained with G-matrix calculations [6, 7].

Table 1. $\Lambda N \rightarrow NN$ decay rate of $^5_\Lambda$He (in units of the free Λ width)

	FREE	*Nonrelativistic*	*Relativistic*
C	0.174	1.7×10^{-3}	0.057
T	0.495	0.488	0.159
PV	0.308	0.244	0.117
TOTAL	0.977	0.694	0.333

Table I compares the decay rate of $^5_\Lambda$He, without including form factors, for the two nonrelativistic correlated potentials: i) the one of Eq. (5), which is the prescription used in refs. [1-3] and has been shown to be consistent with a microscopic interpretation of short range correlations [5], and ii) the one of Eq. (6), which is the nonrelativistic equivalent of the relativistic model used in ref. [4]. This latter prescription for SRC gives twice as much reduction than the standard nonrelativistic model. To illustrate these results we show the tensor integrand in Fig.2. The solid line corresponds to the nonrelativistic prescription. The dashed line, corresponding to the "*relativistic*" scheme of Eq. (6), shows,

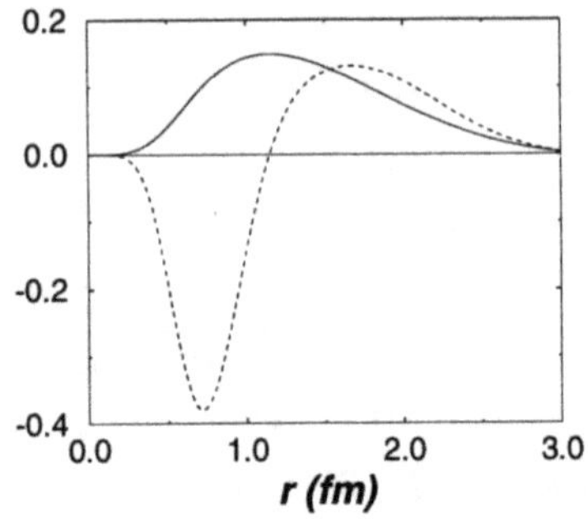

Figure 2. Tensor integrand for the "*relativistic*" (dashed line) and nonrelativistic (solid line) approaches

due to the change of sign in the derivative of $f(r)$, large positive and negative contributions which tend to cancel each other. This gives rise to the bigger reduction of the rates as compared with the nonrelativistic results.

4 Conclusions

We have shown that the discrepancies found in the literature on the effect of SRC in the nonmesonic decay width of Λ hypernuclei, are not due to relativistic effects but to the way the correlation function is implemented. Via a simplified microscopic model we have shown that the correct implementation is the one corresponding to the nonrelativistic calculations.

Acknowledgement. This work has been supported by CICYT contract no. AEN 93-1205 and DGICYT contract no. PB92-0761 (Spain).

References

1. B.H.J. McKellar and B.F. Gibson: Phys. Rev. **C30**, 322 (1984)
2. E. Oset and L.L Salcedo: Nucl. Phys. **A433**, 704 (1985)
3. J.F. Dubach: Nucl. Phys. **A450**, 71c (1986)
4. A. Ramos, C. Bennhold, E. van Meijgaard and B.K. Jennings: Phys. Lett. **B264**, 233 (1991); A. Ramos, E. van Meijgaard, C. Bennhold and B.K. Jennings: Nucl. Phys. **A544**, 703 (1992)
5. A. Parreño, A. Ramos, and E. Oset: Phys. Rev. **C51**, 2477 (1995)
6. H. Bandō: Prog. Theor. Phys. Suppl. No. **81**, 181 (1985)
7. D. Halderson: Phys. Rev. **C48**, 581 (1993)

Few-Body Systems Suppl. 9, 297–301 (1995)

Few-Body Systems

Two-body Mechanisms in Pion Scattering and Pion Photoproduction on the Trinucleon

S. S. Kamalov[1], L. Tiator[2], C. Bennhold[3]

[1] Laboratory of Theoretical Physics, JINR Dubna, Head Post Office Box 79, SU-101000 Moscow, Russia

[2] Institut für Kernphysik, Universität Mainz, 6500 Mainz, Germany

[3] Center of Nuclear Studies, Department of Physics, The George Washington University, Washington, D.C., 20052

Abstract. A breakdown of the Impulse Approximation is studied in pion photoproduction and pion charge exchange on ^{3}He at high momentum transfers. The usual DWIA formalism with Faddeev wave functions which works well for small momentum transfers deviates from experimental measurements by up to two orders of magnitude for $Q^2 > 6\,\mathrm{fm}^{-2}$. It is found that the explicit inclusion of two-body mechanisms, where the photon or pion is absorbed on one nucleon and the pion is emitted from another nucleon removes most of the disagreement with the data.

1 Introduction

Reactions on the trinucleon are an ideal testing ground to search for effects that go beyond the Impulse Approximation (IA) since realistic correlated three-body wave functions are available that are reliable even at high nuclear momentum transfers. Our previous theoretical investigations [1, 2] of pion scattering and pion photoproduction on the trinucleon systems have reached a level where the conventional one-body aspects are treated on a rather accurate level. We obtained a good description of experimental measurements at momentum transfers of $Q^2 < 6\mathrm{fm}^{-2}$. However, at higher momentum transfers large discrepancies appear between measurements and theoretical calculations [1, 2]. In the region of $Q^2 > 8\,\mathrm{fm}^{-2}$, calculations dramatically fail to explain existing data, underestimating them by up to two orders of magnitude.

By starting from pion-nuclear production and absorption operators we study in this work novel two-body mechanisms in pion photoproduction and pion scattering on ^{3}He which do not appear in a standard Distorted Wave Impulse Approximation (DWIA) framework.

2 Formalism

Using the impulse approximation for the pion-nuclear production amplitude one can write

$$T_\pi = \frac{f}{m_\pi} \int d\boldsymbol{r}_\pi \phi^\dagger_\alpha(\boldsymbol{r}_\pi) < f \mid \sum_{j=1}^{A} \boldsymbol{\sigma}\cdot \overleftarrow{\nabla}_\pi \ \tau_\alpha(j)\, \delta(\boldsymbol{r}_\pi - \boldsymbol{r}_j) \mid i >, \qquad (1)$$

where $f/m_\pi = g/2M$ with $g^2/4\pi = 14$, M denotes the nucleon mass, $\boldsymbol{\tau}$ is the nucleon isospin operator, and $\phi_\alpha(\boldsymbol{r}) = \varphi_\alpha \exp(\mathrm{i}\boldsymbol{q}\cdot\boldsymbol{r})$ is the pion wave function with the three isospin components $\alpha = 1, 2, 3$.

The main part of the charged pion photoproduction amplitude - seagull and pion exchange terms - can be obtained by minimal substitution $\boldsymbol{\nabla}_\pi \to \boldsymbol{\nabla}_\pi - \mathrm{i}e\boldsymbol{A}$ (where $\boldsymbol{A} = \boldsymbol{\epsilon}\exp(\mathrm{i}\boldsymbol{k}\cdot\boldsymbol{r})$ is the electromagnetic vector potential with polarization vector $\boldsymbol{\epsilon}$ and photon momentum $\boldsymbol{k}$) in the $\boldsymbol{\sigma}\cdot \overleftarrow{\nabla}_\pi$-operator and in the pion wave function. The more complicated nucleon pole (dispersive or two-step) terms of the pion photoproduction operator is constructed by introducing the electromagnetic field in the nuclear wave function $\mid i >$ and $\mid f >$. The corresponding electromagnetic Hamiltonian can be written as

$$\hat{H}_{\mathrm{em}} = \sum_{n=1}^{A} \frac{\mathrm{i}\hat{e}(n)}{2M}(\boldsymbol{\nabla}_n\cdot\boldsymbol{A} + \boldsymbol{A}\cdot\boldsymbol{\nabla}_n) + \sum_{n=1}^{A}\frac{\hat{\mu}(n)}{2M}\boldsymbol{\sigma}_n\cdot[\boldsymbol{\nabla}\times\boldsymbol{A}], \qquad (2)$$

where the isospin operator for the nucleon charge is $\hat{e} = (1+\tau_3)/2$, the magnetic isospin operator is $\hat{\mu} = \mu_p(1+\tau_3)/2 + \mu_n(1-\tau_3)/2$, and the magnetic moments of the proton, $\mu_p = 2.79$, and the neutron, $\mu_n = -1.91$.

Our analysis indicates that the contribution from the convection part of the electromagnetic Hamiltonian (first term in Eq.(2)) is small due to its nonlocal nature. Therefore, we give only the main expressions for the contributions coming from the magnetic interaction (second term in Eq.(2)). The corresponding dispersive amplitude can be presented in the momentum space as

$$\begin{aligned} T^{(\mathrm{magn}))}_{\pi\gamma} \quad = \quad & \frac{ef\varphi^\dagger_\alpha}{2Mm_\pi} < f \mid \sum_{j,n=1}^{A} \frac{\hat{\tau}_\alpha(j)\,\boldsymbol{\sigma}_j\cdot\boldsymbol{q}\;\hat{\mu}(n)\,\boldsymbol{\sigma}_n\cdot[\boldsymbol{k}\times\boldsymbol{\epsilon}]}{\mathcal{E}_i + E_\gamma - E(\boldsymbol{p}_n+\boldsymbol{k}) - E(\boldsymbol{p}_j) - E(\boldsymbol{p}_k)} + \\ & \frac{\hat{\mu}(n)\,\boldsymbol{\sigma}_n\cdot[\boldsymbol{k}\times\boldsymbol{\epsilon}]\;\hat{\tau}_\alpha(j)\,\boldsymbol{\sigma}_j\cdot\boldsymbol{q}}{\mathcal{E}_i - E_\pi - E(\boldsymbol{p}_n) - E(\boldsymbol{p}_j-\boldsymbol{q}) - E(\boldsymbol{p}_k)} \mid i >, \end{aligned} \qquad (3)$$

where $E(\boldsymbol{p})$ is the nucleon energy in the intermediate state and binding effects are neglected. We note that by performing a multidimensional integration in momentum space we can treat the nucleon momentum dependence in the nuclear propagator exactly without having to resort to a nonrelativistic expansion.

The expression for the dispersive term in Eq.(3) contains matrix elements of one-body as well as two-body operators. The former corresponds to the case $n = j$ and is shown in Fig. 1a-b. This term is identical to the s and u-channels

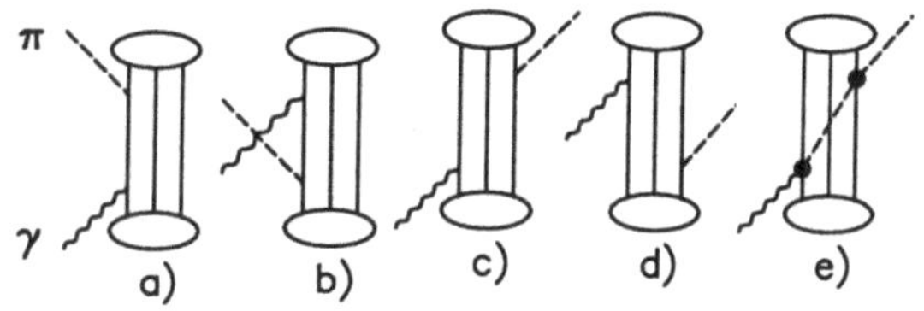

Figure 1. Diagrams for the dispersive and pion rescattering terms in nuclear pion photoproduction

amplitude in pion photoproduction on a single nucleon [3]. The two-body part of Eq.(3) corresponds to the case $n \neq j$ and is shown in Fig.1c-d. This corresponds to a new class of diagrams that do not appear in the IA.

Note that the expression for the dispersive amplitude in pion scattering can be obtained from Eq.(3) by replacing

$$\hat{\mu}(n)\,\boldsymbol{\sigma}_n \cdot [\boldsymbol{k} \times \boldsymbol{\epsilon}]\, e/2M \rightarrow \hat{\tau}_\beta(n)\boldsymbol{\sigma}_n \cdot \boldsymbol{q}'\, \varphi_\beta\, f/m_\pi$$

In Eq.(3) the energy difference $E_\gamma - E_\pi$ is only a few MeV (in the c.m frame), therefore, to leading order the s- and u-channel propagators cancel each other. As had been found before by Levchuk and Shebeko [4], this cancellation is exact for free nucleons in the initial and final states. In our case, however, the nucleons are bound before and after the reaction process, leading to nonzero results. In a study of elastic pion-deuteron scattering Jennings [8] found that the difference between the s- and u-channel contribution is about $6-8\%$ at backward angles; this large cancellation resolved the so-called P_{11} problem. For pion elastic scattering on the trinucleon we find this difference to be closer to 20%, however, as we demonstrate below, this is enough to produce large effect in pion photoproduction and pion scattering in the high momentum transfer region.

3 Results and Discussions

We begin our discussion by considering pion rescattering effects. Within a multiple scattering framework [5] we have studied pion scattering and photoproduction on ^{3}He in detail in refs.[1, 2]. In Fig.2a we show results of our previous work, comparing a PWIA calculation without any pion rescattering with a DWIA computation with full pion-nucleus final state interaction including single charge exchange. The latter one describes the data well up to $Q^2 \approx 6\,\mathrm{fm}^{-2}$. Since the standard multiple scattering framework contains contributions from the trinucleon ground state only, we have estimated the additional contributions from the coupling to the break-up channels using closure approximation. However, comparing the dashed and dash-dotted curves in Fig.2a, it is clear that the disagreement at high momentum transfer can not be improved significantly by contributions from pion rescattering alone.

This situation changes dramatically, once the novel two-body mechanisms, shown in Fig. 1c and d, are taken into account. As shown in Fig. 2a, including

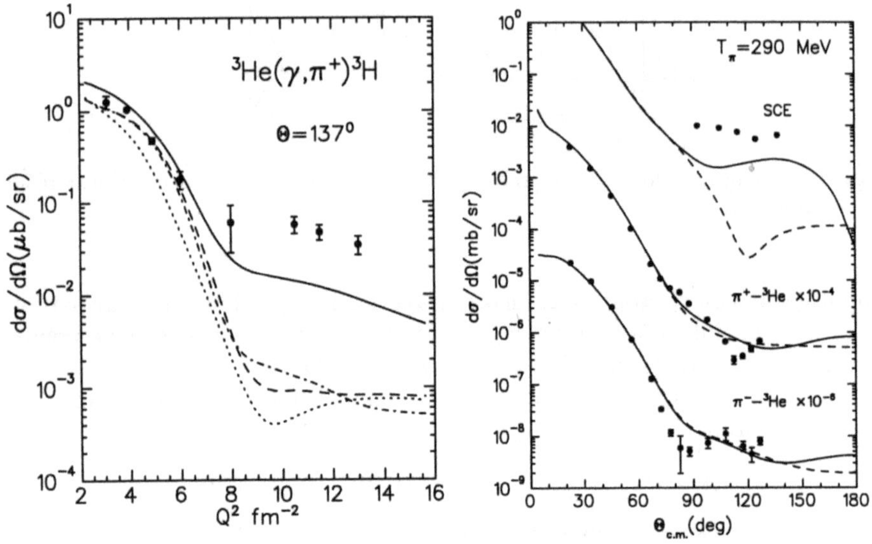

Figure. 2 (a) Differential cross section at $\Theta_{\mathrm{c.m.}} = 137°$ as a function of nuclear momentum transfer Q^2. The dotted (dashed) curves show the PWIA (DWIA) results obtained with Faddeev wave functions [7]. The dash-dotted curve includes the corrections due to the coupling with the break-up channels and the full line shows our complete calculation with the additional novel two-body contribution of Fig. 1c,d. The experimental data are from ref.[6]. (b) Differential cross section for pion scattering on ^{3}He at $T_\pi = 290$ MeV, $\Theta_{\mathrm{c.m.}} = 137°$. The solid and dashed curves are the results obtained with and without novel two-body correction. The experimental data as quoted in ref.[2]

these two-body amplitudes of Eq.(3) within the multiple scattering framework raises the cross section by more than an order of magnitude, thus, these two-body mechanisms in fact become dominant for $Q^2 > 7\ \mathrm{fm}^{-2}$. This effect removes most of the discrepancy between theory and experiment. As mentioned above, even as the leading terms of propagators cancel each other, the next higher-order term (which is about 20% of the leading term) becomes significant in the high momentum transfer region. Our analysis indicates that this large enhancement comes mainly from the isovector magnetic interaction of the two-body operator.

In Fig. 2b we illustrate the importance of the two-body mechanisms in pion scattering on ^{3}He at pion kinetic energie $T_\pi = 290$ MeV. In the elastic channel the corresponding effect is not so dramatic. However, in the pion charge exchange reaction the two-body mechanisms increase the differential cross section at $90° < \theta_\pi < 150°$ by up to two orders of magnitude.

4 Conclusion

In conclusion, we have studied the dramatic disagreement between high-Q (γ, π^+) data on ^{3}He and DWIA calculations that underpredict these data by up to two orders of magnitude. Most of the disagreement with the measurements is removed by explicitly including the dominant two-body mechanisms that go beyond the Impulse Approximation. These two-body terms, where the photon is absorbed on one nucleon and the pion is emitted from another, allow momentum transfer sharing between the two nucleons through the nuclear wave function and dominate the cross section for $Q^2 > 7\,\text{fm}^{-2}$. The most important contribution of these two-body currents are found to be due to the magnetic interaction. All other two-body mechanisms require additional meson exchange interactions between two nucleons and should be considered as higher order terms included in the multiple scattering series. This has been illustrated in our pion rescattering calculation.

From our derivation it is clear that the novel two-body mechanisms are not a special feature of the $^3\text{He}(\gamma, \pi^+)^3\text{H}$ process but play an important role in pion charge exchange on ^{3}He as well. Again we found that the inclusion of the proper two-body terms can resolve the long-standing disagreement of two orders of magnitude between theory and experiment.

Acknowledgement. This work was supported by the Deutsche Forschungsgemeinschaft (SFB201), the U.S. DOE grant DE-FG02-95-ER40907 and the Heisenberg-Landau program.

References

1. S.S. Kamalov, L. Tiator, C. Bennhold: Few-Body Systems **10**, 143 (1991)
2. S.S. Kamalov, L. Tiator, C. Bennhold: Phys. Rev. **C47**, 941 (1993)
3. I. Blomqvist and J.M. Laget: Nucl. Phys. **A280**, 405 (1977)
4. L.G. Levchuk and A.V. Shebeko: Sov. J. Nucl. Phys. **50**, 607 (1989)
5. A.K. Kerman, H. McManus, R.M. Thaler: Ann. Phys. (N.Y.) **8**, 551 (1955)
6. D. Bachelier et al.: Phys. Lett. **B44**, 44 (1973); Nucl. Phys. **A251**, 433 (1975)
7. R.A. Brandenburg, Y.E. Kim, A. Tubis: Phys. Rev. **C12**, 1368 (1975)
8. B.K. Jennings: Phys. Lett. **B205**, 187 (1988)

Few-Body Systems Suppl. 9, 303–306 (1995)

Few-Body Systems

Extended R.G.M. calculations of pion-pion scattering

R. Ceuleneer, C. Semay*

Faculté des Sciences, Université de Mons-Hainaut, B-7000 MONS, Belgium

Abstract. Pion-pion scattering is investigated in the framework of the Resonating Group Method. The pion is described as a quark-antiquark system. Its wavefunction is expanded in extended harmonic oscillator bases. The kernel of the integro-differential equation describing the relative motion of the pions and the associated local effective pion-pion potential are calculated without any simplifying assumption.

Hadron-hadron scattering has extensively been investigated in the framework of the non-relativistic quark model. The Resonating Group Method (RGM) is the natural way to tackle this problem. One-channel RGM calculations lead to a single one-body Schrödinger equation of the form

$$-\frac{1}{2\mu}\nabla_R^2\psi(\boldsymbol{R}) + \int K(\boldsymbol{R},\boldsymbol{R}')\psi(\boldsymbol{R}\,')d^3R' = E\psi(\boldsymbol{R}) \tag{1}$$

where $\psi(\boldsymbol{R})$ describes the relative motion of two complex particles. The precise evaluation of the kernel $K(\boldsymbol{R},\boldsymbol{R}\,')$ is generally very hard. Therefore most RGM calculations rely on drastic approximations which consist, for instance, in oversimplifying the wave functions describing each complex particle taken separately.

To investigate the effects of this type of approximation we have performed an extended RGM calculation of pion-pion scattering. We fully realize that the pion, and accordingly pion-pion scattering, should be treated relativistically since the pion mass ($m_\pi \simeq 140$ MeV) is much smaller than twice the mass of the lightest constituent quarks ($2m_u = 2m_d \simeq 650$ MeV). However $q^2\bar{q}^2$ is the simplest multiquark system which can be separated in two colour singlets and, in this respect, it appears as a convenient testing ground to study the numerical aspects of RGM calculations concerning more complicated multiquark systems, as those that aim at determining the nucleon-nucleon potential from the quark-quark interaction.

*Chercheur qualifié F.N.R.S.

In our calculations the pion is described as a non-relativistic quark-antiquark system using the two-body interaction

$$V_{ij} = -\frac{3}{16}\tilde{\lambda}_i.\tilde{\lambda}_j V(r_{ij}) \tag{2}$$

The matrices $\tilde{\lambda}_i$ are the SU(3) colour generators of the ith particle. $V(r_{ij})$ in GeV is given by

$$V(r_{ij}) = -\frac{0.583}{r_{ij}} + 0.169\, r_{ij} - 0.827 + \frac{0.296}{m_i m_j}\exp(-0.225\, r_{ij}^2)\boldsymbol{s}_i\boldsymbol{s}_j \tag{3}$$

with interparticle distance r_{ij} measured in GeV^{-1}. For masses fixed at the values $m = m_u = m_d = 0.324$ GeV and $m_s = 0.590$ GeV this potential describes quite well the spectra of a great variety of mesons and baryons in extended harmonic oscillator bases [1].

The RGM wavefunction describing a system of two s-wave mesons composed of u and d quarks and their antiparticles is given by

$$\Psi = \Phi(1\bar{1}2\bar{2}) + \Phi(2\bar{2}1\bar{1}) - \Phi(1\bar{2}2\bar{1}) - \Phi(2\bar{1}1\bar{2}) \tag{4}$$

with

$$\Phi(1\bar{1}2\bar{2}) = \sum_c \Phi_c(1\bar{1}2\bar{2})\frac{f_c(R)}{R} \tag{5}$$

and

$$\Phi_c(1\bar{1}2\bar{2}) = \mathcal{C}(1\bar{1})\mathcal{C}(2\bar{2})\phi(r_{1\bar{1}})\phi(r_{2\bar{2}})[Y_l(\hat{R})S(1\bar{1}2\bar{2})]^J_M T(1\bar{1}2\bar{2}) \tag{6}$$

where $\mathcal{C}(1\bar{1})$ and $\mathcal{C}(2\bar{2})$ represent colour singlets. The vectors $\boldsymbol{r}_{1\bar{1}} = \boldsymbol{r}_1 - \boldsymbol{r}_{\bar{1}}$ and $\boldsymbol{r}_{2\bar{2}} = \boldsymbol{r}_2 - \boldsymbol{r}_{\bar{2}}$ are the intrinsic spatial coordinates and $\boldsymbol{R}$ the relative separation of the two clusters

$$\boldsymbol{R} = \frac{1}{2}(\boldsymbol{r}_1 + \boldsymbol{r}_{\bar{1}} - \boldsymbol{r}_2 - \boldsymbol{r}_{\bar{2}}) \quad ; \quad R = |\boldsymbol{R}| \quad ; \quad \hat{R} = \frac{\boldsymbol{R}}{R} \tag{7}$$

The intrinsic wavefunctions $\phi(r_{1\bar{1}})$ and $\phi(r_{2\bar{2}})$ of the clusters are expanded in radial harmonic oscillator bases up to $N\hbar\omega$ oscillator excitation energy as follows

$$\phi(r_{i\bar{i}}) = \sum_{n=0}^{N/2} C(n) R_{n0}(r_{i\bar{i}}) \tag{8}$$

The square bracket in Eq. (6) stands for angular momentum coupling. The spin wavefunction is taken as

$$S(1\bar{1}2\bar{2}) = [[s^{\frac{1}{2}}(1)s^{\frac{1}{2}}(\bar{1})]^{s_{1\bar{1}}}[s^{\frac{1}{2}}(2)s^{\frac{1}{2}}(\bar{2})]^{s_{2\bar{2}}}]^S \tag{9}$$

where $s^{\frac{1}{2}}$ is the spin wavefunction of a single quark. In the same way

$$T(1\bar{1}2\bar{2}) = [[\tau^{\frac{1}{2}}(1)\tau^{\frac{1}{2}}(\bar{1})]^{t_{1\bar{1}}}[\tau^{\frac{1}{2}}(2)\tau^{\frac{1}{2}}(\bar{2})]^{t_{2\bar{2}}}]^T \tag{10}$$

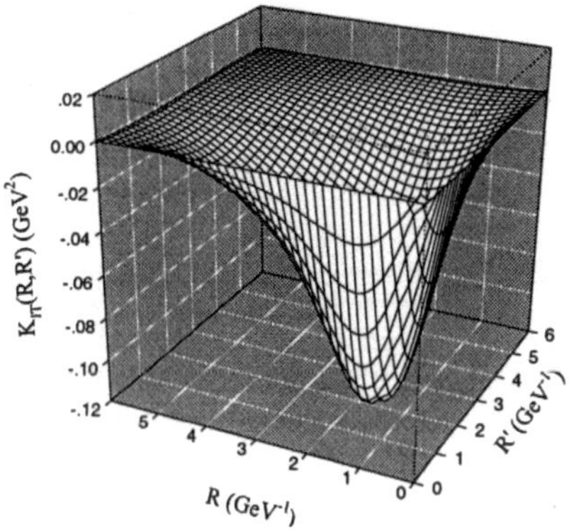

Figure 1. s-wave isoscalar kernel for $N = 0$ and $E = 0.5$ GeV.

denotes the isospin wavefunction.

In one-channel RGM calculations of pion-pion scattering the sum (5) is restricted to a single term, since $s_{1\bar{1}} = s_{2\bar{2}} = 0$ and $t_{1\bar{1}} = t_{2\bar{2}} = 1$, so that a partial radial wave is completely characterized by the quantum numbers $l = J$ and $T = 0, 1, 2$ with $l + T$ even. This partial wave satisfies the equation

$$\left(-\frac{1}{2m}\frac{d^2}{dR^2} + \frac{l(l+1)}{2mR^2} - E\right) f_{lT}(R) + \int_0^\infty K_{lT}(R,R') f_{lT}(R') dR' = 0 \quad (11)$$

deduced from the four-body equation

$$\left\langle \Phi^*_{lT}(1\bar{1}2\bar{2}) \left| -\frac{1}{2m}\sum_{i=1}^{4} \nabla^2_{r_i} + \sum_{i<j}^{4} V_{ij} - \mathcal{E} \right| \Psi \right\rangle = 0 \quad (12)$$

where the brackets are used to denote integration over the spin, isospin, colour and spatial variables keeping R constant. Using usual shell-model techniques this integration can be carried out without any approximation.

Owing to the colour dependence of the two-body interaction the components $\Phi(1\bar{1}2\bar{2})$ and $\Phi(2\bar{2}1\bar{1})$ of Ψ contribute only to the ground state energy $2E_m$ of the mesons, while the remaining components yield a symmetrical kernel $K_{lT}(R,R')$ which vanishes with increasing R. Consequently E represents the relative energy $\mathcal{E} - 2E_m$ of the colliding mesons and, therefore, the mass appearing in the kinetic energy and centrifugal terms of Eq. (11) should be equal to m_π instead of $2m$. This inconsistency, which is often overlooked, is characteristic of RGM calculations. For the two-nucleon system it is of little importance as the nucleon mass amounts to about $3m$. For pion-pion scattering it is hoped that despite the inadequacy of Galilean kinematics, the effective local potential extracted from Eq. (11) might bear a resemblance to the genuine meson-meson interaction. This potential is given by

$$V_{lT}(R) = \frac{1}{f_{lT}(R)}\left(\frac{1}{2m}\frac{d^2}{dR^2} - \frac{l(l+1)}{2mR^2} + E\right) f_{lT}(R) \quad (13)$$

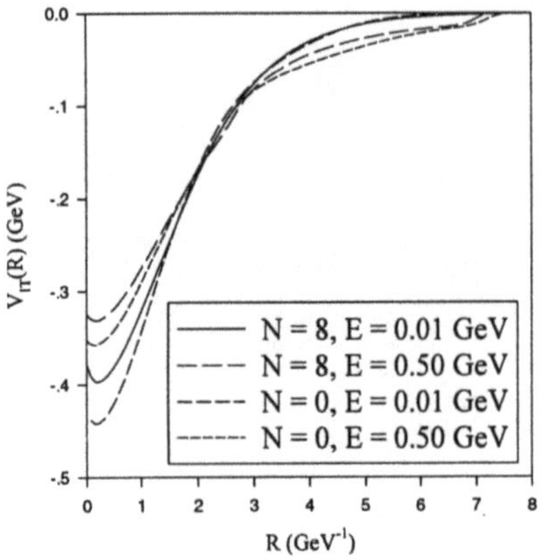

Figure 2. s-wave isoscalar effective potential for indicated values of N and E.

In a first step we have performed a $0\hbar\omega$ calculation using the value 1.60 GeV^{-1} for the oscillator length parameter $b = (\hbar/m\omega)^{\frac{1}{2}}$, which yields the optimum $N = 0$ bound of the theoretical pion mass, namely, 0.195 GeV. The kernels $K_{lT}(R, R')$ and the associated effective potentials were calculated for several l and T values as functions of E; as an example, the s-wave ($l = 0$) isoscalar ($T = 0$) kernel calculated for $E = 0.5$ GeV is displayed in Fig. 1. In a second step we have extended the oscillator basis up to eight quanta of oscillator excitation which, for $b = 1.46$ GeV^{-1}, brings the theoretical pion mass close to the experimental value. The s-wave isoscalar effective potential calculated in the $N = 8$ model space for $E = 0.01$ and 0.5 GeV is compared in Fig. 2 with its $N = 0$ counterpart. It is seen that an important extension of the model space does not modify markedly the $N = 0$ results. We found also that the corresponding variation of the phase shifts is less that 5%. Further analyses using more elaborate two-body forces are in progress.

To conclude we wish to stress that the dependence of the kernel $K_{lT}(R, R')$ upon T is entirely contained in the factor $\langle T(1\bar{1}2\bar{2})|T(1\bar{2}2\bar{1})\rangle$. As this factor is equal to zero for $T = 1$, it is necessary to incorporate isospin-dependent terms in the two-body potential in order to make one-channel RGM calculations suitable for the description of odd waves. Terms of this type might be provided by the meson exchange potentials used recently to derive the spin-orbit component of the nucleon-nucleon interaction [2].

References

1. B. Silvestre-Brac and C. Semay: I.S.N. Grenoble Report 93.69 (1993)

2. A. Valcarce et al.: Phys. Rev. **C51**, 1480 (1995)

Few-Body Systems Suppl. 9, 307–310 (1995)

Few-Body Systems

New experimental results on the $\pi\pi$ interaction in nuclear matter

The CHAOS Collaboration

F. Bonutti[1,2], P. Camerini[1,2], N. Grion[1,2], R. Rui[1,2], P.A. Amaudruz[3], J.F. Brack[3], L. Felawka[3], G.R. Smith[3], G. Hofman[4], M. Kermani[4], S. McFarland [4], K. Raywood[4], M.E. Sevior[4], E.L. Mathie [5], R. Tacik[5], E.F. Gibson[6]

[1] Dipartimento di Fisica dell'Universita' di Trieste, 34127 Trieste, Italy

[2] Istituto Nazionale di Fisica Nucleare, 34127 Trieste, Italy

[3] TRIUMF, Vancouver, B.C., Canada V6T 2A3

[4] Physics Department, University of British Columbia, Vancouver, B.C., Canada V6T 2A6

[5] University of Regina, Regina, Saskatchewan, Canada S4S 0A2

[6] California State University, Sacramento CA 80309, USA

Abstract. Single pion production $A(\pi^+, \pi^+\pi^\pm)$ was investigated in nuclei (^{2}H,^{12}C, ^{40}Ca,^{208}Pb) at an incident pion energy of T_{π^+}=280 MeV. Data were obtained at TRIUMF using the CHAOS spectrometer which detected the two final pions in coincidence. The results presented focus here on two aspects of the $\pi\pi$ reaction in nuclei: the reaction mechanism and the invariant mass behaviour above the $2m_\pi$ threshold.

1 Introduction

Two-pion exchange which couple to the I=J=0,1 quantum numbers account for a considerable fraction of the NN attractive potential at intermediate distances. However, in the nuclear medium, the dynamics of the two correlated pions may be strongly modified [1] by the modifications of the single-pion propagation through its couplings to the nucleons and the Δ-isobar. Recent theoretical calculations on $\pi\pi$ correlations in nuclear matter indicate that the correlated $(\pi\pi)_{I=J=0}$ state builds up considerable strength near the two-pion mass threshold. Furthermore, the strong $\pi\pi$ correlation, which bears the quantum numbers of the σ-meson, may even occur at nuclear densities below the nuclear saturation density [2]. Models of the single pion production reaction, $\pi A \rightarrow \pi\pi A'$, commonly assume that the reaction initiates with the elementary $\pi N \rightarrow \pi\pi N$ process: i.e. with only one nucleon of the nucleus [3].

Two correlated nucleons do not contribute appreciably (~10%) to the pion production cross-section (J. Cohen in ref.[3]). Although these models have a common approach to the production mechanism, they predict very different total cross sections depending on their treatment of pion propagation in nuclear matter (E. Oset in ref.[3]). In order to contribute to the understanding of the dynamics of pion production in the nuclear matter, we have investigated the following reactions:

a) Pion production on deuterium $\pi^+\ {}^2\mathrm{H}\rightarrow \pi^+\pi^- pp$ and $\pi^+\ {}^2\mathrm{H}\rightarrow \pi^+\pi^+ nn$, to understand the reaction mechanism on a nucleon in the absence of nuclear matter. For this nucleus, the pion-induced pion production can be considered a quasi-free process on a nucleon [6].

b) Pion production in light (^{12}C), medium (^{40}Ca), and heavy (^{208}Pb) nuclei, ($A(\pi^+, \pi^+\pi^\pm)A'$) to observe the modification to the $\pi\pi$ interaction due to the strong mean field in nuclear matter.

c) The $\pi^+\pi^+$ reaction channel was also examined to observe the effects of nuclear matter on the I=2 correlated $\pi\pi$ channel.

2 The Experimental Method

The experiment was performed at the TRIUMF Meson Facility on the M11 medium-energy pion channel which, for these measurements, was set to deliver 280 $\pm 4(\sigma)$ MeV positive pions. Final state $\pi^+\pi^-$ and $\pi^+\pi^+$ pairs were analyzed with the CHAOS spectrometer [4] The magnetic spectrometer CHAOS is a cylindrical dipole magnet which identifies charged particle tracks with four co-axial cylindrical wire chambers. CHAOS can achieve a momentum resolution of 1% (σ), however, for these measurements it was 2% (average) since the magnetic field of CHAOS was set as low as 0.5 T in order to detect low energy pions (8 MeV). Particles were mass-identified with three layers of counters: two plastic scintillator counters and one lead-glass Čerenkov counter [5]. This system of counters tightly surrounded the outer wire chamber. With this system and the particle momenta information, almost perfect ($\sim$ 100%) pion-proton discrimination was achieved. Pion-electron discrimination was better than 99% in the momentum range of interest. The pion production reaction was studied in the entire reaction plane, that is $\Delta\Theta \sim 360^o$ and $\Delta\Phi = \pm 7^o$, and the target thicknesses ranged from 0.1 to 0.5 g/cm^2.

3 Experimental Results

In Fig.1, the mass and the (total) energy c undetected proton in the $^2\mathrm{H}(\pi^+, \pi^+\pi^- p)p$ reaction is shown as a basic on the accuracy of the experimental data. The proton mass is reproduc a symmetric distribution

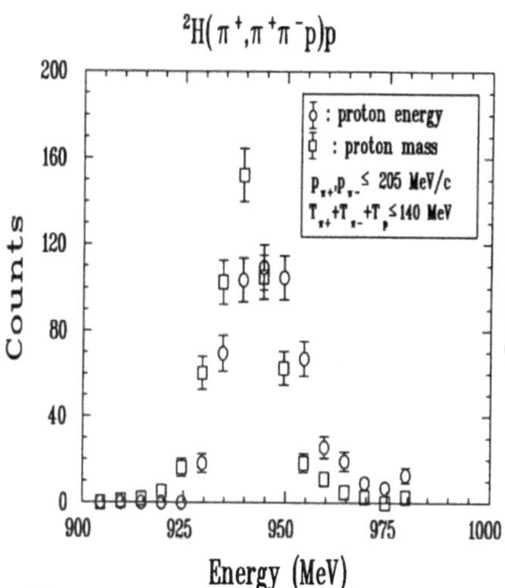

Figure 1. Proton mass (□) and (total) energy (o). The three final particles $\pi^+\pi^-$ and p, are detected in coincidence.

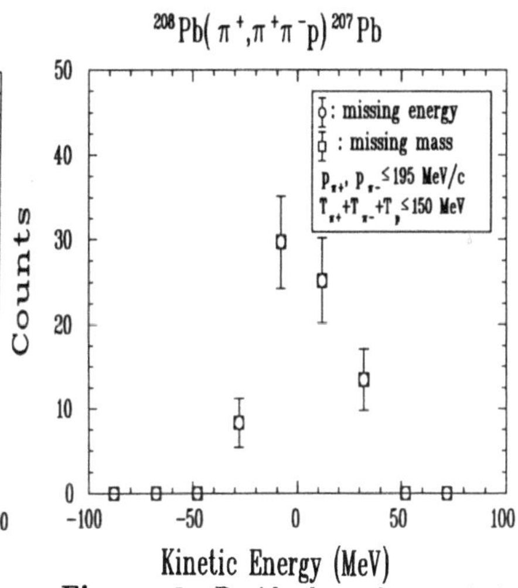

Figure 2. Residual nucleus missing mass (□) and missing (total) energy (o). The ^{208}Pb is assumed to stay in the ground state and the mass subtracted from the energy balance

peaked at 939 MeV and with a $\sigma\sim$ 7MeV, the width being almost entirely due to the pion beam energy spread. These protons had an average kinetic energy of 7 MeV, indicating a spectator role played in the reaction. Figure 2 shows the missing mass and missing (total) energy of the $^{208}Pb(\pi^+,\pi^+\pi^-p)^{207}Pb$ reaction. In this case, the residual nucleus is assumed to lay in the ground state and so the mass is subtracted from the energy balance. The mean value of this mass distribution is $\sim$0 MeV, meaning that the residual nucleus is weakly disturbed by a both the production process and the outgoing particles. The width of the distribution is 18 MeV (σ), which is consistent with the Fermi energy of the interacting nucleon once the beam energy spread is taken into account. Similar results come from the analysis of the other nuclei, ^{12}C and ^{40}Ca, thus supporting the above conclusions. The invariant mass distributions for the $\pi^+\pi^-$ reaction channel are reported in Fig.3. They are normalized to one another at an energy of 350 MeV. The main differences among the distributions is noticeable around the $2m_\pi$ energy threshold. The $\pi^+\pi^-$invariant mass strength depends strongly on the mass number, being almost negligible at threshold for deuterium. A $\pi^+\pi^-$ phase shift analysis is in progress to determine the $\pi\pi$ partial-waves. The behaviour of the measured invariant mass for lead is compared with the prediction (full line) of the model described in ref.[2] (G.Chanfray and Z.Aouissat). The model accounts for the behaviour of two strongly interacting pions in the I=J=0 channel embedded in a nuclear environment of density $\rho\leq\rho_0$ (the nuclear saturation density). The agreement is remarkable in the low-energy region, where the deuterium exhibits little strength. Instead, the $\pi^+\pi^+$ invariant mass distributions for ^{2}H,^{12}C, ^{40}Ca,^{208}Pb display

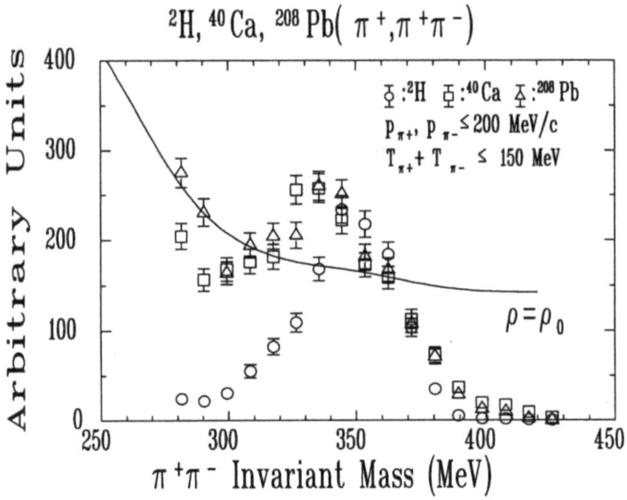

Figure 3. $\pi^+\pi^-$ invariant mass distribution for ^{2}H (o), ^{40}Ca (□) and ^{208}Pb (△). Model calculations (full line) are from ref.[2] (G.Chanfray and Z.Aouissat) which apply to the ^{208}Pb data.

a different behaviour: their shapes roughly overlap in the energy interval 280-420 MeV (not reported here) and follow the predictions of phase-space. The present data support the picture of a σ-like resonance structure for two correlated pions embedded in the nuclear medium. However, this statement is not conclusive since our $(\pi, \pi\pi)$ data require a more exhaustive analysis.

References

1. V. Bernard et al.: Phys. Rev. Lett. **9**, 966 (1987); T. Hatsuda and T. Kunihiro: Progr. Theor. Phys. Suppl. **91**, 284 (1987); C.V. Christov et al.: Nucl. Phys. **A510**, 689 (1990)

2. G. Chanfray et al.: Phys. Lett. **B3**, 325 (1991); V. Mull et al.: Phys.Lett. **B286** , 13 (1992); Z. Aouissat et al.: Nucl. Phys. **A581**, 471 (1995)

3. J.M. Eisenberg: Phys.Lett. **B93**, 12 (1980); R.M. Rockmore: Phys.Rev. **C27**, 2150 (1983); R.M. Rockmore: Phys.Rev. **C29**, 1534 (1984); E. Oset et al.: Nucl. Phys. **A454**, 637 (1986); J. Cohen and J.M. Eisenberg: Nucl. Phys. **A395**, 389 (1983)

4. G.R. Smith et al.: Nucl. Instr. Meth. in Phys. Res. (in print)

5. F. Bonutti et al.: Nucl. Instr. and Meth. in Phys. Res. **A350**, 136 (1994)

6. R. Rui et al.: Nucl. Phys. **A517**, 455 (1990); V. Sossi et al.: Nucl. Phys. **A548**, 562 (1990)

Few-Body Systems Suppl. 9, 311–314 (1995)

Few-Body Systems

Hidden Chiral Symmetry and Low Energy Theorem

J. Smejkal, E. Truhlík*

Institute of Nuclear Physics, Czech Academy of Sciences,
CZ-250 68 Řež n. Prague, Czech Republic

Abstract.
The low-energy theorem for pion electroweak production amplitude is derived within the framework of the hidden chiral symmetry approach.

1 Introduction

Processes with π, ρ and A_1 mesons were intensively studied soon after formulating current algebra [1]. The aim was to extend the validity of the electroweak interaction with hadrons from threshold towards higher energies. It was understood subsequently that an effective method for the reproduction of the current algebra results and for the extension to intermediate energies region is the phenomenological Lagrangian method based on the $SU(2)_L \times SU(2)_R$ group [2, 3, 4], where the vector mesons ρ and A_1 are considered as the so-called massive Yang-Mills (YM) [5] gauge fields.

The effective Lagrangians [2, 3] of the $\pi\rho A_1$ system contain generally terms with three and four field derivatives which leads to the appearance of arbitrary parameters. This arbitrariness was eliminated in [4] by the requirement that the effective Lagrangian of the $\pi\rho A_1$ system would contain no more than two field derivatives in each term.

The hard pion Lagrangian [4] was used in [6, 7, 8] as a starting point for constructing the electroweak nuclear meson exchange currents (MECs). It was found that the theory predicts the existence of a transverse MEC $j_\mu^{A_1\pi}$ of the pion range, a part of which is fixed by the low-energy theorem. Then question arises, how model dependent is the rest of the current $j_\mu^{A_1\pi}$ [9].

Actually, the MECs are closely related to the corresponding pion production amplitudes [6, 7, 10]. Consequently, the model dependence of MECs can be equivalently studied on these amplitudes.

*Supported by grants Nos. 148410 (Grant Agency of the Czech Academy of Sciences) and 202/94/0370 (Grant Agency of the Czech Republic).

Another approach to study the problems discussed above is the hidden local symmetry (HLS) scheme [11, 12]. In this scheme, the vector mesons (gauge bosons) are generated dynamically.

Our aim is to compare the content of the low-energy theorems in both approaches and to study the model dependence of the amplitudes.

2 Lagrangians and Currents

We constructed Lagrangian for the meson-nucleon system within the framework of the HLS method [11] as an analogue of the corresponding YM Lagrangian. We found that the difference between these Lagrangians contains only various contact terms (see [9] for more details). Electroweak currents [9], generated from the Lagrangians by the Gell-Mann-Levy method, were subsequently used for the construction of the pion production amplitudes.

3 Low-Energy Theorems

We first considered the pion production from the nucleon by the axial-vector current:

$$N(p_1) + J^j_{\mathrm{A},\lambda}(k) \longrightarrow N(p_2) + \pi^n(q) \ . \tag{1}$$

The Born part of the amplitude is the same in both approaches. Consequently, we will be interested in the non-Born part $\overline{M}_{\mathrm{A},\lambda}$ only.

In the limit $k = 0$, we can write,

$$\begin{aligned} \overline{M}_{\mathrm{A},\lambda}(q^2, k = 0) \;=\; & \mathrm{i}\,\overline{u}(p_2)\,\{\, a^{(+)}_{nj}\,\gamma(q^2)\,q_\lambda \;+ \\ & a^{(-)}_{nj}\,[\,\mathrm{i}\,\beta'(q^2)\,\sigma_{\lambda\nu}\,q_\nu \;-\; 2\mathrm{i}\,\alpha'(q^2)\,M\gamma_\lambda\,]\}\,u(p_1) \ , \end{aligned} \tag{2}$$

where

$$a^{(\pm)}_{nj} \;=\; \frac{1}{4}\,[\,\tau_n, \tau_j\,]_\pm \ . \tag{3}$$

Apart from the Δ-isobar current contribution we obtain (see [6]):

$$\begin{aligned} \alpha'^{\,\mathrm{YM}}(q^2) \;&=\; \frac{1}{2Mf_\pi}\left[1 - g_{\mathrm{A}}^2 - \frac{q^2}{q^2 + m_\rho^2}\right] \ , \qquad \gamma^{\,\mathrm{YM}}(q^2) \;=\; 0 \ , \\ \beta'^{\,\mathrm{YM}}(q^2) \;&=\; -\,\frac{\kappa_{\mathrm{V}}}{2Mf_\pi}\,\frac{m_\rho^2}{q^2 + m_\rho^2} \ . \end{aligned} \tag{4}$$

On the other side, we have from current algebras and PCAC ([13]),

$$\alpha'(0) \;=\; \frac{1}{2Mf_\pi}\,(1 - g_{\mathrm{A}}^2) \ . \tag{5}$$

As we can see, this equation is fulfilled by $\alpha'(0)$ from Eq. (4).

Analogously, we can obtain from the HLS Lagrangian,

$$\alpha'^{\,\mathrm{HLS}}(q^2) = \frac{1}{2Mf_\pi}\left[1 - g_{\mathrm{A}}^2 - \frac{3}{2}\frac{q^2}{q^2+m_\rho^2}\right] , \qquad \gamma^{\,\mathrm{HLS}}(q^2) = 0 ,$$

$$\beta'^{\,\mathrm{HLS}}(q^2) = -\frac{\kappa_{\mathrm{V}}}{2Mf_\pi}\left[\frac{m_\rho^2}{q^2+m_\rho^2} + \frac{1}{2}\frac{q^2}{q^2+m_\rho^2}\right] . \tag{6}$$

Generally, the functions $\beta'^{\,\mathrm{HLS}}(q^2)$ and $\alpha'^{\,\mathrm{HLS}}(q^2)$ differ from the analogous functions of Eqs. (4), but they have the same value for $q^2 = 0$. Consequently, Eq. (5) is fulfilled also in this case.

In the limit $q = 0$, we can write the decomposition similar to Eq. (2):

$$\begin{aligned}\overline{M}_{\mathrm{A},\lambda}(q=0,k^2) &= \mathrm{i}\,\overline{u}(p_2)\,\{\,a_{nj}^{(+)}\,\rho(k^2)\,k_\lambda \,+ \\ &\quad a_{nj}^{(-)}\,[\,\mathrm{i}\,\lambda(k^2)\,\sigma_{\lambda\nu}\,k_\nu \,-\, 2\mathrm{i}\,\alpha'(k^2)\,M\gamma_\lambda\,]\}\,u(p_1) \; .\end{aligned} \tag{7}$$

In this case, current algebras and PCAC give us directly limits for the decomposition functions (7) (see [6]),

$$\rho(k^2) = -\frac{1}{f_\pi}\,g_{\mathrm{A}}\,g_{\mathrm{P}}(k^2) , \qquad \lambda(k^2) = \frac{1}{f_\pi}\,F_2^{\mathrm{V}}(k^2) ,$$

$$\alpha'(k^2) = \frac{1}{2Mf_\pi}\,[\,F_1^{\mathrm{V}}(k^2) \,-\, g_{\mathrm{A}}\,g_{\mathrm{A}}(k^2)\,] \; . \tag{8}$$

The decomposition functions are expressed in (8) in terms of form factors F_1^{V}, F_2^{V}, g_{A} and g_{P}. It is possible to show that both sets of model functions obey Eqs. (8).

We further considered the pion production from the nucleon by the vector current:

$$N(p_1) \,+\, J^j_{\mathrm{V},\lambda}(k) \;\longrightarrow\; N(p_2) \,+\, \pi^n(q) \; . \tag{9}$$

Now we can parametrize the corresponding non-Born amplitude in the limit $q = 0$ as follows [14],

$$\begin{aligned}\overline{M}_{\mathrm{V},\lambda}(q=0,k^2) &= \overline{u}(p_2)\,\{\,a_{nj}^{(+)}\,\varphi(k^2)\,\gamma_5\,\sigma_{\lambda\mu}\,k_\mu \,+ \\ &\quad a_{nj}^{(-)}\,[\,B(k^2)\,\gamma_5\,\gamma_\lambda \,-\, \mathrm{i}\,\kappa(k^2)\,\gamma_5\,k_\lambda\,]\}\,u(p_1) \; .\end{aligned} \tag{10}$$

Current algebras and PCAC give us for the form factors $\varphi(k^2)$, $B(k^2)$ and $\kappa(k^2)$ equations:

$$\varphi(k^2) = -\frac{g_{\mathrm{A}}}{2Mf_\pi}\,F_2^{\mathrm{V}}(k^2) , \qquad B(k^2) = -\frac{1}{f_\pi}\,[\,g_{\mathrm{A}}(k^2) \,-\, g_{\mathrm{A}}\,F_1^{\mathrm{V}}(k^2)\,] ,$$

$$\kappa(k^2) = \frac{2M}{f_\pi}\,\frac{1}{k^2}\,[\,g_{\mathrm{A}}(k^2) \,-\, g_{\mathrm{A}}\,F_\pi^{\mathrm{V}}(k^2)\,] \; . \tag{11}$$

It can be shown that the form factors φ, B and κ obtained in both models fulfil equations given above.

As to the non-Born amplitude in the limit $k = 0$, it satisfies the following equation for both models:

$$\overline{M}_{\mathrm{V},\lambda}(q^2, k=0) = 0 \; . \tag{12}$$

4 Conclusions

Although both models (YM and HLS) give different amplitudes of the pion production from the nucleon by the vector or axial-vector currents in general, all requirements folowing from current algebra and PCAC are fulfilled.

We have found that the difference between the amplitudes in both models is given by terms containing the product $k \cdot q$. These terms will influence the calculations of observables at higher energies. Their effect on the differential cross sections of the reaction

$$e + d \longrightarrow e' + np \tag{13}$$

was studied in [9] and it was shown that it is not negligible.

We conclude that both model Lagrangians are admissible for use in describing the nuclear phenomena at intermediate energies. The advantage of the HLS approach is that it is based on a formally firm ground.

References

1. H.J. Schnitzer, S. Weinberg: Phys. Rev. **164**, 1828 (1967); J.G. Gerstein, H.J. Schnitzer: Phys. Rev. **170**, 1638 (1968)
2. J. Schwinger: Phys. Lett. **24B**, 473 (1967); J. Wess, B. Zumino: Phys. Rev. **163**, 1727 (1967)
3. P. Chang, F. Gürsey: Phys. Rev. **164**, 1752 (1967); **169**, 1397 (1968); S. Weinberg: Phys. Rev. Lett. **18**, 507 (1967); Phys. Rev. **166**, 1568 (1968); B.W. Lee, H.T. Nieh: Phys. Rev. **166**, 1507 (1968)
4. V.I. Ogievetsky, B.M. Zupnik: Nucl. Phys. **B24**, 612 (1970)
5. C.N. Yang, R. Mills: Phys. Rev. **96**, 191 (1954)
6. E. Ivanov, E. Truhlík: Nucl. Phys. **A316**, 437 (1979)
7. J. Adam, Jr., E. Truhlík: Czech. J. Phys. **B34**, 1157 (1984)
8. E. Truhlík, J. Adam, Jr.: Nucl. Phys. **A492**, 529 (1989)
9. J. Smejkal, S. Platchkov, E. Truhlík: in Proceedings of the International Conference "*Few-Body XV*", Peñiscola, Spain (1995)
10. E. Truhlík: Czech. J. Phys. **43**, 467 (1993)
11. M. Bando, T. Kugo, K. Yamawaki: Phys. Reports **164**, 217 (1988)
12. U.-G. Meißner: Phys. Reports **161**, 213 (1988)
13. S.L. Adler: Phys. Rev. **140**, 736 (1965); W.I. Weisberger: Phys. Rev. **143**, 1302 (1966)
14. E. Ivanov, E. Truhlík: Fiz. Elem. Chastits At. Yadra **12**, 492 (1981); (english translation) Sov. J. Part. Nucl. **12**, 198 (1981)

Few-Body Systems Suppl. 9, 315–318 (1995)

Few-Body Systems

High Resolution Measurements of (γ, N) at Intermediate Energy. How Important are Meson Exchange Current Effects ?

J.R.M. Annand[1*], J-O. Adler[2], B-E. Andersson,[2] S.A. Bulychjev[2†], G.I. Crawford[1], P.D. Harty[1], L. Isaksson[2], J.C. McGeorge[1], G.J. Miller[1], J. Ryckebusch[3], H. Ruijter[2], B. Schrøder[2]

[1] Department of Physics and Astronomy, University of Glasgow, G12 8QQ, SCOTLAND.

[2] Institute of Physics, University of Lund, Sölvegatan 14, S-223 62 Lund, SWEDEN.

[3] Institute for Nuclear Physics, Proeftuinstraat 86, B-9000, Gent, BELGIUM.

Abstract. Tagged photon measurements of (γ, N) reaction cross sections are compared with the predictions of various, recent microscopic models. The implications for the different theoretical descriptions are discussed, notably the dependence of the (γ, N) cross section on meson exchange currents and final-state rescattering.

At intermediate energy, $E_\gamma \sim 50 - 150$ MeV, (γ, N) reactions provide a relatively clean and potentially illuminating probe of nuclear structure. At these energies the photon interacts with one and two-body nuclear currents, probing in the first instance high-momentum components of the single-particle wave-function, due to the large momentum mismatch between incoming photon and outgoing nucleon. This was an early goal of (γ, p) measurements [1], but after the first bremsstrahlung measurements of (γ, n) [2] it became clear that $\sigma_{\gamma,n} \sim \sigma_{\gamma,p}$, which is inconsistent with the quasi-free knockout (QFK) picture. Absorption on a correlated $p-n$ pair obviates the need for a high initial-state momentum and quasi-deuteron (QD) processes feed the (γ, N) channels when one nucleon is reabsorbed. At the present energies and momentum transfers the coupling to one-pion-exchange currents would be expected to be the dominant component of two-nucleon absorption.

*Corresponding Author: tel. +44 141 330 6428, E-mail jrma@np.ph.gla.ac.uk
†Present Address: ITEP, Moscow 117259, Russia

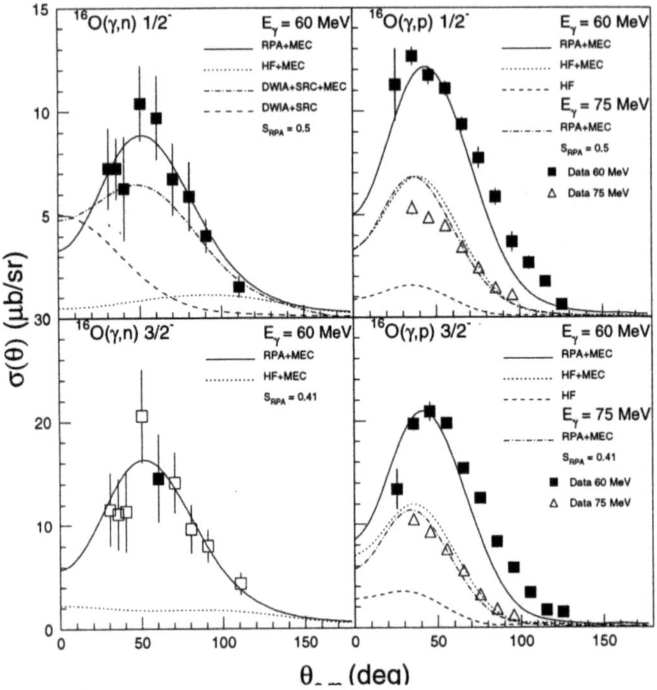

Figure 1. Measured differential cross sections and theoretical calculations for the *1h*, A=15 states.

The experiments are taking place at the tagged photon facility of MAX-lab in Lund, Sweden, which offers very good energy resolution ($\Delta E_\gamma = 300$ keV) and a well-known beam intensity. Targets of mass A = 4–40 have been measured, but this contribution will concentrate mainly on ^{16}O for which the most complete set of data has been taken [3, 4] and for which several theoretical calculations are available. Of the low-lying states in the A = 15 system resolved and measured in the present work, the $\frac{1}{2}^-$ (ground) and $\frac{3}{2}^-$ (6.2 MeV) states have substantial hole (*1h*) components, which may be populated by both QFK and QD processes. The $\frac{1}{2}^+, \frac{5}{2}^+$ doublet at $\sim$ 5.2 MeV is largely two-hole-one-particle (*2h1p*) in character and thus is not accessible via simple knockout.

Differential cross sections for ^{16}O(γ, N) reactions to the $\frac{1}{2}^-$ and $\frac{3}{2}^-$ states are displayed in Fig.1 where they are compared with the results of coupled-channels calculations, which were performed within a consistent, continuum Hartree-Fock-RPA framework [5]. As the states were assumed to be pure *1h*, the calculations have been multiplied by spectroscopic factors (S_{RPA}, Fig.1), extracted from an equivalent analysis of ^{16}O$(e, e'p)$ data. In the description of the initial photoabsorption mechanism the current operator, obtained by minimal

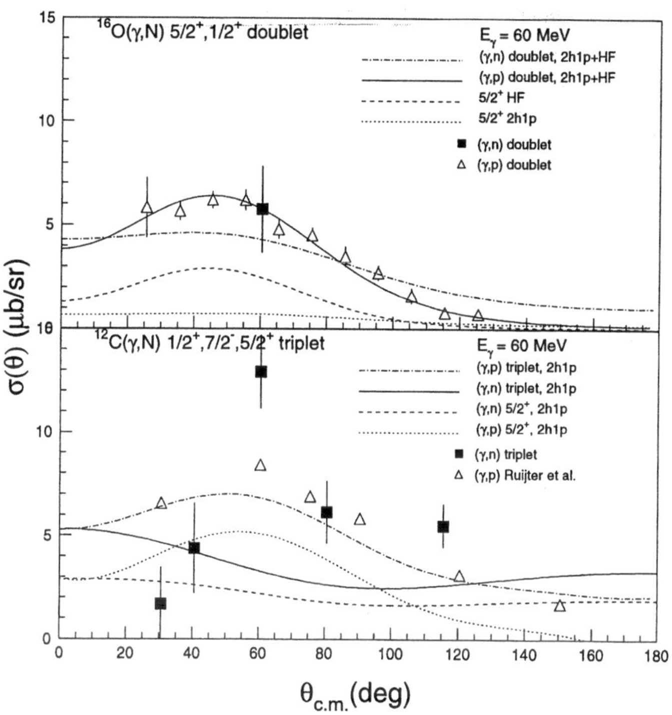

Figure 2. Measurements and calculations for (γ, N) reactions to largely *2h1p* states

substitution in the Hamiltonian, includes both one-body and two-body terms. The latter was essentially determined by the momentum-dependent-force component of the extended Skyrme-type effective $N-N$ interaction also used for the coupled channels calculations. The full calculation is labelled RPA+MEC in Fig.1, but to assess the importance of final-state rescattering and coupling to collective modes of the nucleus, equivalent direct-knockout calculations were made using the same Skyrme interaction, with one (HF, Fig.1) and two-body (HF+MEC, Fig.1) current terms included.

The other calculations shown in Fig.1 use an extension of the Pavia DWIA approach [6]. Here short-range correlation effects are crudely modelled in the nuclear wavefunction and the current operator has a two-body part, derived from the one-pion-exchange potential, where only the seagull term has been retained. Calculations with and without the two-body term are labelled DWIA+SRC+MEC and DWIA+SRC in Fig.1 respectively. The extended DWIA calculation gives a fair description of the (γ, n_0) cross section, and predicts a dominant MEC effect, in disagreement with the HF-RPA calculation. However the former lacks the degree of self-consistency of the latter and is extremely sensitive to the choice of optical potential.

Differential cross sections for the largely *2h1p* states excited in $^{16}O(\gamma, N)$ and $^{12}C(\gamma, N)$ (the unresolved triplet of states at $\sim$ 7 MeV excitation) are displayed in Fig.2. The calculations come from a direct knockout model [7] (RPA handles only *1h* states) where MEC effects are assumed to be fully responsible for the observed cross section, these being implemented through a non-relativistic reduction of two-body currents derived from the one-pion-exchange potential. Seagull and pion-in-flight terms are included. In the case of $^{12}C(\gamma, N)$ the difference in shape of the (γ, n) [8] and (γ, p) [9] differential cross sections is not reproduced by the MEC calculations [7]. which reproduce the integrated triplet strength in (γ, p) quite well, but fail for (γ, n). The $\frac{5}{2}^+$ is predicted to be the dominant member, rather at odds with (γ, p) measurements [9] where the observed centroid of the unresolved peak sits at $\sim$ 6.8 MeV excitation. $^{12}C(\gamma, p\gamma')$ measurements performed in Lund in June 1995 will attempt to resolve the states and thus shed more light on the problem.

In the case of $^{16}O(\gamma, N)$ reactions to the $\frac{5}{2}^+, \frac{1}{2}^+$ doublet the explicit MEC calculations [3] (*2h1p* Fig.2, note $\sigma_{\frac{1}{2}^+} \ll \sigma_{\frac{5}{2}^+}$) fall far below the measured differential cross sections. However the ground state of ^{16}O is impure, having an admixture of the *2s1d*-shell from which the doublet states can be reached by QFK or 2N absorption where one nucleon returns to its original orbital. Processes of this type were calculated (HF Fig.2) using the model described above (HF+MEC in Fig.1), and when added coherently to the MEC calculation (*2h1p*+HF Fig.2) a reasonable description of the data is obtained. Clearly a consistent description of (γ, N) to the largely *2h1p* states in the A=11 and A=15 systems remains to be achieved.

References

1. D.J.S. Findlay, R.O. Owens: Nucl. Phys. **A292**, 53 (1977)
2. H. Göringer, B. Schoch, G. Lührs: Nucl. Phys. **A384**, 414 (1982)
3. G.J. Miller et al. : Nucl. Phys. **A586**, 125 (1995)
4. B.-E. Andersson et al. : Phys. Rev. C **51**, 2553 (1995)
5. J. Ryckebusch et al. : Nucl.Phys. **A476**, 237 (1988)
6. G. Benenti et al. : Nucl. Phys. **A574**, 716 (1994)
7. J. Ryckebusch et al. : Phys. Rev. C **46**, R829 (1992)
8. J.R.M. Annand et al. : Phys. Rev. Lett. **71**, 2703 (1993)
9. H. Ruijter et al. : Private Communication

Few-Body Systems Suppl. 9, 319–323 (1995)

Few-Body Systems

A zero-degree spectrometer in CELSIUS and the $d(d,2\pi)^4$He reaction

Chr. Bargholtz[1], K. Fransson[1], L. Holmberg[1], K. Lindh[1], D. Protic[2], L. Sandberg[1], I. Sitnikova[3], P.-E. Tegnér[1], P. Thörngren Engblom[1], G. Weiss[1], K. Wilhelmsen Rolander[1]

[1] Department of Physics, Stockholm University, Box 6730, S-113 85 Stockholm, Sweden

[2] Institut für Kernphysik, Forschungszentrum Jülich, Postfach 1913, D-52425 Jülich, Germany

[3] Department of Nuclear Physics, St. Petersburg University, 198904 St. Petersburg, Russia

Abstract.

For the realization of near threshold studies a small-size spectrometer has been developed by the Nuclear Physics Division at Stockholm University in conjunction with Institut für Kernphysik in Jülich. A particle telescope is mounted inside the CELSIUS vacuum chamber in the bend following the cluster-jet target. It is possible to vary the measuring position within the dipole field to cover different magnetic rigidities which makes it a versatile tool for studies of threshold reactions. The first aim has been to study two-pion production and we present preliminary results from measurements of the $d(d,2\pi)^4$He reaction.

1 Introduction

The search for resonance-like two-pion states has been going on for the last three decades. Still remains to be clarified, both experimentally and theoretically, the origin of the enhancement in the cross-section for meson production in the isospin zero channel, the so called ABC (Abashian, Booth and Crowe) effect first discovered in the reaction $p(d,X)^3$He [1, 2].

The model with two intermediate Δ resonances decaying back to back, or both forward, in the center of mass, reproduces the peaks at maximum and minimum missing mass, respectively, at 0° in the laboratory, but fails at larger angles [3, 4, 5]. Among other proposed mechanisms is the excitation of two gluon condensates [6].

It has been suggested that one possible way to settle this issue is to measure the invariant mass spectrum from the inclusive reaction $d(d,X)^4$He at different beam energies [7]. In 1976 Banaigs et al. [8] reported on extensive measurements of this reaction at kinetic energies ranging from 0.8 to 2.4 GeV, showing that the production of ^{4}He was dominated by the ABC peak for kinetic energies up to 1.94 GeV. At higher energies the effect started to disappear. Curves representing the cross section obtained from phase space showed no resemblance with the experimental results. We have done a series of experiments concerning the same reaction at a beam energy of 570 MeV, corresponding to 29 MeV above the production threshold for two neutral pions in the center of mass system.

Three runs have been carried through. The first one in June 1992 had the character of an exploratory measurement. The second one in December 1992 yielded an integrated luminosity of $4.2 \cdot 10^{34}\text{cm}^{-2}$, too low for any conclusions to be drawn about a possible ABC effect. The total cross section obtained for the $d(d,2\pi)^4$He reaction was approximately 40 nb with an estimated uncertainty of 50 %, assuming isotropy in the c.m. system. This spring more data were gathered in an experiment lasting for three weeks. The total integrated luminosity now amounts to at least $2 \cdot 10^{35}\text{cm}^{-2}$. The analysis of these data is under way.

In Section 2 the experimental arrangement and method are described and in Section 3 the preliminary results are discussed.

2 Experimental Design

2.1 Site of Experiment

The CELSIUS synchrotron and storage ring [9] at The Svedberg Laboratory in Uppsala, Sweden, is well suited for near threshold experiments in the energy range of light meson production. This is due to the possibility of obtaining high precision in energy without loss of luminosity, by using a thin internal target and a cooled circulating beam.

Various beams are available. The maximum energy is 1.36 GeV for protons, but can be raised to 1.75 GeV with new power supplies. Light-ion beams are produced by the cold cathode type ion source, while heavy ions are obtained from an ECR (Electron Cyclotron Resonance) source.

The ions are accelerated to their injection energy in the Gustaf Werner cyclotron and enter the ring at one of the four straight sections. Opposite to the injection straight section the electron cooling system is situated. The target facilities are a cluster jet and a recently installed pellet target at the two remaining straight sections.

In the fourth quadrant, following the cluster-jet target, a solid-state telescope has been installed inside the second dipole magnet. At this position particles with approximately half the rigidity of the beam are deflected out of the beam just enough to reach the detectors.

The number of particles stored in each cycle in the present experiment varied between 10^9 and 10^{10}. After good injections the beam current was about 4-5 mA, which combined with the thickness of the deuterium target provided by the cluster jet, 10^{14} atoms/cm^2, yielded a luminosity of 10^{30} cm^{-2} s^{-1}.

2.2 The Solid-State Telescope

The telescope consists of two ΔE detectors, one position sensitive and one structureless, stacked together with one thick stopping E detector, all of them made of high-purity germanium.

The position sensitive element is a 1.7 mm thick strip detector of a matrix kind with parallel readout for the individual strips. The sensitive area is 64×36 mm^2. The position resolution is 1 mm horizontally and 2 mm vertically. Each of the strips are connected by UHV feed-throughs to charge-sensitive preamplifiers. The angular resolution is better than 2 mrad taking into account the emittance of the electron cooled beam.

The thicknesses of the other two detectors are 1.1 and 14.5 mm respectively. ^{4}He-ions with an energy of 347 MeV are fully stopped. The detectors are cooled by circulating liquid nitrogen.

The susceptibility to radiation damage of germanium detectors made it necessary to develop techniques of annealing in site without breaking the vacuum. Annealing can be done repeatedly without deterioration of the detector properties [10]. During the last run in April of this year, warming to temperatures around $50 - 100°$ was needed approximately twice a week.

In order to minimize the particle exposure of the telescope it is inserted into and withdrawn from the region close to the circulating beam in phase with the CELSIUS cycle, i.e. during injection, acceleration and cooling of the beam the telescope is placed in a 'parking position' and then it is moved to the measuring position, where it remains until shortly before the dumping of the beam. No degradation of the beam intensity was seen during the fifteen minutes long cycles.

2.3 Experimental Method

Particles with different charge-to-mass ratio are well separated in $\Delta E - E$ plots. In Fig.1(a) the ridges represent in turn protons, deuterons, ^{3}He and ^{4}He particles. These ridges are, except for the collection of ^{4}He-particles, background originating from scattering in the vacuum chamber. They can however be exploited for calibration and/or monitoring of the effective thicknesses of the detectors (i.e. depletion depths).

The summed energy of a detected ^{4}He particle and its position coordinates in the detector in combination with results from ray-trace calculations yield the momentum (p_α) and angle of emission (Θ) at the target. The missing mass M_x can then be calculated according to the formula

$$M_x^2 = E_{\text{tot}}^2 + m_\alpha^2 - 2E_{\text{tot}}E_\alpha - p_d^2 + 2p_d p_\alpha \cdot cos\Theta \qquad (1)$$

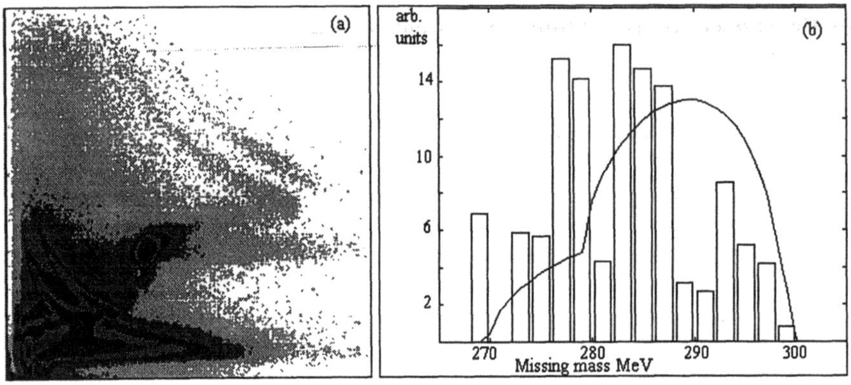

Figure 1. (a) A $\Delta E - E$ plot of the data being analyzed. Except for the ridges from stopped particles, there is a peak from the reaction $d(d, N\pi)^3$He. These ^{3}He-ions pass through the detectors. Nuclear reactions in the E detector are seen as horizontal ridges. (b) Preliminary experimental missing-mass spectrum, corrected for the acceptance of the detector. The solid line represents a calculation including effects of phase space and final-state interactions between the two pions.

3 Discussion

The missing-mass spectrum based on the limited data set from December 1992 is shown in Fig.1. If the tendency that is seen in earlier experiments persists in the analysis of the latest data, with a peak at minimum missing mass, a dip in the middle of the spectrum and a somewhat smaller peak at maximum missing mass, the argument for the ABC anomaly being inherent in the production mechanism is strengthened.

Qualitatively the observed structure of the missing mass spectrum is reproduced by a matrix element proportional to $\overline{k_1} \cdot \overline{k_2}$, where $\overline{k_1}$ and $\overline{k_2}$ are the π-momentum three-vectors in the c.m. system [11, 12]. This corresponds to a total J=0 obtained by having the π-mesons both in p wave relative to the ^{4}He, but in s wave relative to each other. Schepkin has suggested that the other possibility of having J=2, in the exit channel, which would cancel most of the dip mentioned above, could be suppressed due to the Pauli exclusion principle in the entrance channel. When the wavefunctions of the two deuterons overlap, the spins would tend to be antiparallel.

To continue the studies of two-pion production near threshold an experiment concerning the reaction ^{14}N$(d, 2\pi)^{16}$O is scheduled for the spring of 1996. This reaction could give information on the issue of the dressing of pions in nuclear matter [13] and the possible medium modifications of the strength function in the $T = 0$ channel.

The authors wish to thank the personnel of CELSIUS and of the detector laboratory at Institut für Kernphysik in Jülich, whose efforts made these experiments possible. The Swedish Natural Science Research Council and the

Royal Swedish Academy of Sciences have supported part of the work.

References

1. A. Abashian, N.E. Booth, K.M. Crowe: Phys. Rev. Lett. **5**, 258 (1960)
2. A. Abashian, N.E. Booth, K.M. Crowe: Phys. Rev. Lett. **7**, 35 (1961)
3. T. Risser and M.D. Shuster: Phys. Lett. **B43**, 68 (1973)
4. I. Bar-Nir, T. Risser, M.D. Shuster: Nucl. Phys. **B87**, 109 (1975)
5. F. Plouin et al.: Nucl. Phys. **A302**, 413 (1978)
6. J. Bordes et al.: Phys. Lett. **B223**, 251 (1989)
7. A. Codino and F. Plouin: Preprint LNS/Ph/94 -06 1994
8. J. Banaigs et al.: Nucl. Phys. **B105**, 52 (1976)
9. C. Ekström et al.: Phys. Scr. **T22**, 256 (1988)
10. D.L. Friesel, B.S. Flanders, R.H. Pehl: Nucl. Inst. Meth. **207**, 403 (1983)
11. C. Wilkin: Private Communication
12. M. Schepkin: Private Communication
13. P. Schuck, W. Nörenberg, G. Chanfray: Z. Phys. A **330**, 119 (1988)

Few-Body Systems Suppl. 9, 324–338 (1995)

Few-Body Systems

Physics Program and Experimental Equipment at CEBAF

Volker D. Burkert

CEBAF, 12000 Jefferson Avenue, Newport News, Virginia 23606, USA

Abstract. The initial complement of experimental instrumentation and the physics program at CEBAF are discussed. Using the power of the electromagnetic and neutral weak interaction, the structure of light quark mesons, baryons and light nuclei will be studied utilizing high duty cycle electron and photon beams with energies up to 4 GeV.

1 Introduction

Electron scattering as a probe of the internal structure of nucleons and nuclei has been employed for several decades, mostly in inclusive reactions and using low duty cycle machines. Experiments in the deep inelastic regime revealed the quark substructure of the nucleon, and more recently showed that the spin structure of the nucleon is more complicated than originally anticipated.

A more detailed understanding of the structure of nucleons and nuclei requires the measurement of more exclusive channels. For example, the study of the excited states of the nucleon requires the identification of spin, parity, and isospin of a state, which can only be accomplished by studying the resonance decay channels, and therefore requires exclusive measurements. In the past, low duty cycle machines have limited exclusive experiments to a few processes, mostly single pion production, and to restricted kinematics.

The construction of high current, high duty cycle electron accelerators has changed this situation in a significant way. Electromagnetic processes may now be studied with statistical sensitivities comparable to hadronic reactions. This brings to bear the full capability of the electromagnetic interaction as a probe of the internal structure of hadrons and nuclei.

2 The Accelerator and the Initial Experimental Equipment

The CEBAF electron accelerator is based on superconducting rf cavities operated in a continuous wave (cw) mode. A schematic of the machine is shown

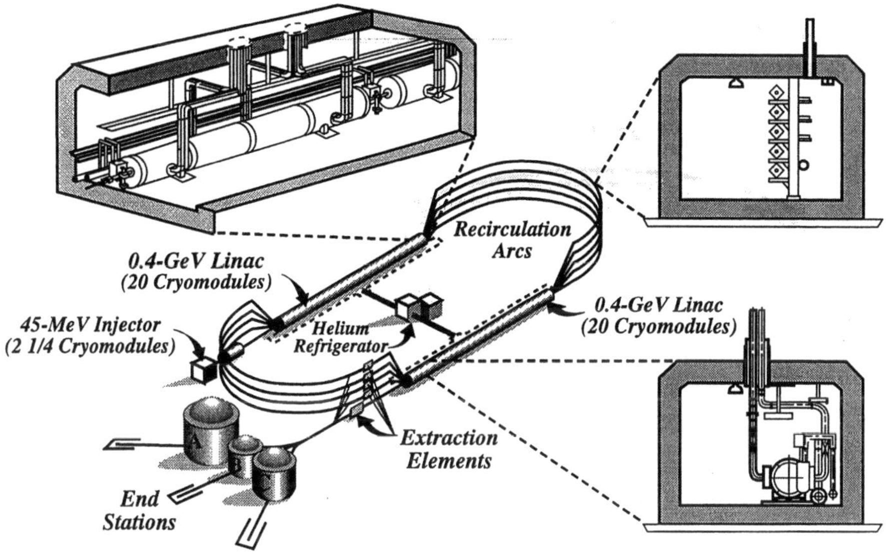

Figure 1. Schematics of the CEBAF accelerator.

in Fig. 1. Two parallel linacs in a "race track" configuration boost the beam energy by 800 MeV for each turn. The beam is recirculated five times to reach an initial maximum energy of 4 GeV. The heart of the machine are the five-cell niobium cavities, which have a minimum gradient of 5 MeV per meter. The cavities perform significantly better than the specifications, therefore providing the technical basis for a future energy upgrade. The machine can deliver electrons to 3 experimental areas (Hall A, B, C) at either the same energy, or at multiples of 1/5 of the end energy. Due to the virtual lack of synchrotron radiation, the energy spread in the beam is $\Delta E/E \leq 10^{-4}$. Beams can be extracted at each recirculation, thus allowing the operation of the experimental halls with simultaneous beams of different, though correlated, energies. The 1.5 GHz rf structure allows simulatanous beams to be delivered to the halls at a frequency of 500 MHz. The micro bunches can also be loaded with different electron densities, which provides the basis for operating the experimental areas with currents spanning a large dynamic range. In addition, a polarized electron gun can be operated in parallel with the standard thermionic unpolarized gun.

The halls are equipped with spectrometers for complementary experimental programs (Fig. 2). Hall C, which is already fully instrumented, contains two magnetic spectrometers of medium resolution with $\delta p/p \leq 10^{-3}$ but different maximum momenta: the High Momentum Spectrometer (HMS) has a maximum momentum of 7 GeV/c, and the Short Orbit Spectrometer (SOS) of 1.8 GeV/c, respectively.

Hall A will house two high resolution spectrometers (HRS) with $\delta p/p \leq 10^{-4}$ and a maximum momentum of 4 GeV/c, instrumented for electron and hadron detection, respectively. The spectrometers are expected to be operational in spring 1996.

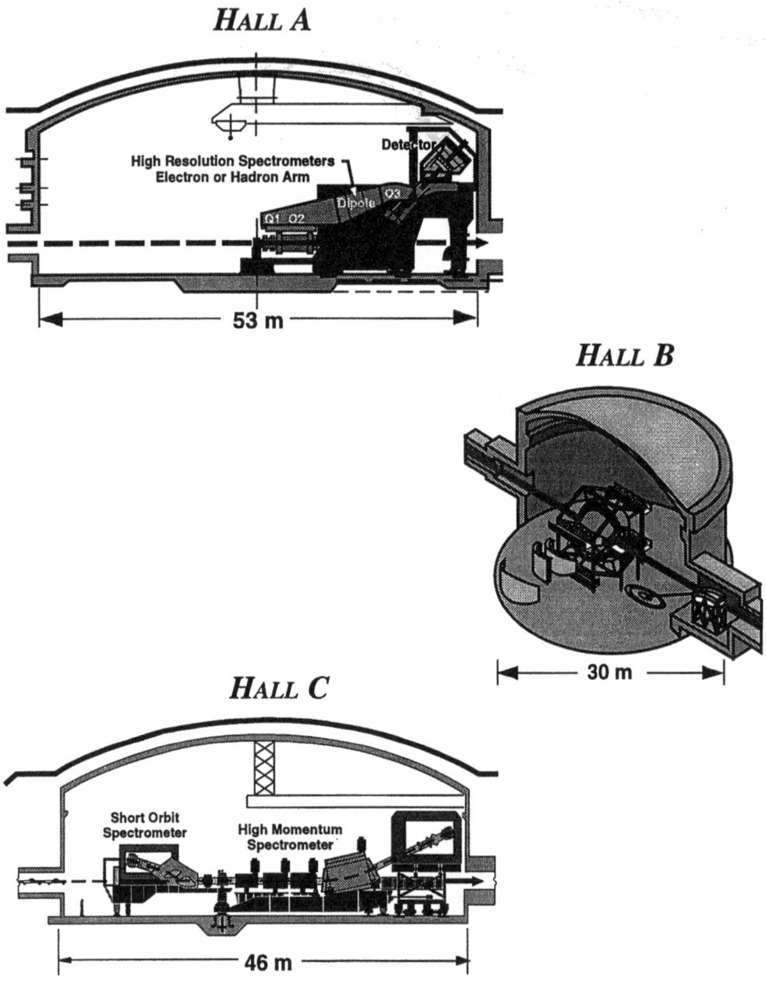

Figure 2. The initial equipment in the experimental halls.

Hall B will house the CEBAF Large Acceptance Spectrometer (CLAS) and a tagged photon facility. CLAS is based on a multi-gap magnet with six superconducting coils, symmetrically arranged to generate an approximately toroidal magnetic field distribution. Each of the six sectors is instrumented with drift chambers, time-of-flight counters, Cerenkov counters for electron identification, and electromagnetic calorimetry for photon and neutron detection. Completion of hall B is expected in the fall of 1996.

3 Structure of the Nucleon.

The interest in the structure of the ground state nucleon, its excited states, and their role in nuclear properties has dramatically increased in recent years and is one of the major motivations in support of the experimental program at CEBAF. The structure of the nucleon may be probed in elastic electron nucleon scattering and in inelastic reactions induced by electrons or photons. These experiments measure the charge and current distribution of the nucleons and the transition currents to their excited states. Knowledge of these quantities allows testing of models describing the nucleon structure at low and intermediate energy and momentum transfer. With increasing momentum transfer Q^2, the transition from the non-perturbative regime to the perturbative regime can be studied, where simple quark counting rules and power law behavior may apply [1].

More than half of all approved experiments at CEBAF address questions related to the structure of light baryons and mesons some of which either require measurement of spin observables, or the sensitivity to fundamental quantities is increased significantly in spin observables. Polarized electron beams, polarized targets, and proton and neutron recoil polarimeters will be important tools in these studies.

3.1 Electromagnetic Form Factors.

In elastic electron nucleon scattering the hadronic current may be specified by the electric and magnetic form factors $G_{\mathrm{E}}(Q^2)$ and $G_{\mathrm{M}}(Q^2)$. The usual technique for measuring the elastic form factors is the Rosenbluth separation, where one makes use of the different angular dependence of the electric and the magnetic term in the unpolarized elastic cross section to separate $|G_{\mathrm{E}}|$ and $|G_{\mathrm{M}}|$.

$$\frac{d\sigma}{d\Omega} = \left(\frac{d\sigma}{d\Omega}\right)_{\mathrm{M}} \frac{E'}{E}[(G_{\mathrm{E}}^2 + \tau G_{\mathrm{M}}^2) + 2\tau(1+\tau)G_{\mathrm{M}}^2 \mathrm{tg}^2\frac{\theta}{2}] \tag{1}$$

where $\tau = \frac{Q^2}{4M^2}$. This technique ceases to be useful, when either $G_{\mathrm{E}}^2 \ll G_{\mathrm{M}}^2$, or at high values of Q^2, where the magnetic contribution dominates both the angular dependent and the angular independent term. Unlike for the proton, the Rosenbluth separation of G_{E}^n from G_{M}^n for a neutron target is difficult even at low Q^2, because of the small size of G_{E}^n compared to G_{M}^n. At $Q^2 < 1$ GeV^2, G_{E}^n has been extracted from elastic electron-deuteron scattering data assuming a model for the deuteron structure [2].

A model-independent determination of G_{E}^n can be obtained by measuring the polarization asymmetry

$$A_{en} = \frac{2\tau\cos\theta\, v'_{\mathrm{T}} + 2\sqrt{2\tau(1+\tau)}\cdot(G_{\mathrm{E}}^n/G_{\mathrm{M}}^n)\sin\theta\cos\phi\, v'_{\mathrm{TL}}}{v_{\mathrm{L}}\,(1+\tau)(G_{\mathrm{E}}^n/G_{\mathrm{M}}^n)^2 + 2\tau\, v_{\mathrm{T}}} \tag{2}$$

where v_{L}, v_{T}, v'_{T}, v'_{TL} are known kinematic quantities and ϕ and θ define the orientation of the nucleon spin relative to the scattering plane. From the

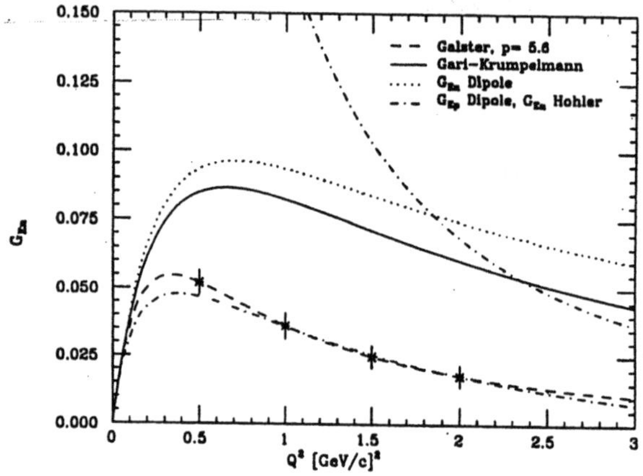

Figure 3. Projected data for a measurement of G_E^n using a polarized ND_3 solid state target.

asymmetry, the ratio G_E^n/G_M^n can be extracted. Knowing G_M^n the electric form factor can be determined. One method uses a polarized deuterium or ^3He target, either as an ultra thin gas target in an electron storage ring, or a solid state target ND_3, or a dense ^3He gas target, in an external electron beam. If the polarization asymmetry is measured using vector polarized deuterium, it will be necessary to measure the recoil neutron in coincidence with the scattered electron to eliminate the much larger contributions from the polarized proton in the deuteron. The binding of the neutron in the deuteron has negligible effect on the polarization asymmetry and on G_E^n/G_M^n, as long as the recoil neutron is emitted at small angles with respect to the direction of the virtual photon [3].

A second method uses an unpolarized deuterium target, and the polarization of the recoiling neutron is measured in a second scattering experiment. Both methods will be employed at CEBAF [4] to measure G_E^n/G_M^n for Q^2 up to 2-3 GeV^2 (Fig. 3). In both cases, the scattered electron will be detected in the hall C HMS spectrometer, and the neutron will be detected in a narrow cone around the direction of the virtual photon. One experiment uses a polarized ND_3 solid state target, while the other one uses a plastic scintillator based polarimeter to measures the polarization of the recoil neutron.

The polarization techniques can also be employed to measure the electric form factor G_E^p of the proton. Most promising in this respect is the recoil polarization techniques using an unpolarized hydrogen target [5]. With a 4 GeV beam G_E^p can be measured for Q^2 up to 5 GeV^2, with statistical errors of less than 5%.

Our knowledge of the magnetic form factor of the neutron is also unsatisfactory. Quasi-elastic electron scattering off deuterons appears to be the most promising way to determine G_M^n at high momentum transfers. Two experiment will measure G_M^n over a large Q^2 range [6]. A common problem is how to determine the neutron detection efficiency accurately. A large acceptance detector such as CLAS offers the possibility to determine the efficiency by measuring the reaction $p(e, e'\pi^+)n$ using a hydrogen target, simultaneously. The neutron kinematics is completely constrained in this reaction, and can be used to determine the neutron detection efficiency. The ratio of quasi-free $e-n$ and $e-p$ coincidences allows to determine G_M^n using the known proton form factors as normalization.

3.2 *Baryon Resonance Transitions.*

A large number of resonances, attributed to the excitation of the nucleon have been observed in hadron scattering, the $\Delta(1232)$ being the most prominent one. Electromagnetic excitation of the resonances addresses fundamental questions about the interaction of quarks and gluons in confined systems. Specifically, one would like to study how the transition between the 3-quark ground state and excited states is mediated. Measurement of the Q^2 evolution of the transition form factors provides information about the wave function of the excited state. At high momentum transfer, one may observe the transition from the non-perturbative regime of QCD to the perturbative regime. A complete program to study nucleon resonance transitions involves measurement of polarization observables.

The lowest mass resonant state, the $\Delta(1232)$ is of special interest. In SU(6) symmetric quark models, this transition is explained by a simple quark spin-flip in the $L_{3Q} = 0$ ground state, corresponding to a magnetic dipole transition M_{1+}. In QCD based models, which include color magnetic interactions arising from the one-gluon exchange, the $\Delta(1232)$ acquires an $L_{3Q} = 2$ component, leading to small electric and scalar contributions (e.g. $|E_{1+}/M_{1+}| \simeq 0.01$ at $Q^2 = 0$). The ratio $|E_{1+}/M_{1+}|$ is predicted to be weakly dependent on Q^2.

Experiments at CEBAF are in preparation to measure the electromagnetic transition amplitudes for the nucleon to the Δ [7], as well as to many higher mass resonances [8], over a large Q^2 range, using both unpolarized pion and eta electroproduction, as well as polarized electron beams and polarized targets or recoil polarimeters [9]. For the $N\Delta$ transition one obtains information about the terms

$$M_{1+}\ ,\ \mathrm{Re}(E_{1+}M^*_{1+})\ ,\ \mathrm{Re}(S_{1+}M^*_{1+})\ ,\ \mathrm{Im}(E_{1+}M^*_{1+})\ ,\ \mathrm{Im}(S_{1+}M^*_{1+}).$$

Projected data the $\Delta(1232)$ are shown in Fig. 4. The imaginary parts of the bilinear terms can be measured using polarization degrees of freedom.

Of topical interest are the transition form factors to the $N(1440)$, a candidate for a state with a large gluonic content [10]. Gluonic excitations of baryons are not distinguished by exotic quantum numbers from ordinary baryons. Electroproduction of theses states may be the only available tool in the search for signatures of these states, as their transition form factors are expected to have very different Q^2 dependence [10].

The QCD motivated extension of the non-relativistic quark model [11] predicts many states which have not been observed in πN reactions. Several of these states are predicted to couple strongly to photons (real or virtual) and may thus be searched for in photoproduction or electroproduction experiments.

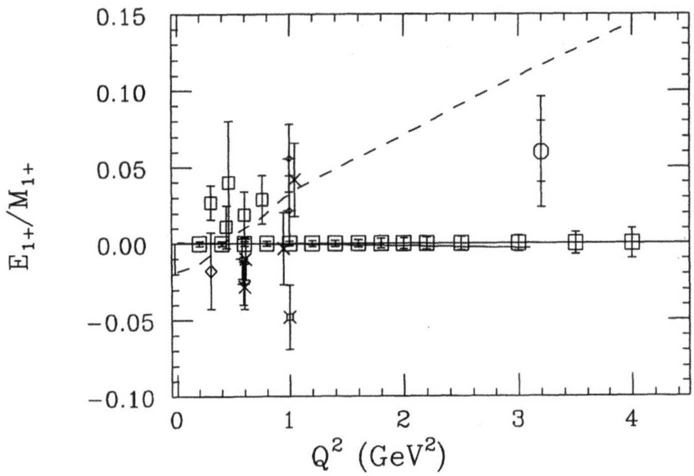

Figure 4. Data for E_{1+}/M_{1+} compared to quark model calculations. The error bars on the horizontal axis are projections of experiment E-89-037.

Search for some of these states in multi-pion and vector meson production are in preparation at CEBAF [12]. Virtual Compton scattering off nucleons $p(e,e'p)\gamma$ is also sensitive to the excitation of nucleon resonances and will be measured in an experiment using the HRS spectrometers [13]. One advantage is the absence of final state interaction, while the low rate makes it difficult to achieve sufficient kinematical coverage for a complete partial wave analysis. At small momentum transfers t to the nucleon, the reaction $ep \rightarrow en\pi^+$ is sensitive to the pion charge form factor and can directly be extracted from the longitudinal response function:

$$\sigma_{\rm L} \sim -t \cdot g^2_{\pi NN}(t)\frac{F^2_\pi}{(t-m^2_\pi)^2} \tag{3}$$

where $g_{\pi NN}$ represents the πNN coupling. Much improved knowledge of $F_\pi(Q^2)$ over a large Q^2 range is expected from this experiment [14].

3.3 Parity Violation Experiments.

At low and medium energy ($Q^2 \ll M^2_{Z^o}$) neutral current interactions, the parity violating contributions arise from the interference between the one-photon exchange and the neutral weak boson Z^o exchange graphs.

Parity violation in $\vec{e}p \rightarrow ep$ probes the strange vector current $\bar{s}\gamma_\mu s$. Sizeable $s\bar{s}$ contributions in the proton are suggested by the results of the polarized structure function experiments. The parity violating asymmetry in elastic ep scattering is given by:

$$A_{\vec{e}p} \simeq \frac{G_{\rm F}Q^2}{\pi\alpha\sqrt{2}}\frac{\epsilon G^p_{\rm E}G^Z_{\rm E}+\tau G^p_{\rm M}G^Z_{\rm M}}{\epsilon(G^p_{\rm E})^2+\tau(G^p_{\rm M})^2} \tag{4}$$

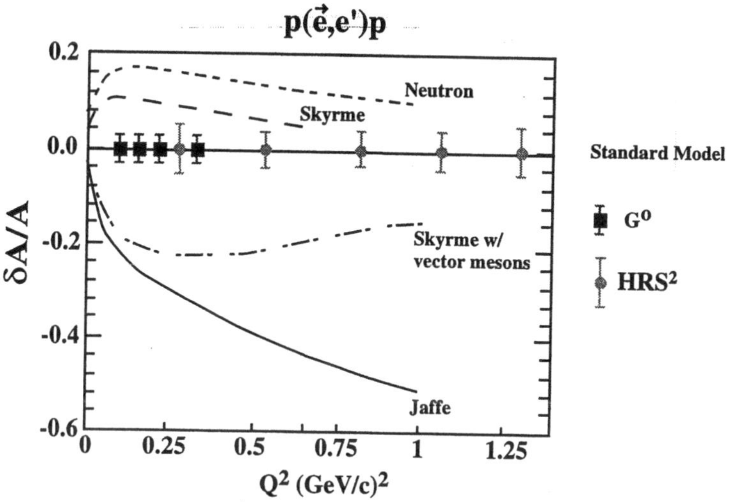

Figure 5. Projected data for the polarized electron asymmetry $p(\vec{e}, e')p$.

In the Standard Model of particle physics:

$$G^Z_{\mathrm{E,M}} = (\frac{1}{2} - \sin^2\theta_{\mathrm{W}})G^p_{\mathrm{E,M}} - \frac{1}{4}(G^n_{\mathrm{E,M}} + G^s_{\mathrm{E,M}}) \tag{5}$$

where $G^{p,n}_{\mathrm{E,M}}$ are the usual electromagnetic form factors, and $G^Z_{\mathrm{E,M}}$ are the neutral weak form factors. At large electron scattering angles the first term dominates. G^Z_{E}, G^Z_{M} can be separated by varying the electron kinematics and G^s_{E}, G^s_{M} can be determined using the known electromagnetic form factors.

Three experiments are in preparation at CEBAF to study parity violation in elastic electron scattering experiments [17]. Projected data of one of the experiments are shown in Fig. 5.

3.4 Q^2 Evolution of the Nucleon Spin Structure

The study of the spin structure of the nucleon has focussed on the short distance behavior where the measurements may be interpreted in terms of the parton structure of the nucleon. Studies of the Q^2 evolution of the spin structure functions g_1 and g_2 down to small Q^2 will give important constraints on the models aimed at describing the nucleon structure at larger distances. At $Q^2 = 0$, the Gerasimov-Drell-Hearn (GDH) sum rule constrains the slope of Γ_1 to the anomalous magnetic moment of the target:

$$\frac{2M}{Q^2}\Gamma_1 \to -\frac{1}{4}\kappa^2 \tag{6}$$

The negative value is in conflict with a simple extrapolation from the deep inelastic regime.

The GDH sum rule and its Q^2 evolution have never been studied experimentally. The only empirical information comes from the analysis of pion photo-

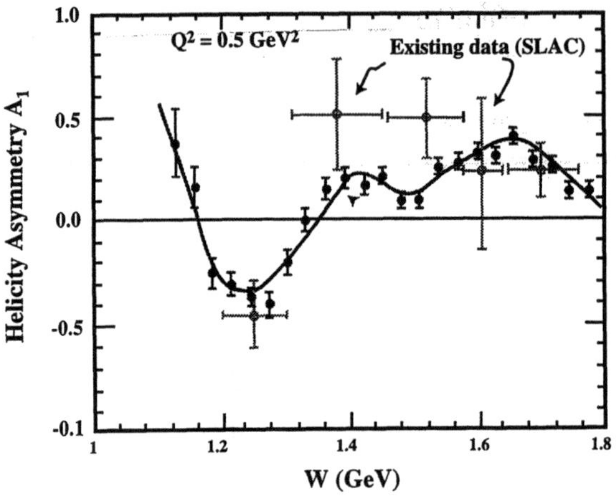

Figure 6. Projected data (full circle) on the double polarization asymmetry $\vec{p}(\vec{e}, e')X$ at fixed Q^2.

and electroproduction experiments [15, 16]. There are plans to measure the sum rule at $Q^2 = 0$ for the protons and neutrons [18] using a circularly polarized photon beam and a polarized solid state target. The Q^2 evolution will be studied in three experiments [19] using polarized NH_3, ND_3, and ^3He targets, respectively. With information on the proton and neutron, the evolution of the Bjorken integral $\Gamma_1^p - \Gamma_1^n$ can be studied down to $Q^2 = 0$, where it should be constraint by the respective difference of the GDH sum rules. Fig. 6 shows projected data on the double polarization asymmetry using a NH_3 target.

3.5 Strangeness Production

Electromagnetic production of strange particles $\gamma p \to K^+Y$ ($Y = \Lambda, \Lambda^*, \Sigma, \Sigma^*$) have been poorly studied in the past. Consequently, the production mechanism is not well understood. In a diagrammatic approach the process is sensitive to the $K\Lambda N$ and $K\Lambda N^*$ coupling constants. Coupling constants extracted from photoproduction data and from hadronic data disagree. Calculations indicate that the Λ polarization is very sensitive to specific ingredients of the model, in particular on assumptions about the coupling constants. Measurement of the Λ recoil polarization in photoproduction reactions will thus yield independent information about the hadronic coupling constants.

An efficient experimental program to study polarization degrees of freedom in the $K\Lambda$ channel benefit greatly from the use of large acceptance detectors with nearly 4π solid angle coverage. For example, the Λ polarization can be inferred from an analysis of $\Lambda \to \pi^- p$ decay. Using a longitudinally polarized electron beam, circularly polarized bremsstrahlungs photons can be generated, and the polarization transfer reaction $\vec{\gamma} p \to K\vec{\Lambda}$ can be studied as well.

Very little electroproduction data is available on $\gamma_v p \to K^+\Lambda$. Similar to the π^+ production, the longitudinal cross section for this reaction may be related to

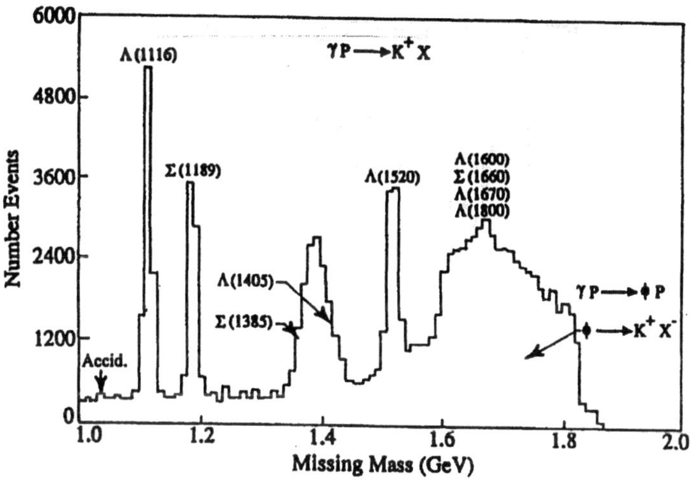

Figure 7. Missing mass distribution for $\gamma p \to K^+X$.

the K^+ elastic form factor [20], although the model dependence is larger than for the π^+ form factor. An intriguing problem is presented by the $f_0(975)$. The state does not fit into the standard $q\bar{q}$ scheme for mesons. It is considered a candidate for an exotic $q\bar{q}q\bar{q}$ state, or a $(q\bar{q})(q\bar{q})$ molecule. If the $f_0(975)$ is a weakly bound mesonic molecule a strong t dependence may be expected in electromagnetic production. f_0 production will be studied both in the $\pi\pi$ and $K\bar{K}$ channels [21]. CLAS experiments will cover hyperon photoproduction on the proton [22] and deuteron [23], and hyperon electroproduction [24]. Figure 7 shows the expected missing mass distribution in $\gamma p \to K^+X$ for a proton target. The Λ, Σ^0, $\Lambda(1520)$, and a combination of $\Lambda(1405)$ and $\Sigma^0(1385)$ can be isolated.

Electroproduction of ϕ mesons probes the production mechanism as a function of the distance scale, and therefore can probe deviations from the dominantly diffractive behavior observed in photoproduction as a function of Q^2.

4 Study of Light Nuclei

Nuclei containing only a few nucleons ($\lesssim 4$) provide the best hope for an accurate description of a nuclear system. Their study therefore provides a laboratory for the study of nucleon-nucleon interaction at short distances, relativistic effects, meson exchange currents, and possible quark effects.

4.1 Elastic Form Factors of Light Nuclei

Elastic scattering on the deuteron is described by three independent form factors. The standard Rosenbluth techniques allows to determine the magnetic form factor and a combination of charged and quadrupole (electric) form factors. Additional polarization measurements are required for a complete deter-

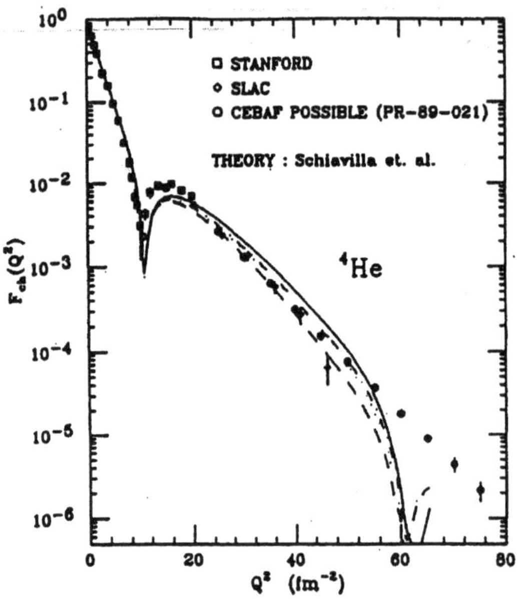

Figure 8. Charge form factor of ^{4}He. Projected data of experiment E-89-021 are shown as full squares.

mination of all form factors. The magnetic and electric contributions will be measured using the spectrometers in hall A [25], while a specialized setup in hall C is required to measure the tensor polarization [26], which, at small scattering angles, is given by

$$T_{20} \simeq -\sqrt{2}\frac{\frac{3}{2}\tau G_Q(\frac{2}{3}\tau G_Q + 2G_C)}{G_C^2 + \frac{8}{9}\tau^2 G_Q^2} \tag{7}$$

Combined with the unpolarized measurements, the charged and quadrupole form factors G_C, G_Q can be determined, separately.

The charge and magnetic form factors of ^{3}H and ^{3}He have been measured up to ~ 1 GeV^2 and the charged form factor of ^{4}He is known up to ~ 1.8 GeV^2 at SLAC. Experiments at CEBAF will extend these measurements to $Q^2 \simeq 2.3$ GeV^2 and $Q^2 \simeq 3$ GeV^2, respectively [25]. Fig. 8 shows existing and projected data on ^{4}He. Deviations from the diffractive pattern seen at lower Q^2, could be a signature for quark effects.

4.2 Deuteron Disintegration

Results of experiments at SLAC on the photodisintegration of the deuteron (Fig. 9) seem to be consistent with a simple constituent counting rule behavior of the differential cross section at photon energies > 1.3 GeV, and $\theta_p = 90^o$:

$$\frac{d\sigma}{dt} \sim s^{-11} \tag{8}$$

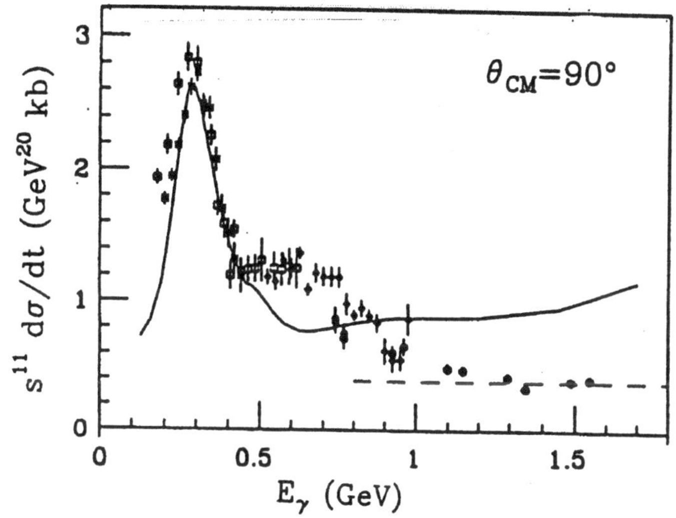

Figure 9. Deuteron photodisintegration cross section multiplied by s^{11}. The dashed line indicates the energy dependence expected from the quark counting rules (not normalized).

Using the HMS spectrometer the reaction will be measured [27] up to $E_\gamma = 4$ GeV. Additional information will come from a study of the proton recoil polarization [28] which is very sensitive to the reaction mechanism. At asymptotic energies QCD makes the simple prediction $P_N \to 0$. Any non-zero results would indicate that the asymptotic regime has not been reached, and the observed scaling behavior of the differential cross section must have a different origin. Information about the reaction mechanism will also come from the measurement of the angular dependence and the comparison of $d(\gamma, p)n$ and $d(\gamma, p)\Delta^0$ using CLAS [29]. Electrodisintegration of the deuteron $d(e, e'p)n$ allows probing the isoscalar and isovector parts of the hadronic current at varying distance. The HRS spectrometers in hall A will be used to separate the longitudinal and transverse contributions for protons emitted along $\boldsymbol{q}$. A measurement of the recoil protons in the electron scattering plane but left and right of $\boldsymbol{q}$, and for electrons scattered at forward and backward angles, will separate three response functions. This will allow stringent tests of microscopic models for this reaction [30]. Additional information about the deuteron structure and the reaction mechanism will be obtained [31] from measurement of polarization observables in coplanar $d(\vec{e}, e'\vec{p})n$ reactions. An experiment using in CLAS [32] will study $d(e, e'p)n$ for high momentum protons emitted at backward angles relative to $\boldsymbol{q}$ and probe the short distance structure of the deuteron wave function.

4.3 Hadron Emission from Nuclei

An important program in the past has been the measurement of single nucleon knockout processes such as $A(e, e'p)(A-1)^*$, where the remnant nucleus is

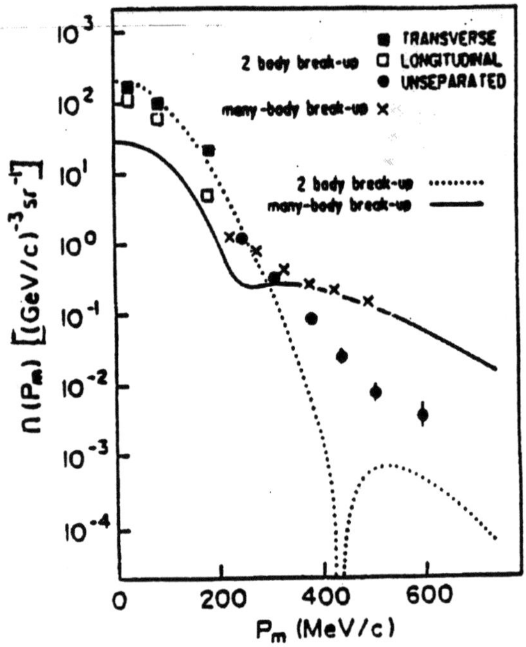

Figure 10. Nucleon momentum density distribution in ^{4}He versus missing momentum. For momenta greater than 350 MeV/c, multi-nucleon break-up becomes the dominant process.

left in the ground state or an excited state. Single nucleon densities have been measured up $\sim$ 600 MeV/c for the lightest nuclei, and up to $\sim$ 300 MeV/c for heavier nuclei. At CEBAF energies missing momenta up to 1 GeV/c can be probed for ^{4}He using the hall A HRS [33], and the three in-plane response functions can be measured. Experiments on ^{4}He have shown that momenta $>$ 350 MeV/c largely reside in multi-body breakup channels (Fig. 10), and their investigation becomes increasingly important at higher energy transfer. Using CLAS, measurements for multi-hadron final states will be carried out [34] on nuclei ranging from ^{3}He to ^{56}Fe.

4.4 Properties of Hadrons in Nuclei

An important question in nuclear physics is how the properties of hadrons are modified in the nuclear medium. For example, the mass and decay width of vector mesons are predicted to change significantly if they are produced inside a nucleus. Masses and widths of vector mesons can be studied most effectively using the e^+e^- decay, which, in contrast to the hadronic decay, is not be affected by final state interactions. Measurements of ρ and ω production are especially interesting as the short-lived ρ will decay mostly inside the nucleus, while the ω will decay outside the nucleus, and no effect should be observed in the latter case (Fig. 11). Photoproduction of e^+e^- pairs on various nuclei will be measured

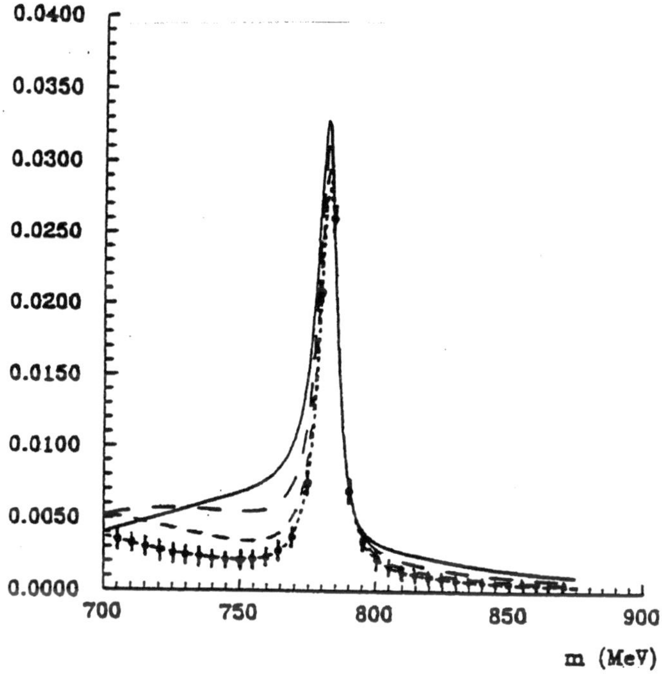

Figure 11. Projected data of experiment E-94-002. The curves represent different assumption about the ρ and ω mass shifts in nuclei.

using CLAS [35].

Hadronic production of strange particles in nuclei has been studied for decades at proton machines. The electromagnetic probe has important features, which makes its use complementary to hadronic studies. First, it allows to deposit hyperons deep inside the nucleus. Second, it strongly populates unnatural parity states starting from a 0^+ nuclear target. A program at CEBAF will start with the study of the lightest hypernucleus produced in ${}^4\mathrm{He}(e, e'K^+)^4_\Lambda\mathrm{He}$ using the HMS and SOS spectrometers [36]. Other experiments will require a specialized spectrometer setup with high resolution [37].

5 Summary and Outlook.

The CEBAF electron accelerator will be a powerful tool in probing the internal structure of baryons and light nuclei with unprecedented precision. The initial experimental equipment has been designed for complementary experimental programs. Already from the initial experimental program we expect significant insight into the manifestation of QCD in the confinement regime. We can also hope that limitations of the hadronic picture of light nuclei will be revealed as nuclear wave functions are probed at very short distances.

As of May 1995, the accelerator has reached 5 recirculations and a 4 GeV beam was transferred to hall C. First physics experiments are expected to run in fall 1995. Completion of the hall A equipment is expected for spring 1996, and the instrumentation in hall B should be operational by the end of 1996.

There is an excellent possibility that the maximum energy of the accelerator will be upgraded to about 6 GeV in 1997. A further energy upgrade into the 10 GeV energy regime has been proposed [38], and appears feasible with relatively modest financial investment. This would be accomplished by installing additional accelerating cavities, and by replacing cavities with low field gradients with better performing cavities.

References

1. S.J. Brodsky and G.P. Lepage: Phys. Rev. **D24**, 2848 (1981)
2. S. Platchkov et al. : Nucl.Phys. **A510**, 740 (1990)
3. H. Arenhövel, W. Leidemann, and E.L. Tomusiak: Z. Phys. **A331**, 123 (1988)
4. CEBAF experiments E-93-026, E-93-038
5. CEBAF experiment E-93-027.
6. CEBAF experiments E-93-024, E-94-017
7. CEBAF experiments E-89-037, E-89-042, E-94-014
8. CEBAF experiments E-89-038, E-89-039, E-91-002
9. CEBAF experiments E-91-011, E-93-036
10. Z.P. Li, V. Burkert, Zh. Li: Phys. Rev. **D46**, 70 (1992)
11. N. Isgur and G. Karl: Phys. Lett. **72B**, 109 (1977); Phys. Rev. **D23**, 817 (1981)
12. CEBAF experiments E-91-024, E-93-006, E-93-033, E-94-109
13. CEBAF experiment E-93-050
14. CEBAF experiment E-93-021
15. I. Karliner: Phys. Rev. **D7**, 2717 (1973)
16. V. Burkert and Zh. Li: Phys. Rev. **D47**, 46 (1993)
17. CEBAF experiments E-91-004, E-91-010, E-91-017
18. CEBAF experiment E-91-015, PR-94-117
19. CEBAF experiment E-91-023, E-93-009, E-94-010
20. CEBAF experiment E-93-018
21. CEBAF experiment E-89-043
22. CEBAF experiments E-89-004, E-89-024
23. CEBAF experiment E-89-045
24. CEBAF experiment E-93-030
25. CEBAF experiments E-91-026, E-89-021
26. CEBAF experiment E-94-018
27. CEBAF experiment E-89-012
28. CEBAF experiment E-89-019
29. CEBAF experiment E-93-017
30. CEBAF experiment E-94-004
31. CEBAF experiment E-89-028
32. CEBAF experiments E-93-043, E-94-102
33. CEBAF experiment E-89-044
34. CEBAF experiments E-89-027, E-89-031, E-89-032, E-89-036
35. CEBAF experiment E-94-002
36. CEBAF experiment E-91-016
37. CEBAF experiment E-89-009
38. CEBAF at Higher Energies A White Paper, February 1995

Few-Body Systems Suppl. 9, 339–348 (1995)

Few-Body Systems

Results from the Real Photon Programme at MAMI

J. Ahrens * †

Institut für Kernphysik, Universität Mainz, D-55099 Mainz, Germany

Abstract. The accelerator MAMI (Mainzer Mikrotron) with its excellent beam properties and its duty factor of 100% allows to apply coincidence techniques in intermediate energy nuclear physics, which has led to a variety of results with high precision. The real photon collaboration (A2) uses energy tagged bremsstrahlung photons with energies up to 800 MeV. We have results on total photon absorption (proton, deuteron, ^{3}He and U), on Compton scattering (proton, ^{4}He and ^{12}C) and on meson production, especially on π°– and η–photoproduction. Double pion production on the proton and the deuteron was measured. The two and three body breakup of ^{3}He was investigated as well as the reactions (γ,NN) and $(\gamma,\pi^\pm N)$ on complex nuclei, etc. Some results are shown and discussed.

1 Introduction

The 100% duty factor accelerator MAMI [1] has gone into operation for experiments in 1990. Since then the international A2–Collaboration has used the Glasgow bremsstrahlung tagger [2] to make experiments with monochromatic, real photons with energies up to 800 MeV, with the aim to investigate the properties of the nucleons in different environments, i.e. as free and as bound particles. In the available energy range several baryon resonances, $P_{33}(1232)$, $P_{11}(1440)$, $D_{13}(1520)$ and $S_{11}(1535)$, can be selectively excited by different reaction channels. The institutions participating in the collaboration are:
CEN Saclay, DAPNIA/SPN, Gif sur Yvette, France,
Department of Physics and Astronomy, Glasgow University, Scotland,
Department of Physics and Astronomy, University of Edinburgh, Scotland,
INFN, Laboratori di Frascati, Italy,
INFN, Sezione di Genova, Italy,
INFN, Sezione di Pavia, Italy,
Institute for Nuclear Research, Moscow, Russia,

*for the A2-Collaboration

†Supported by DFG under SFB 201 and funding agencies of the collaborating groups

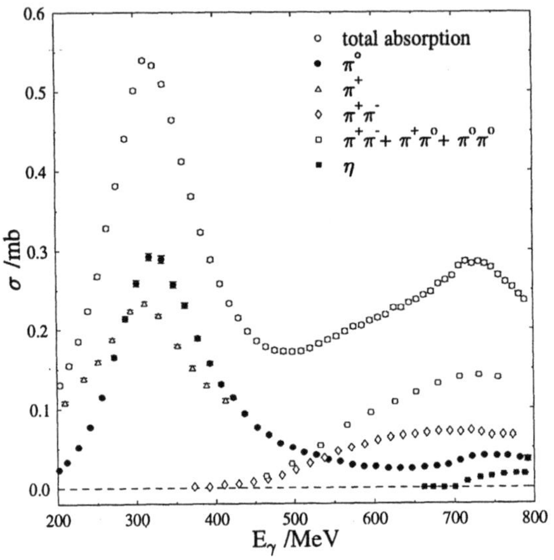

Figure 1. Total cross sections for photoproduction on the proton as measured in the framework of the A2–Collaboration. Data with open symbols have been obtained with the detector DAPHNE, and those with full symbols with TAPS (see following section)

Institut für Kernphysik, Universität Mainz, Germany,
Institut für Strahlungsphysik, Universität Stuttgart, Germany,
Physikalisches Institut, Universität Bonn, Germany
Physikalisches Institut, Universität Tübingen, Germany
II. Physikalisches Institut, Universität Gießen, Germany
II. Physikalisches Institut, Universität Göttingen, Germany.

They have all contributed to the equipment used. Thus a variety of detectors has been developed and put into operation to fulfill the requirements of our complex research programme. As an example of our present results total cross sections for the proton are shown in Fig. 1. All data have been measured within our collaboration.

2 Experimental set-up

2.1 The tagged photon facility

The electron accelerator MAMI provides 100% duty factor beams with energies between 180 and 855 MeV in steps of 15 MeV, with a maximum current of 100 μA and excellent beam emittance and stability. There are electron sources available for longitudinally polarized electrons. If only low currents are needed, strained GaAs can be used in these sources that have provided up to 80% degree of polarization. The energy of bremsstrahlung photons is given by the Glasgow tagger, which is a wide band magnetic electron spectrometer with 352 electron

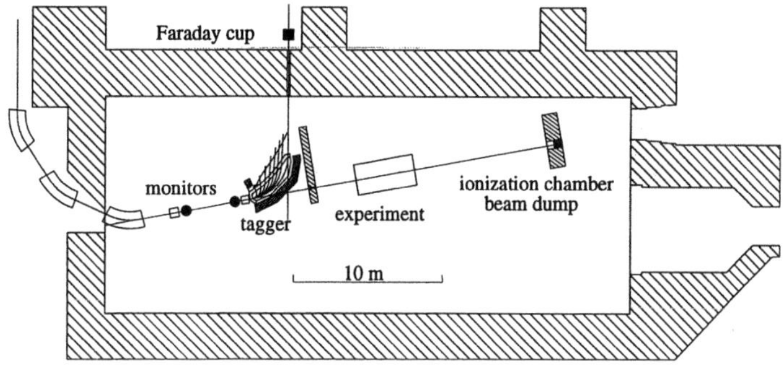

Figure 2. The experimental area available for measurements with real photons. In front of the tagger there are two groups of monitors (each equipped with scintillation screens, wire scanners and position sensitive pickup cavities) 3.5 m apart that allow the monitoring and on-line surveillance of the beam. The main electron beam is dumped in a neighbouring hall. In the well shielded space for experiments the appropriate detectors can be installed.

detectors that cover a photon energy range from 50 to 800 MeV in steps of 2 MeV for incoming electrons of 855 MeV. The tagging technique limits the photon flux available on the reaction target to roughly 10^8/s in the selected energy range. In order to obtain linearly polarized photons a diamond is used as a radiator [3]. Circularly polarized photons are produced by helicity transfer from longitudinally polarized electrons. Since the tagger can only be operated with low currents the above mentioned high degree of electron polarization can be used. Figure 2 shows the installation in the tagger–hall.

The use of polarized beams requires good means for monitoring and polarimetry. When using coherent bremsstrahlung the characteristic shape of the spectrum is continuously seen in the tagger electron detectors. This shape agrees very well with calculations. In the photon beam the ratio of coherent to incoherent bremsstrahlung can be sensitively influenced by collimation. For that case it is harder to perform good calculations. We have used in a first attempt coherent π° photoproduction on ^{4}He in order to determine the degree of linear polarization of the photons (coherent π°–production from spin = 0 nuclei has 100% analyzing power for linearly polarized photons). Figure 3 shows the comparison between measurement and calculation for preliminary results for the relative intensity spectrum of the photon beam and for the degree of polarization.

2.2 Detectors

In the following a list of available detectors will be given with short descriptions of their properties. The names in parentheses indicate their origin.

• CATS [4] (Mainz, Moscow, Göttingen) is a photon spectrometer mainly meant for the measurement of Compton scattering. It consists of a big segmented NaI

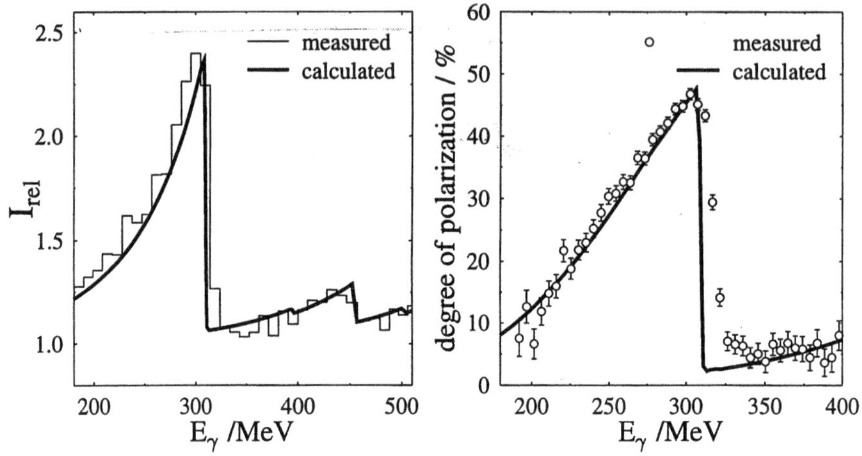

Figure 3. Comparisons of measurement and calculation for the relative intensity spectrum of coherent bremsstrahlung photons after collimation (left) and the degree of polarization (right) from coherent π°–production on ^{4}He.

spectrometer (19″ in diameter and 25″ long) with an excellent energy resolution of 1.5% in the region 50 to 600 MeV, and a 2π–array of 61 BaF_2–crystals (TAPS-type, see below) surrounding the target at the side opposite to the NaI in order to measure or suppress π°–mesons.

• COPP [5] (Genova, Frascati) is an array of photon and proton detectors to measure Compton scattering from the proton. Photons are detected in lead-glass blocks and protons in plastic scintillators. COPP is built in azimuthal symmetry and thus suited for measurements with linearly polarized photons.

• DAPHNE [6] (Saclay, Pavia): this is mainly a charged particle tracking detector with a large acceptance of solid angle. It has cylindrical symmetry and consits of three layers of wire chambers and several layers of plastic scintillators and convertors.

• LARA (Frascati, Genova, Göttingen, Mainz, Moscow, Tübingen) is built similar to COPP, but has a large acceptance of polar angle. The detection of protons was achieved by wire chambers and time-of-flight (TOF, see below) plastic scintillators.

• PPAC [7] (Stuttgart) is an almost 4π–array of 14 wire chambers for the detection of fission products. The detector was used to measure the total fission cross section of actinides as a substitute for the total photon absorption cross section.

• PIP–TOF [8] (Edinburgh, Glasgow, Tübingen) has been developed to investigate reactions with two hadrons in the final state. It consists of a $\Delta E - \Delta E - E - E - E - E$ plastic scintillator hodoscope (PIP) and a set of time-of-flight detectors (TOF) composed of bars of plastic scintillator (3 m long, 20 cm wide and 5 cm thick) with which a total area of 60 m^2 can be covered.

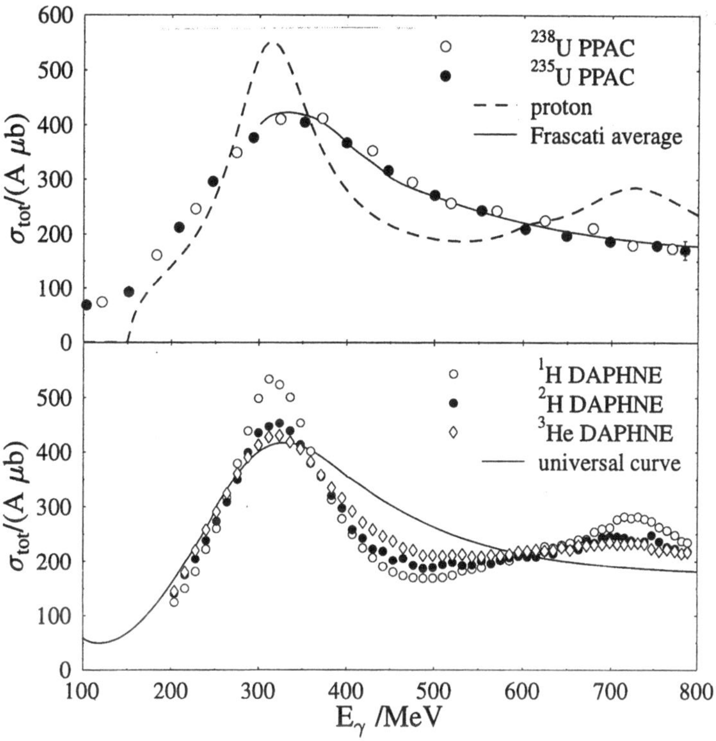

Figure 4. Photon total absorption cross sections per nucleon. Top: results for ^{235}U and ^{238}U in comparison with the proton and an average from data obtained in Frascati [12] (the latter data have been measured up to 1.2 GeV). Bottom: results for light nuclei in comparison with the universal curve (see text).

• TAPS [9] (Gießen, TAPS–Collaboration) consists of walls of BaF_2 crystals of hexagonal shape (smallest diameter 5.9 cm, 25 cm long). The purpose of this detector is to investigate mainly reactions that have two or more photons as reaction products, i.e. single and double π°– and η–production.

3 Experimental results

3.1 Total photon absorption

Figure 4 shows results obtained with the detectors PPAC [10] and DAPHNE [11]. As has been known before for the region of the Δ–resonance [13] the cross sections per nucleon ranging from ^{6}Li to ^{238}U have, within their errors, the same shape. Now this has been found also for the region of the second group of resonances. The cross sections for light nuclei show a steady approach to this behaviour. The fact that there is no structure to be seen in the universal curve above the Δ–resonance has not yet been fully understood. Fermi motion alone should not have this drastic effect.

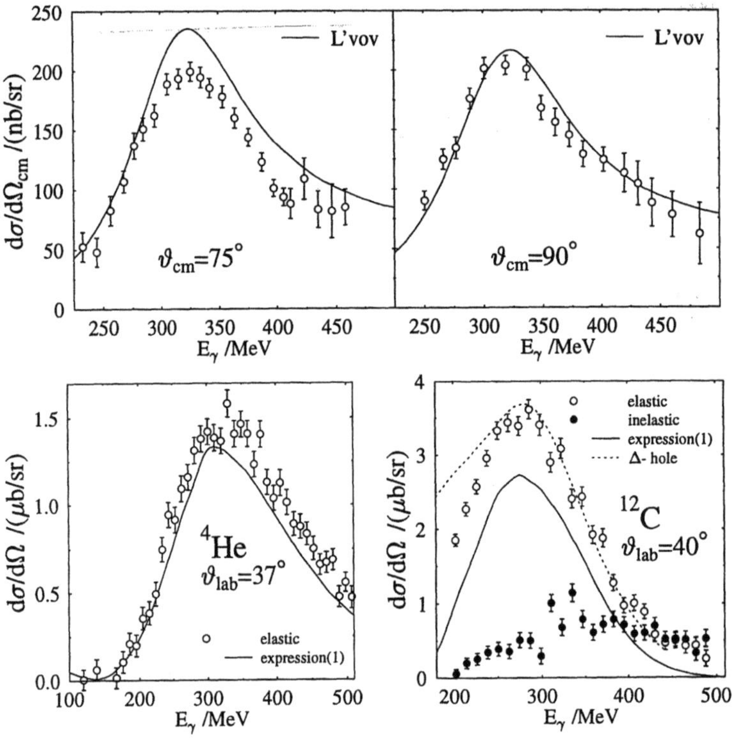

Figure 5. Results of Compton scattering on the proton (top) and on ^{4}He and ^{12}C (bottom). For the proton results of predictions by A.L'vov [16] are shown for comparison. In the case of ^{12}C the excellent resolution of CATS has allowed us to separate inelastic from elastic scattering [15]. The Δ-hole-calculation is from [17]. The full curves shown for the complex nuclei are based on formula (1).

3.2 Compton scattering

We have measured Compton scattering with CATS and COPP on the proton [14], on ^{4}He and on ^{12}C [15]. The results are shown in Fig. 5. Our proton results are compared with calculations by A. L'vov [16] which are based on dispersion relations and the use of the multipole parametrization as given in [18]. It seems that the calculation overestimates the effect. Since we see a similar behaviour in π^0–production, this may be a hint that the M_{1+}–amplitude has been overestimated in [18].

For complex nuclei the cross section can be roughly estimated by the formula

$$\frac{d\sigma}{d\Omega}(\vartheta, E_\gamma) = \frac{d\sigma}{d\Omega}(0, E_\gamma)\frac{1+\cos^2(\vartheta)}{2}F^2(q) \tag{1}$$

which is based on the forward cross section calculated from the total absorption

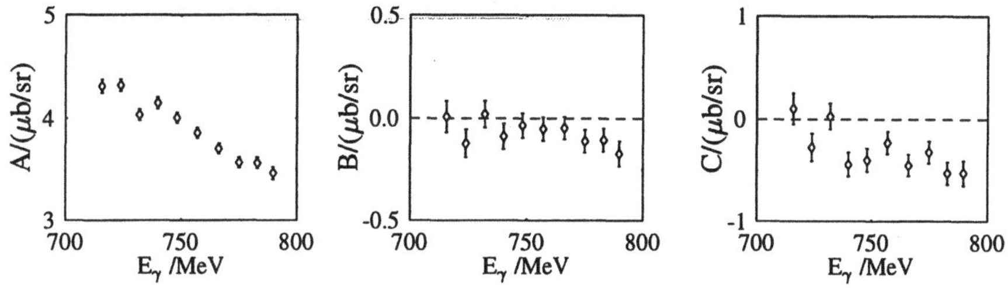

Figure 6. Coefficients to the angular distribution of the η-mesons.

cross section and the assumption that the angular distribution is given by impulse approximation. $F(q)$ is the formfactor as measured in electron scattering.

3.3 Meson–photoproduction

We have measured the photoproduction of mesons on the proton and on complex nuclei. Only part of the data can be shown here.

3.3.1 η-production on the proton

Due to the energy range of MAMI we have measured η-production at threshold with the detector TAPS. The data have been fully analyzed and published [19]. From these data (they are shown with other results in Figs. 1 and 8). We were able to determine the η–mass as m_η=(547.12±0.06±0.25), where the last error is systematical. This value nicely confirms the latest determination as found in the reaction $d\ p \rightarrow {}^3\mathrm{He}\ \eta$. In addition we could confirm the literature value of the branching ratio for the decays $\eta \rightarrow 2\gamma$ and $\eta \rightarrow 3\pi^\circ$. From the angular distribution of the η-mesons we extracted with

$$\frac{d\sigma}{d\Omega} = \frac{q_{\text{meson}}}{k}(A + B\cos(\vartheta) + C\cos^2(\vartheta)) \tag{2}$$

the coefficients which are given in Fig. 6 from a fit to the data. The strong role of the S_{11}-resonance is shown by the dominance of A.

3.3.2 Photon asymmetry of the single pion production on the proton

The single pion production on the proton, $p(\gamma,\pi^\circ)p$ and $p(\gamma,\pi^+)n$, in the region of the Δ–resonance has been measured with linearly polarized photons and the detector DAPHNE. The aim is to determine the $E_{1+}^{3/2}$–amplitude, the resonant part of which gives the quadrupole deformation of the $N-\Delta$ transition. Part of the measurements have been analyzed. Figure 7 shows the results. In the case of π°–production the ratio E_{1+}/M_{1+} can be read from $C_\parallel/A_\parallel$ since the M_{1-}–amplitude is there small.

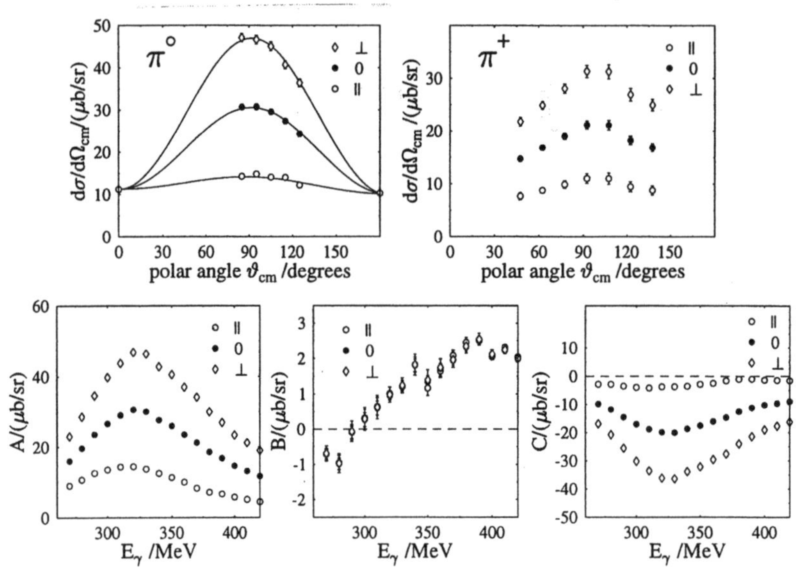

Figure 7. Differential cross sections for the π°– and π^{+}–production on the proton at 320 MeV (top) and coefficients (equivalent to formula (2)) for π°–production (bottom).

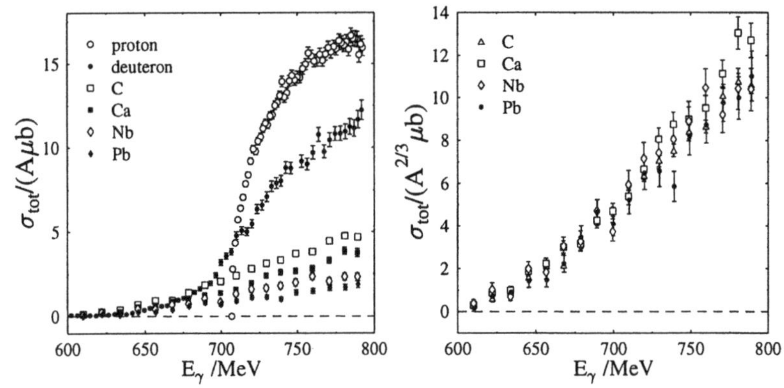

Figure 8. η–production on nuclei. Left: total cross section per nucleon for different nuclei. Right: cross sections/$A^{2/3}$ for complex nuclei.

3.3.3 η–photoproduction on nuclei

The η–production on nuclei has been measured also in order to learn about medium effects. Figure 8 shows the results. Comparison with the proton cross section shows that η-mesons have been produced on nuclei predominantly in quasifree processes. The clear $A^{2/3}$ dependence of the cross sections for complex

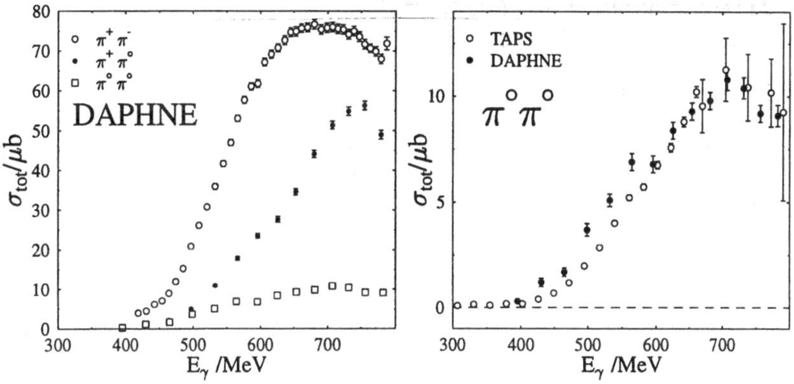

Figure 9. Double pion production on the proton. Left: the three possible reactions as measured with DAPHNE. Right: comparison of results from DAPHNE and TAPS for the production of two neutral pions. This cross section is very small below 400 MeV.

nuclei (a fitting procedure gave $A^{0.67\pm0.01}$) shows that only η-mesons produced at the surface have a chance to leave the nucleus. This indicates a small mean free path of the order of 1 fm.

3.3.4 Double pion production on the proton

Measurements with DAPHNE [20] and TAPS have allowed to extract the cross sections for double pion production on the proton, $p(\gamma,\pi^+\pi^-)p$, $p(\gamma,\pi^+\pi^\circ)n$ and $p(\gamma,\pi^\circ\pi^\circ)p$. Figure 9 shows the results. Model calculations for these processes are presently only able to describe the $\pi^+\pi^-$-channel, but not the other two sufficiently well. The $\pi^\circ\pi^\circ$-channel is of special interest since there the contact term is strongly suppressed and one thus ought to have direct access to the resonances.

4 Outlook

In the future we plan to extensively use polarized photons. We have started activities to experimentally test the Gerasimov–Drell–Hearn sum rule [21] and to determine the spin polarizability of the nucleons. In these experiments circularly polarized photons will react with longitudinally polarized nucleons.

References

1. H. Herminghaus et al. : Nucl. Instr. Meth. **A138**, 1 (1976)
2. I. Anthony et al. : Nucl. Instr. Meth. **A301**, 230 (1991)
3. D. Lohmann, J. Peise et al. : Nucl. Instr. Meth. **A343**, 494 (1994)

4. J. Ahrens et al.: In: *Proc. Int. Workshop on Future Detectors for Photonuclear Experiments*, ed. D. Branford, Edinburgh 1991

5. G.P. Capitani et al. : INFN Report, INFN/BE 89/6

6. G. Audit et al. : Nucl. Instr. Meth. **A301**, 473 (1991)

7. W. Wilke et al. : Nucl. Instr. Meth. **A272**, 785 (1988)

8. I.J.D. MacGregor: In: *Proc. Int. Workshop on Future Detectors for Photonuclear Experiments*, ed. D. Branford, Edinburgh 1991;
T. Hehl: ibidem

9. R. Novotny: IEEE Trans. Nucl. Sci., **38**, 379 (1991)

10. Th. Frommhold et al. : Phys. Lett. **B295**, 28 (1992)

11. M. MacCormick et al. : submitted to Phys. Rev. **C**

12. N. Bianchi et al. : Phys. Lett. **B299**, 219 (1993); Phys. Lett. **B309**, 5 (1993)

13. J. Ahrens: Nucl. Phys. **A446**, 229c (1985)

14. M. Sanzone: In: *Proc. Perspectives in Nucl. Phys. at Intermediate Energies*, ed. S. Boffi et al. , p. 455, Trieste 1993 ;
J. Peise et al. : to be published in Phys. Lett. **B**

15. F. Wissmann et al. : Phys. Lett. **B335**, 119 (1994)

16. A. L'vov, V.A. Petrun'kin, Lec. Notes in Phys. **365**, 123 (1990)

17. B. Koerfgen, F. Osterfeld: Phys. Rev. **C50**, 1637 (1994)

18. R.A. Arndt et al. : Phys. Rev. **C42**, 1853 (1990);
R.A. Arndt: Private Communication

19. B. Krusche et al. : Phys. Rev. Lett. **74**, 3736 (1995);
B. Krusche et al. : Z. Phys. **A351**, 237 (1995);
B. Krusche et al. : submitted to Phys. Lett. **B**

20. A. Braghieri et al. : submitted to Phys. Lett. **B**;
L.M. Murphy, J.M. Laget: submitted to Phys. Lett. **B**

21. S.B. Gerasimov: Sov. J. Nucl. Phys. **2**, 430 (1966);
S.D. Drell, A.C. Hearn: Phys. Rev. Lett. **16**, 908 (1966);
M. Anselmino et al. : Sov. J. Nucl. Phys. **49**, 136 (1989);
D. Drechsel: Prog. Part. Nucl. Phys. **34**, 181 (1995)

Few-Body Systems Suppl. 9, 349–354 (1995)

Few-Body Systems

Polarized Photon Scattering from ^{4}He

D. Moricciani[1], V. Bellini[2,3], M. Capogni[1], A. Caracappa[4], L. Casano[1], R.M. Chasteler[5,6], A. D'Angelo[1], F. Ghio[7], B. Girolami[7], S. Hoblit[8], L. Hu[1], M. Khandaker[9], O.C. Kistner[4], L.H. Kramer[10], C.M. Laymon[5,6], A.I. L'vov[11], B. Marks[5,12], L. Miceli[4], V.A. Petrunkin[11], B.J. Rice[5,6], A.M. Sandorfi[4], C. Schaerf[1,13], C.E. Thorn[4], D.R. Tilley[5,12], H.R. Weller[5,6]

[1] INFN-Roma 2, Via della Ricerca Scientifica 1, I-00133, Rome, Italy

[2] Physics Department, University of Catania, Corso Italia 57, I-95129, Catania, Italy

[3] INFN-LNS, Via S. Sofia 44, I-95125, Catania, Italy

[4] Physics Department, Brookhaven National Laboratory, Upton N.Y. 11973, USA

[5] Triangle Universities Nuclear Laboratory, Duke University, Box 90308, Durham, NC 27708-0305

[6] Physics Department , Duke University, Box 90305, Durham, NC 27708-0305

[7] Istituto Superiore di Sanitá and INFN-ISS, Viale Regina Elena 299, I-00161, Rome, Italy

[8] Physics Department, University of Virginia, Charlottesville, VA 22901

[9] Physics Department, Virginia Polytechnic Institute & State University, Blacksburg, VA 24061

[10] Laboratory for Nuclear Science, MIT, Cambridge, MA, USA

[11] Lebedev Physical Institute, Leninsky Prospect 53, 117924, Moscow, Russia

[12] Physics Department, North Carolina State University, Raleigh, NC 27695-8202

[13] Physics Department, University of Rome "Tor Vergata", Via della Ricerca Scientifica 1, I-00133, Rome, Italy

Abstract. We have measured, for the first time, the differential cross section of and the parameter of asymmetry for the reactions $^4\mathrm{He}(\vec{\gamma},\gamma)^4\mathrm{He}$ with linearly polarized photon. The LEGS polarized and tagged gamma-ray beam of Brookhaven National Laboratory has been used in the energy region 180–310 MeV. Data are colleted for five laboratory angles 30°, 45°, 72.5°, 110° and 130°.

1 Introduction

Photonuclear scattering reaction off nuclei, in the intermediate energy range, are a powerfull tool to investigate the creation and the propagation of the Δ Isobar in the nuclear media.

At energies above pion threshold, the $\gamma - \mathcal{N}$ amplitude dominates the elastic photon scattering reaction [1, 2] while the nuclear media effects manifest themselves in term of MEC and Δ–Isobar contributions.

The last contribution play a important role at energies near at the Δ resonance and nuclear medium corrections on the Δ propagation in the nucleus have also to be taken into account. These corrections are due to kinematical effects like the Fermi motion of the nucleons and to dynamical effect like: Pauli blocking of the Δ decay, $\Delta - \mathcal{N}$ binding, multiple scattering of the pions from the Δ decays and coupling of the Δ to π channels. All this effect produce a shift and broadening of the Δ peak [3, 4, 5, 6].

The differential cross section for the coherent Compton scattering of linerly polarized photon on ^{4}He is given by the following relation:

$$\frac{d\sigma}{d\Omega}(E_\gamma, \vartheta_\gamma^{\rm cm}, \varphi_\gamma^{\rm cm}) = W_0(E_\gamma, \vartheta_\gamma^{\rm cm}) + P\, W_1(E_\gamma, \vartheta_\gamma^{\rm cm}) \cos(2\varphi) \qquad (1)$$

where E_γ is the energy of the incoming photon in the laboratory frame, $\vartheta_\gamma^{\rm cm}$ and φ are the polar and azimuthal angle of the scattered photon in the center of mass frame. W_0 is the unpolarized structure function, W_1 is the polarization dependent structure funcion and P is the degree of linear polarization of the incoming photon.

We present the first experiment which permitted the measurement of the W_1 structure function. In the past year coherent Compton scattering on nuclei have been very difficult to perform, because of the necessity of using tagged and polarized photon beams with enought intensity and energy resolution to select the elastically scattered photons. The existing data are available for unpolarized bremsstrahlung photon beam at average incoming photon energies of 180 MeV [7] and 187, 235, 280 MeV [9] and 320 [8]; in this last case only for the forward scattering angle.

2 Experimental Apparatus

The experiment was carried out using the LEGS[1] polarized and tagged photon beam, produced by the backscattering of Laser light on the 2.5 GeV electrons circulating in the NSLS at the Brookhaven National Laboratory [10, 11]. The incoming photons, tagged with an energy resolution of 5 MeV (FWHM), impinged on a 10 cm liquid ^{4}He target with density $\varrho = 0.140 \pm 0.001$ g/cm^3.

The scattered photons were detected by a high resolution, total absorption cylindrical NaI(Tl) 19"x 19" scintillator detector, this detector was rounded

[1] Laser Electron Gamma Source

by 12 plastic scintillators (annulus). The annulus detectors was used as a veto counter for cosmic rays and as an electromagnetic shower leakage detector. A 2.5 cm plastic scintillator placed in front of the detector, was used in anticoincidence to reject charged particles. A 21 cm diameter lead collimator was used to define the detector geometrical acceptance ($\Delta\Omega = 0.13$ sr) and the whole apparatus was externally shielded with lead.

Eight NaI detectors (bars) were placed around the target, covering a large solid angle, with a window in the direction of the scattering detector; these bars were used as vetoes against the competing processes (neutral pion photoproduction, photon scattering with ^{4}He break-up, π^0 photoproduction with ^{4}He break-up).

For the first time it was possible to obtain the complete separation of the coherent photon scattering events from the background reactions.

Measurements of the two structure functions Eq. (2), were performed at six polar angles equal to 31°, 45°, 72.5°, 90°, 110° and 130° for the scattered photons in the laboratory frame.

The true coherent scattering events N_{ev} were identified in a scatter plot where the difference between the measured and the theoretical Compton gamma energy was drawn versus the measured energy in the main detector.

3 Cross Sections

The experimental results for the two structure functions W_0 and W_1 were determined through the following relations:

$$W_0 = \frac{d\sigma_{\parallel} P_{\perp} + d\sigma_{\perp} P_{\parallel}}{P_{\parallel} + P_{\perp}} \qquad W_1 = \frac{d\sigma_{\parallel} - d\sigma_{\perp}}{P_{\parallel} + P_{\perp}} \tag{2}$$

where $P_{\parallel}$ and $P_{\perp}$ are the degrees of the incoming photons linear polarization in the directions parallel and perpendicular to the scattering plane.

The results are shown in Figs. (1,2). Comparisons with existing data are possible for the unpolarized differential cross-section at slightly different values of the incoming photon energy.

At energies close to the Δ resonance, general good agreement is found with data from refs. [8, 9]; for the two lowest energy bins the difference in the incoming photon energy between our data and those of refs. [7, 9] may account for the discrepancy in the angular distribution.

Unpolarized differential cross sections are also compared with predictions based on a Δ-hole model from ref. [5] (dotted curve). The solid curves in Figs. (1,2) are the predictions from L'vov and Petrun'kin [1, 2]. They use relativistic dispersion relations to obtain a reliable γ-nucleon scattering amplitude [13, 14]. The γ-nucleus cross section is calculated in the Impulse Approximation with Fermi smearing of the incoming photon energy; E1-MEC currents were also included through a Siegert-like procedure of minimal substitution and also by explicitly taking into account the emission and the absorption of virtual pions

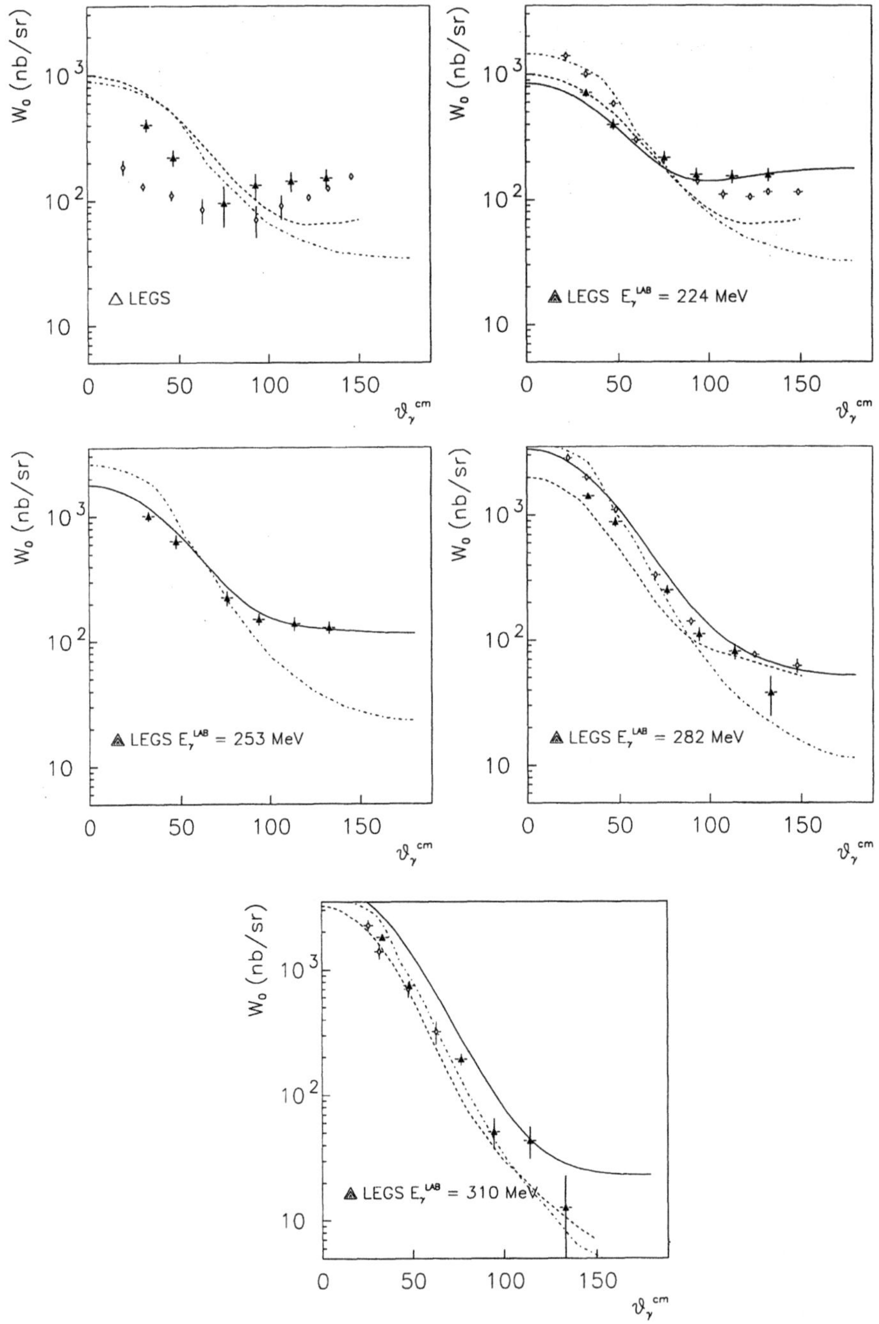

Figure 1. $W_0(\vartheta_\gamma^{cm})$.

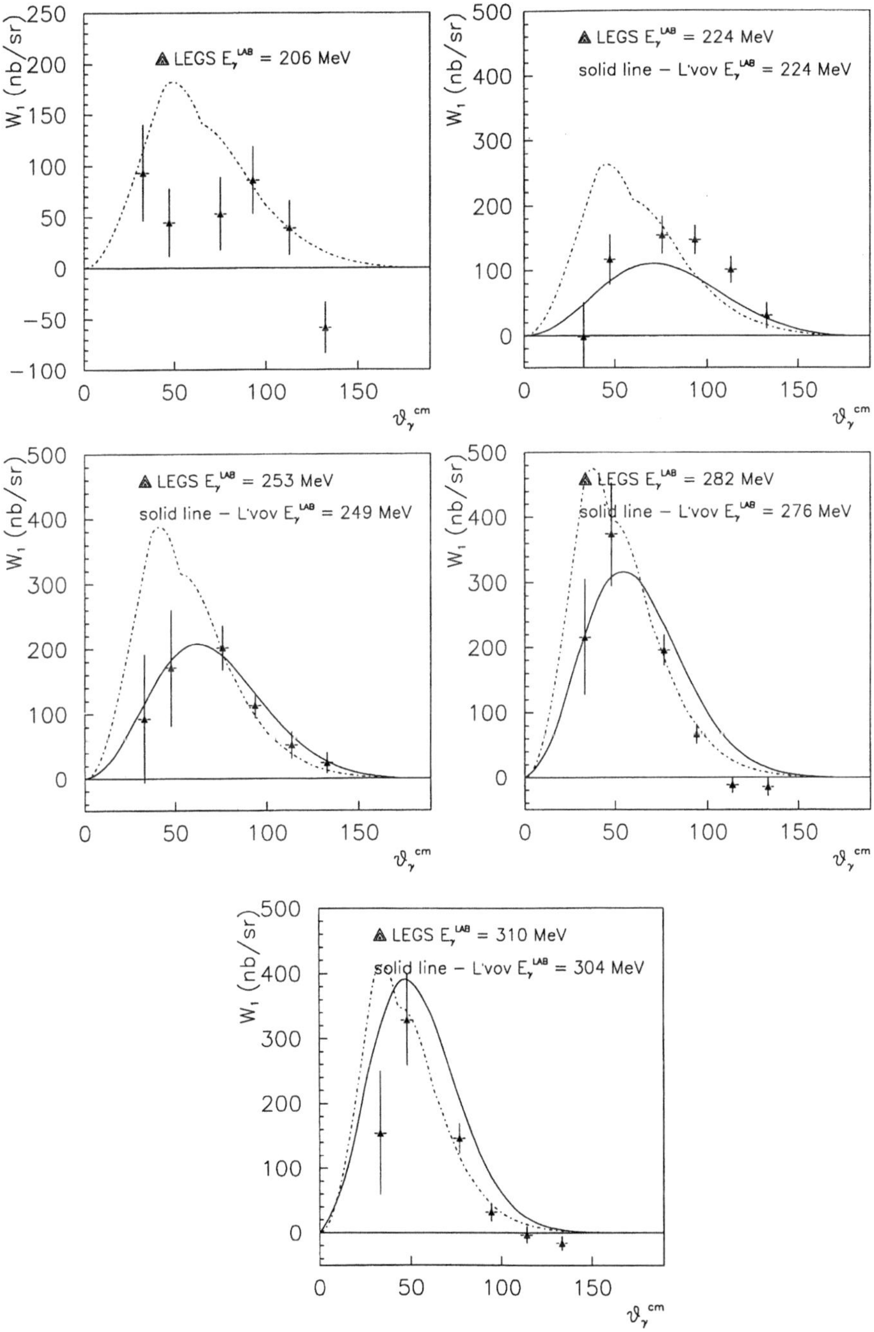

Figure 2. $W_1(\vartheta_\gamma^{cm})$.

at the $\gamma N\Delta$ vertex [15]. M1-MEC contributions, including those related with the Δ spreading, are not included in the calculation.

The dash-dotted curve is simple predictions including only the electric-dipole contribution to the cross section. It is given by the following relation:

$$\frac{d\sigma}{d\Omega}(E_\gamma, \vartheta_\gamma^{\rm cm}, \varphi) = |F(q^2)|^2 \frac{d\sigma}{d\Omega}(E_\gamma, 0)\Big(\frac{1+\cos^2\vartheta}{2} + P\,\frac{\sin^2\vartheta\cos 2\varphi}{2}\Big) \quad (3)$$

where $F(q^2)$ is the elastic form factor of the ^{4}He. Good agreement is found between our results and the predictions by L'vov and Petrunkin, except at the Δ−resonance peak, where the theoretical calculation overestimates the experimental strenght. The Δ−hole model, on the contrary, provides a good description of the unpolarized cross section at the Δ−peak, but important discrepancies are found with experimental results at lower photon energy. These results indicate the need of correctly including the Δ modifications effects in a prediction where non-resonant terms are also properly taken into account.

References

1. A.I. L'vov and V.A. Petrun'kin: Lect. Notes in Physics **365**, 123 (1990)
2. V.A. Petrun'kin and A.I. L'vov: In: *Proc. of the 8th Int. Seminar on Electromagnetic Interactions of Nuclei at Low and Medium Energies* Moscow, Dec. 1991, published by Institute for Nuclear Research, Moscow, 109 (1992)
3. W. Weise: Nuclear Physics **A358**, 163 (1981)
4. E. Oset and W. Weise: Nuclear Physics **A368**, 375 (1981)
5. J.H. Koch, E.J. Moniz and N. Ohtsuka: Annals of Physics **154**, 99 (1984)
6. J. Vesper, D. Drechsel, N. Ohtsuka: Nuclear Physics **A466**, 652 (1987)
7. E.J. Austin et al. : Physical Review Letters **57**, 972 (1986)
8. E.J. Austin et al. : Physical Review Letters **61**, 1922 (1988)
9. D. Delli Carpini et al. : Physical Review **C 43**, 1525 (1991)
10. A.M. Sandorfi et al. : IEEE Trans. **fNS-30**, B 3093 (1983)
11. C.E. Thorn et al. : Nucl. Instr. Meth. **A285**, 447 (1989)
12. J.H. Koch et al. : Annals of Physics **154**, 99 (1984)
13. A.I. L'vov: Sov. J. Nucl. Phys. **34**, 597 (1981)
14. A.I. L'vov: Sov. J. Nucl. Phys. **42**, 583 (1985)
15. A.I. L'vov: Topics in Atomic Science and Technique, ser. Gen and Nucl. Phys. **2/35/**, 51 (1986)

Few-Body Systems Suppl. 9, 355–368 (1995)

Few-Body Systems

Strangeness Electro- and Photo-Production at CEBAF

Reinhard A. Schumacher*

Department of Physics, Carnegie Mellon University, Pittsburgh, PA 15213, U.S.A.

Abstract. Strange-particle physics can extend our understanding of nuclear and nucleonic structure and reaction mechanisms. Numerous experiments at CEBAF will investigate strangeness in elementary production and nuclear reactions. The established program includes the study of hyperon electromagnetic decays, the elementary photo- and electro-production of hyperons on nucleons and light nuclei, and the electroproduction of light hypernuclear species. There are relevant experiments planned in all three experimental halls at CEBAF; advances made possible by these new facilities are discussed.

1 Introduction

The strangeness-related physics program at CEBAF has been evolving for several years. However, no experimental data have been obtained since I reviewed this subject last year [1]. This talk is thus an evolutionary update of that previous report, including several new proposals and letters of intent which have appeared since then.

A sketch of an idealized electromagnetic quark-pair creation is shown in Fig. 1. When the electron scatters, it produces a virtual photon with four-momentum $q = (\nu, \boldsymbol{q})$, leading to an interaction that results in an $s\bar{s}$ pair, which hadronizes into a kaon and a hyperon. While it would be wonderful if we could compute such processes directly from QCD, at CEBAF energies one cannot ignore the many baryonic resonances, or non-perturbative degrees of freedom, which play a role in arriving at the final state [2]. Thus the current state of the art is much more along the lines of Fig. 2, which shows some of the tree–level Feynman diagrams that describe the elementary production process. As we shall see, the mechanism of strangeness production is not yet well in hand even at this level of description, and this motivates new experiments

* *E-mail address:* reinhard@ernest.phys.cmu.edu

at CEBAF. In passing we note that total cross sections for real strangeness photoproduction are on the order of a few microbarns, quite small fractions of the total photoproduction cross section.

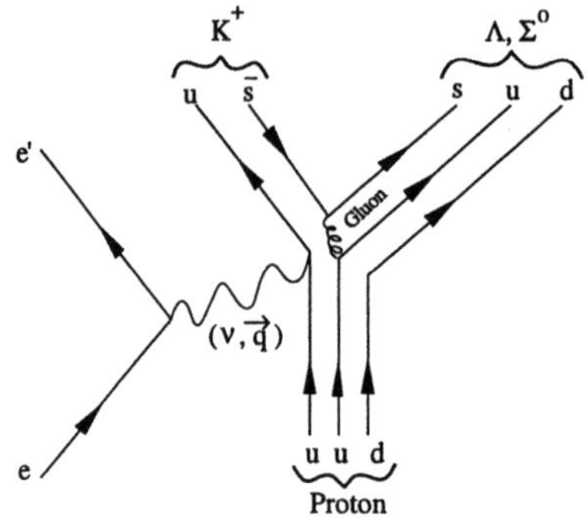

Figure 1. Quark-line strangeness production picture, ignoring non-perturbative effects. Note that the photon need not couple directly to the strange quark pair.

The CEBAF accelerator is now complete; 4 GeV beam was achieved in May 1995. The approved experiments will be carried out in Halls A, B and C [4].

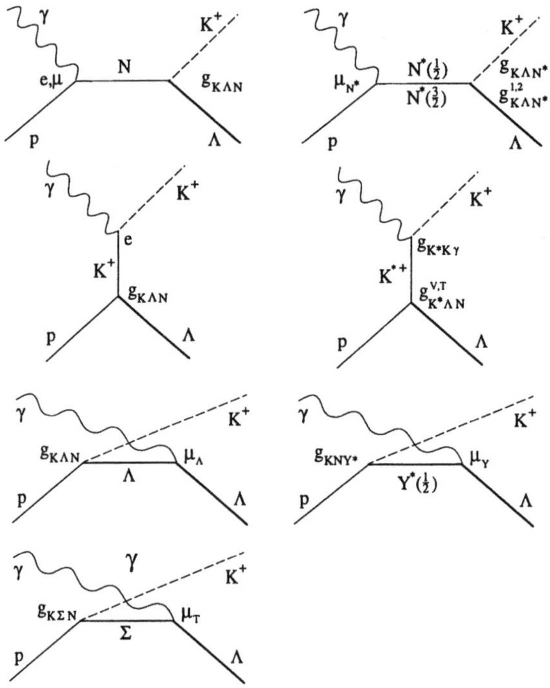

Figure 2. Resonance diagrams for the process $\gamma + p \rightarrow K^+ + \Lambda$. Born terms (left side) and resonant terms (right side) contribute. After ref. [3].

Table 1. Approved experiments at CEBAF involving strangeness, as of July 1995 [5].

Title of Experiment	Exp #	Hall	Contact person	Beam Days
Radiative Decays of the Low-Lying Hyperons	89-024	B	G. Mutchler Rice Univ.	65
Electromagnetic Production of Hyperons	89-004	B	R. Schumacher Carnegie Mellon	65
Measurement of the Structure Functions for Kaon Electroproduction	93-030	B	M. Mestayer CEBAF	50
L/T Cross Section Separation in $p(e,e'K^+)\Lambda(\Sigma^0)$ for $0.5<Q^2<2.0\ (GeV/c)^2$, W> 1.7 GeV , and $t_{min}>0.1\ (GeV/c)^2$.	93-018	C	O.K. Baker Hampton Univ.	15
Measurements of the Electroproduction of the Λ(gnd), $\Lambda^*(1520)$, and $f_0(975)$ via the K^+K^-p and $K^+\pi^-p$ Final States	89-043	B	L. Dennis Florida State	48
Study of Kaon Photo-production on Deuterium	89-045	B	B. Mecking CEBAF	23
Quasi-free Strangeness Production in Nuclei	91-014	B	C. Hyde-Wright Old Dominion	25
Electroproduction of Kaons and Light Hypernuclei	91-016	C	B. Zeidman Argonne	21
Investigation of the Spin Dependence of the ΛN Effective Interaction in the p Shell	89-009	C	E. Hungerford Houston	25
Measurement of the Polarizarion of the $\phi(1020)$ in Electroproduction	93-022	B	E. Smith CEBAF	15
G^0: Measurement of the Flavor Singlet Form Factors of the Proton	91-017	C	D. Beck Univ. of Illinois	46+X
Measurement of Strange Quark Effects Using Parity-Violating Elastic Scattering from ^{4}He at Q^2=0.6 GeV^2	91-010	A	P. Souder Syracuse Univ.	42
Parity Violation on Elastic Scattering from the Proton & ^{4}He	91-004	A	E. Beise Univ. of Maryland	85

Hall B contains the Large Acceptance Spectrometer (CLAS), which is a toroidal magnet design optimized to track charged particles from 8 to 140 degrees for momenta from about 250 MeV/c to 4 GeV/c. An electromagnetic calorimeter and gas Cherenkov detector provide electron, photon, and π^0 detection for angles below 45°. For electroproduction, the luminosity limits expected to be about $10^{34}\text{cm}^{-2}\text{sec}^{-1}$. Momentum resolution is expected to be about 1%. For strangeness production, K/p separation will be done by time-of-flight over a 4 meter flight path, and is expected to work up to about 2 GeV/c. For real photon experiments a photon tagging system operates from 20% to 95% of the

Table 2. Conditionally approved experiments and Letters-of-Intent involving strangeness, as of July 1995 [5].

Title of Experiment	ID #	Hall	Contact person	Beam Days
High Resolution 1p Shell Hypernuclear Spectroscopy	PR-94-107	A	F. Garibaldi Sanita/INFN	24
Electroproduction of Kaons up to $Q^2 = 3$ $(GeV/c)^2$	PR-94-108	A	P. Markowitz Hampton Univ.	
Measurement of K^0 Electroproduction	LOI-94-106	B	R. Magahiz Carnegie Mellon	
Electroproduction of K* Mesons	LOI-94-104	B	D.Doughty CEBAF	
High Resolution Electroproduction of Light Hypernuclei (Helium 4)	LOI-94-103	A	T. Saito	
Heavy Hypernuclei Lifetime, Direct Measurement at CEBAF	LOI-94-101	A	A.T.Margarian	

bremsstrahlung endpoint energy, with about 5 MeV energy resolution. Completion of CLAS is scheduled for late in 1996.

In Hall C there are two principal spectrometers. The Short Orbit Spectrometer (SOS) is suitable for detecting kaons up to 1.5 GeV/c with a momentum resolution of 0.1% FWHM and 7 msr acceptance. The High Momentum Spectrometer (HMS) will detect electrons up to 6 GeV/c, with similar momentum resolution and acceptance. These spectrometers are in their early data taking phases now. For high resolution hypernuclear physics, there are long-range plans for dedicated spectrometers for low momentum electrons and kaons, as well as a dipole to separate forward-scattered electrons and kaons from the beam (see Sect. 4).

In Hall A there are two identical high resolution spectrometers. Recently there have been proposals to use these spectrometers, both for studying elementary production, and when outfitted with septum magnets to reach small angles, for doing hypernuclear physics.

Table 1 is a list of all experiments at CEBAF which involve strangeness. In this talk I will outline these experiments, except those which probe the strange sea using the parity violation in electro-weak scattering of electrons. Table 2 lists recently proposed experiments which were conditionally approved (by PAC9), or are at the Letter of Intent stage.

2 Elementary Production

The field of photo– and electro–production of strange particles benefits from the substantial body of theory and experiment that exists for single pion production. Essentially all of the formalism for these studies is taken from the pion production area. The general formalism for pseudoscalar meson electro– and photo– production can be found in many places, for example refs. [3,6-9]. One finds there are many minor variants among authors in defining notation, cross sections, coordinate systems, and amplitudes.

The relevant kinematic quantities for one-photon exchange are the 4-momentum of the virtual photon $q = (\nu, \boldsymbol{q})$ and the direction (θ_m, ϕ) of the meson relative to $\boldsymbol{q}$ in the electron scattering plane, as shown in Fig. 3. The photon's energy is $\nu = E - E'$, and its 3- momentum is $\boldsymbol{q}$. The electron vertex is perfectly well known, which means that the polarization of the virtual photon is determined purely from the electron kinematics. The polarization has three components, since the photon is off shell, two transverse and one longitudinal. The polarization parameter ϵ, defined as

$$\epsilon = \left(1 - 2\frac{\boldsymbol{q}^2}{q^2}\tan^2\frac{\theta_e}{2}\right)^{-1} \tag{1}$$

ranges over ± 1. It gives the degree of linear transverse polarization along x ($\epsilon = +1$) or y ($\epsilon = -1$), and is directly proportional to the longitudinal polarization along z. The reaction factorizes into a virtual photon cross section and a virtual photon flux. The virtual photon flux into a phase space element $dE'd\Omega_e$ is given by

$$\Gamma(E', \Omega_e) = \frac{\alpha}{4\pi^2}\frac{E'}{E}\frac{W^2 - m_n^2}{m_n}\frac{1}{-q^2}\frac{1}{1-\epsilon} \tag{2}$$

where W is the invariant photon-nucleon energy and m_n is the nucleon mass.

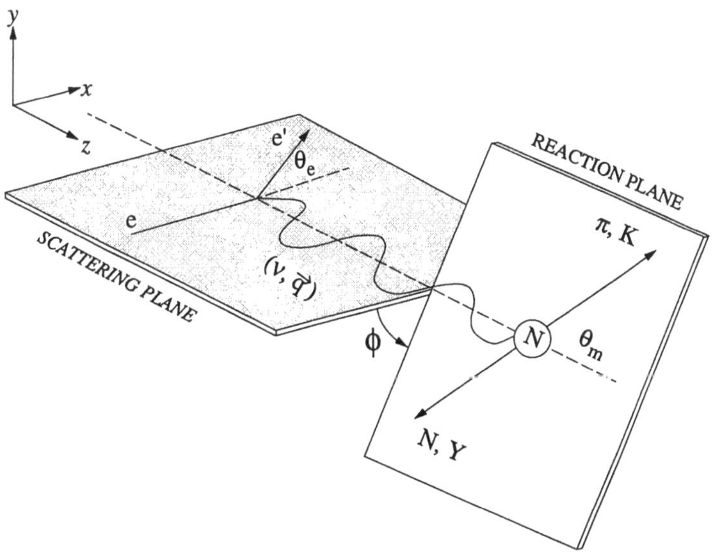

Figure 3. Kinematic variables for meson electroproduction.

When kaons are produced with unpolarized electron beams and only the electron and meson are detected, then there are four kinematically separable pieces to the cross section which can be written in the form

$$\frac{d\sigma}{dE' d\Omega_e d\Omega_m} = \Gamma(E', \Omega_e)$$
$$\times \left(\frac{d\sigma_{\mathrm{T}}}{d\Omega_m} + \epsilon \frac{d\sigma_{\mathrm{L}}}{d\Omega_m} + \epsilon \cos(2\phi) \frac{d\sigma_{\mathrm{TT}}}{d\Omega_m} + \sqrt{\epsilon(\epsilon+1)/2} \cos(\phi) \frac{d\sigma_{\mathrm{LT}}}{d\Omega_m} \right) \quad (3)$$

The components of the cross section are each related to six complex gauge and Lorentz invariant amplitudes derived from the hadronic currents, $\boldsymbol{J}$. Various equivalent sets of amplitudes are used in the literature: helicity, transversity, and extended CGLN basis amplitudes are used, and all can be written as linear combinations of each other. In any representation, σ_{T} is the unpolarized transverse cross section, which in rectangular coordinates is proportional to $\int d\Omega_m [< J_x^* J_x + J_y^* J_y >]$ (ref. [7]). The longitudinal cross section, σ_{L}, goes as $\int d\Omega_m < J_z^* J_z >$. In inclusive experiments, where the meson is not detected, these two terms are extracted from data at fixed q^2 and W, but for differing ϵ, and doing the famous Rosenbluth separation. The transverse-transverse cross section σ_{TT} arises from the interference of the two components of transverse current, and is proportional to $\int d\Omega_m [< J_x^* J_x - J_y^* J_y >]$. The photon vertex dictates that this term is proportional to $\cos\phi$, where $\phi = 0$ corresponds to the meson being detected on the "$+x$" side in the electron scattering plane. This term can also be accessed with linearly polarized real photons. The σ_{LT} term is due to the interference of the longitudinal and transverse parts of the current and is proportional to $\int d\Omega_m [< J_x^* J_z - J_z^* J_x >]$.

Because the longitudinal photons transfer no angular momentum, σ_{L} is sensitive to the exchange of $0^{\pm}$ exchanges in the t-channel, and hence is sensitive to pions, diquarks, or other spin-zero objects in the nucleon. At low q^2 and near minimum t, the longitudinal cross section is dominant in pion production [7]. With suitable extrapolations it has been possible to extract the pion form factor from pion electroproduction data [10]. The analogous work for kaons has not been done, but will be the subject of several CEBAF experiments (see Sect. 2.2)

Computing the relevant amplitudes directly from QCD-inspired models has not been attempted until very recently for the strangeness-producing channels. More typically, models have been developed using hadronic and mesonic degrees of freedom. Figure 2 shows the diagrams that are needed. The lowest order Born terms involve proton, lambda, and kaon exchange in the s, u, and t-channels, respectively. Other important terms involve the exchange of the Σ, the spin-1 kaon resonances, the spin 1/2 and spin 3/2 nucleon resonances, and spin 1/2 hyperon resonances. The coupling constants can be extracted from the data and compared either with values predicted by SU(6), or with values obtained from other strangeness-producing reactions. The most elaborate theoretical treatment of kaon electroproduction, which includes crossing symmetry and s and t channel duality, is that of Cotanch and coworkers [11]. In photoproduction, where only four complex amplitudes are needed, recent work has been done by Adelseck and Saghai [12], Adelseck, Bennhold & Wright [3]; very recently a chiral quark model approach has been presented by Li [13].

2.1 On the proton - photoproduction

The elementary photoproduction reactions on the proton are $\gamma + p \rightarrow K^+ + \Lambda$, $\gamma + p \rightarrow K^+ + \Sigma^\circ$, and $\gamma + p \rightarrow K^\circ + \Sigma^+$. Total cross sections exist for all cases, differential cross section data exist for the Λ and Σ°, and very sparse hyperon and target polarization data exist for the Λ. While being restricted to $q^2 = 0$, real photon experiments using tagged or untagged bremsstrahlung beams have been more numerous than electroproduction experiments because of their relative simplicity. Figure 4 gives an example of the existing Λ differential cross section data and polarization data [14, 3]. About a dozen theoretical models of these reactions have been published over the past thirty years, all based on the diagrammatic techniques mentioned earlier. The relatively poor quality of the data has led to a wide range of "best" fits, such that even the principal Born couplings $g_{K\Lambda N}$ and $g_{K\Sigma N}$ (see Fig. 2) vary over a wide range, differing even in sign in some cases [15]. Unlike pion photoproduction where one strong resonance, the $\Delta(1232)$, dominates the reaction, for hyperons the resonance structure is just not well known.

The contribution of CEBAF to this field will be to provide data for all three elementary channels, and to emphasize the polarization variables which show large sensitivity to the model parameters. Experiment E89-004 will use CLAS to obtain such data, using the detector to examine the self-analyzing weak decays of the hyperons to an extent impossible in earlier measurements. Figure 4, for example, shows an old analysis of Renard [14] which illustrates the sensitivity of that model to Λ polarization data: the hatched region shows the range of predictions due to reasonable variations of the couplings $g_{K\Lambda N}$ and $g_{K\Sigma N}$. The polarization shown is that of the hyperon perpendicular to the (γ, K) plane (albeit with the sign opposite to most modern authors' usage). The Σ° polarization information will be accessible in CLAS because this hyperon decays 100% to $\Lambda\gamma$. The polarization of the Λ, turns out to be $-1/3$ of the Σ° polarization, thus preserving in diluted form the polarization produced in the reaction. In another category of polarization measurements, CLAS will be able to operate with circularly polarized photon beams produced by bremsstrahlung from polarized electrons. Discussions are underway to extend E89-004 to include such measurements.

Real photoproduction of the Σ^+ and Σ° and Λ are of equal importance, since all three reactions ought to be described within the same theoretical framework. An interesting feature of Σ^+ production is the absence of simple t-channel exchange because the photon does not couple to the K°. This results in a backward peak in the predicted differential cross section, which would be easy to find experimentally [16]. By detecting $K^\circ \rightarrow \pi^+\pi^-$ this reaction will be accessible in CLAS.

For the Σ^+ there exist only a few total cross section data points. But Mart et al. [17] have recently shown how existing models can fail badly when applied to the sparse data in this channel, typically overpredicting the cross section by an order of magnitude, thus underscoring the need for significant data in this area. Z. Li[13] has recently presented a chiral quark model of strangeness

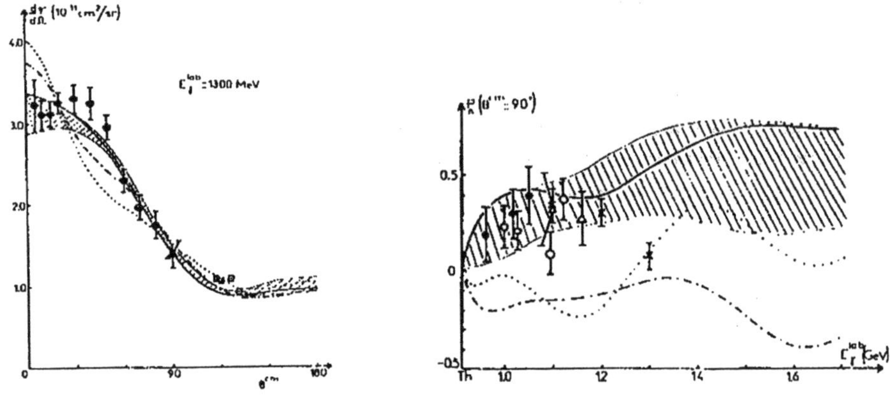

Figure 4. (a) Differential cross sections for $\gamma + p \to K^+ + \Lambda$ as a function of kaon c.m. angle, from ref. [14]. The shaded region corresponds to $g_{K\Lambda N}/\sqrt{4\pi}$ varying from 1.1 to 2.8. (b) Λ polarization data for $p(\gamma, K^+)\Lambda$ for kaon c.m. angles of $90° \pm 5°$.

photoproduction which overcomes the guesswork and data-fitting of traditional hadronic models. In his model the mesons are treated as Goldstone bosons, but couplings to known resonances are computed exactly, and the low energy theorem is fulfilled. The Λ data are reproduced well, and problems in other models with unitarity at higher energies and blow-up of the Σ^+ cross section are gone.

2.2 On the proton - electroproduction

Several CEBAF experiments will study the $(e, e'K)$ reaction on the proton. Each of them plans to separate the four components of the cross section in Eq. (3). Experiment 93-030 will use CLAS to obtain charged kaon production data over a wide kinematic range to maximize the ability to do the L/T separations. While systematic uncertainties inherent in a large acceptance detector are a possible limitation, the acceptance also makes extraction of the interference structure functions potentially easier than in a conventional spectrometer. A recent Letter of Intent, 94-106, would extend these studies in CLAS to include the neutral kaon channels. K^o's have essentially never been measured in electroproduction, and CLAS is ideally suited to their detection.

In Hall C, Experiment 93-018 uses the SOS and HMS spectrometers to do the strange-particle analog of the single pion electroproduction done in the past. σ_{LT} will be obtained over a small range by moving the SOS spectrometer left and right in the lepton scattering plane. The goals are the same as for the Hall B experiments. Using the Hall A spectrometers has also been proposed (PR94-108) for extending the electroproduction measurements from 2 to 3 $(\mathrm{GeV}/c)^2$, but this will require 6 GeV beam and successful running of the Hall C experiments.

The resonance structure of $(e, e'K)$ is unknown, as mentioned above, and the reaction is not dominated by a single strong resonance. As one example of the physics to be addressed, Fig. 5 shows the Q^2 dependence ($Q^2 = -q^2$) of the electroproduction of Λ and Σ^o for a fixed value of the total c.m. energy W. For Λ production the cross section first rises from the real photon point at $Q^2 = 0$, while for Σ^o production the drop is monotonic with Q^2. This is interpreted as an indication that longitudinal photon contributions are important for Λ production but not Σ production. A rough longitudinal-transverse separation was obtained [18] by measuring at different ϵ for the same W and Q^2, supporting the conclusion that a large longitudinal component contributes to Λ production only. A suggested explanation [19] is that longitudinal photons contribute strongly to kaon exchange in the t-channel, but this process is expected to be larger for Λ production than for Σ production since $g_{K\Lambda N} > g_{K\Sigma N}$, and hence the difference in Q dependence.

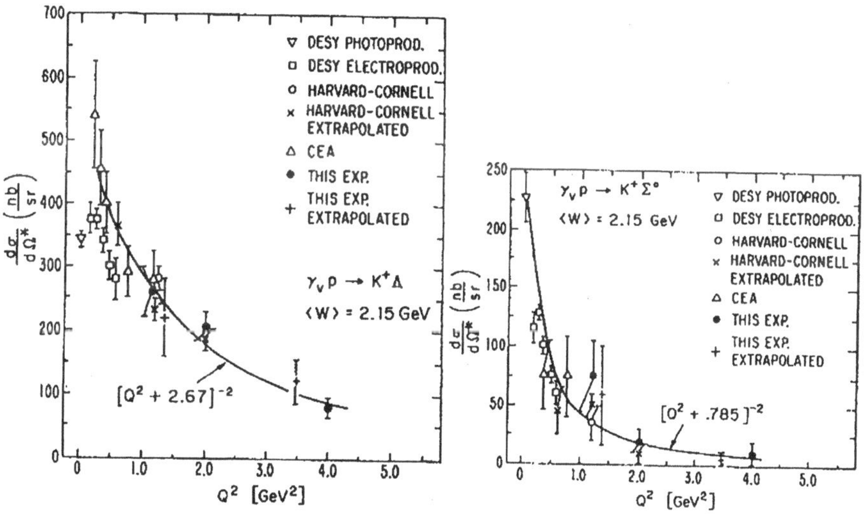

Figure 5. Q^2 dependence for $\gamma_{(v)} + p \to K^+ + \{\Lambda, \Sigma^o\}$. The trends away from the real photon point at $Q^2 = 0$ differ strongly. Data are from ref.[23] and others cited therein. $d\sigma/d\Omega^*$ is the sum of longitudinal and unpolarized transverse cross sections in the c.m. of the photon and proton.

The Q^2 dependence of Σ^o production is much steeper than that for Λ production. In quark-parton models [20-22] this feature was interpreted as consistent with the decrease of the ratio $F_1^{\gamma n}/F_1^{\gamma p}$, the deep inelastic electron-nucleon structure functions of the neutron and proton, as Bjorken $x = Q^2/2M\nu$ goes to 1. In this limit the production of forward-going kaons off u-quarks tends to leave behind an isospin 0 pair $u-d$ quarks, from which the production of $I=1$ baryons (Σ) is suppressed in favor of $I=0$ baryons (Λ).

The t dependence of the longitudinal cross section can also be interpreted in terms of the kaon form factor, similar to work done on the pion form factor. Experiment 93-018 will emphasize this aspect of their analysis.

2.3 *On the deuteron - photoproduction*

After the elementary production mechanism on the proton is established, it only makes sense to extend our knowledge using a deuteron target to get the three reactions on the neutron: $\gamma+n \rightarrow K^{+}+\Sigma^{-}$; $\gamma+n \rightarrow K^{\circ}+\Lambda$; $\gamma+n \rightarrow K^{\circ}+\Sigma^{\circ}$. A set of data including all isospin channels will naturally provide the most rigorous test of our understanding. This will be done by CEBAF experiment 89-045. No data of this kind exist to date. Besides extracting the elementary cross sections [24], it may be possible to explore YN interaction effects; since the photon and the K^{+} both interact "weakly," the deviations from quasi-free behavior in the (γ, K) reaction will be due to the interaction of the final state hyperon–nucleon system [25]. This experiment will also scan the cross section as a function of missing mass in the region between the Λ and the Σ°. Here the amplitudes for production of the two hyperons will interfere, producing a cusp that has been predicted to be visible [26]. This channel-coupling cusp has been seen in $D(\pi^{+}, K^{+})$ and $D(K^{-}, \pi^{-})$ reactions [27] but electromagnetic studies of the shape of the cusp will reveal the relative phase of the production amplitudes. Experiment 91-016 will also hunt for this cusp in electroproduction in Hall C.

Another goal of E89-045 in Hall B is to look for narrow structures near the ΣN threshold which might be interpreted as $S=-1$ dibaryons [28]. The spin-one $D1$ should be 20-40 MeV above the cusp while the spin-zero $D0$ should be $20-40$ MeV below the cusp. The $D1$ can be reached from the deuteron using reactions with small spin-flip probability: (π, K), (K, π), and (γ, K). But the (γ, K) reaction is special because only it can also populate the spin-flipped $D0$ state. This state should be narrower than the $D1$, and therefore might be worth a look despite failure to detect the $D1$ in the mesonic reactions [29].

2.4 *On heavier nuclei - inclusive kaon photoproduction*

Once the elementary amplitudes for kaon photoproduction are known, it is interesting to know if the inclusive nuclear response can be described in a simple quasi–free picture, *i.e.* incoherent production on individual nucleons. From inclusive $A(e, e')$ and $(e, e'p)$ studies it is known that medium modifications such as two-nucleon currents can substantially alter quasi-free behavior. There may be analogous effects in strangeness production, though no ernest theoretical estimates have been made.

Experiment 91-014 will study the relative importance of the one-nucleon (γ, K) versus either two-nucleon (γ, K) currents or KY final state interactions (FSI). The targets will be ^{3}He, ^{4}He, and ^{12}C. Due to strangeness conservation, the total $A(\gamma, K)$ cross sections will be insensitive to FSI, since either a K^{+} or a K° must emerge once either has been created. The A dependence of kaon production for D:^{4}He:C thus will be sensitive to the presence of medium modifications of the elementary amplitudes. The ratio of K^+/K° yields for isoscalar nuclei will be measured and compared to deuterium. This ratio would probe either two-nucleon currents or charge-exchange final state interactions. Finally, the differential ^{3}He data will be a detailed test of the quasi-free model, because good wavefunctions exist and calculations should be most reliable.

3 Radiative Hyperon Decays

Radiative decays are in general useful for testing quark models of baryons. Their widths are sensitive to the detailed quark wave functions used in the models, with predictions presently varying by a factor of sixty. In an elementary quark model certain radiative transitions are forbidden: *e.g.* the $I = 0$ $\Lambda(1405)$ cannot decay to the two lower-lying $I = 1$ Σ states because a single photon, described by one-body operators, cannot connect states that involve changing two quarks. There is some belief that the $\Lambda(1405)$ consists at least partly of a KN molecular state, owing to the proximity of the KN threshold at 1437 MeV. Because the $\Lambda(1405)$ and the $\Sigma(1385)$ are below this threshold they cannot be studied using stopped K^- on liquid hydrogen.

The CLAS spectrometer is suited to making these states using the reaction $\gamma p \rightarrow K^+Y^*$. In Experiment 89-024, the incident real photons are tagged, the kaons are detected and together used to construct the Y^* missing mass; in the cases of the $\Lambda(1405)$ and the $\Sigma(1385)$ there is an overlap because the states are 50 and 36 MeV wide, while the $\Lambda(1520)$ is well separated. The radiative decays to the ground state Λ will be detected by following the decay chain $Y^* \rightarrow \gamma\Lambda$ followed by $\Lambda \rightarrow \pi^- p$. Detection in the EGN calorimeter of the photon, in addition to the charged particles, greatly suppresses the background due to π° decays. Detection of decays with the $\Sigma(1193)$ as an intermediate state will be more difficult because both decay photons must be detected.

4 Hypernuclear Production

No data exist on electromagnetic production of hypernuclear states. The reasons are clear: the large momentum transfers (comparable to the (π, K) reaction) and small predicted cross sections (on the order of 10 nb/sr for even the lowest momentum transfers) make such reactions challenging to measure. The 5 MeV photon energy resolution of the CLAS photon tagging system make hypernuclear photoproduction experiments in Hall B unlikely. In Halls A and C, however, the $A(e, e'K)_\Lambda A$ reaction will be studied. Detecting the electrons close to zero degrees minimizes the momentum transfer, and the count rate for populating specific states will be on the order of several counts per day, similar to (π, K) studies. There are two main benefits in doing $(e, e'K)$ hypernuclear spectroscopy. The first is that up to an order of magnitude improvement in resolution may be possible over present hadronic data. The second is that while the (π, K) reaction populates high-spin natural-parity states, the $(e, e'K)$ reaction also populates the unnatural parity states, starting from 0^+ nuclei, via the spin-flip interaction. CEBAF can therefore provide data complimentary to the hadronic production reactions.

The goal of all the proposed hypernuclear experiments is measuring the Λ-nucleus spin-orbit splitting, which is related to the ΛN spin-orbit interaction. The strength of this interaction is known to be small [30], with splittings of under 0.5 MeV, in constrast to the nucleon spin-orbit interaction which is of order several MeV . Thus, while gamma-ray studies have seen transitions between widely spaced levels, the fine structure has not been seen. Experiment 89-009 will study the spectroscopy of the p-shell in Hall C; the plan is to use a 1.8 GeV electron beam, a dedicated Enge split-pole magnet to detect scattered electrons at zero degrees, and the SOS spectrometer with about 1.2 MeV resolution for the kaons. A "splitting magnet" must be added at the target to peel off the electrons and kaons to large enough angles to enter the spectrometers. A survey of p-shell nuclei is planned to map the overall response and to evaluate rates and backgrounds. As a future development this program hopes to develop a dedicated kaon spectrometer with resolution in the range of 120 keV.

The conditionally approved proposal 94-107 plans a similar program of p-shell studies in Hall A, with a goal of 300 to 400 keV resolution [31]. There, too, the target must be moved upstream, septum magnet placed in front of each spectrometer; full approval awaits a review of technical feasibility.

Experiment 91-016 will examine electroproduction of kaons, again in Hall C, on the proton, deuterium, ^{3}He and ^{4}He, but without special efforts to obtain high resolution. The physics goals are examination of the $\Lambda - \Sigma$ cusp in deuterium (see Sect. 2.3) and examination of bound state angular distributions in the helium isotopes. Other goals include dibaryon hunting near the $\Sigma - N$ cusp, and searching for narrow Σ states. The measurements are essentially exploratory to obtain cross sections and count rates. This experiment will use the "standard" Hall C configuration, combining the SOS for kaons (same as experiment 89-009) with the HMS for the electrons. Both spectrometers will be at angles greater than 12 degrees, and the missing mass-resolution is expected to be about 3 MeV.

5 Hidden Strangeness

Hidden strangeness means that the strangeness quantum number is zero in the initial and final states, but strange quarks manifest their presence in some particular property of a system. We will not discuss the parity violation experiments here, but mention instead plans to measure $\phi(1020)$ electroproduction as a method of measuring the strange quark content of the nucleon. In a hard scattering process, an $s\bar{s}$ component of the proton may be knocked out to create a final state ϕ. In a model due to Henley an coworkers [32], such contributions may be of the same order as the diffractive processes described using vector meson dominance if the strange quark sea of the nucleon in the range of 10% to 20%.

Experiment 93-022 plans to look for such ϕ knock-out via electroproduction with the CLAS spectrometer. The key to separating the diffractive from the knockout interactions is to examine the self-analyzing decay of the ϕ's. Henley et al. predict that the knockout mechanism of interest goes via pseudo-scalar (0^-) exchange, different from production that involves scalar (0^+) exchange. In the rest frame of the decaying ϕ's the kaons are produced with an angular distribution, W, as

$$W(\phi) = \frac{3}{8\pi}\sin^2\theta(1 \pm \epsilon\cos 2\psi) \tag{4}$$

where θ is the kaon polar angle and ψ is the azimuthal angle with respect to the lepton plane. Pseudo-scalar and scalar exchange are characterized by the + and the − sign, respectively. By measuring this decay distribution, the experiment plans to achieve a sensitivity of a 5% admixture of pseudo-scalar exchange, at which level the OZI- suppressed π and η exchange diagrams are also expected to contribute. Any larger measured amount of 0^- exchange would be a possible signature of strange quark components in the nucleon.

6 Conclusion

This broad overview has only skimmed the contents of strangeness–related proposals at CEBAF. I hope, however, that it has convinced you that an important and exciting aspect of CEBAF's overall program is in the area of strangeness physics. In another year or two I hope there will actually be some new results to report.

References

1. R. A. Schumacher: Nucl. Phys. **A585**, 63c (1995)
2. V. Burkert: these *Proceedings*
3. R.A.Adelseck, C. Bennhold, and L.E.Wright: Phys. Rev. **C32**, 1681 (1985)
4. *CEBAF Conceptual Design Report*, April 1990
5. Copies of proposals are available upon request from the CEBAF Users' Liaison Office, at 804-249-7536
6. N. Dombey: *Hadronic Interactions of Electrons and Photons*, Ed. J. Cumming and H. Osborn, Acad. Press 1971
7. J. M. Laget: Can. J. Phys. **62**, 1046 (1984)
8. C. W. Akerlof et al.: Phys Rev **163**, 1482 (1967)
9. R.C.E. Devenish and D. H. Leith: Phys. Rev. **D5**, 47 (1972)
10. P. Brauel et al.: Z. Phys. **C3**, 101 (1979)

11. Robert A. Williams, Cheung-Ryong Ji, and Stephen R. Cotanch: Phys. Rev. **C46**, 1617 (1992); Stephen Cotanch and Shian Hsiao: Nucl. Phys. **A450**, 419c (1986)

12. R.A.Adelseck and B. Saghai: Phys. Rev. **C42**, 108 (1990)

13. Zhenping Li: to be published in Phys. Rev. **C**

14. Y. Renard: Nucl. Phys. **B40**, 499 (1972); F. M. Renard and Y. Renard: Nucl. Phys. **B25**, 491 (1971)

15. R.A. Schumacher: *Particle Production Near Threshold* (AIP Conf. Proc. 41), p. 378, H. Nann, E.Stephenson, Eds. 1990

16. C. Bennhold: private communication

17. T. Mart, C. Bennhold, and C. E. Hyde-Wright: Phys. Rev. **C51**, R1074 (1995); see also these *Proceedings*

18. C. J. Bebek et al.: Phys Rev **D15**, 3082 (1977)

19. T. Azemoon et al.: Nucl. Phys. **B95**, 77 (1975)

20. F.E. Close: Nucl. Phys. **B73**, 410 (1974)

21. O. Nachtmann: Nucl. Phys. **B74**, 422 (1974)

22. J. Cleymans and F.E.Close: Nucl. Phys. **B85**, 429 (1975)

23. C. J. Bebek et al.: Phys Rev **D15**, 594 (1977)

24. Xiaodong Li, L.E. Wright and C. Bennhold: Phys. Rev. **C45** 2011 (1992)

25. R.A.Adelseck and L.E.Wright: Phys. Rev. **C38**, 1965 (1988); Ralf Anton Adelseck: PhD thesis Ohio University 1988

26. S. R. Cotanch: *Strangeness Production Studies at CEBAF* (Proc. of the Conf. on Medium and High-Energy Physics), p. 666, Taipei 1989

27. C. Pigot et al.; Nucl. Phys. **B249**, 172 (1985); T. H. Tan: Phys. Rev. Lett. **23**, 395 (1985)

28. C.B.Dover: Nucl. Phys. **A450**, 95c (1986)

29. H. Piekarz et al.: Nucl. Phys. **A450**, 85c (1986); H. Piekarz et al.: **A479**, 263c (1988)

30. D.J. Millener, A. Gal, C. B. Dover, and R. H. Dalitz: Phys. Rev. **C31**, 499 (1985)

31. S. Frullani: these *Proceedings*

32. E.M. Henley et al.: Phys Lett. **B281**, 178 (1991)

33. C. B. Dover and P. M. Fishbane: Phys Rev Lett. **64**, 3115 (1990)

Few-Body Systems Suppl. 9, 369–373 (1995)

Few-Body Systems

Kaon Photo- and Electroproduction on Nucleons

T. Mart,[1] C. Bennhold[2]

[1] Institut für Kernphysik, Johannes Gutenberg-Universität, 55099 Mainz, Germany,

[2] Center of Nuclear Studies, Department of Physics, The George Washington University, Washington, D. C. 20052, USA

Abstract. We extend previous models of kaon photo- and electroproduction in order to include all six isospin channels. It is found that the inclusion of the few available data for the reactions $\gamma p \rightarrow K^0\Sigma^+$ and $\gamma n \rightarrow K^+\Sigma^-$ in the fit leads to drastically reduced Born coupling constants g_Λ and g_Σ. The result suggests the need to include hadronic form factors in a gauge invariant fashion. It is also shown that the K^0 form factor can be seen in $K^0\Lambda$ electroproduction.

1 Introduction

The investigation of kaon photo- and electroproduction has the potential to become an important aspect in meson physics since many important features of the electromagnetic and hadronic interaction involving strangeness can be studied through these processes. Interest in this topic has increased mainly through the construction of new high-duty continuous electron beam machine such as ELSA in Bonn or CEBAF. Recently, a new set of photoproduction data has appeared from the SAPHIR Collaboration in Bonn [1]. On the other hand, the theoretical description of the photo- and electroproduction of kaon is still far from settled. The lack of experimental data, incomplete knowledge of the hadronic coupling constants (CC), and the many resonances that can contribute to the reaction can be blamed for a proliferating number of models. Meanwhile, however, a basic understanding of these elementary reactions is required in order to predict cross sections for the photoproduction of hypernuclei.

In this paper we develop extensions of previous models in order to include all six isospin channels of kaon production on the nucleon. For this purpose we employ the few available total cross section data for the charged Σ-photoproduction reactions, $\gamma p \rightarrow K^0\Sigma^+$and $\gamma n \rightarrow K^+\Sigma^-$. Unfortunately, the presently available data are too few to allow quantitative conclusions. Future kaon data from Bonn and CEBAF that include all production channels are therefore expected to solve this problem. Since the kaon form factor (FF) can be accessed through kaon electroproduction, we present here also the effect of different K^0 FF in K^0 electroproduction.

2 The Elementary Kaon Production Operator

We choose an isobaric model [2] to describe the photo- and electroproduction processes. To relate all kaon productions channels, we employ the isospin formalism and use some information on resonance decay widths. Our model for KY-production consists of the standard Born terms along with the intermediate K^*-exchange. Furthermore, we have incorporated the N^* resonances $S_{11}(1650)$ and $P_{11}(1710)$ as well as the Δ resonances $S_{31}(1900)$ and $P_{31}(1910)$ which can only contribute to Σ photoproduction. For $K\Lambda$-production, we excluded the Δ resonances due to isospin conservation. Our choice of resonances was guided by our goal to draw qualitative conclusions about the behavior of CC with a simple model that contains as few parameters as needed to achieve a reasonable χ^2.

2.1 Charged Σ Photoproduction

Table 2.1 shows the extracted CC from previous works along with the present calculation. In Fig. 1 we compare the total cross section obtained from different models for all possible isospin channels in kaon photoproduction. One of the more recent models in K^+ photoproduction [3] which yields CC in agreement with hadronic determinations and SU(3) values (set II) is shown for the $K\Lambda$ channels. Since the model fits only the data below 1.5 GeV, it diverges rapidly at higher energies. A more advanced model for $K\Lambda$ production (set I) from Ref. [4] is also shown for both Λ channels. Since both Born couplings g_Λ and g_Σ in this model are already small, the total cross section diverges slower as the energy increases. The simplest model (set III) for $K\Sigma$ is taken from Ref. [5]. Since it contains the largest g_Σ, the cross sections drastically increase as the function of energy. The different predictions of our new model with set IV and set V of the CC illustrate the same point. Fitting all $K^+\Sigma^0$ photo- and electroproduction data (set IV) leads to very large discrepancies with the $K^+\Sigma^-$ and $K^0\Sigma^+$ total cross section data. Including those data into the fit yields a coupling strength g_Σ (set V) that differs by almost a factor of 10 from the coupling constant in set IV. Thus, fitting all $K\Sigma$ data simultaneously reduces the Born couplings to very small values, almost eliminating the Born terms. The same pattern, although less drastic, can be seen in set VI and VII, where we included the new Bonn data and fitted both $K\Lambda$ and $K\Sigma$ channels simultaneously.

2.2 Inclusion of a Hadronic Form Factor

Until now no analysis has ever included a form factor at the hadronic vertex. The above result suggests the need to include a hadronic FF, since such a FF is expected to reduce the divergent Born terms at higher energies. However, it is well known that the FF contributions to different diagrams violates gauge

Table 2.1. The main CC set I and II fit the $K\Lambda$ data. Set III, IV, and V are from $K\Sigma$ studies, set V includes charged Σ data, while set IV does not. Set VI is generated by fitting to all $K\Lambda$ and $K\Sigma$ but excluding the charged Σ data, set VII includes the charged Σ, and set VIII shows the result after we used the hadronic FF. (*) means not reported, while (†) shows our present calculation

Set	data included	$\frac{g_\Sigma}{\sqrt{4\pi}}$	$\frac{g_\Lambda}{\sqrt{4\pi}}$	N	$\frac{\chi^2}{N}$	Ref.
I	$p(\gamma,K^+)\Lambda$, $p(e,e'K^+)\Lambda$	0.23	−2.38	(*)	(*)	[4]
II	$p(\gamma,K^+)\Lambda$	1.18	−4.17	117	1.3	[3]
III	$p(\gamma,K^+)\Sigma^0$	2.72	−1.84	86	3.15	[5]
IV	$p(\gamma,K^+)\Sigma^0$, $p(e,e'K^+)\Sigma^0$	1.30	−0.84	182	2.67	[2]
V	$p(\gamma,K^+)\Sigma^0$, $p(e,e'K^+)\Sigma^0$ $p(\gamma,K^0)\Sigma^+$, $n(\gamma,K^+)\Sigma^-$	0.13	0.51	190	5.30	[2]
VI	$p(\gamma,K^+)\Lambda$, $p(e,e'K^+)\Lambda$, $p(\gamma,K^+)\Sigma^0$, $p(e,e'K^+)\Sigma^0$	0.97	−2.07	663	6.67	(†)
VII	$p(\gamma,K^+)\Lambda$, $p(e,e'K^+)\Lambda$, $p(\gamma,K^+)\Sigma^0$, $p(e,e'K^+)\Sigma^0$, $p(\gamma,K^0)\Sigma^+$, $n(\gamma,K^+)\Sigma^-$	0.53	−2.10	671	7.18	(†)
VIII	$p(\gamma,K^+)\Lambda$, $p(e,e'K^+)\Lambda$, $p(\gamma,K^+)\Sigma^0$, $p(e,e'K^+)\Sigma^0$, $p(\gamma,K^0)\Sigma^+$, $n(\gamma,K^+)\Sigma^-$	1.27	−4.09	671	5.48	(†)

invariance. Hence, one may consider two ways to include a hadronic FF, either by multiplying the whole amplitude with a FF or through minimal substitution which was motivated by Ohta [6]. To obtain a qualitative understanding we use the first method by multiplying the amplitudes with a monopole FF $F(\Lambda,t) = (\Lambda^2 - m_K^2)/(\Lambda^2 - t)$. The result is quite dramatic. As shown in set VIII the main CC increase to values consistent with SU(3) and the χ^2 is reduced significantly, even though we used the entire data set.

3 The K^0 Form Factor

Similar to the K^+ FF, the K^0 FF can be accessed through kaon electroproduction. Among the SU(3) pseudoscalar mesons, the K^0 is the only neutral system that can have a nonvanishing electromagnetic FF at finite photon four-momentum, k^2. The FF for π^0 and η vanish since the quark masses are the same and the charges opposite, while for the K^0 the non zero FF is a direct consequence of the unequal quark masses, and, therefore, is strongly dependent on the value of the strange quark mass. We have employed two models [7], both of which reproduce charge radii and weak decay constants, to calculate the K^0 FF, the light-cone quark (LCQ) model with Gaussian radial (harmonic oscillator parameter $\beta = 320$ MeV) and the quark-meson vertex (QMV) model with

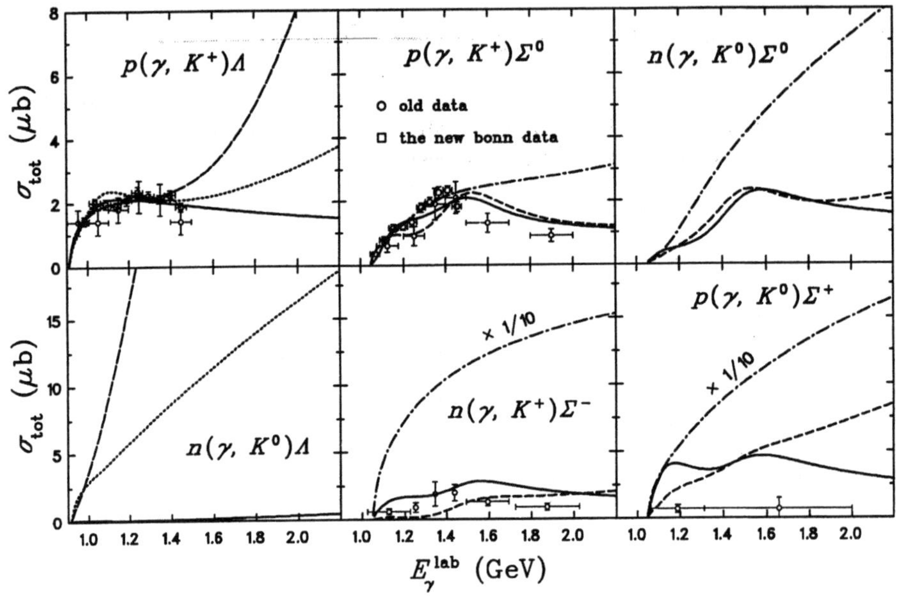

Figure 1. Total cross sections in kaon photoproduction. The dotted (dashed) curve in $K\Lambda$ represents the model with CC of set I (II) of Table 2.1. The dash-dotted (dashed) curve in $K\Sigma$ comes from set III (V), while the solid curve fits all data in the $K\Lambda$ and $K\Sigma$ channels (set VIII). The new experimental data are from [1].

a monopole-like quark vertex function ($\Lambda = 800$ MeV).

We have examined the sensitivity of the $p(e,e'K^0)\Sigma^+$, $n(e,e'K^0)\Sigma^0$ and $n(e,e'K^0)\Lambda$ reactions to the different K^0 form factor models and found only very small changes in the observables of the two Σ production reactions. This can be understood from the fact that the leading Born coupling constant, g_Σ, which multiplies the K^0 t-channel pole term, in most models is much smaller than the g_Λ coupling constant which governs the $K\Lambda$ Born terms. In particular, we used the model from Ref. [4] (set I in Table 2.1) to investigate the FF behavior. The result is demonstrated in Fig. 2. Clearly both models give a considerable effect on the longitudinal cross section, in fact, they increase the cross section by up to 30% at certain kinematics, while the transverse cross section remains insensitive.

4 Conclusion

Kaon photo- and electroproduction have been studied simultaneously for all six possible isospin channels. We have demonstrated that existing models can only partially describe the process. The hadronic FF has been shown can improve the models by enhancing the two main CC and reducing the χ^2/N. A complete data set in all isospin channels is urgently needed to improve the models, especially for charged Σ and $K^0\Lambda$ photoproduction. We have also shown that the K^0 FF

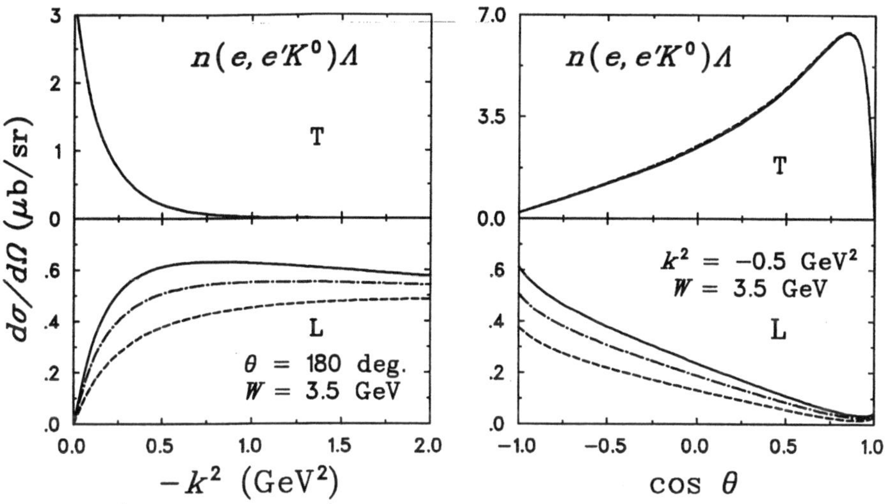

Figure 2. Transverse and longitudinal cross sections for the K^0 electroproduction. The solid (dash-dotted, dashed) line shows the results with the K^0 FF obtained with the LCQ model (QMV model, no K^0 FF).

can have a significant effect on the longitudinal $K^0\Lambda$ electroproduction cross section for certain kinematics, hence a measurement of this FF might be a challenging project for CEBAF in the future.

Acknowledgement. The work of TM is supported by Deutscher Akademischer Austauschdienst and Deutsche Forschungsgemeinschaft (SFB 201) while the work of CB is supported by the US DOE grant no. DE-FG02-95-ER40907.

References

1. M. Bockhorst et al.: Z. Phys. **C63**, 37 (1994)
2. T. Mart, C. Bennhold, and C.E. Hyde-Wright: Phys. Rev. **C51**, R1074 (1995); T. Mart and C. Bennhold: Nucl. Phys. **A585**, 369c (1995)
3. R.A. Adelseck and B. Saghai: Phys. Rev. **C42**, 108 (1990)
4. R.A. Williams, C.R. Ji, and S.R. Cotanch: Phys. Rev. **C46**, 1617 (1992)
5. C. Bennhold: Phys. Rev. **C39**, 1944 (1989)
6. K. Ohta: Phys. Rev. **C40**, 1335 (1989); R.L. Workman, H.W.L. Naus, and S.J. Pollock: Phys. Rev. **C45**, 2511 (1992)
7. C. Bennhold, H. Ito, and T. Mart: to be published

Few-Body Systems Suppl. 9, 374–378 (1995)

Few-Body Systems

Strangeness Studies off Proton and Nuclei in CEBAF Hall A

E. Cisbani [1], S. Frullani [1], F. Garibaldi [1], M. Iodice [1], G.M. Urciuoli [1], R. De Leo [2], A. Leone [3], R. Perrino [3], C.C. Chang [4], P. Markowitz [4], M. Sotona [5], O.K. Baker [6], T. Saito [7]

[1] Istituto Superiore di Sanità, Physics Laboratory - Istituto Nazionale di Fisica Nucleare, Sezione Sanità, Viale Regina Elena 299, I-00161 Roma, Italy

[2] University of Bari, Physics Department - Istituto Nazionale di Fisica Nucleare, Sezione di Bari, Via Amendola 173, I-70126 Bari, Italy

[3] University of Lecce, Physics Department - Istituto Nazionale di Fisica Nucleare, Sezione di Lecce, Via per Arnesano, Lecce, Italy

[4] University of Maryland, College Park, Maryland, U.S.A.

[5] Nuclear Physics Institute, 25068 Řež, Czech Republic

[6] Hampton University, Hampton, Virginia, U.S.A.

[7] Tohoku University, Japan

Abstract. Plans for experimental activity on kaon electro-production using CEBAF Hall A facilities also upgraded with small scattering angle capability are presented. The kinematics coverage allowed by the possible detection of high momentum kaons as well as the high resolution of the spectrometers give unique opportunity for experimental measurements.

1 Introduction

Experimental data on associated hyperon-kaon photo- and electro-production have not been upgraded since about twenty years due to the unavailability of suited accelerators. On the contrary, in the last years, the planned constructions of new facilities has increased the interest in this subject both from experimental and theoretical point of view. While photo-production data have already been started to be produced at ELSA in Bonn and will be also produced at GRAAL in Grenoble as well as in CEBAF, this last laboratory will be the only place where progress can be done in electroproduction process. The three CEBAF Halls have complementary instrumentation, in particular strangeness production measurements can profit of the Hall A facilities for high luminosity experiments (up to $10^{38}cm^{-2}s^{-1}$) in which particles identification (pions, kaons, protons) up to 4 GeV/c, good control of systematic errors or high resolution or forward angle capability (it is foreseen to decrease the minimum scattering angle from 12.5 to 6 degrees) are required.

2 Off proton

In the low energy region various semiphenomenological approaches describing the elementary process in terms of mesons and baryons are used (see [1]). The semiphenomenological parameters of the theory (strong coupling constants, transition magnetic moments) are fitted to the available photoproduction data, making use of transition amplitudes based on the tree level Feynman diagrams. Phenomenological χ^2 parameter fits are not unique due to a large number of coupling constants and form factors associated with the diagrams considered and to the large experimental errors and restricted kinematics coverage of the available data. In addition, in most models the values of the leading coupling constants of the vertex proton, kaon and hyperon are in serious disagreement with the broken SU(3) predictions. The new facilities will allow measurements with high quality data for production cross sections as well as for hyperon polarization, the broad kinematics interval covered will give more severe constraints to the models. Measurements will approach the region where description of the reaction in terms of quark level mechanisms is attempted.

2.1 Elementary unpolarized cross section

The elecroproduction cross-section, when polarization is not taken into account, is determined by 4 response functions [2]: transverse R_{T}, longitudinal R_{L}, longitudinal-transverse interference term R_{LT} and transverse-transverse interference term R_{TT}. The last two terms are multiplied in the expression of the cross section by $\cos\phi$ and $\cos 2\phi$ respectively (ϕ being the angle between scattering and production planes). R_{LT} can be separated with two measurements of the cross section at ϕ=0 and ϕ=180 degrees. To separate R_{TT}, the experimental apparatus has to have out-of-plane capabilities. For the special case in which the kaon is detected in the direction of the virtual photon the two interference terms are zero and the remaining part of the electro-production cross section is given by :

$$\frac{d\sigma_{\mathrm{U}}}{d\Omega_k} = \frac{q_k}{k_\gamma}(R_{\mathrm{T}} + \epsilon_{\mathrm{L}} R_{\mathrm{L}}) \tag{1}$$

where q_k and k_γ are the momentum of kaon and virtual photon. σ_{U} may be further divided into the transverse and the longitudinal parts as far as the cross section is measured at two different values of linear virtual photon polarization ϵ_{L}. The measurement of separated structure functions gives much more constraints to a satisfactory theoretical description of the reaction. For this purpose, for a given value of W (total energy of the photon-hadron system),

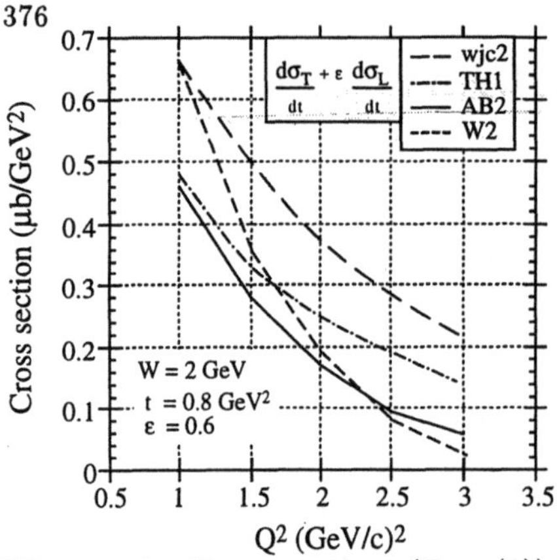

Figure 1. Cross section (Eq. (1)) computed with different theoretical models that are referred in [3]. Precise measurements at level of a few per cent, as foreseen in CEBAF, will put severe constraints to the models.

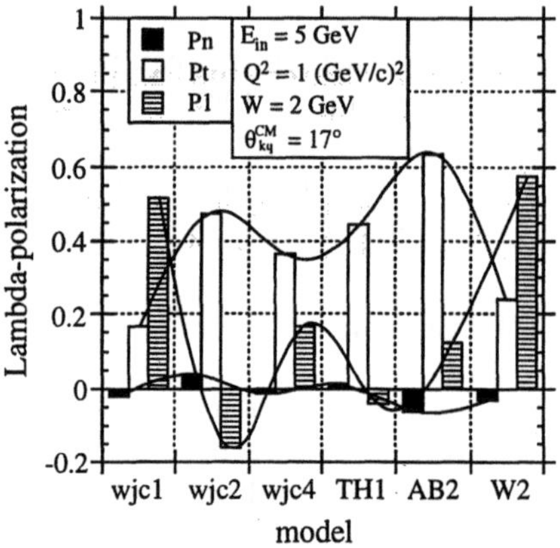

Figure 2. Polarization components of Λ^0 in the reaction induced by polarized electrons according to different models referred in [3]. Polarization measurements will discriminate among models.

q_μ (4-momentum of the virtual photon) and t (squared 4-momentum transfer to the hyperon, connected with the angle between the emitted kaon and the 3-momentum of the virtual photon), that are the variables on which the four structure functions depend, more independent measurements of the cross section must be carried out. CEBAF will provide precise sets of measurements of the basic reactions $p(e,e'K^+)\Lambda^0, \Sigma^0$ in a wide kinematics range. In Hall C approved experiment will cover the $Q^2(=-q_\mu^2)$ interval 0.5 - 1.5 $(\mathrm{GeV}/c)^2$, while in Hall A a conditionally approved experiment will cover up to 3.0 $(\mathrm{GeV}/c)^2$, as regards W the limits will be 1.9 and 2.2 GeV for the two experiments respectively and for t the coverage will be up to - 3 GeV^2. In Fig. 1 values of cross section (Eq.(1)), computed for some models in the kinematics range that will be explored, are shown and, with projected experimental errors of the order of a few per cent, it is clear how the foreseen measurements will constrain theoretical models.

2.2 Hyperon polarization measurements

When the polarizations of the incident electron and of the detected hyperon are considered, the cross section depends on 18 response functions. The studies with polarized electrons and measurement of hyperon polarization add therefore more selectivity against models. The cross section for an incoming electron with helicity h and polarization of the outgoing hyperon $\boldsymbol{\Pi}$ in a system of reference having versors $\boldsymbol{s}_R$ is :

$$\sigma^h(\hat{\boldsymbol{s}}_R) = 1/2\sigma_0(1 + \boldsymbol{\Pi}_1 \cdot \hat{\boldsymbol{s}}_R + h(A + \boldsymbol{\Pi}_2 \cdot \hat{\boldsymbol{s}}_R)) \qquad (2)$$

where σ_0 is the unpolarized cross section, A is the electron analyzing power and $\boldsymbol{\Pi}_1$, $\boldsymbol{\Pi}_2$ are the electron helicity independent and dependent parts of the

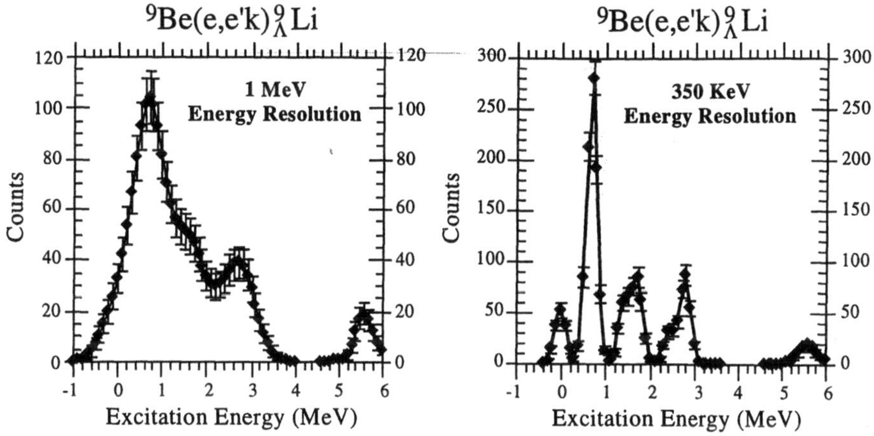

Figure 3. The expected results of the Hypernuclear Li-9 spectrum resulting from the reaction $^9\mathrm{Be}(e,e'K^+)^9\mathrm{Li}_\Lambda$. The effect of resolution on distinction of the populated levels shows how our knowledge of hypernuclear transitions can improve if high resolution is achieved. In the high resolution spectrum the splitting of spin doublets starts to be detected. These results are computed with shell model nuclear and hypernuclear wave functions deduced with effective NN interaction of Utrecht group and effective ΛN interaction derived from Nijmegen soft core hyperon-nucleon interaction (for discussion see [1]).

hyperon polarization ($\boldsymbol{\Pi} = \boldsymbol{\Pi}_1 + h\boldsymbol{\Pi}_2$). The weak hyperon decay is a self analyzing process for the measurement of the hyperon polarization, indeed the daughter baryon of a spin 1/2 hyperon has an angular distribution

$$\frac{dN}{d\Omega} = \frac{1}{4\pi}(1 + \alpha\Pi\cos\psi) \tag{3}$$

where ψ is the angle between the hyperon polarization and the daughter baryon direction in the rest frame of the hyperon and α is the strength of the decay asymmetry (0.642 for Λ^0 decay).Considering a system of axes for polarization component with $\boldsymbol{l}$ along the hyperon direction, $\boldsymbol{n}$ perpendicular to the reaction plane and $\boldsymbol{t} = \boldsymbol{n} \times \boldsymbol{l}$, the daughter baryon angular distribution in the laboratory frame can be written as :

$$\frac{dN}{d\Omega} = \frac{1}{4\pi}(1 + \alpha(\Pi_l\cos\theta + \Pi_n\sin\theta\cos\chi + \Pi_t\sin\theta\sin\chi)) \tag{4}$$

where θ is the polar angle of the baryon respect to the hyperon direction and χ is the azimuthal angle. From Eq.(4) it is possible to see that right/left and up/down asymmetries measure $\boldsymbol{t}$ and $\boldsymbol{n}$ polarization components while the polar angular distribution measures the longitudinal component.To perform polarization studies triple coincidences are required and additional equipment is needed, then these measurements will be proposed as a second phase of the Hall A strangeness electro-production program.

3 Off nuclei

Our present knowledge of hypernuclear states comes essentially from the strangeness exchange reaction $K^- + n \Longrightarrow \pi^- + \Lambda^0$, the dominating transitions are $\Delta l = 0$ and $\Delta s = 0$ and the substitutional states, lying as a rule in the continuum, are predominantly populated . In the associated production reaction $\pi^+ + n \Longrightarrow K^+ + \Lambda^0$ the favored transitions are $\Delta l = 1, 2$ and $\Delta s = 0$. This means that also $p_N \Rightarrow s_\Lambda$ transitions are stimulated but, due to the lack of sufficiently strong spin flip, only lower members of the hypernuclear bound states are populated. The energy resolution of all these process is typically of the order of 2 MeV or more, and with this value is out of any possibility the detection of spin doublet splitting so important for a better understanding of ΛN interaction. Only the electromagnetic production of strangeness with CEBAF beam and spectrometers quality can afford the possibility to obtain high resolution data on hypernuclear spectra. Moreover, due to the strong spin flip contribution, both members of the spin doublet may be populated. The weaker interaction of the electromagnetic probe respect to the two other probes can be, in part, overcome with the high intensity of the electron beam and with a proper choice of scattering angles both for scattered electron and outgoing kaon. This require the possibility to perform the experiment with a set-up in which the scattering angles can be as low as 6 degrees: to this end two septa have to be added to the already existing Hall A facilities. Taking into account electron beam and detected particles spreading, an overall resolution of 350 keV for the hypernuclear spectra can be obtained. In fig. 3 it is shown how the richness of the spectrum increases going from a resolution of 1 MeV to 350 keV. This possibility places CEBAF Hall A hypernuclear activity in the front end of the field.

References

1. J. Adam Jr., J. Mares, O. Richter, M. Sotona and S. Frullani: Czech. J. Phys. **42**, 1167 (1992)

2. D. Drechsel and L. Tiator: J. Phys. **G8**, 449 (1992)

3. [wjc1,wjc2] R.A. Williams, Ch.-R. Ji and S. R. Cotanch : Phys. Rev. **D41**, 1449 (1990); [wjc4] R.A. Williams, Ch.-R. Ji and S. R. Cotanch : Phys. Rev. **D46**, 1617 (1992); [TH1] H. Thom : Phys. Rev. **151**, 1322 (1966); [AB2] R. A. Adelseck, C. Bennhold and L. E. Wright: Phys. Rev. **C32**, 1681 (1985); [W2] R. L. Workman: Phys. Rev. **C44**, 552 (1991)

Few-Body Systems Suppl. 9, 379–383 (1995)

Few-Body Systems

Electroproductions of Light Λ- and Σ-Hypernuclei

S. Shinmura

Department of Applied Mathematics, Faculty of Engineering, Gifu University, Yanagido 1-1, Gifu 501-11, Japan

Abstract. Theoretical estimations of production cross sections of light Λ- and Σ-hypernuclei in the $(e, e'K^+)$ reactions are presented. Unnatural-parity states and stretched states are favorably excited. The cross sections for Λ-hypernuclei are sufficiently measurable. If only $s1/2$-state of Σ is bound, it may be difficult to observe signals of Σ-hypernuclei. If ${}^4_\Sigma$H(spin=1) exists, the signal can be observed in the reaction.

1 Introduction

In recent years, the (K^-, π) and (π, K^+) experiments have been performed intensively and have clarified properties of many hypernuclear states. However, these states are only a part of possible hypernuclear states. These reactions are suitable to excite substitutional states and excited states without a deep nucleon-hole. As alternative tools, the photo- and electro-productions, that is, hypernuclear productions in (γ, K^+) and $(e, e'K^+)$ reactions, have been discussed[1-11]. Since γ, e and K^+ interact weakly with nuclear medium, they can excite easily hypernuclei with a deep hole or a deep hyperon. Further, these reactions are dominated by spin-flip amplitudes and we can expect excitation of unnatural-parity states. In this talk, I report our theoretical estimations [12] for the electroproductions of Λ- and Σ-hypernuclei and discuss possibility to observe these hypernuclei.

2 Formalism

Elementary processes $N(e, e'K^+)Y$ ($Y = \Lambda$ or Σ) are assumed to consist of one-photon-exchange and $N(\gamma, K^+)Y$. As a model of $N(\gamma, K^+)Y$, we employ a sum of one-particle-exchange diagrams as shown in Fig. 1, which are generally used in theoretical studies [1,3-7,9-17]. We use the coupling constants determined in ref. [1] and ref. [6] for Y=Λ and Σ, respectively. Their values are determined to reproduce the experimental data at around 1 GeV region. In impulse approximation, only protons in target nuclei contribute to

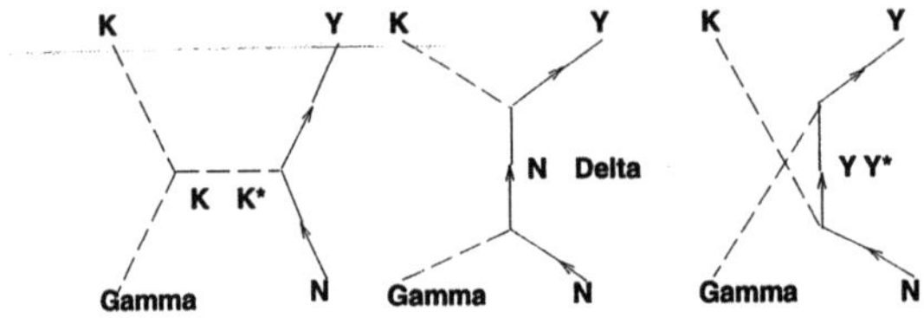

Figure 1. Model of $N(\gamma, K^+)Y$ vertex.

Λ-hypernuclear productions. On the other hand, for Σ production, our model satisfies $n(e, e'K^+)\Sigma^- \gg p(e, e'K^+)\Sigma^0$. Therefore, neutrons mainly contribute to Σ-hypernuclear productions. This is very interesting feature. Using these elementary processes, we calculate hypernuclear productions in RIA. Detailed expressions of the production cross sections are given in ref. [12]

To perform relativistic calculations, we use a relativistic single particle model of nuclei and hypernuclei [4, 18]. The scalar and vector potentials are determined so as to desctribe the phenomenological single-particle energies. The Woods-Saxon shape is assumed except for $A = 4$ cases, where the Gaussian shape is assumed. Potential parameters and calculated single-particle energies are given in Table 2.1. For Σ, we assume shallow potentials which bind only $s1/2\Sigma$ in p-shell hypernuclei.

Table 2.1. Potential parameters and single particle energies. For ${}^{12}_{\Lambda}$B, ${}^{12}_{\Sigma}$B and ${}^{4}_{\Sigma}$H*, the same potentials as ${}^{16}_{\Lambda}$N, ${}^{16}_{\Sigma}$N and ${}^{4}_{\Sigma}$H are assumed, respectively.

	^{4}He	^{12}C	^{16}O	${}^{4}_{\Lambda}$H	${}^{4}_{\Lambda}$H*	${}^{16}_{\Lambda}$N	${}^{4}_{\Sigma}$H	${}^{16}_{\Sigma}$N
S_0	-286.0	-288.0	-426.0	-142.5	-137.2	-187.0	-243.0	-187.0
V_0	200.0	215.0	340.0	100.0	100.0	155.0	200.0	177.0
a	-	0.7	0.7	-	-	0.7	-	0.7
r_0	1.2	1.2	1.055	1.0	1.0	1.15	1.0	1.15
$s1/2$	-20.1	-34.9	-40.3	-2.1	-1.0	-13.3	-3.0	-5.1
$p3/2$	-	-15.3	-18.3	-	-	-2.7	-	-
$p1/2$	-	-	-12.0	-	-	-1.7	-	-

3 Numerical results

To clarify what kinds of hypernuclear states are favorably excited, we calculate the cross sections for a common kinematical condition, that is, p_e=3.0 GeV/c, $p_{e'}$=1.2 GeV/c, $(\theta_{e'},\phi_{e'})$=(6°, 0°) and (θ_K,ϕ_K)=(10°,180°), where, (θ_K,ϕ_K) was determined so as to minimize the momentum transfer. Results are given in Table 3.1 and 3.2. We find that the stretched states are favorably excited. This can be explained by large momentum transfers in the reactions. For $[(p3/2)_p^{-1}, (p3/2)_\Lambda]_{J_f=3}$, we obtain cross sections of 2.485 and 2.667 nb/sr^2/GeV for ${}^{16}_{\Lambda}$N and ${}^{12}_{\Lambda}$B, respectively. For Σ-hypernuclei, $[(p3/2)_N^{-1}, (s1/2)_\Sigma]_{J_f=2}$ has large cross sections, 0.314 and 0.420 nb/sr^2/GeV,

Table 3.1. Production cross sections of Λ-hypernuclei in nb/sr^2/GeV

hole	Λ	J_f	${}^{16}_{\Lambda}$N	${}^{12}_{\Lambda}$B	${}^{4}_{\Lambda}$H	J_f	${}^{16}_{\Lambda}$N	${}^{12}_{\Lambda}$B	${}^{4}_{\Lambda}$H*
$s1/2$	$s1/2$	0	0.002	0.003	0.020	1	0.393	0.644	2.260
	$p3/2$	1	0.203	0.223		2	0.631	0.709	
	$p1/2$	0	0.002	0.003		1	0.369	0.354	
$p3/2$	$s1/2$	1	0.526	0.724		2	1.576	2.185	
	$p3/2$	0	0.000	0.000		1	0.388	0.390	
		2	0.013	0.020		3	2.485	2.667	
	$p1/2$	1	0.021	0.024		2	1.329	1.252	
$p1/2$	$s1/2$	0	0.033			1	0.844		
	$p3/2$	1	0.023			2	1.320		
	$p1/2$	0	0.002			1	0.555		

Table 3.2. Production cross sections of Σ-hypernuclei in nb/sr^2/GeV

hole	Σ	J_f	${}^{16}_{\Sigma}$N	${}^{12}_{\Sigma}$B	${}^{4}_{\Sigma}$H	J_f	${}^{16}_{\Sigma}$N	${}^{12}_{\Sigma}$B	${}^{4}_{\Sigma}$H*
$s1/2$	$s1/2$	0	0.001	0.001	0.054	1	0.013	0.029	4.021
$p3/2$		1	0.077	0.112		2	0.314	0.420	
$p1/2$		0	0.055			1	0.093		

which are about one fifth of 1.576 and 2.185 for corresponding Λ-hypernuclear states.

If the single-particle potential for Σ is the same as that for Λ, we obtain the cross sections of order of nb. Unnatural-parity states are strongly excited. In fact, the cross sections 2.260 and 4.021 nb/sr^2/GeV for ${}^{4}_{\Lambda}$H* and ${}^{4}_{\Sigma}$H* are larger than those for ${}^{4}_{\Lambda}$H and ${}^{4}_{\Sigma}$H by two orders of magnitude. This is due to large spin-flip amplitudes in the elemetary processes. Angular distributions for

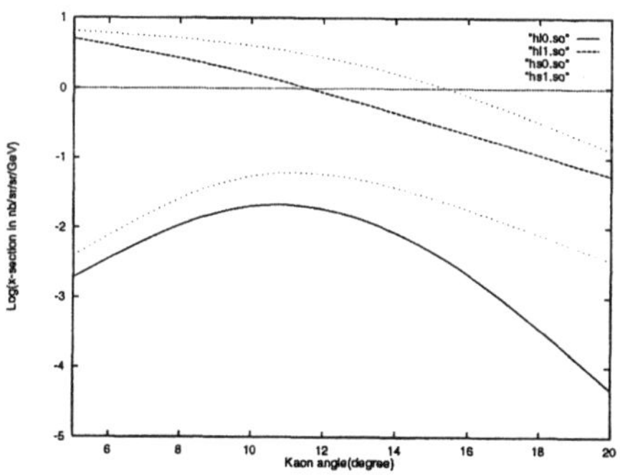

Figure 2. Angular distributions for ^{4}He$(e, e'K^+){}^{4}_{\Lambda}$H, ${}^{4}_{\Lambda}$H*, ${}^{4}_{\Sigma}$H and ${}^{4}_{\Sigma}$H*,which are denoted by lower-solid, upper-solid, lower-dashed and upper-dashed lines, respectively.

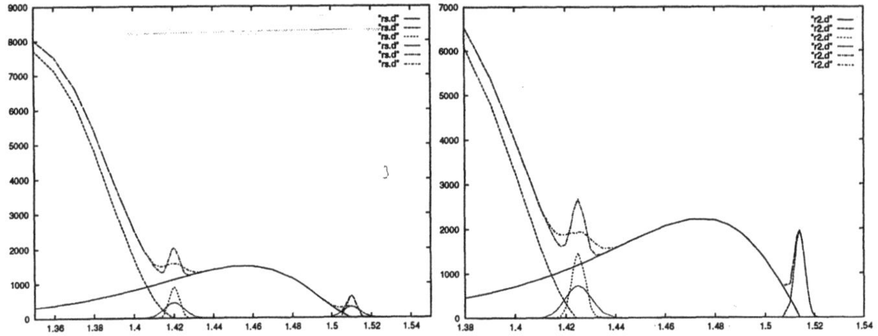

Figure 3. Kaon spectrum in ^{4}He$(e, e'K^+)$ at $\theta_K = 10°$(a) and $\theta_K = 5°$(b). Peaks Λ and Σ are $^4_\Lambda$H*(Γ=4MeV) and $^4_\Sigma$H*(Γ=5,10 MeV) productions, respectively

these states have a typical character of the spin-flip process, as shown in Fig.2.

To examine whether signals of hypernuclei are measurable, we compare them to quasifree hyperon productions(QHP), which are the largest background process. For ^{4}He target, we obtain results as shown in Fig.3. We find the signal of $^4_\Lambda$H* is sufficiently measurable. For $^4_\Sigma$H*, the result depends on the width. If the width is smaller than 10MeV, we can observe the signal. QHP increase with target mass number. Therefore, for ^{16}O and ^{12}C targets, the signals of Σ-hypernuclear states may not be measurable.

For higher incident energies, we calculate the cross sections for $[(p3/2)_p^{-1}(p3/2)_\Lambda]_{J_f=3}$ which is the most favorably excited in the reaction. We obtain a result as given in Table 3.3. For given $p_{e'}$ and $\theta_{e'}$, θ_K is determined so as to minimize the momentum transfer. For larger $p_{e'}$, that is, for smaller momentum of intermediate γ, we get smaller cross section. For $(p_e - p_{e'})$ ~2GeV, we obtain the largest cross section. In this case, K^+ momentum is about 1.7 GeV/c.

Table 3.3. Production cross sections of $[(p3/2)_p^{-1}(p3/2)_\Lambda]_{J_f=3}$ for higher incident energies in nb/sr^2/GeV($\theta_{e'} = 6°$)

p_e=4.0GeV/c				p_e=5.0GeV/c			
$p_{e'}$	p_K	θ_K	Cross section	$p_{e'}$	p_K	θ_K	Cross section
1.6	2.16	4	6.57	2.6	2.16	6	7.45
2.0	1.75	6	7.39	3.0	1.75	9	7.81
2.4	1.33	9	6.76	3.4	1.33	12	5.69
2.8	0.89	14	3.15	3.8	0.88	18	1.81
3.0	0.64	17	0.94	4.0	0.64	22	0.39
3.2	0.36	22	0.03	4.2	0.35	28	0.01

4 Summary

We calculated the electroproduction cross sections of light Λ- and Σ-hypernuclei in the relativistic impulse approximation. We obtain measurable cross sections for stretched states and unnatural-parity states of Λ-hypernuclei. The cross sections for Σ-hypernuclei are generally smaller than those for Λ-hypernculei, if single-particle potential for Σ is shallow. If p-state Σ is bound, we get measureble cross sections. Further, If ${}^4_\Sigma$H(spin=1) exists, we can observe the signal. The $(e, e'K^+)$ reaction is a promising tool, at least for Λ-hypernuclear productions, and is complementary to other types of reactions.

References

1. S. S. Hsiao and S. R. Cotanch: Phys. Rev. **C28**, 1668 (1983)
2. A. M. Bernstein, T. W. Donnelly and G. N. Epstein: Nucl. Phys. **A358**, 195c (1981)
3. A. S. Rosenthal et al.: Ann. Phys. **184**, 33 (1988)
4. C. Bennhold and L. E. Wright: Phys. Rev. **C39**, 927 (1989)
5. H. Tanabe, M. Kohno and C. Bennhold: Phys. Rev. **C39**, 741 (1989)
6. C. Bennhold: Phys. Rev. **C39**, 1944 (1989)
7. C. Bennhold: Nucl. Phys. **A547**, 79c (1992)
8. M. Sotona et al.: Nucl. Phys. **A547**, 63c (1992)
9. J. Cohen: Phys. Rev. **C32**, 543 (1985)
10. J. Cohen, M. W. Price and G. E. Walker: Phys. Lett. **B188**, 393 (1987)
11. J. Cohen: Int. J. Mod. Phys. **A4**, 1 (1989)
12. S. Shinmura: Prog. Theor. Phys. **92**, 571 (1994)
13. H. Thom: Phys. Rev. **151**, 1322 (1966)
14. R. A. Adelseck and L. E. Wright: Phys. Rev. **C38**, 1965 (1988)
15. R. A. Adelseck, C. Bennhold and L. E. Wright: Phys. Rev. **C32**, 1681 (1985)
16. H. Kadowaki et al.: Prog. Theor. Phys. **79**, 263 (1988)
17. R. A. Adelseck and B. Saghai: Phys. Rev. **C42**, 108 (1990)
18. A. Ramos et al.: Nucl. Phys. **A544**, 703 (1992)

Few-Body Systems Suppl. 9, 384–398 (1995)

Few-Body Systems

Achievements and Challenges in 3*N*- and 4*N*-Systems

W. Glöckle[1], H. Witała[2], H. Kamada[3], D. Hüber[1], J. Golak[2]

[1] Institut für Theoretische Physik II, Ruhr Universität Bochum, D-44780 Bochum, Germany

[2] Institute of Physics, Jagellonian University, PL-30059 Cracow, Poland

[3] Paul Scherrer Institut (PSI), CH-5232 Villingen, Switzerland

Abstract. Recent results on rigorous solutions of the 3*N* Faddeev equations but also open problems and suggestions for new experiments in 3*N* scattering are presented. For ^{4}He binding energies and properties of the ground state wavefunction for various realistic *NN* forces are given based on precise solutions of the 4*N* Faddeev-Yakubovsky equations.

The treatment of three- and four-nucleon systems has seen great progress in recent years and we report on recent results and challenges in 3*N* scattering and on bound state calculations for ^{4}He, all based on realistic nuclear forces.

1 3*N* Scattering

One of the early stated reasons to study three-nucleon (3*N*) scattering was, to possibly distinguish between different nucleon-nucleon (*NN*) forces and thus to single out the one with the "correct" off-shell behaviour. In the past this aim could not be achieved since the forces were not well enough tuned to the *NN* data (they were also too simple in their operator structure) and there were severe limitations in solving the 3*N* scattering equations. Both has changed now. There exists now several so called realistic *NN* forces, which are very well adjusted to the most recent phase-shift analysis (PSA) for *np* and *pp* data up to 350 MeV [1] and the 3*N* Faddeev equations can be solved [2] with very high accuracy which is far better than the experimental error bars. Thus the old question mentioned can be adressed now much better.

We shall report here on recent results of the Bochum-Cracow collaboration. The work is performed in momentum space and in a partial wave representation. We choose the Faddeev equations in the form

$$T|\phi> = tP|\phi> + tPG_0T|\phi> \tag{1}$$

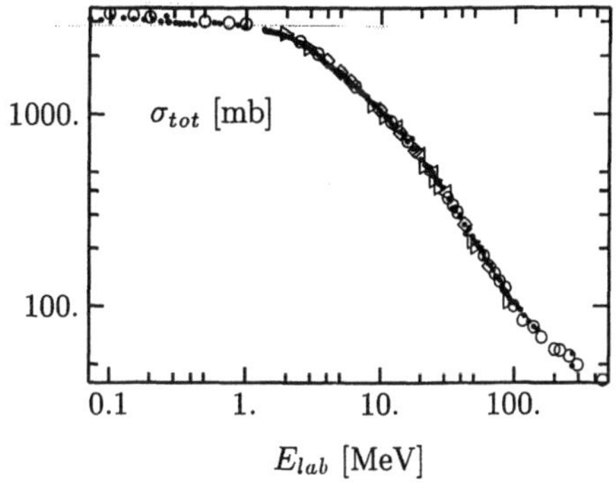

Figure 1. The total *nd* cross section, data (•), theory (open symbols).

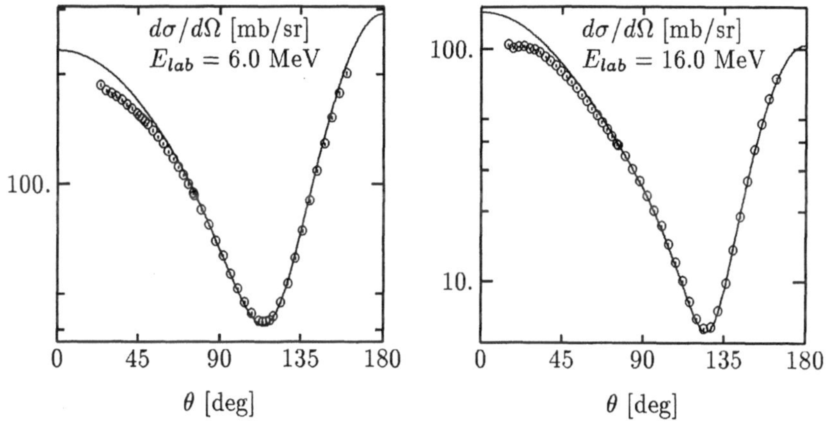

Figure 2. The differential cross section for elastic *Nd* scattering. *pd* data from [14].

which contain the NN off-shell t-operator t, the free $3N$ propagator G_0 and the sum P of a cyclical and anticyclical permutation of 3 objects. That equation generates the multiple scattering series for the $3N$ breakup process initiated by the scattering of a nucleon on a deuteron. The corresponding initial channel state is ϕ, a product of a deuteron state and a momentum eigenstate of relative motion of the projectile nucleon and the deuteron. We refer for all technical details to the original literature [2, 3, 4]. Two recent benchmark calculations [5, 6] demonstrate the maturity of the present day techniques to solve the $3N$ scattering equations very precisely. In [5] our results were compared to solutions of the Faddeev equations in configuration space with a quite different mathematical appearence and in [6] to the pair correlated hyperspherical harmonic basis method, which is again quite different from our way of solving $3N$ scattering. In both cases the agreement among the results of the different techniques was perfect. If one adds a three nucleon force ($3N$F) a generalised set of equations [7] is

$$T|\phi> \quad = \quad tP|\phi> + tG_0T_4|\phi> + tPG_0T|\phi>$$

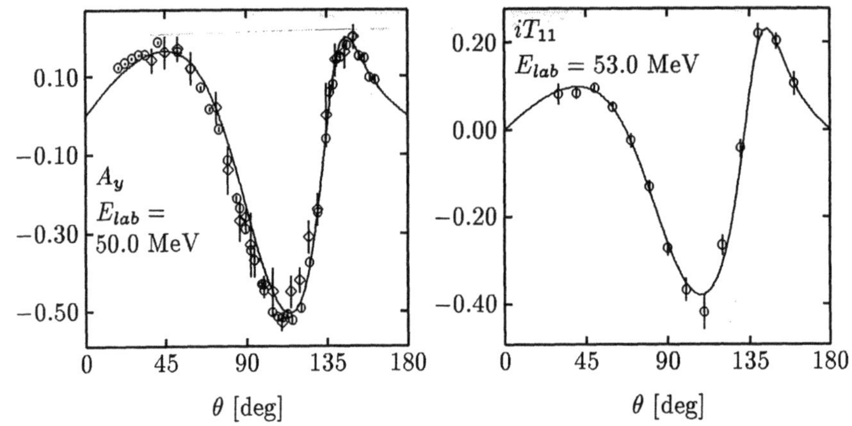

Figure 3. The neutron and deuteron vector analyzing powers. *pd* data for A_y are from [15], *nd* data from [16]. *pd* data for iT_{11} are from [17].

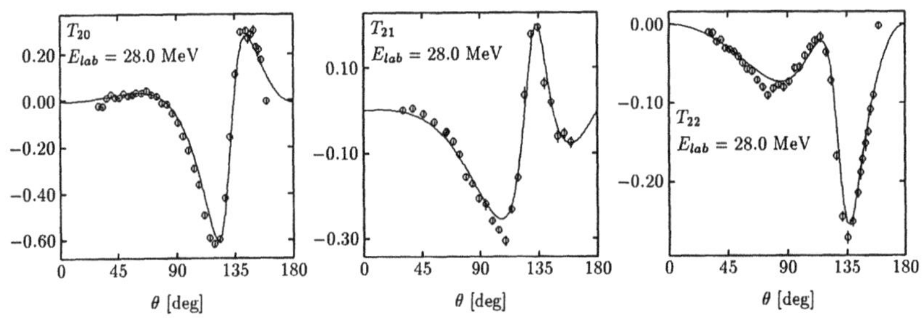

Figure 4. The deuteron tensor analyzing powers. *pd* data are from [18].

$$T_4|\phi> \;=\; (1+P)t_4|\phi> + (1+P)t_4G_0T|\phi> \qquad (2)$$

with t_4 being driven by the three-nucleon force ($3NF$) V_4:

$$t_4 = t_4 + V_4G_0t_4 \qquad (3)$$

Then the operators for the breakup process and elastic Nd scattering are

$$\begin{aligned} U_0 &= (1+P)T + T_4 \\ U &= PG_0^{-1} + PT + T_4 \end{aligned} \qquad (4)$$

respectively.

We have analysed very many $3N$ scattering data, using the most modern NN forces, AV18 [8], Nijm93, NijmI, NijmII [9], and also somewhat older ones, Bonn B [10], Paris [11], Nijm78 [12], and AV14 [13]. Overall we find a very good description of the data using these NN forces only and the results are very stable with respect to exchanging NN forces. This is especially true for the first four potentials mentioned, the newest ones, which are perfectly well adjusted to the most recent PSA. Deviations in the past between different NN force predictions were often caused by different on-shell properties, which can appear magnified

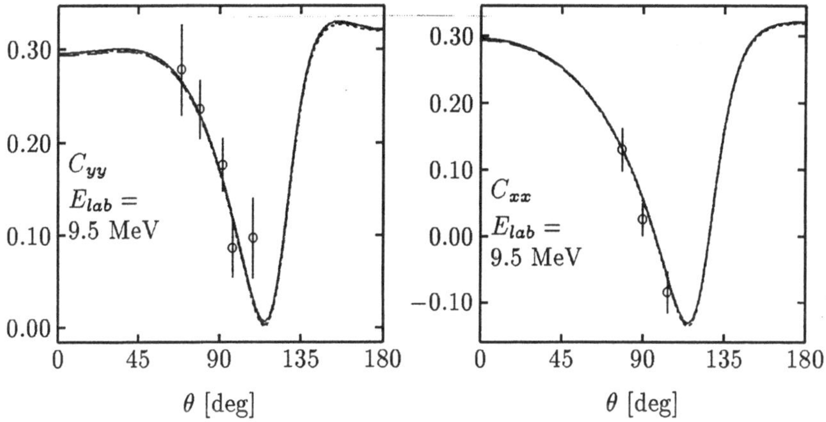

Figure 5. The vector spin correlation coefficients. *pd* data are from [19].

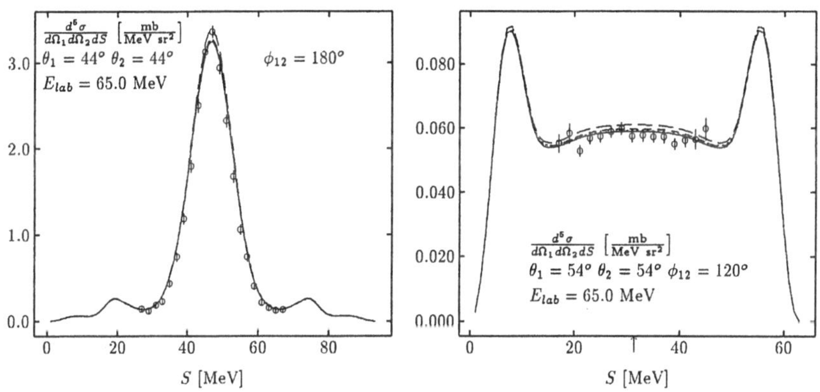

Figure 6. Two breakup cross sections with good agreement between experiment and theory. *pd* data are from for the QFS configuration (left hand side) and from for the space star configuration (right hand side).

in certain $3N$ observables. Thus right now we cannot see significant differences in the predictions (except for certain breakup configurations) of these newest NN forces, though they have quite different radial shapes and some of them are purely local while others have nonlocalities built in.

We illustrate the very many cases of agreement between theory and experiment by a few examples in Figs. (1-6) and refer to [24, 25, 26] for a larger overview. All the curves shown are calculated with the most modern NN forces mentioned above.

We would like to emphasize again the stability of that theoretical picture against exchanges of NN forces. This however is only true if they are fairly well phase equivalent. Clearcut examples, which react quite sensitively on on-shell differences are the spin transfer coefficient $K_y^{y'}$ [27] and the analyzing power A_y in elastic nd scattering [28] or the final state interaction (FSI) peak in the nd breakup process [29].

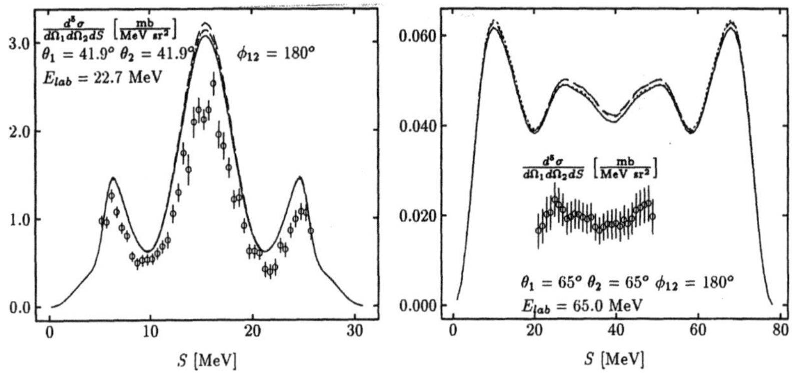

Figure 7. Two breakup cross sections with disagreement between experiment and theory. *pd* data are from [22] for the QFS configuration (left hand side) and from [23] for a nameless configuration (right hand side).

Despite the existence of quite a few data specific ones are missing and can be expected to provide more information. The total breakup cross section data are old and inaccurate [30]. Since we have a perfect agreement with the total *nd* cross section (of the order of 1%) and also the angular distribution in elastic *nd* scattering is described very well, one has to conclude, that also the total breakup cross section should agree equally well with theory. The explicit verification would put a highly welcome constraint on the still sparkling situation for specific breakup cross sections, where we have both, cases of spectacular agreement and cases of strong disagreement. The latter ones should be remeasured. For instance some of recent data for *nd* breakup at 13 MeV taken at TUNL [31] do not confirm older data measured at Erlangen [32], which were in striking disagreement with our theory [33], and now they agree with our predictions. There are more cases, which are waiting for a reconfirmation or a modification. We display two of them in Fig. 7.

An absolutely ideal test for the potential energy of three nucleons would be a complete "4π" survey of all breakup cross sections. This has been done in the past [34], but the theoretical analysis at that time is insufficient on present day standards. In order to stimulate renewed experimental effort we are performing presently sensitivity studies comparing the cross section predictions of the most modern *NN* forces among each other. In steps of 5 degrees we evaluate the breakup cross sections for 13 MeV as a function of the two detector positions and find that the largest deviations among all most modern *NN* force predictions are at most 10%. All of the 8-10% deviations occur in *nn* and *np* FSI peaks and can be traced back to different *nn* and *np* scattering lenghts, especially for AV18, which we use without electromagnetic terms. Figure 8 displays in the space of the angular positions of the two detectors θ_1, θ_2, and $\phi_{12} = \phi_1 - \phi_2$, the region, where all the potential predictions coincide within 3% and are $\geq 1 \frac{\text{mb}}{\text{MeV sr}^2}$. Precise data there would be very welcome and discrepancies there would challenge very much present day nuclear forces.

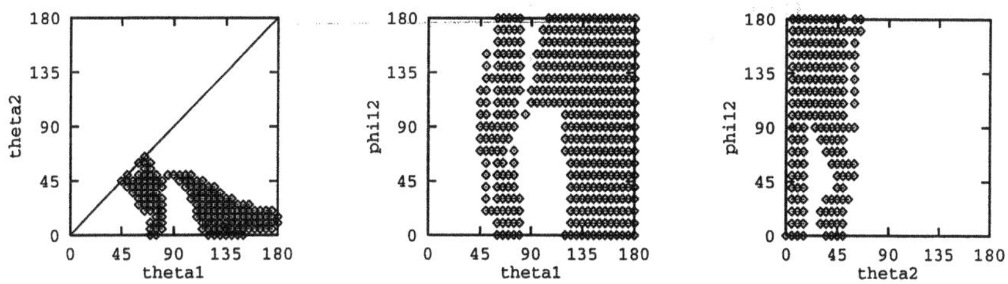

Figure 8. The insensitive breakup configurations at $E_{\rm lab} = 13$ MeV (see text).

In elastic scattering more and precise spin transfer coefficients from the nucleon to the tensor polarised deuteron would be very useful, since they are quite sensitive to ^3P_j *NN* force components [24, 35]. Data for spin transfer coefficients from the nucleon to the nucleon are totally missing above 30 MeV, as far as we know. They are sensitive to $^3S_1 - ^3D_1$ force components, but also to *P*- and *D*-wave forces. Though a nice set of tensor analysing powers [36] exist for a wide range of energies the quality of the data should be improved in order to test theory harder. It would be interesting to see whether the spectacular agreement seen for instance at 22.7 MeV [24, 37] persists also at higher energies.

The discrepancies in some breakup cross sections are not the only ones. Recent tensor analysing power data [38] measured at $E_{d,\rm lab} = 52.1$ MeV for specific breakup configurations are in striking disagreement with theory and certainly more data, also at neighbouring angles, are needed to clarify the situation. Similarily a tensor analysing power measurement [39] at $E_{d,\rm lab} = 94.5$ MeV in the symmetric constant relative energy geometry exhibits a peak which is absent in theory and calls for a renewed measurement. Last not least we should point to the most challenging puzzle right now, the low energy *nd* analysing power A_y [40]. The present day situation is displayed in Fig. 9. The most recent *NN* force predictions lie close together and underestimate the data by about 30%. That observable is extremely sensitive to ^3P_j *NN* force components. If the most recent *NN* phase-shift parameters in the ^3P_j states are the true and final ones, one envisages in that observable a candidate which magnifies off-shell or $3N$ force effects drastically.

The extraction of the *nn* scattering length from the *nn* FSI peak in the *nd* breakup process is an old issue and still inconclusive. For a recent study see [29]. We found that the decisive quantity, the *nn* FSI peak height depends significantly on the choice of the *NN* force and is thus model dependent. Interestingly enough, if one regards that peak height as a function of the production angle of the two neutrons, the model dependence vanishes at a certain angle, which is for instance 43 degree at 13.0 MeV. This appears to be an optimal case to keep the model dependence on a minimum and a measurement for extracting a_{nn} at that angle appears to be promising.

If one increases the nucleon laboratory energy to 150 MeV and higher not only a technical challenge arises due to the increase of necessary partial waves

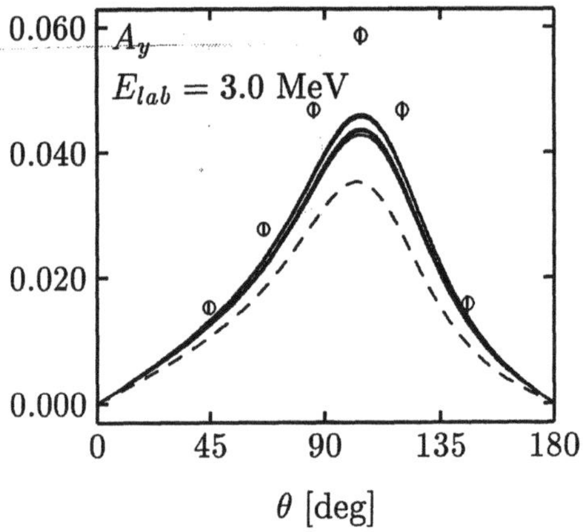

Figure 9. A_y in elastic *nd* scattering. Solid curves are different *NN* force predictions, dashed curve Bonn B + full TM 3*N*F, *nd* data from [41].

and *NN* force components but also one has to expect that relativistic effects will start to play a noticeable role. A first indication can already be seen in the angular distribution for elastic *nd* scattering, pointed out in [26]. We are presently investigating, whether the onset of relativistic effects is responsible for that discrepancy. We also study the question for which breakup configurations the deuteron can be considered to be a nucleon target (neutron or proton) and conversely, where rescattering is predominant.

Another challenge is the full inclusion of the *pp* Coulomb force together with realistic *NN* forces. This is still an unsolved problem above the breakup threshold. For simple separable *S*-wave Yamaguchi forces first results appeared [42]. Below that threshold the Pisa group succeeded to achieve precise solutions [43] using the pair correlated hyperspherical harmonic basis method. This opens now the door to compare their sophisticated calculations with equally precise data of various spin observables in elastic *pd* scattering. This appears to be very promising to also test 3*N*F effects, on which we would like to comment now.

We solved [44] the set (2) using the Bonn B *NN* force together with the Tucson-Melbourne (TM) 3*N*F [45] consisting of $\pi-\pi$, $\pi-\rho$, and $\rho-\rho$ exchanges. Since ^{3}H is overbound for this force combination the resulting 3*N*F effects in the 3*N* continuum are likely overestimated. Nevertheless the effects evaluated at 14 and 3 MeV turned out to be rather small in general, except at the lower energy. At 3 MeV nucleon lab energy, for instance C_{xx}, C_{yy}, $K_y^{z'z'}$ show very strong 3*N*F effects, while the same observables are hardly affected at 14 MeV. On the other hand $K_y^{y'}$ shows 3*N*F effects at both energies and the differential cross section and the tensor analysing powers in elastic *Nd* scattering show essentially no 3*N*F effect at all. Examples for that behaviour are displayed in Fig. 10. Breakup cross sections at both energies exhibit only very small 3*N*F effects. Small shifts in the QFS and FSI peak heights of about 5% are the strongest cases we found. Interestingly these shifts in the FSI peak height

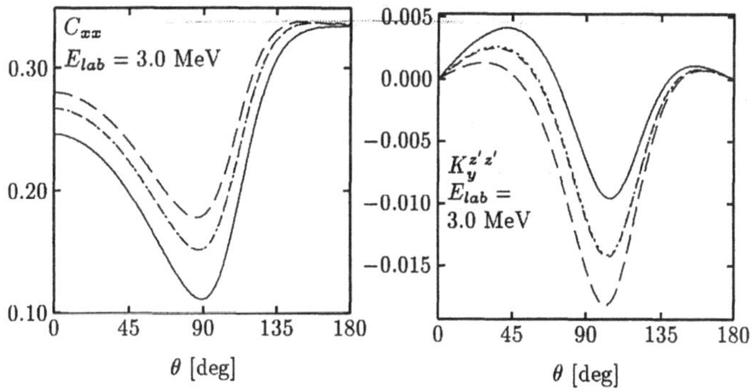

Figure 10. C_{xx} and $K_y^{z'z'}$ for elastic *nd* scattering at 3 MeV. Bonn B (solid curve), Bonn B + π-π TM 3*N*F (dashed) and Bonn B + full TM 3*N*F (short dashed and dotted curves).

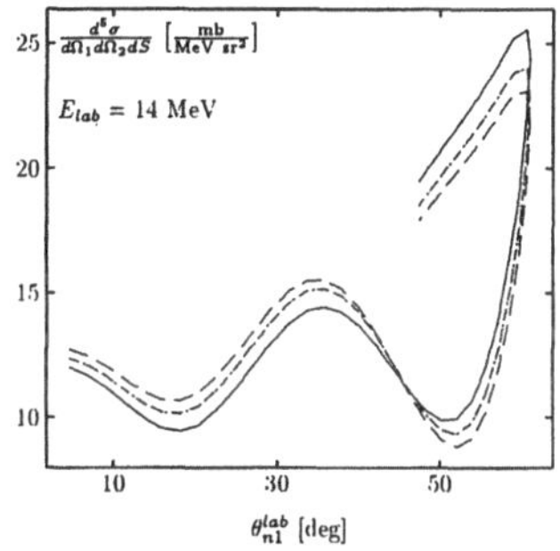

Figure 11. Breakup cross section in the *nn* FSI peak as a function of a neutron lab angle. Description as in Fig. 10.

vanish at the same "magic angle", where also different *NN* force predictions coincide [29], see Fig. 11.

At very low energies, below about 10 MeV, one can find 3*N* scattering observables which scale with the triton binding energy [25]. This feature has been revealed by pairing different *NN* forces with the TM 3*N*F and adjusting the 3*N*F in each case to the correct ^{3}H binding energy. While the *NN* force predictions for these scaling observables deviate from each other, they coincide once the proper 3*N*F is switched on. There are however also other observables, which do not scale, like A_y. The effect of the TM 3*N*F on A_y even increases the discrepancy to the data and aggravates the puzzle, as exhibited in Fig. 9.

We pointed out interesting challenges in the 3*N* continuum, but overall one can state, that the *NN* force picture only provides a very good description of most data.

2 The 4*N*-Bound State

3*N*- and 4*N*-bound states have attracted vivid attention since the early days of nuclear physics. While the Faddeev equations have been solved precisely for realistic nuclear forces since quite a few years, [46], it is only very recently

[47, 48] that the Faddeev - Yakubovsky (FY) equations for four nucleons could be solved with comparable rigorousness. Though the FY-equations have been written down about 30 years ago [49] they are much less familiar than the three-body Faddeev equations [50]. Therefore we would like to sketch briefly their derivation incorporating immediately the simplification due to the identity of the four nucleons. We start from the Schrödinger equation in integral form

$$\Psi = G_0 \sum_{ij} V_{ij} \Psi \tag{5}$$

and define the Faddeev components as for three particles

$$\Psi \equiv \sum_{ij} \psi_{ij} \tag{6}$$

with

$$\psi_{ij} \equiv G_0 V_{ij} \Psi = G_0 V_{ij} \sum_{kl} \psi_{kl} \tag{7}$$

Summing the pair force V_{ij} to infinite order into the NN t-matrix t_{ij} yields

$$\psi_{ij} = G_0 t_{ij} \sum_{kl \neq ij} \psi_{kl} \tag{8}$$

These are six coupled equations. There is only one independent amplitude ψ_{ij} (ij arbitrary), the other result by particle permutations. Take for instance

$$\psi_{12} = G_0 t_{12} (\psi_{23} + \psi_{31} + \psi_{14} + \psi_{24} + \psi_{34}) \tag{9}$$

According to Yakubovsky ψ_{12} is split into three subamplitudes, nowadays called Yakubovsky components, which allow to treat seperately 3-body subclusters and $2+2$ subclusters:

$$\begin{aligned} \psi_1 &\equiv G_0 t_{12} (\psi_{23} + \psi_{31}) \\ \psi_2 &\equiv G_0 t_{12} \psi_{34} \end{aligned} \tag{10}$$

and $-P_{34}\psi_1 = G_0 t_{12}(\psi_{14} + \psi_{24})$. Introducing $P \equiv P_{12}P_{23} + P_{13}P_{23}$, which permutes the particles in the subcluster (123) only, we get

$$\psi_1 = G_0 t_{12} P \psi_{12} = G_0 t_{12} P (\psi_1 - P_{34}\psi_1 + \psi_2) \tag{11}$$

Further let $\tilde{P} \equiv P_{13}P_{24}$ interchange the two two-body subclusters (12), (34). Then

$$\psi_2 = G_0 t_{12} \tilde{P} \psi_{12} = G_0 t_{12} \tilde{P} (\psi_1 - P_{34}\psi_1 + \psi_2) \tag{12}$$

Now we can sum up all forces within the 3-body subcluster (123) and within the $2+2$ subclusters (12), (34):

$$\begin{aligned} (1 - G_0 t_{12} P)\psi_1 &= G_0 t_{12} P (-P_{34}\psi_1 + \psi_2) \\ (1 - G_0 t_{12} \tilde{P})\psi_2 &= G_0 t_{12} \tilde{P} (1 - P_{34})\psi_1 \end{aligned} \tag{13}$$

The inversion of the operators on the left hand side yields

$$\begin{aligned} \psi_1 &= -G_0TPP_{34}\psi_1 + G_0TP\psi_2 \\ \psi_2 &= G_0\tilde{T}\tilde{P}(1-P_{34})\psi_1 \end{aligned} \tag{14}$$

where T and $\tilde{T}$ obey

$$\begin{aligned} T &= t_{12} + TPG_0t_{12} \\ \tilde{T} &= t_{12} + \tilde{T}\tilde{P}G_0t_{12} \end{aligned} \tag{15}$$

Equations (14) are the two coupled FY-equations and Eqs. (15) determine the subcluster problems (123) and (12), (34), respectively. According to our decomposition the total state is

$$\Psi = (1 + P - P_{34}P + \tilde{P})((1 - P_{34})\psi_1 + \psi_2) \tag{16}$$

We solve the set (14)-(15) in momentum space and in a partial wave representation [47]. Allowing the forces to act in two-nucleon states up to total angular momentum $j = 4$ one reaches convergence for 398 combinations of discrete quantum numbers (orbital angular momenta, spins and isospins), usually called channels. More physically spoken, the $3N$ subclusters can be in the states of total angular momenta and parities $1/2^{\pm}$, $3/2^{\pm}$, $5/2^{+}$ and the two-body subclusters in the states $0^{\pm}$ to $4^{\pm}$. The huge matrix problem after discretisation of the two types of three Jacobi momenta is handled by a Lanczos type method. Comparative studies with other techniques [51, 52] underline the reliability of the result. For instance our result for AV14, which is 24.73 MeV, compares favourably well with 24.68 MeV of the Pisa group. We display in table 2.1 our results for various NN forces. For the sake of completeness we also included ^3H binding energies for the most recent NN forces taking charge-independence breaking into account. This has not yet been included in the ^4He binding energy results, but can easily be done and can effect the binding energy by as much as 1 MeV. Since we neglect the pp Coulomb force one has to compare to 29 MeV for ^4He. Like in ^3H there is underbinding. Note the larger binding energy for Bonn B, which is likely due to its inherent nonlocality. In any case the missing binding energy for both nuclei is only of the order of a few percent of the total potential energy in these nuclei. That discrepancy is naturally taken care off by the $3N$F's studied up to now (like the TM $3N$F or the Ruhrpot [53]). The Ruhrpot seems to be the first theoretical framework, where NN and $3N$F's are derived consistently. Much more theoretical work, however, has still to be done to achieve a consistent picture of $2N$ and $3N$ forces.

Besides the often presented nucleon momentum distributions and the averaged NN correlation functions, one can regard state dependent NN correlation functions and the most probable geometrical nucleon sites to characterize the bound states. In Fig. 12 we display for d, ^3H and ^4He

$$C_{lsjt}(r) \equiv \langle \Psi | \delta(r - r_{ij}) P_{lsjt} | \Psi \rangle \tag{17}$$

Potential	p_d	$E_{^3\mathrm{H}}$	Potential	p_d	$E_{^3\mathrm{H}}$	$E_{^4\mathrm{He}}$
Bonn B (cd)	4.99	7.92	Bonn B	4.99	8.14	27.07
Nijm II (cd)	5.65	7.64	Nijm78	5.39	7.63	25.10
Nijm I (cd)	5.68	7.73	Paris	5.77	7.46	24.32
Nijm93 (cd)	5.76	7.66	Ruhrpot	5.85	7.64	24.60
AV18 (cd)	5.78	7.65	AV14	6.08	7.68	24.73
			AV8	6.08	7.79	25.31

Table 2.1. Our ^{3}H and ^{4}He binding energies based on $j \leq 4$ *NN* forces.

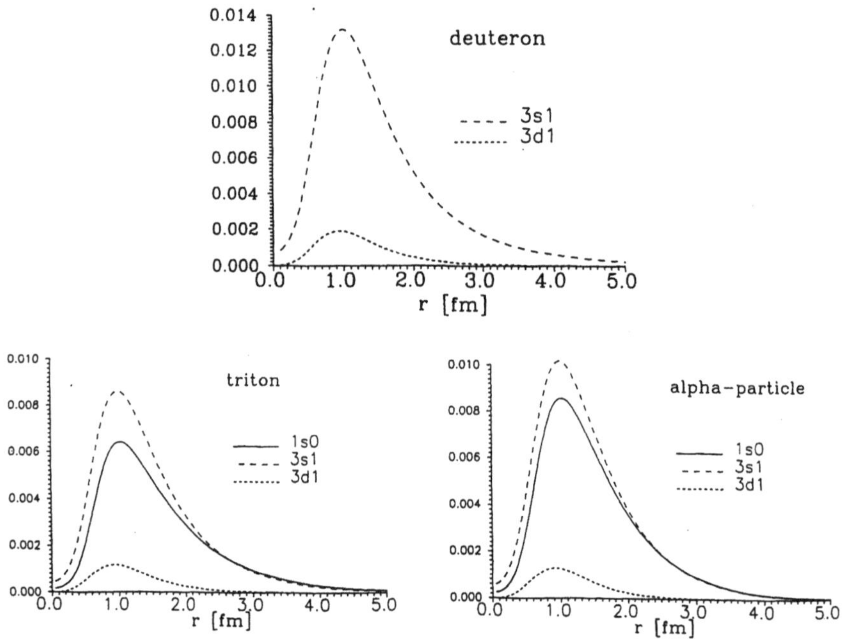

Figure 12. State dependent two-nucleon correlation functions.

where P_{lsjt} projects onto the *NN* states with $lsjt$ = 0001, 0110, 2110, respectively. The correlation functions for *P*- and *D*-wave states are less populated by far more than a factor 10. We see a great similarity among the three nuclei, with the $^1S_0$ correlations of course missing in the deuteron.

The most probable sites of the nucleons are displayed in Fig. 13 for ^{3}H and ^{4}He. We see essentially an equiliteral triangle and a tetrahedron. Because of the pure $t = 1$ forces the distances between identical particles are somewhat larger than between neutrons and protons. Also for ^{4}He the pair distances are slightly larger than for ^{3}H. The pair correlation functions together with these most probable geometrical locations demonstrate, that nucleons are mostly well seperated from each other and therefore strong medium effects of nucleon properties, like changes of electromagnetic nucleon form factors, should not be expected. Of course with lower probabilities one should see cases of close encounter using sufficiently short wavelength probes.

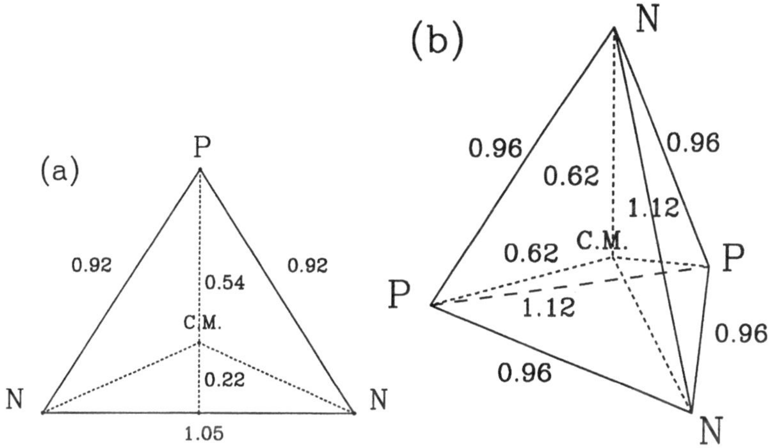

Figure 13. The most probable geometrical nucleon sites for (a) ^{3}H and (b) ^{4}He.

The inclusion of $3NF$'s into the Yakubovsky scheme requires some care [54] and first restricted results have been achieved. On the other hand the Pisa group [55] got already converged results by their method and also results have been achieved by the GFMC method [56]

It will be an exciting challenge to determine the complex resonance positions for ^{4}He by exact few-body techniques and realistic forces. Since geometrically the four nucleons will be in different configurations in comparison to the ground state, one has to expect, that $3NF$'s will act differently in the various "excited states" of ^{4}He.

Overall the results achieved up to now for bound and scattering states and based on realistic nuclear forces are very promising and will and already find applications in studying responses to external electroweak probes [57, 58].

Acknowledgement. The work was supported by the Deutsche Forschungsgemeinschaft and the Polish Commitee for Scientific Research under Grant No. 2 P302 104 06. The numerical calculations have been performed on the Cray Y-MP of the Höchstleistungsrechenzentrum Jülich, Germany.

References

1. V.G.J. Stoks et al. : Phys. Rev. **C48**, 792 (1993)

2. H. Witała, Th. Cornelius, W. Glöckle: Few-Body Systems **3**, 2 (1988)

3. W. Glöckle: *The Quantum-Mechanical Few-Body Problem.* Berlin - Heidelberg: Springer 1983

4. W. Glöckle: Lecture Notes in Physics **273**, 3 (1987)

5. J. L. Friar et al. : Phys. Rev. **C51**, 2356 (1995)

6. D. Hüber et al. : Phys. Rev. **C51**, 1100 (1995)

7. D. Hüber, H. Witała, W. Glöckle: Few-Body Systems **14**, 171 (1993)

8. R. B. Wiringa, V.G.J. Stoks, R. Schiavilla: Phys. Rev. **C51**, 38 (1995)

9. V.G.J. Stoks et al. : Phys. Rev. **C49**, 2950 (1994)

10. R. Machleidt, K. Holinde, Ch. Elster: Phys. Rep. **149**, 1 (1987);
R. Machleidt: Adv. Nucl. Phys. **19**, 189 (1989)

11. M. Lacombe et al. : Phys. Rev. **C21**, 861 (1980)

12. M.M. Nagels, T.A. Rijken, J.J. de Swart: Phys. Rev. **D17**, 768 (1978)

13. R. B. Wiringa, R.A. Smith, T.L. Ainsworth: Phys. Rev. **C29**, 1207 (1984)

14. K. Sagara et al. : Phys. Rev. **C50**, 576 (1994)

15. N.S.P. King et al. : Phys. Lett. **69B**, 151 (1977)

16. J.W. Watson et al. : Phys. Rev. **C25**, 2219 (1982);
J.L. Romero et al. : Phys. Rev. **C25**, 2214 (1982)

17. H. Witała et al. : Few-Body Systems **15**, 67 (1993)

18. K. Hatanaka et al. : Nucl. Phys. **A428**, 77 (1984)

19. J. Chauvin, D. Garreta, M. Fruneau: Nucl. Phys. **A247**, 335 (1975)

20. K. Bodek: Private Communication

21. J. Zejma: Ph-D thesis, Cracow 1995 (unpublished)

22. M. Zadro et al. : Il Nuovo Cimento **107A**, 185 (1994)

23. J. Krug: Private Communication

24. W. Glöckle, H. Witała, Th. Cornelius: In: *Proceedings of the 12th International IUPAP Conference on Few-Body Problems in Physics*, ed. H. W. Fearing, Vancouver, Nucl. Phys. **A508**, 115c (1990);
W. Glöckle, H. Witała, Th. Cornelius: In: *Proceedings of the 25th Zakopane School on Physics, Vol 2: Selected Topics in Nuclear Structure*, p. 300, eds. J. Styczen and Z. Stachura, Singapore: World Scientific 1990;
W. Glöckle, H. Kamada, H. Witała: Physicalla Mag. **14**, 271 (1992);
W. Glöckle, H. Witała: Kerntechnik **57**, 220 (1992);

H. Kamada, H. Witała, W. Glöckle: In: *Proceedings of the International Workshop on Few-Body Problems in Low-Energy Physics*, eds. A. A. Chumbalov, Zh. B. Kurmanov, N. Zh. Takibayev, Alma-Ata, Republic of Kazakhstan, Journal of Nuclear Physics **56**, 29 (1993);
W. Glöckle, H. Kamada, H. Witała: In: *Proceedings of the National Conference on Physics of Few-Body and Quark-Hadronic Systems*, p. 75, eds. V. Boldshev, V. Kotlyar, A. Shebeko, KFTI/Kharkov, Ukraine 1994;
W. Glöckle, H. Kamada, H. Witała, In: *Contemporary Topics in Medium Energy Physics*, p. 243, eds. K. Goeke, W. -Y. Pauchy Hwang, J. Speth, New York: Plenum Press 1994;

25. W. Glöckle, H. Witała, H. Kamada, D. Hüber, J. Golak: In: *Proceedings of the 14th International Conference on Few-Body Problems in Physics*, Williamsburg 1994;

26. W. Glöckle, H. Kamada, H. Witała, D. Hüber, J. Golak, K. Miyagawa, S. Ishikawa: In: *Proceedings of the 15th European Conference on Few-Body Problems in Physics*, Peñiscola, Spain 1995

27. L. Sydow et al. : Nucl. Phys. **A567**, 55 (1994)

28. H. Witała, W. Glöckle: Nucl. Phys. **A528**, 48 (1991)

29. H. Witała, D. Hüber, W. Glöckle, W. Tornow, D. E. Gonzalez Trotter: accepted for publication in Few-Body Systems

30. M. Holmberg: Nucl. Phys. A129, 327 (1969);
H. Carton et al. : Phys. Rev. **123**, 213 (1961);
G. Pauletta et al. : Nucl. Phys. **A255**, 257 (1975);
N. Koori: J. Phys. Soc. Japan **32**, 306 (1972)

31. W. Tornow: Private Communication

32. J. Strate et al. : J. Phys. G: Nucl. Phys. **L229**, 14 (1988)

33. H. Witała, Th. Cornelius, W. Glöckle: Few-Body Systems **5**, 89 (1988)

34. G.J.F. Blommestijn et al. : Nucl. Phys. **A365**, 202 (1981)

35. H. Witała, W. Glöckle, Th. Cornelius: Nucl. Phys. **A496**, 446 (1989)

36. H. Witała et al. : Few-Body Systems **15**, 67 (1993)

37. H. Witała, W. Glöckle: Few-Body Systems, Suppl. **6**, 291 (1992)

38. L. M. Qin et al. : Nucl. Phys. **A587**, 252 (1995)

39. H. Witała et al. : submitted for publication

40. H. Witała, D. Hüber, W. Glöckle: Phys. Rev. **C49**, R14 (1994)

41. J.E. McAnninch et al. : Phys. Lett. **B307**, 13 (1993)

42. E.O. Alt, M. Rauh: Few-Body Systems, Suppl **7**, 160 (1994)

43. A. Kievsky, M. Viviani, S. Rosati: preprint Pisa 1995

44. H. Witała, D. Hüber, W. Glöckle, J. Golak, A. Stadler, J. Adam Jr.: submitted for publication

45. S.A. Coon, M.T. Pena: Phys. Rev. **C48**, 2559 (1993)

46. Y. Wu, S. Ishikawa, T. Sasakawa: Few-Body Systems **15**, 145 (1993)

47. H. Kamada, W. Glöckle: Nucl. Phys. **A548**, 205 (1992); Few-Body Systems, Suppl **7**, 217 (1994)

48. W. Glöckle, H. Kamada: Phys. Rev. Lett. **71**, 971 (1993)

49. O.A. Yakubovsky: Sov. J. Nucl. Phys. **5**, 937 (1967)

50. L.D. Faddeev: Sov. Phys. JETP **12**, 1014 (1961)

51. M. Viviani, A. Kievsky, S. Rosati: Few-Body Systems **18**, 25 (1995)

52. J. Carlson, R. Schiavilla: Few-Body Systems, Suppl **7**, 349 (1994)

53. D. Plümper, J. Flender, M.F. Gari: Phys. Rev. **C49**, 2370 (1994)

54. W. Glöckle, H. Kamada: Nucl. Phys. **A560**, 541 (1993)

55. M. Viviani, A. Kievsky, S. Rosati: Few-Body Systems **18**, 25 (1995)

56. J. Carlson, R. Schiavilla: Few-Body Systems, Suppl **7**, 349 (1994)

57. R. Schiavilla, R.B. Wiringa, J. Carlson: Phys. Rev. Lett. **70**, 3856 (1993)

58. S. Ishikawa, H. Kamada, W. Glöckle, J. Golak, H. Witała: Nuovo Cimento **A107**, 305 (1994);
W. Glöckle, J. Golak, H. Kamada, S. Ishikawa, H. Witała: In: *Proceedings of the Workshop on Electron-Nucleus Scattering*, p. 66, EIPC, Italy, 1993, eds. O. Fabrocini, R. Schiavilla, Singapore: World Scientific 1994;
J. Golak, H. Kamada, H. Witała, W. Glöckle, S. Ishikawa: Phys. Rev. **C51**, 1638 (1995) ;
J. Golak, H. Witała, H. Kamada, D. Hüber, W. Glöckle, S. Ishikawa: to appear in Phys. Rev. **C**;
S. Ishikawa, H. Kamada, W. Glöckle, J. Golak, H. Witała: Phys. Lett. **B339**, 293 (1994)

Few-Body Systems Suppl. 9, 399–403 (1995)

Few-Body Systems

S-Matrix Parameters for Elastic Neutron-Deuteron Scattering above the Breakup Threshold

D. Hüber[1], J. Golak[2], H. Witała[2], W. Glöckle[1], H. Kamada[3]

[1] Institut für Theoretische Physik II, Ruhr Universität Bochum, D-44780 Bochum, Germany

[2] Institute of Physics, Jagellonian University, PL-30059 Cracow, Poland

[3] Paul Scherrer Institut (PSI), CH-5232 Villingen, Switzerland

Abstract. Complex eigenphases and mixing parameters for elastic nd scattering above the breakup threshold are calculated. Faddeev equations are solved precisely using the Bonn B NN potential. We find peculiar energy variations in the mixing parameter $\eta^{\frac{3}{2}-}$ but show that they have no effect on observables and are simply an artifact of the S-matrix parametrization. It is found, that the nd analyzing power A_y depends most sensitively on the three eigenphases $\delta^{\frac{1}{2}}_{\frac{3}{2},1}$, $\delta^{\frac{3}{2}}_{\frac{3}{2},1}$ and $\delta^{\frac{5}{2}}_{\frac{3}{2},1}$, which again are predominantly generated by the ^3P_j NN force components. Therefore a determination of those eigenphases from experimental data would help to constrain the ^3P_j NN forces.

1 Introduction

Little is known about the complex interaction between a nucleon and a composite particle and how it arises from the nucleon-nucleon (NN) interaction. The three nucleon ($3N$) system offers a first nontrivial case to study the interaction between a nucleon and the most simple composite particle, the deuteron, in a quantitative manner. In a first investigation [1] we calculated eigenphases and mixing parameters for elastic nd scattering below the breakup threshold. In a second study [2] we extended the investigations of [1] to energies above the breakup threshold, where all the eigenphases and mixing parameters become complex. In [2] we used the realistic Bonn B [3] NN force. The Faddeev equations are solved precisely in momentum space as described in [4]. The details of the definition and parametrization of the S-matrix are given in [1], and the new features linked to complex eigenphases and mixing parameters are described in [2].

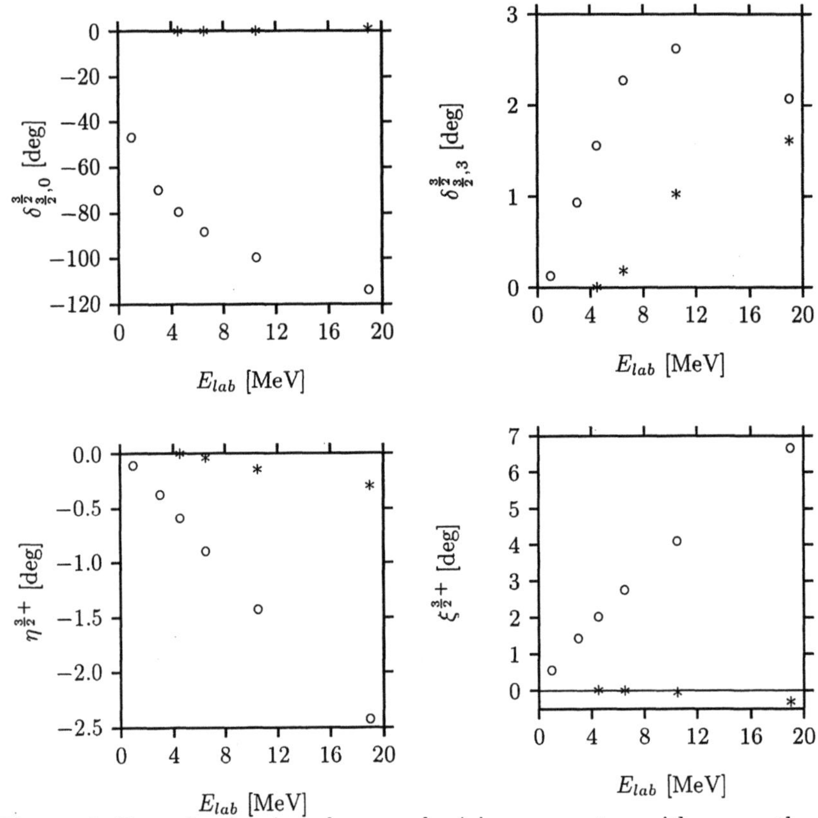

Figure 1. Examples for eigenphases and mixing parameters with a smooth energy behaviour. Open circles are the real parts and stars are the imaginary parts. The notation is $\delta^{J}_{\Sigma\lambda}$, where Σ is the channel spin and λ the orbital angular momentum of the nucleon.

2 Results

The results presented here are obtained by taking the NN force into account up to a maximal total two-body angular momentum of $j_{\max} = 3$. The $j = 3$ force components are needed only in the state of total $3N$ angular momentum and parity $J^{\Pi} = 1/2^{+}$ and there especially for $\eta^{\frac{1}{2}+}$, where they contribute about 3.5%.

Most eigenphases and mixing parameters vary smoothly with energy. Examples are shown in Fig. 1. There are parameters with a negligibly small imaginary part like $\delta^{\frac{3}{2}}_{\frac{3}{2},0}$ and $\xi^{\frac{3}{2}+}$ and ones with big imaginary parts like $\delta^{\frac{3}{2}}_{\frac{3}{2},3}$ and $\eta^{\frac{1}{2}+}$. The latter result throws doubts on a phase-shift analysis (PSA) in the past [5], where imaginary parts have been neglected.

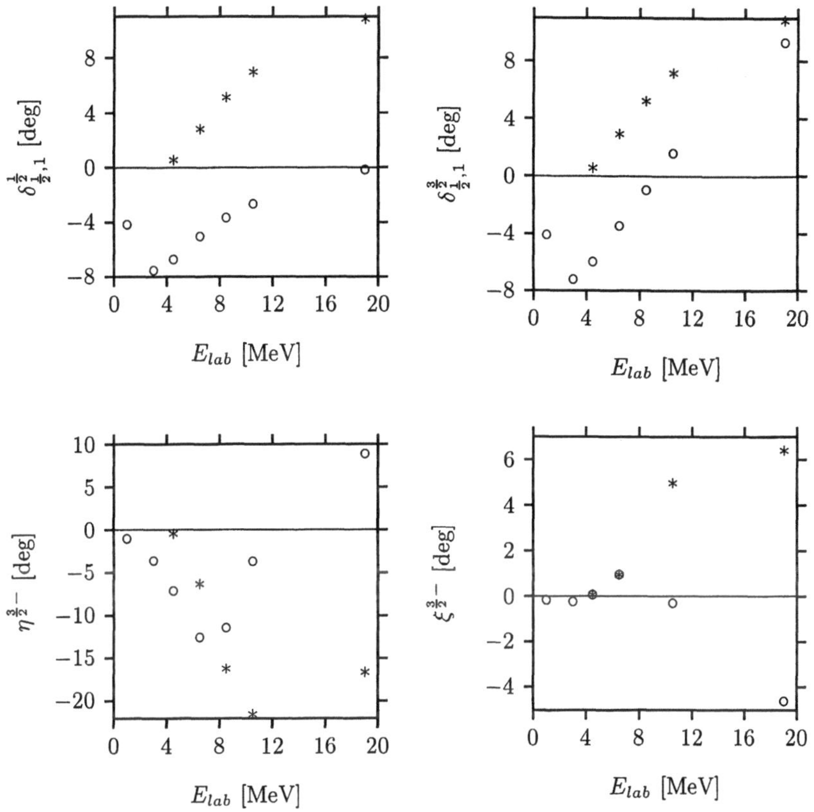

Figure 2. The eigenphases and mixing parameters which show strong variations with energy. Open circles are the real parts and stars are the imaginary parts.

Also a very recent PSA [6] at $E_{\text{lab}} = 3$ MeV used too stringent assumptions. The authors of [6] treated only the low-λ phase-shift parameters ($\lambda \leq 2$) as free parameters and took the higher ones just from a model for the nucleon exchange process. This is only qualitatively true, but fails quantitatively, as our calculations at $E_{\text{lab}} = 3$ MeV show. Comparing the phase-shift parameters using only the nucleon exchange process to the results of the full calculation we found for some mixing parameters relatively big differences: $\eta^{\frac{5}{2}+}$, $\xi^{\frac{5}{2}+}$, $\epsilon^{\frac{5}{2}-}$, $\epsilon^{\frac{7}{2}-}$ and $\epsilon^{\frac{9}{2}+}$ deviate from the full calulation between 6 and 13%. Thus those quantities should be treated as free fit parameters in a PSA, contrary to what has been done in [6].

But not all phase-shift parameters vary smoothly with energy. The two eigenphases $\delta^{\frac{1}{2}}_{\frac{1}{2},1}$ and $\delta^{\frac{3}{2}}_{\frac{1}{2},1}$ show a strong minimum in the real part at the threshold energy $-\frac{3}{2}E_d \approx 3.34$ MeV, where E_d is the deuteron binding energy. The mixing parameter $\xi^{\frac{3}{2}-}$ shows small oscillations in the real part and $\eta^{\frac{3}{2}-}$ even

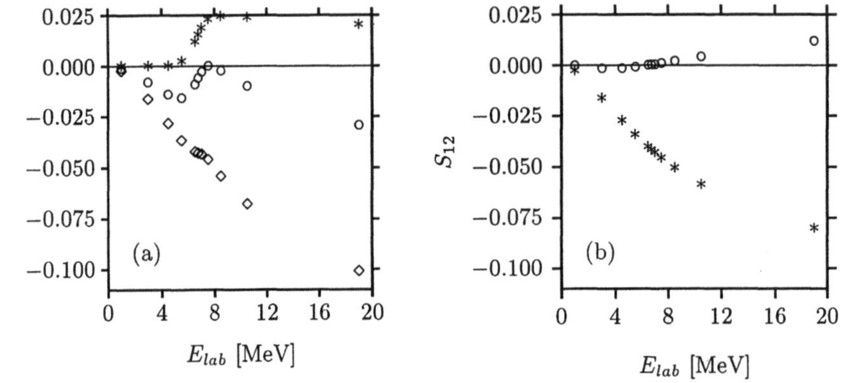

Figure 3. a) Im S_{12} within the approximation used in Eq. 1 ($\diamond$) together with its two contributing terms, Re $\eta^{\frac{3}{2}-}$ Re $(\delta^{\frac{3}{2}}_{\frac{3}{2},3} - \delta^{\frac{3}{2}}_{\frac{1}{2},1})$ ($\circ$) and Im $\eta^{\frac{3}{2}-}$ Im $(\delta^{\frac{3}{2}}_{\frac{3}{2},3} - \delta^{\frac{3}{2}}_{\frac{1}{2},1})$ ($\star$), respectively. b) Full calculation of S_{12}, real part ($\circ$) and imaginary part ($\star$).

strong variations above the breakup threshold. This is exhibited in Fig. 2.

The strong variations of $\eta^{\frac{3}{2}-}$ are of special interest. Do they also lead to a strong energy dependence in the S-matrix and in the observables? This is not the case. Thus the strong energy dependence of $\eta^{\frac{3}{2}-}$ must be an artifact of the parametrization of the S-matrix. To demonstrate this we expand the S-matrix up to first order in the small eigenphases and mixing parameters. Within this approximation it is found that only the S_{12} element of the 3×3 S-matrix depends on $\eta^{\frac{3}{2}-}$. The imaginary part of S_{12}, which is one order of magnitude larger than the real part (see Fig. 3b), reads in this approximation

$$\text{Im } S_{12} \approx 2\,(\text{Re } \eta^{\frac{3}{2}-} \text{ Re } (\delta^{\frac{3}{2}}_{\frac{3}{2},3} - \delta^{\frac{3}{2}}_{\frac{1}{2},1}) - \text{Im } \eta^{\frac{3}{2}-} \text{ Im } (\delta^{\frac{3}{2}}_{\frac{3}{2},3} - \delta^{\frac{3}{2}}_{\frac{1}{2},1})) \qquad (1)$$

In Fig. 3a we plotted the imaginary part of S_{12} according to Eq. (1) together with its two contributions of the right hand side of Eq. (1). It can be seen from Fig. 3, that these two contributions have a strong energy dependence which cancels in the difference, so that Im S_{12} becomes smooth with energy. The good quality of the approximation used in Eq. (1) can be seen by comparing it to the full calculation in Fig. 3b.

Next it is interesting to see the dependence of the *nd* elastic scattering observables on the nd phase-shifts and again their dependence on the *NN* force components. Because of the pending A_y puzzle [7] we concentrate here on A_y. We find that A_y is sensitive to only three eigenphases, namely $\delta^{\frac{1}{2}}_{\frac{3}{2},1}$, $\delta^{\frac{3}{2}}_{\frac{3}{2},1}$ and $\delta^{\frac{5}{2}}_{\frac{3}{2},1}$. The sensitivity to these three eigenphases is very strong. The relative change in the eigenphases is enhanced by about a factor of 10 in A_y. Thus a change of less than 1% in one of these eigenphases leads still to significant effects in A_y. Each of these three nd eigenphases is sensitive only to one NN force component: $\delta^{\frac{1}{2}}_{\frac{3}{2},1}$ to $^3P_0$, $\delta^{\frac{3}{2}}_{\frac{3}{2},1}$ to $^3P_1$ and $\delta^{\frac{5}{2}}_{\frac{3}{2},1}$ to $^3P_2 - ^3F_2$, respectively. An experimental determination of the three nd eigenphases $\delta^{\frac{1}{2}}_{\frac{3}{2},1}$, $\delta^{\frac{3}{2}}_{\frac{3}{2},1}$ and $\delta^{\frac{5}{2}}_{\frac{3}{2},1}$ in a PSA would possibly provide valuable information on the ^3P_j NN force phases and is worthwhile to do. Of course one cannot rule out the effect of a $3N$ force, but all $3N$ forces used up to now failed in describing A_y [7,8].

Finaly we would like to point out that a change of a NN force component influences always more than one eigenphase or mixing parameter.

Acknowledgement. This work was supported by the EC under Grant No. CI1*-CT91-0894 (D.H.), the Polish Committee for Scientific Research under Grant No. 2 P302 104 06 (J.G.,H.W.) and the Deutsche Forschungsgemeinschaft (H.K.). The numerical calculations have been performed on the Cray Y-MP of the Höchstleistungsrechenzentrum in Jülich, Germany.

References

1. D. Hüber, W. Glöckle, J. Golak, H. Witała, H. Kamada, A. Kievsky, S. Rosati, M. Viviani: Phys. Rev. **C51**, 1100 (1995)

2. D. Hüber, J. Golak, H. Witała, W. Glöckle, H. Kamada: accepted for publication in Few-Body Systems

3. R. Machleidt: Adv. Nucl. Phys. **19**, 189 (1989)

4. H. Witała, Th. Cornelius, W. Glöckle: Few-Body Systems **3**, 123 (1988)

5. P.A. Schmelzbach, W. Grüebler, R.E. White, V. König, R. Risler, P. Marmier: Nucl. Phys. **A197**, 273 (1972)

6. L.D. Knutson, L.O. Lamm, J.E. McAninch: Phys. Rev. Lett. **71**, 3762 (1993)

7. H. Witała, D. Hüber, W. Glöckle: Phys. Rev. **C49**, R14 (1994);

8. D. Hüber, H. Witała, W. Glöckle: Few-Body Systems **14**, 171 (1993);
D. Hüber, H. Witała, H. Kamada, W. Glöckle: Contribution to the 15th European Conference on Few-Body Physics, Peñiscola, Spain, June 1995;
M. Viviani, A. Kievsky, S. Rosati: Contribution to the 15th European Conference on Few-Body Physics, Peñiscola, Spain, June 1995;
K. Chielewski, J. Haidenbauer, U. Meyer, S. Nemoto, P.U. Sauer, N.W. Schellingerhout: Contribution to the 15th European Conference on Few-Body Physics, Peñiscola, Spain , June 1995

Few-Body Systems Suppl. 9, 405–409 (1995)

Few-Body Systems

Variational Calculations for Continuum States in Few–Nucleon Systems

A. Kievsky

Istituto Nazionale di Fisica Nucleare, Sezione di Pisa, I-56100 Pisa, Italy

Abstract. Deuteron analyzing powers at 3.0 MeV for *pd* scattering and scattering lengths for $p - {}^3$He and $n - {}^3$H are calculated. The method consists in an expansion of the wave function in terms of the Correlated Hyperspherical Harmonic basis. The reactance matrix for each process is calculated by means of the Kohn variational principle. The results are compared with the new high-precision measurements allowing for a detailed check of the potential models investigated.

Low energy reactions in few–nucleon systems are a natural stage to test our knowledge about the NN interaction. Theoretical predictions for different observables in such reactions provide information on the validity of the description of the nucleus in terms of nucleonic degrees of freedom. On the other hand, they establish our capability to solve the multidimensional Schroedinger equation. Typical processes in the three–nucleon system are the *pd* and *nd* scattering as well as photodisintegration and capture reactions. In the four–nucleon systems, there is a rich variety of scattering processes since different clusters with two or three nucleons appear in the initial or final state.

At present, a complete solution of the Schroedinger equation for nuclear systems with $A = 3, 4$ is not available. For $A = 3$ the solution of the Faddeev equations provides an exact description of the system, but the method encounters some difficulties when the Coulomb interaction is switched on. For $A = 4$ the corresponding Faddeev–Yakubovsky equations have been solved only for the alpha particle ground state [1].

Recently a variational approach has been developed to describe the bound state of the three– and four–nucleon systems [2, 3]. The method consists in a channel decomposition of the wave function (as in the Faddeev technique) and the radial dependence is constructed by expanding the corresponding amplitudes in terms of the Correlated Hyperspherical Harmonic (CHH) basis [2]. The flexibility of the basis to reproduce the minor details of the wave function has been tested by comparison with other accurate techniques [1, 4, 5].

The extension of the method to scattering processes is given in ref. [6]

for nd and pd elastic scattering below the break–up threshold. The reactance matrix has been calculated by means of the Kohn variational principle and the corresponding phase–shift and mixing parameters have been compared with those given by the Faddeev technique in momentum space [7]. In general, the reactance matrix is calculated for a given energy and for all possible quantum numbers L, S, J, π (L is the relative angular momentum between the clusters, S is the total spin, J the total angular momentum, and π the parity) with $L \leq L_{\rm max}$. $L_{\rm max}$ is fixed in such a way that, for $L > L_{\rm max}$, the corresponding phase–shifts are negligible. Therefore, the transition matrix M can be obtained and the cross section and polarization observables are calculated by a simple trace operation [8]. In ref. [9] the CHH expansion has been used to calculate the nd and pd cross section and the proton and neutron analyzing powers. The laboratory energy was fixed at 3.0 MeV to compare with the new and high–quality measurements of refs. [10, 11, 12]. In addition, the theoretical phase–shift and mixing parameters have been compared with the phase shift analysis determined from the data in ref. [10].

In ref. [9] four potential models have been investigated: the Argonne AV14 and AV18 interactions in correspondence with the three–nucleon interactions (TNI) of Brazil (BR) and Urbana (UR), fixed to reproduce the binding energy of the three–nucleon system. The conclusion was that the charge dependent interaction AV18 substantially improves the agreement with the data but still some differences subsist. The inclusion of the TNI modifies very little the theoretical curves. Here we use the CHH expansion to evaluate the deuteron analyzing powers for pd scattering at 3.0 MeV lab energy. The tensor analyzing power iT_{11} is given in Fig.1 for the four potential models investigated. A behavior similar to that of the proton A_y is observed. All the potential models underestimate the data but the inclusion of charge dependence terms in the hamiltonian produces a better agreement. The TNI terms give a small contribution moving the curve in the right direction. The tensor analyzing powers T_{20}, T_{21} and T_{22} are given in Figs. 2-4, for the AV18 and AV18+UR potential models. It is interesting to notice that an improvement is obtained for the T_{22} when the TNI is included but the opposite situation happens for the T_{21} at angles near 90°. In conclusion, polarization observables provide a powerful tool to evaluate NN interactions as well as TNI terms. Further investigation are necessary in order to adjust the nuclear interaction not only to reproduce the binding energies but also scattering observables in light nuclei.

There is a rich variety of reactions at low energy involving four nucleons. If the total energy of the process is such that only two fragments can be found in the initial or final state, the method described above can be used to obtain the reactance matrix. Typical reactions to be studied are $n(p)-{}^3$H, $n(p)-{}^3$He and $d-d$ scattering, that can be recombined (depending on the incident energy) to form different final states. Theoretical calculations, if realistic interactions are considered, are complicated and lengthy. Zero energy processes are the simplest ones to be studied and they will be presented here as a first step in the calculations of all the mentioned reactions. The first one studied is the $p-{}^3$He zero energy scattering, characterized by the quantum numbers $J=0^+, T=1$.

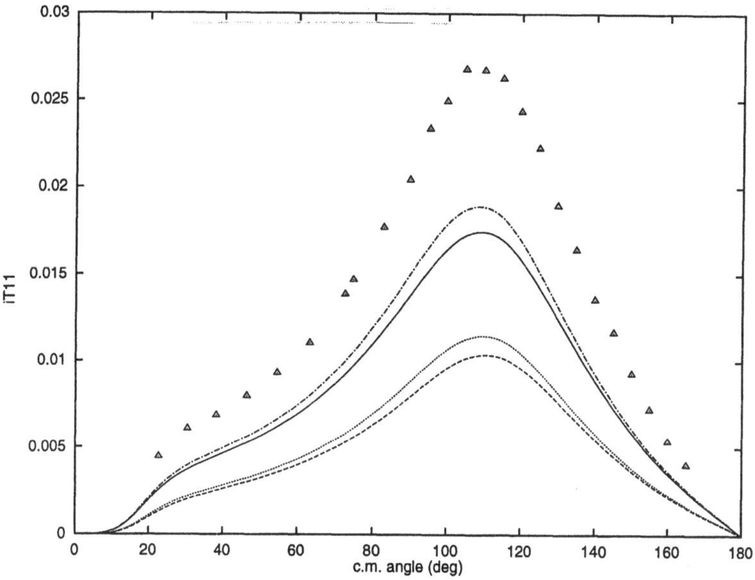

Figure 1. The deuteron analyzing power iT_{11} calculated with the four potential models: AV18 (solid line), AV14 (dotted line), AV14+BR (dashed line), and AV18+UR (dashed-dotted line). The experimental points are from ref. [10].

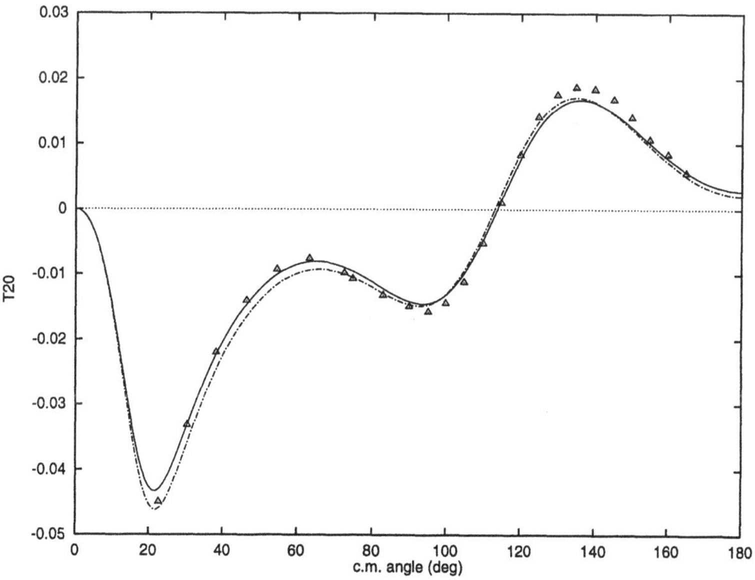

Figure 2. The deuteron analyzing power T_{20} calculated with the AV18 (solid line) and AV18+UR (dashed-dotted line) potential models. The experimental points are from ref. [10].

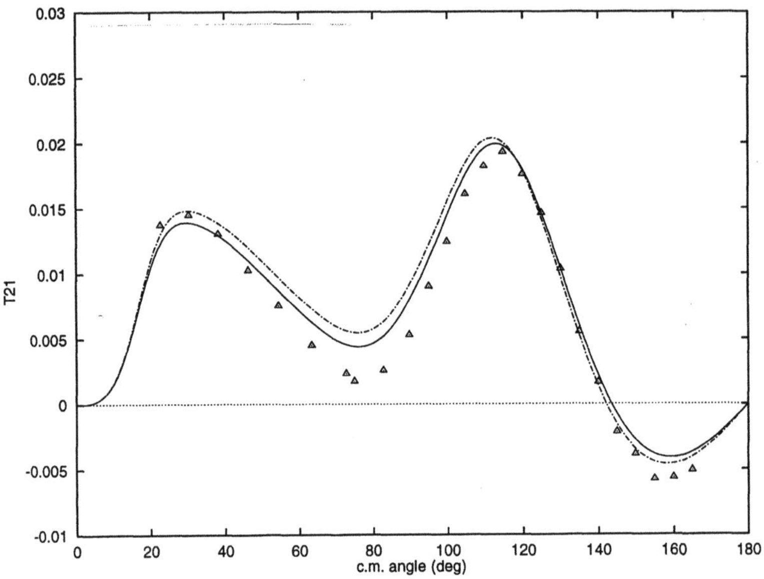

Figure 3. The deuteron analyzing power T_{21} calculated with the AV18 (solid line) and AV18+UR (dashed–dotted line) potential models. The experimental points are from ref. [10].

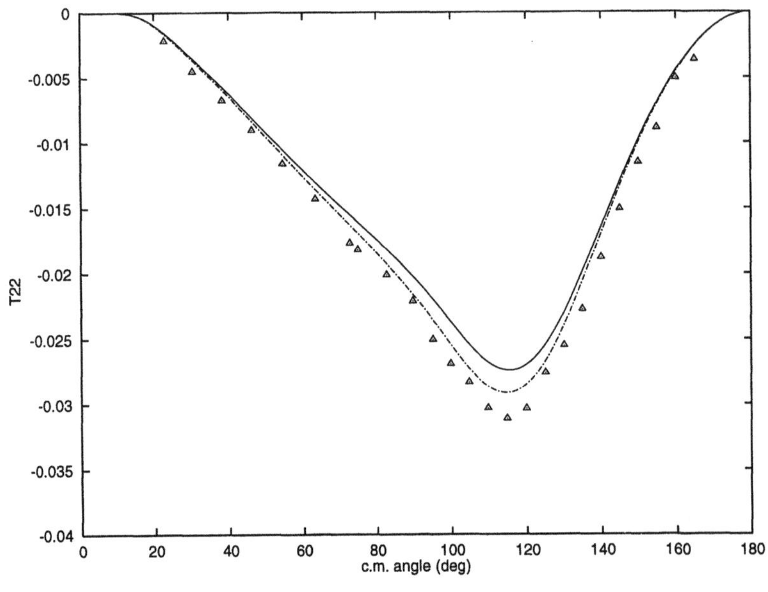

Figure 4. The deuteron analyzing power T_{22} calculated with the AV18 (solid line) and AV18+UR (dashed–dotted line) potential models. The experimental points are from ref. [10].

The singlet scattering lenght obtained is ${}^1a = 11.2\,\mathrm{fm}$ in correspondence to the AV14+UR potential model. The experimental value obtained by extrapolating the low energy data is $10.5 \pm 1.4\,\mathrm{fm}$ [13] in good agreement with the theoretical prediction. For the $n-{}^3\mathrm{H}$ zero energy reaction our prediction (in a three channel calculation) for the singlet scattering lenght is ${}^1a = 4.12\,\mathrm{fm}$. In this case the experimental value is $4.98 \pm 0.12\,\mathrm{fm}$ but new measurements are necessary in order to get a definite value [14]. The study of the reactions for different J, T, π values is in progress.

Acknowledgement. The author would like to thank S.Rosati, M.Viviani, D.Hüber and L.Knutson for useful discussions.

References

1. H. Kamada, W. Glöckle: Few–Body Syst., Suppl. **7**, 217 (1994)
2. A. Kievsky, S. Rosati, M. Viviani: Nucl. Phys. **A551**, 241 (1993)
3. M. Viviani, A. Kievsky, S. Rosati: Few–Body Syst. **18**, 25 (1995)
4. C.R. Chen et al. : Phys. Rev. **C31**, 266 (1985)
5. H. Kameyama, M. Kamimura, Y. Kukushima: Phys. Rev. **C40**, 974 (1989)
6. A. Kievsky, M. Viviani, S. Rosati: Nucl. Phys. **A577**, 511 (1994)
7. D. Huber et al. : Phys. Rev. **C51**, 1100 (1995)
8. R.G. Seyler: Nucl. Phys. **A124**, 253 (1969)
9. A. Kievsky, M. Viviani, S. Rosati: Phys. Rev. **C** (in print)
10. L.D. Knutson et al.: Phys. Rev. Lett. **71**, 3762 (1993)
11. K. Sagara et al. : Phys. Rev. **C50**, 576 (1994)
12. J.E. McAninch et al. : Phys. Rev. **C50**, 589 (1994)
13. P.E. Tegner, C. Bargholtz: Astrophysical J. **272**, 311 (1983)
14. H. Rauch et al. : Phys.Let. **165B**, 39 (1985)

Few-Body Systems Suppl. 9, 410–414 (1995)

Few-Body Systems

Off-Shell *NN* Potential and Nuclear Binding

R. Machleidt*, F. Sammarruca, Y. Song

Department of Physics, University of Idaho, Moscow, ID 83843, U. S. A.

Abstract. A new, high-precision, charge-dependent Bonn nucleon-nucleon potential ('CD-Bonn') has been constructed that is exactly phase-equivalent to the Nijmegen phase shift analysis and the new high-quality Nijmegen potentials. This *non-local* CD-Bonn potential predicts **8.0 MeV binding energy for the triton** (in a charge-dependent 34-channel Faddeev calculation) which is about 0.4 MeV more than the predictions by the local (phase-equivalent) Nijmegen potential. We pin down origin and size of the nonlocality in the Bonn potential, in analytic and numeric form. The nonlocality is due to the use of the correct relativistic off-shell Feynman amplitude of one-boson-exchange avoiding the commonly used on-shell approximations which yield the local potentials. Adding the relativistic effects from the relativistic nucleon propagators in the Faddeev equations, brings the CD-Bonn result up to 8.2 MeV triton binding. This leaves a difference of only about 0.3 MeV to experiment, which may possibly be explained by refinements in the treatment of relativity and the inclusion of other nonlocalities (e. g., those due to the composite nature of hadrons). Nuclear matter calculations further illustrate our points.

1 Introduction

One of the most fundamental goals of theoretical nuclear physics is to explain the properties of atomic nuclei in terms of the basic interactions between the nucleons. In spite of several decades of hard work, this goal is still a challenge today. One of the most serious problems encountered is a lack of binding energy when conventional two-nucleon potentials are used in nuclear structure calculations [1]. To fix this problem, it has been suggested to introduce a phenomenological three-nucleon ($3N$) force which is adjusted such as to provide the missing binding in nuclei—and a program pursuing consistently this point of view is well under way [2].

However, before opening the Pandora's box of $3N$ forces, it may be more reasonable to put the conventional nucleon-nucleon (NN) potentials under some scrutiny. As it turns out, most NN potentials applied in nuclear structure are

**E-mail address:* machleid@phys.uidaho.edu

local because this is more convenient when calculations are performed in configuration space. Notice, however, that numerical convenience is not really a fundamental theoretical argument for the alleged local nature of the nuclear force. One the contrary, any deeper insight into any more fundamental mechanism underlying the nuclear force is bound to yield a non-local interaction. Major sources for non-locality are the composite nature of hadrons and relativity. It is the purpose of this contribution to investigate the non-locality of the nuclear force as implied by relativity (relativistic meson-exchange) and its impact on nuclear structure predictions.

2 Relativistic Meson-Exchange and Non-Locality

Most nucleon-nucleon (*NN*) potentials are based more or less on meson exchange. The corresponding Feynman diagrams are non-local expressions due to relativity and energy-dependence. However, for convenience, some practitioners introduce specific approximations (e. g., the 'on-shell' approximation) which reduce the Feyman diagrams to local functions changing the potential, particularly, off-shell. Recent examples for such local *NN* potentials are the Nijmegen-II [3], Reid'93 [3], and Argonne V_{18} [4] potentials; the Nijmegen-I potential [3] has some non-locality in the central force but is local otherwise. In Fig. 1, left half, we display the half off-shell central forces of the Nijm-I, -II, and the CD-Bonn [5] potentials; the latter uses the unapproximated Feynman amplitudes.

As will be explained below, more important for nuclear structure is the off-shell tensor force. The local approximation of the pion tensor potential which is used in all common potentials except Bonn, has the familiar form,

$$^{\pi}\tilde{V}_{\rm T}(r) = -\frac{g_\pi^2}{4\pi}\left(\frac{m_\pi}{2M}\right)^2\left(1+\frac{3}{m_\pi r}+\frac{3}{(m_\pi r)^2}\right)\frac{e^{-m_\pi r}}{r}S_{12}\,. \qquad (1)$$

The transformation of this local potential into momentum space yields for the $^3S_1$-$^3D_1$ amplitude

$$^{\pi}\tilde{V}_{02}^{110}(q_0,k) = -\frac{g_\pi^2}{4\pi}\frac{\sqrt{8}}{4\pi M^2 q_0 k}[q_0^2 Q_2(z) - 2q_0 k Q_1(z) + k^2 Q_0(z)] \qquad (2)$$

with Q_L the Legendre function of the 2. kind and $z \equiv (q_0^2 + k^2 + m_\pi^2)/(2q_0 k)$. However, the original $^3S_1$-$^3D_1$ transition potential as it results from the relativistic one-pion-exchange Feynman amplitude is

$$\begin{aligned} ^{\pi}V_{02}^{110}(q_0,k) &= -\frac{g_\pi^2}{4\pi}\frac{\sqrt{8}}{4\pi M^2 q_0 k}[(E_{q_0}-M)(E_k+M)Q_2(z) - 2q_0 k Q_1(z) \\ &\quad +(E_{q_0}+M)(E_k-M)Q_0(z)] \end{aligned} \qquad (3)$$

with $E_{q_0} \equiv \sqrt{M^2+q_0^2}$ and $E_k \equiv \sqrt{M^2+k^2}$. Expanding these roots in terms of q_0^2/M and k^2/M yields

$$^{\pi}V_{02}^{110}(q_0,k) \approx -\frac{g_\pi^2}{4\pi}\frac{\sqrt{8}}{4\pi M^2 q_0 k}\left[\left(q_0^2 + \frac{q_0^2 k^2}{4M^2} - \frac{q_0^4}{4M^2}\cdots\right)Q_2(z) - 2q_0 k Q_1(z)\right.$$

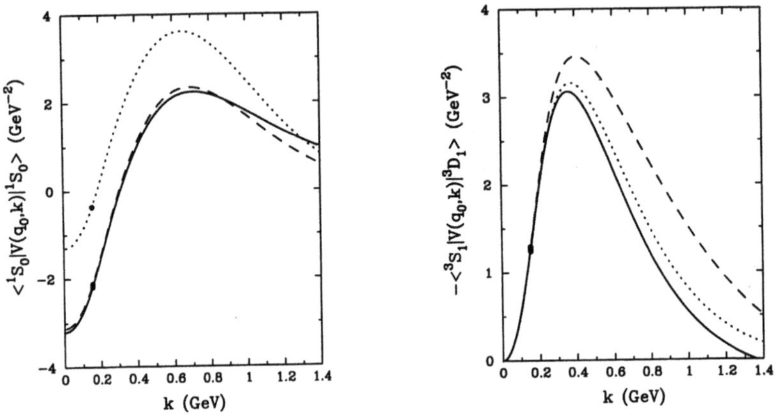

Figure 1. Half off-shell potentials $V(q_0, k)$ for CD-Bonn (solid line), Nijm-I (dashed), and Nijm-II (dotted). The momentum q_0 is held fixed at 153 MeV which corresponds to a lab. energy of 50 MeV. The solid dots denote the on-shell points ($k = q_0$). **Left:** potentials in the ${}^1S_0$ state. **Right:** ${}^3S_1$-${}^3D_1$ transition potentials.

$$+\left(k^2 + \frac{q_0^2 k^2}{4M^2} - \frac{k^4}{4M^2} \cdots \right) Q_0(z)\Bigg] \; . \tag{4}$$

Keeping terms up to momentum squared, leads to the local approximation Eq. (2). The largest term in the next order is $-k^2/(4M^2)Q_0(z)$ which damps the tensor potential off-shell.

In Fig. 1, right half, we show the ${}^3S_1$-${}^3D_1$ potential matrix element, $-V_{02}^{110}(q_0, k)$, for various potentials. It is seen that, particularly for large off-shell momenta, the CD-Bonn potential is substantially smaller than the local potentials indicating a weaker *off-shell* tensor force.

3 Off-Shell Potential and Nuclear Structure

Predictions for *NN* observables may be based upon the on-shell two-nucleon t-matrix derived from a given *NN* potential V. The off-shell *NN* t-matrix is input for momentum-space Faddeev calculations of the three-nucleon system. The calculation of the t-matrix always involves the *NN* potential on- and off-(the-energy-)shell. We illustrate this for the example of the partial-wave t-matrix, $t_{L'L}^{JST}$, in the ${}^3S_1$ two-nucleon state, which is given by

$$\begin{aligned} t_{00}^{110}(q', q; E) &= V_{00}^{110}(q', q) - \int_0^\infty k^2 dk V_{00}^{110}(q', k) \frac{M}{k^2 - ME - i\epsilon} t_{00}^{110}(k, q; E) \\ &\quad - \int_0^\infty k^2 dk V_{02}^{110}(q', k) \frac{M}{k^2 - ME - i\epsilon} t_{20}^{110}(k, q; E) \; . \end{aligned} \tag{5}$$

Table 3.1. Recent high-precision *NN* potentials and some of their predictions

	CD-Bonn	**Nijm-I**	**Nijm-II**
Characteristics	non-local	non-loc. central local tensor	local
χ^2**/datum**	1.03	1.03	1.03
Deuteron D-state prob. (%)	4.83	5.66	5.64
Triton binding energy (MeV)			
non-relativistic calculation	8.00	7.72	7.62
relativistic calculation	8.19	–	–
Nuclear matter energy (MeV)			
partial-wave contributions			
at $k_F = 1.35$ fm^{-1}: $^1S_0$	-16.76	-16.73	-16.11
$^3S_1$	-18.95	-17.55	-17.05
$^3P_0$	-3.09	-3.11	-3.06
$^3P_1$	9.85	9.73	9.72
$^3P_2$	-7.02	-7.00	-7.01
Total at $k_F = 1.35$ fm^{-1}	−13.60	−12.40	−11.02
Saturation energy	−17.45	−14.41	−11.68

For free-space *NN* scattering, $E = q_0^2/M$ with M the nucleon mass and q_0 the c.m. on-shell momentum which is related to the lab. energy by $E_{\text{lab}} = 2q_0^2/M$. Notice that in the integral terms the potential contributes essentially off-shell. In view of Fig. 1, it is clear that local potentials produce much larger integral terms in Eq. (5) than non-local ones.

Nuclear structure calculations can be based upon a t- or Brueckner g-matrix which is defined very similarly to t in Eq. (5). In the three-body Faddeev equations, the t-matrix is fully off-shell and E is negative; the negative E reduces the magnitude of the integral terms. The larger the terms, the larger the quenching. Since the integral terms are attractive, the quenching produces a repulsive effect. Thus, large off-shell potentials, implying large integral terms, yield less attraction in three- and many-body systems. This explains essentially why local potentials predict less binding energy than the non-local CD-Bonn potential.

4 Effect of Non-Locality on Nuclear Binding

To accurately pin down the exact effect of off-shell differences between potentials on nuclear structure predictions, it is important to use potentials that are identical in their on-shell preditions for the two-nucleon system and fit the experimental *NN* data exactly. Therefore we have improved the Bonn potential [5] such as to fit the *NN* data below 350 MeV as precisely as the recently developed high-quality local Nijmegen [3] and Argonne [4] potentials. The new charge-dependent, high-precision Bonn potential [5], dubbed CD-Bonn, reproduces the 4301 pp and np data below 350 MeV with a χ^2/datum of almost one. Our results are summarized in Table 1. The non-local CD-Bonn potential predicts 8.00 MeV for the triton binding energy in a 34-channel charge-dependent Faddeev calculation. The corresponding result for local, high-precision potentials is 7.62 MeV. Thus, about half of the discrepancy between local-potential predictions and the experimental value of 8.48 MeV can be explained by non-locality in the NN interaction. Performing a relativistic Faddeev calculation [6] with CD-Bonn brings the result further up to 8.19 MeV.

We have also performed standard Brueckner nuclear matter calculations since they reveal in a very transparent way which partial waves are mainly responsible for differences in binding energy predictions (cf. Table 1).

5 Conclusions

Based upon any more fundamental view point, the nuclear force is non-local in nature. Non-locality of the NN potential may have a significant impact on microscopic nuclear structure predictions. Here, we have shown that the non-locality implied by relativity increases the predictions for the binding energy of nuclei—bringing them closer to their experimental values.

In view of the fact that our current calculations do not even take the more important source of non-locality due to the composite nature of hadrons into account, it is conceivable that nuclear binding can be explained quantitatively without any significant contribution from nuclear many-body forces.

Acknowledgement. Supported in part by NSF-Grant PHY-9211607.

References

1. J.L. Friar et al. : Phys. Lett. **B311**, 4 (1993);
   B.D. Day and R. B. Wiringa: Phys. Rev. C **32**, 1057 (1985)
2. B.S. Pudliner et al. : Phys. Rev. Lett. **74**, 4396 (1995)
3. V.G.J. Stoks et al. : Phys. Rev. C **49**, 2950 (1994)
4. R.B. Wiringa et al. : Phys. Rev. C **51**, 38 (1995)
5. R. Machleidt: Adv. Nucl. Phys. **19**, 189 (1989);
   to be published
6. F. Sammarruca, D.P. Xu, and R. Machleidt: Phys. Rev. C **46**, 1636 (1992)

Few-Body Systems Suppl. 9, 415–428 (1995)

Few-Body Systems

Electron Scattering from the Deuteron Using the Gross Equation

J.W. Van Orden[1,3], N. Devine[3], F. Gross[2,3]

[1] Department of Physics, Old Dominion University, Norfolk, Virginia 23529, USA

[2] Department of Physics, College of William and Mary, Williamsburg, Virginia 23185, USA

[3] The Continuous Electron Beam Accelerator Facility, 12000 Jefferson Ave., Newport News, Virginia 23606, USA

Abstract. The elastic electromagnetic form factors for the deuteron are calculated in the context of a one-boson-exchange model using the Gross or Spectator equation [1]. The formalism is manifestly covariant and gauge invariant. Results are shown for the impulse approximation and for $\rho\pi\gamma$ exchange currents. The impulse approximation results are quite close to the available data which suggests that only a relatively small exchange current contribution is required. It is shown that by using a soft form factor for the exchange current, the model provides a very good representation of the data.

1 Introduction

With CEBAF now coming on line, it will be routine to probe nuclear systems with electron scattering where the energy and momentum transfers will be well in excess of the nucleon mass. Under such circumstances, the usual nonrelativistic description of the nucleus is no longer reliable. It is, therefore, necessary to develop relativistically covariant models of the nuclear system. In addition, it is important to maintain gauge invariance in such calculations if they are to be reliable. This is a difficult problem and at present this program can be carried out completely in the deuteron. Here, we will present a calculation of the electromagnetic form factors of the deuteron in the context of the Gross equation [1]. This calculation is manifestly Lorentz covariant and has been constructed to be gauge invariant. We will show that by carefully constraining the interaction model to fit the nucleon-nucleon phase shifts and by using only minimal exchange currents we are capable of obtaining an excellent description of the available data.

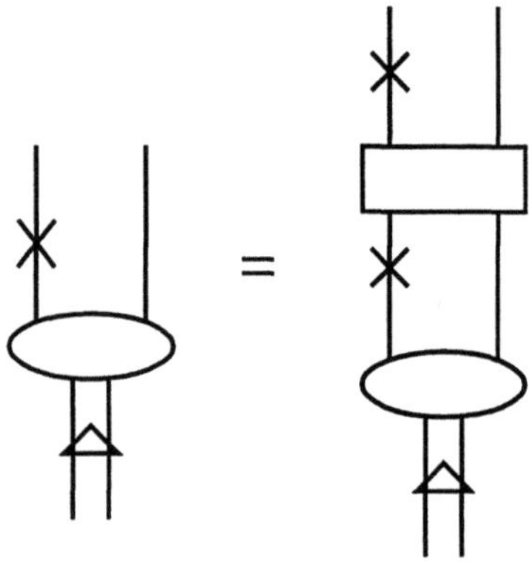

Figure 1. Feynman diagrams representing the homogeneous integral equation for th Gross wave function.

2 The Model

The Gross equation [1] can best be understood as an example of a quasipoten tial equation [2, 3]. The common characteristic of all quasipotential equation is the replacement of the free intermediate-state two-nucleon propagator in th Bethe-Salpeter equation by a new propagator that includes a delta-functio constraining the relative energy of the intermediate states. This reduces th four-dimensional Bethe-Salpeter equation to a three-dimensional integral equa tion. In the case of the Gross equation the relative energy constraint is obtaine by keeping only the positive energy pole of one of the nucleons in performin energy loop integrals. This is illustrated by Fig. 1 which shows the Feynma diagrams representing the homogeneous Gross wave equation. The ovals rep resent deuteron vertex functions, the single lines nucleon propagators and th rectangular box the interaction kernel. The crosses on nucleon propagators rep resent particles placed on their positive energy mass-shells. In this case we hav chosen to always place the left-hand nucleon (particle 1) on shell. Note tha in order to obtain a consistent integral equation, the external propagator fo particle 1 is placed on shell. Note also that in defining the wave function, the ex ternal nucleon lines represent nucleon propagators rather than simply spinors The on-shell nucleon propagator simply becomes a positive energy projectio operator. As will be shown below, calculation of current matrix elements re quires the deuteron vertex function with both external nucleons off mass shell This is represented by the Feynman diagrams in Fig. 2. The dots on the verte function and the kernel represent attachment points for nucleon propagators Note that on the right hand side of this equation the deuteron vertex has on nucleon on mass shell and this vertex along with the on- and off-shell propa gators is identical to the Gross wave function as defined by Fig. 1. Therefore the off-shell deuteron vertex function can be obtained by quadrature using th

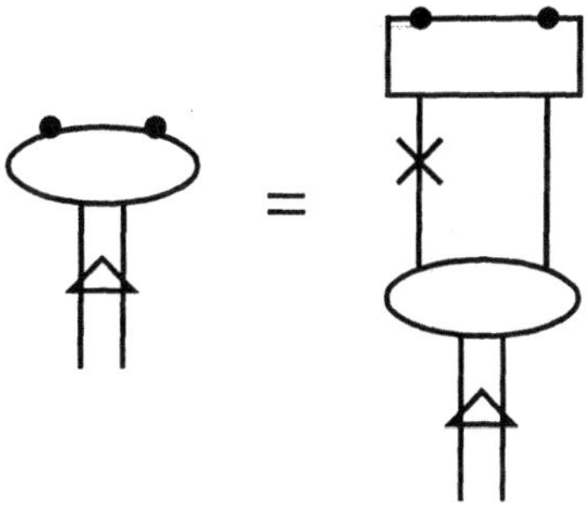

Figure 2. Feynman diagrams representing the deuteron vertex function.

Gross wave function and the kernel with three external momenta off mass shell.

This vertex function for the deuteron can be written as [4]

$$(\Gamma(p,P)\cdot\xi_{\lambda_d}(P)\mathcal{C})_{ab} \tag{1}$$

where P is the four-momentum of the deuteron, p is the relative momentum of the external nucleons, $\xi_{\lambda_d}(P)$ is the polarization four-vector for the deuteron, $\mathcal{C}$ is the Dirac charge conjugation matrix , the subscripts a and b are indices in the Dirac spinor space, and Γ^μ can be determined by basic symmetry arguments to have the general form:

$$\begin{aligned}\Gamma^\mu(p,P) &= f_1(p^2,p\cdot P)\gamma^\mu + g_1(p^2,p\cdot P)\frac{p^\mu}{m} \\ &+\frac{\not{p}_2-m}{m}\left(f_2(p^2,p\cdot P)\gamma^\mu + g_2(p^2,p\cdot P)\frac{p^\mu}{m}\right) \\ &+\left(f_2(p^2,-p\cdot P)\gamma^\mu + g_2(p^2,-p\cdot P)\frac{p^\mu}{m}\right)\frac{\not{p}_1+m}{m} \\ &+\frac{\not{p}_2-m}{m}\left(f_4(p^2,p\cdot P)\gamma^\mu + g_4(p^2,p\cdot P)\frac{p^\mu}{m}\right)\frac{\not{p}_1+m}{m}\end{aligned} \tag{2}$$

where $p_1 = \frac{P}{2}+p$ and $p_2 = \frac{P}{2}-p$. A generalization of the Pauli symmetry requires that

$$\Gamma^\mu(p,P) = -\mathcal{C}\Gamma^{\mu T}(-p,P)\mathcal{C}^{-1}. \tag{3}$$

Note that for given values of p and P, the vertex function depends upon eight scalar functions, two pairs of which are related by inversion of p. In the case of the constrained spectator vertex function where particle 1 is on its positive energy mass shell, only four of these functions contribute. Therefore, the general off-shell vertex function can be represented by eight radial functions while the constrained constrained Gross wave functions have only four radial wave functions.

Since the on-shell constraint used in construction the Gross wave function and the fully-off-shell vertex function is not symmetric for the intermediate state nucleons, satisfying the generalized Pauli symmetry given by Eq. (3) requires some care. It has been shown in ref. [5] that the Pauli symmetry can be satisfied by requiring that the kernel of the integral equation be antisymmetrized. This has the unwanted effect of introducing unphysical singularities into the wave and scattering integral equations. But it has been shown that the quantitative effects of this defect are small and that the solution of the integral equations can be achieved at the expense of some technical complexity.

The Gross wave functions are normalized in the deuteron rest frame, where $P = (M_d, \mathbf{0})$, according to [5]

$$\begin{aligned} 1 &= \int \frac{d^3p}{(2\pi)^3}\frac{m}{E_p}\bar{\psi}(p,P)\gamma^0\psi(p,P) \\ &- \int \frac{d^3p'}{(2\pi)^3}\int \frac{d^3p'}{(2\pi)^3}\bar{\psi}\frac{\partial}{\partial M_d}V(p',p;P)\psi(p,P) \end{aligned} \tag{4}$$

where M_d is the deuteron mass, $\psi(p,P)$ is the Gross wave function and $V(p',p;P)$ is the interaction kernel. The derivative term accounts for possible energy dependence of the kernel. Although we have written this normalization expression as it appears in the deuteron rest frame where it is of a relatively simple form, the normalization can be written in a covariant form. Indeed, as a check of some elements of the numerical calculation of the deuteron form factors, we have verified that the covariant normalization condition is satisfied in a variety of Lorentz frames.

Reference [5] describes in considerable detail the application of the spectator equation to nucleon-nucleon scattering and the deuteron bound state. Four models are presented for the NN interaction. Each model has been fitted to the NN phase shift data of Arndt and Roper, SP89 [6] with the aid of the error matrix obtained in the phase shift fit. These fits are constrained such that the deuteron bound state mass is correct. The resulting phase shift calculations are then compared to the data base and typically obtain a χ^2 per datum of approximately 2 for energies from 0 MeV to 225 MeV of laboratory kinetic energy. This is quite good for a one-boson-exchange model.

An unusual feature of these calculations is that they allow for off-shell meson-nucleon couplings not usually present in such models. For example, the basic form of the πNN coupling is chosen to be [4, 5]:

$$g_\pi\left[\lambda_\pi\gamma^5 + (1-\lambda_\pi)\frac{(\not{p}-\not{p}')}{2m}\gamma^5\right] \tag{5}$$

where λ_π is a parameter which extrapolates between pseudoscalar and pseudovector coupling. Note that this coupling is independent of λ_π when the nucleons are on mass shell, so models with differing values of λ_π differ in their off-shell content. All meson-nucleon interactions also include factorable form factors.

The meson-nucleon-nucleon form factors depend upon the invariant masses of the three virtual particles connected to the interaction vertex. That is, if p and p' are the initial and final nucleon four-momenta, and $\ell = p - p'$ is the meson four-momentum, the general form of the form factor is $F(p'^2, p^2, \ell^2)$. For simplicity, we assume that the form factor can be written in a factorable form [5, 7]

$$F(p'^2, p^2, \ell^2) = h(p'^2)h(p^2)f(\ell^2) \tag{6}$$

where the meson form factor is taken to be

$$f(\ell^2) = \frac{(\Lambda_\mu^2 - \mu^2)^2 + \Lambda_\mu^4}{(\Lambda_\mu^2 - \ell^2)^2 + \Lambda_\mu^4} \tag{7}$$

and the nucleon form factor is

$$h(p^2) = \frac{2(\Lambda_n^2 - m^2)^2}{(\Lambda_n^2 - p^2)^2 + (\Lambda_n^2 - m^2)^2} \tag{8}$$

where Λ_μ and Λ_n are meson and nucleon form factor masses. The presence of the nucleon form factor is an unusual feature of the calculations presented here. This form factor allows the four-momenta of the nucleons to be controlled such that the contributions from highly-virtual nucleons can be limited.

The interaction model used in the calculations shown here is a variation on model IIB of ref. [5]. The parameters of the model have been adjusted to fit the Nijmegen energy dependent np phase shifts [11]. The results are compared to the data base in SAID [6] to give a χ^2 per datum of 1.89 for energies of 1 to 250 MeV and of 2.53 for 1 to 350 MeV. This model uses a one boson-exchange kernel containing six mesons: π, η, σ, σ_1, ω and ρ. The σ_1 meson is a scalar-isovector companion to the σ with a mass comparable to the σ mass. The pion mixing parameter was fixed at $\lambda_\pi = 0$ for pure pseudovector coupling. A total of thirteen parameters were adjusted in the fitting procedure.

The deuteron wave functions for this model are shown in Fig. 3. There are four wave functions, the usual S and D waves that appear in the nonrelativistic treatment of the deuteron and singlet and triplet P waves of relativistic origin. The contributions to the normalization of the wave function from these components are: 92.979% for the S wave, 5.015% for the D wave, 0.049% for the triplet P wave and 0.009% for the singlet P wave. Note that this gives a total of only about 98%. The remaining 2% is associated with the derivative term in (4). This is almost entirely the result of energy dependence introduced into the kernel by the nucleon form factor (8). Note that the signs of the singlet and triplet P waves are opposite for this model.

3 Current Matrix Elements and the Gross Equation

There are two problems associated with the calculation of the electromagnetic current matrix elements using these wave functions. The first of these is common to all Bethe-Salpeter based calculations of the deuteron electromagnetic

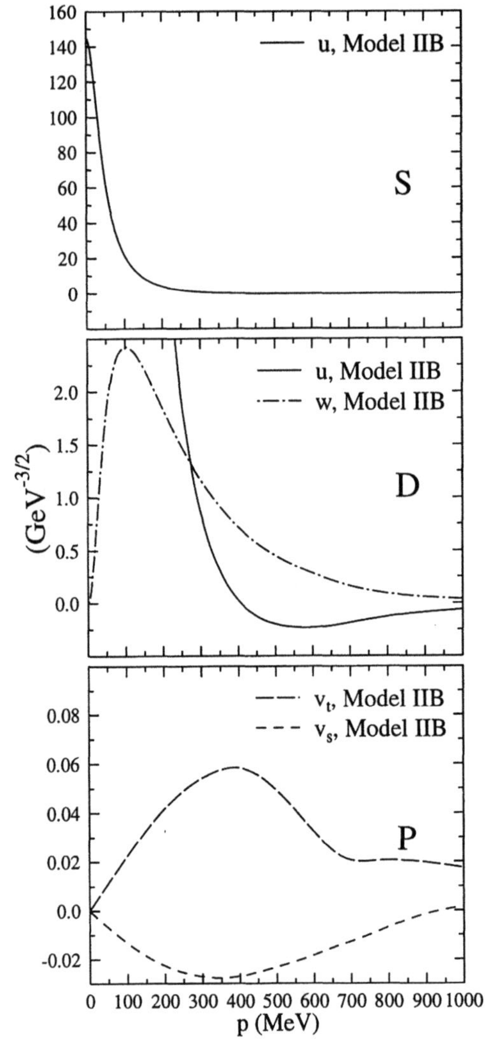

Figure 3. Gross deuteron wave functions for Model IIB. The S wave is denoted by u, the D wave by w and the singlet and triplet P waves by v_t and v_s.

properties. This is associated with the use of form factors at the strong and electromagnetic vertices in such models. If the couplings were truly pointlike, the construction of the appropriate one- and two-body electromagnetic current operators would straightforward. The single-nucleon current operator could be taken directly from the effective lagrangian of the model. The two-nucleon current keeping two-body contributions could be calculated from all two-nucleon irreducible diagrams constructed by attaching a virtual photon to all nucleon or meson lines with charges or magnetic moments in the diagrams representing the irreducible Bethe-Salpeter current. These diagrams can be easily shown to satisfy the necessary one- and two-body Ward-Takahashi identities [8, 7].

The complication comes from the introduction of form factors at the strong and electromagnetic vertices to account for the finite sizes of the nucleons and mesons. Simply multiplying the various strong and electromagnetic vertices with form factors will violate the Ward-Takahashi identities. This problem is discussed in detail in ref. [7] for models containing factorable form factors such as is defined by (6). Using the arguments of this reference it is possible to construct a minimal description of the single-nucleon current operator given by [3]:

$$\begin{aligned} J^{(i)\mu}(p',p) &= F_1(Q^2)f_0(p'^2,p^2)\,\gamma^\mu + \frac{F_2(Q^2)}{2m}h_0(p'^2,p^2)\,i\sigma^{\mu\nu}q_\nu \\ &\quad +F_3(Q^2)g_0(p'^2,p^2)\,\frac{\not p' - m}{2m}\,\gamma^\mu\,\frac{\not p - m}{2m} \end{aligned} \tag{9}$$

where

$$f_0(p'^2,p^2) \equiv \frac{h(p^2)}{h(p'^2)}\,\frac{m^2-p'^2}{p^2-p'^2} + \frac{h(p'^2)}{h(p^2)}\,\frac{m^2-p^2}{p'^2-p^2}, \tag{10}$$

$$g_0(p'^2,p^2) \equiv \left(\frac{h(p^2)}{h(p'^2)} - \frac{h(p'^2)}{h(p^2)}\right)\frac{4m^2}{p'^2-p^2} \tag{11}$$

and $h_0(p'^2,p^2)$ is an arbitrary function subject only to the constraint that $h_0(m^2,m^2)=1$. In the calculations presented here, this function is chosen to be $h_0(p'^2,p^2)=f_0(p'^2,p^2)$, for simplicity.

For elastic scattering from the deuteron, only isoscalar two-body exchange currents can contribute. The only possible isoscalar contributions for the one-boson-exchange model used here are of the type $\rho\pi\gamma$, $\omega\eta\gamma$, $\omega\sigma\gamma$, etc. These currents have couplings that are individually gauge invariant and therefore require no complicated modification of the off-shell behavior of the vertex functions and form factors in order to maintain gauge invariance.

The second problem is associated with the use of a quasipotential formalism. The quasipotential prescriptions impose constraints on the phase space available to nucleons propagating in intermediate states. In general this constraint must be applied consistently to the calculation of the current matrix elements which will lead to an effective current which differs from the Bethe-Salpeter current operator. The procedure for obtaining the effective current operator for a general quasipotential reduction has not yet been sufficiently studied. However, the correct form of the matrix element for the spectator or

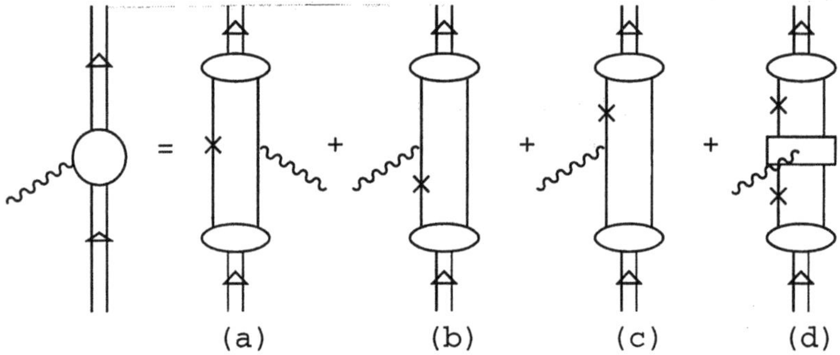

Figure 4. Feynman diagrams representing the Gross current matrix element.

Gross equation has been described in ref. [7], and a description of the matrix element for the Blankenbecler-Sugar equation is described in ref. [9]. In the case of the Gross equation, the correct expression for the current matrix element can be obtained by keeping only the positive energy nucleon poles for particle 1 in the evaluation of the energy loop integrals of the Bethe-Salpeter current matrix element. For the elastic matrix elements this leads to the Feynman diagrams displayed in Fig. 4. Diagram 4a, where the virtual photon is absorbed on particle 2, has particle 1 constrained on shell for both the initial and final state vertex functions and can be written in the form of a matrix element of the single-nucleon current operator between two Gross wave functions. However, if the virtual photon is absorbed on particle 1, the positive energy pole can be picked up for the propagator before the absorption of the virtual photon or the one after. This leads to diagrams 4b and 4c. In these diagrams only the initial or final vertex function is on shell and the other must be off shell. These two diagrams do not have the simple form associated with the nonrelativistic impulse approximation as does diagram 4a. The equation for the off-shell vertex function represented by Fig. 2 can be used to write the matrix elements for these diagrams in terms of the constrained Gross wave function. The resulting diagrams may be viewed as interaction current contributions which are necessary to accommodate the on-shell constraint. It should be noted that only by calculating diagrams 4a, 4b and 4c can the proper normalization of the charge be recovered from the charge form factor in the limit $Q^2 \rightarrow 0$. Diagrams involving two-body Bethe-Salpeter currents will have two internal energy loops which can be constrained independently to give diagram 4d. Note that the meson exchange currents must be symmetrized. Also, it should be noted that the Bethe-Salpeter and Gross two-body meson exchange currents will only be the same at the one-boson-exchange level.

Prior to ref. [7], it was assumed that the proper form of the Gross current matrix element was described by diagram 4a along with a symmetric diagram where the photon attaches to particle 1 and particle 2 is placed on mass shell

[10]. Because of the symmetry of the matrix element, the contribution of the second diagram is equivalent to diagram 4a. Thus this approximation is equivalent to simply calculating 2×diagram 4a. Since the form of this approximation looks like a matrix element of a single-nucleon current between spectator wave functions, it is referred to as the relativistic impulse approximation (RIA). Since the combination of diagrams 4a, 4b and 4c are related to the Bethe-Salpeter relativistic impulse approximation but represent a complete gauge-invariant description of the Gross one-body current matrix elements we will refer to it here as the complete impulse approximation (CIA).

In calculating the diagrams of Fig. 4 it is possible to proceed in two basically different ways. The scalar coefficient functions for half-off-shell and fully-off-shell vertex functions as represented in Eq. (2) can be determined and the matrix element can be evaluated as a trace in the Dirac matrix space. Alternately the various elements of the Feynman diagrams can be projected onto positive and negative energy plane wave Dirac spinors and the wave and vertex functions expanded in a partial wave basis in the deuteron rest frames. It is then necessary to boost the wave and vertex functions to the Breit frame where the calculations are performed. The various elements of the CIA have been calculated independently in both ways and are in agreement. The exchange current contributions have only been calculated using the spinor expansion due to their greater complexity.

4 Results

Figure 5 shows the structure functions $A(Q^2)$ and $B(Q^2)$ as calculated with our model for variations on the impulse approximation. The relativistic impulse approximation of Hummel and Tjon [12] is shown for reference and is labelled "Tjon, RIA". This calculation uses the Höhler single nucleon electromagnetic form factors [13] while our calculations use the dipole parameterization of Galster [14]. Three versions of the impulse approximation are calculated using our model. The calculation labelled "RIA" uses the old definition of the RIA with a current matrix element calculated as twice the result of diagram 4a and using only an on-shell form of the single nucleon current operator obtained by setting $f_0 = h_0 = 1$ and $g_0 = 0$ in (9). The curve labelled "RIA, off shell" is the same as the first but with the full form of the off-shell single-nucleon current operator, and the curve labelled "CIA" is the complete impulse approximation corresponding to diagrams 4a, 4b and 4c with the completely off-shell single-nucleon current operator. For $A(Q^2)$,the RIA and the RIA of Hummel and Tjon are in reasonable agreement except at large Q^2. Use of the off-shell current operator increases the size of $A(Q^2)$ to more closely describe the data above $Q^2 = 1$ GeV^2. The CIA decreases the size of $A(Q^2)$ from 1 to 5 GeV^2 moving away from the data but remaining above the RIA.

For $B(Q^2)$, the minimum of the RIA is at larger Q^2 than is the calculation of Hummel and Tjon. This appears to be the result of dynamical differences in the interaction models used. In particular the position of the minimum of

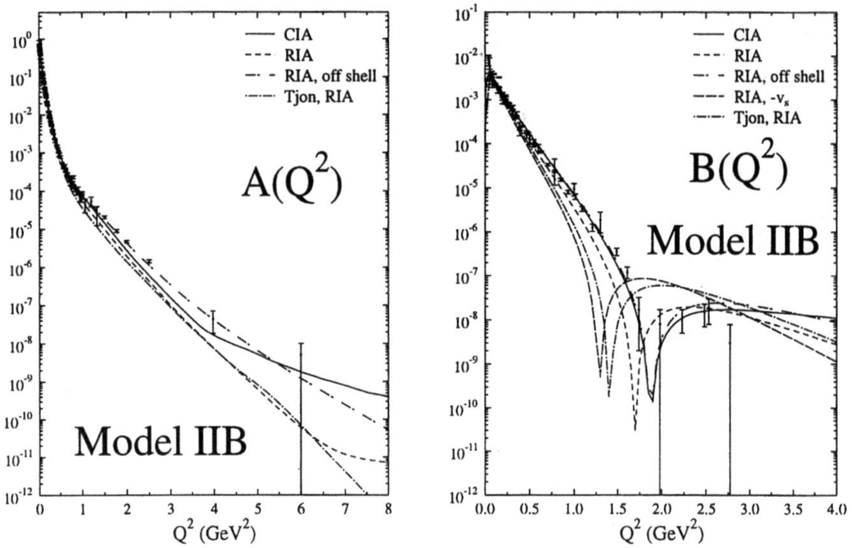

Figure 5. A and B structure functions calculated in the impulse approximation.

$B(Q^2)$ is particularly sensitive to the sign of the singlet P wave v_s as can be seen by simply changing the sign of this wave function component in the calculation of the RIA as is shown in the curve labelled "RIA, $-v_s$". The effect of this change is to produce a large downward shift in the position of the minimum. This causes little change in $A(Q^2)$ or T_{20}, however. The use of the off-shell current operator in the RIA moves the minimum to even larger values of Q^2. The CIA is very close to the off-shell RIA and both are in remarkably close agreement to the data. These curves show that only small contributions from the exchange currents are required to bring the CIA into good agreement with the data.

As mentioned above, the isoscalar exchange currents are all independently gauge invariant and are thus not needed to obtain a gauge invariant calculation of the form factors and structure constant. However, the $\rho\pi\gamma$ exchange current is related to the AAV anomaly [15] and the coupling and size of the contribution to the form factors at $Q^2 = 0$ is reasonably well constrained. The form factor for the $\rho\pi\gamma$ vertex is not known however and is a source of uncertainty in the calculations. The coupling constant for the $\omega\eta\gamma$ can be extracted from existing data but with less accuracy than in the previous case. Again the form factor is not known. Since a σ meson of 500 MeV does not exist except as a possible 2π-exchange enhancement, the existence and role of the $\omega\sigma\gamma$ exchange current is unclear. If we assume that $\omega \rightarrow \gamma + 2\pi$ is related to our $\omega \rightarrow \gamma\sigma$, then coupling constant for the $\omega\sigma\gamma$ can be extracted from existing data. Again the

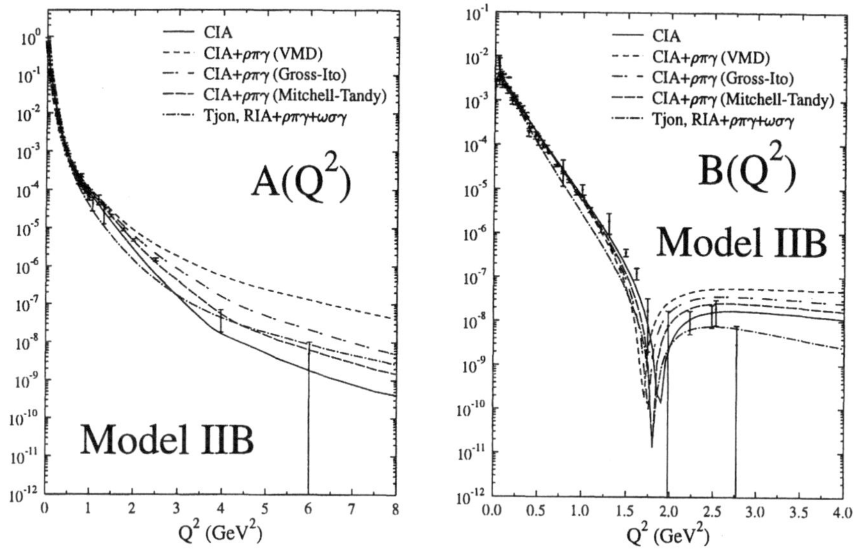

Figure 6. A and B structure functions with exchange currents.

form factor is not known. The couplings and form factors for the other possible exchange currents can be predicted by quark models, but are not otherwise constrained. The contributions of the exchange currents to the elastic form factors of the deuteron have been calculated by Hummel and Tjon [12] in an approximate fashion using the Blankenbecler-Sugar equation. The form factors for all contributions were taken to be given by the vector dominance model (VMD). It was found that $\rho\pi\gamma$ and $\omega\sigma\gamma$ exchange currents were needed to obtain any agreement with the data and that contributions of the $\omega\eta\gamma$ exchange currents were small. These calculations for $A(Q^2)$ and $B(Q^2)$ are shown in Fig. 6 labelled as "Tjon, RIA+$\rho\pi\gamma$ + $\omega\sigma\gamma$". The CIA calculation is also shown for reference to the previous figures. Calculations of the contributions of the $\rho\pi\gamma$ exchange current are shown for our model as calculated with the VMD form factors (labelled "CIA+$\rho\pi\gamma$ (VMD)"), and quark model form factors as calculated by Gross and Ito [16] (labelled "CIA+$\rho\pi\gamma$ (Gross-Ito)") and by Mitchell and Tandy [17] (labelled "CIA+$\rho\pi\gamma$ (Mitchell-Tandy)"). Both of the quark model form factors are softer than the VMD. The $\rho\pi\gamma$ exchange currents tend to increase the size of $A(Q^2)$ and to move the minimum of $B(Q^2)$ to lower Q^2. In both cases the VMD form factors produce much too large an effect while the softer quark model form factors give smaller effects. Indeed the calculation with the Mitchell-Tandy form factor is remarkably close to the data. Contributions from $\omega\sigma\gamma$ and $\omega\eta\gamma$ have also been calculated in our model but the results are still under examination.

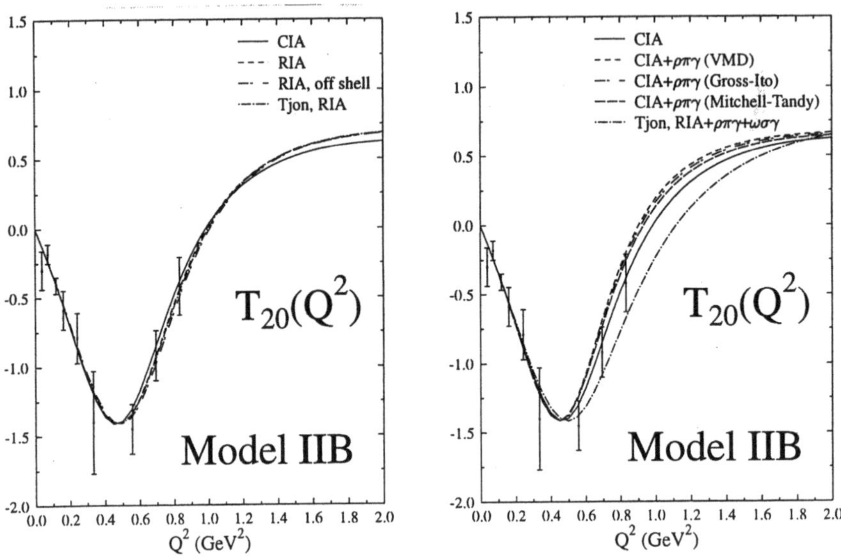

Figure 7. The tensor polarization T_{20}.

Figure 7 shows the quantity

$$T_{20}(Q^2) = \tilde{t}_{20}(Q^2) = -\frac{\sqrt{2}Q^2}{3M_d^2}\frac{G_C G_Q + (Q^2/12M_d^2)G_Q^2}{G_C^2 + (Q^4/18M_d^4)G_Q^2} \tag{12}$$

for the various calculations. This quantity is relatively insensitive to variations in the impulse approximation and all such calculations are consistent with the data. A somewhat greater sensitivity is seen for the exchange current contributions. The quality of the data is not yet sufficient to distinguish among the various models, however.

Figure 8 shows the charge and quadrupole form factors $G_C(Q^2)$ and $G_Q(Q^2)$, comparing the impulse approximation and exchange current calculations of this work and Tjon, et al. to the data. The position of the zero in $G_C(Q^2)$ is quite sensitive to the underlying dynamics of the impulse approximation and to the exchange currents. The complete exchange current calculation of Hummel and Tjon seems to give a zero within the range required by the data, while our exchange current calculation with the Mitchell-Tandy form factor seems give the zero at too small a value of Q^2. Our calculation seems to be a little too large at smaller Q^2. As should be expected from the T_{20} calculations, the quadrupole form factor $G_Q(Q^2)$ is less sensitive to the dynamics. Our calculations seem to be slightly too large at larger Q^2 while those of Hummel and Tjon are slightly too small.

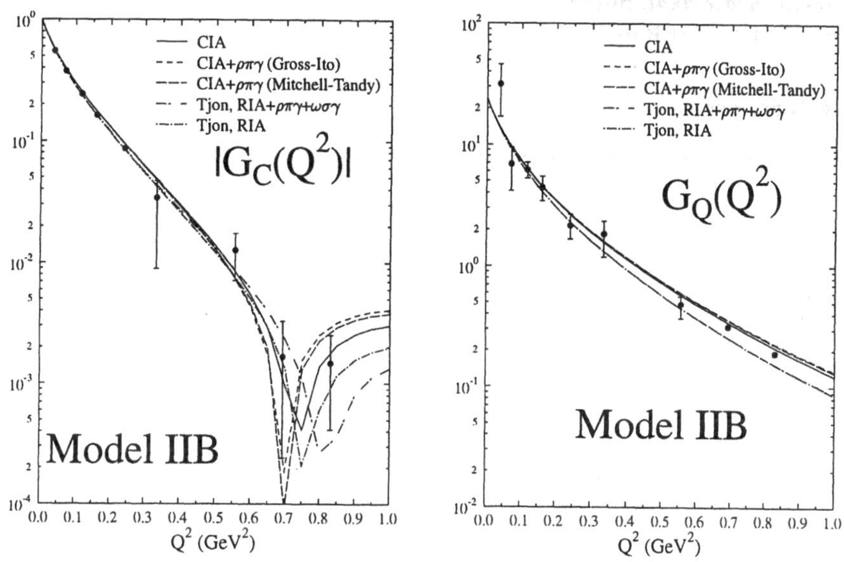

Figure 8. The charge and quadrupole form factors G_C and G_Q.

5 Summary

We have constructed a complete, relativistically covariant and gauge invariant model of elastic electron scattering from the deuteron using the Gross equation. The calculation includes the complete impulse approximation and $\rho\pi\gamma$ exchange currents. We find that the structure function $B(Q^2)$ is extremely sensitive to the presence of small P-wave components of the deuteron wave function of relativistic origin. By using a soft $\rho\pi\gamma$ electromagnetic form factor we have been able to obtain an excellent description of the data.

This, however, should be viewed as a preliminary result since we have not yet studied the sensitivity of the model to the choice of single-nucleon electromagnetic form factors and to the arbitrary decision to place $h_0 = f_0$ in the representation of the off-shell single-nucleon current operator. There are also some additional problems with the interaction model IIB. In particular, the deuteron quadrupole moment and the asymptotic D/S ratio are too small for this model while the magnetic moment is too large. We are presently looking for more accurate models. The Gross equation is also being applied to the calculation of the triton binding energy [18] and we expect that this will result in some additional constraints on acceptable interaction models. It is possible that our best interaction models may produce different results for the deuteron form factors.

Acknowledgement. Supported by D.O.E. contracts #DE-AC05-84ER40150 and #DE-FG05-94ER40832

References

1. F. Gross: Phys. Rev. **186**, 1448 (1969); Phys. Rev. D **10**, 223 (1974); Phys. Rev. C **26**, 2203 (1982)
2. For an introduction to quasipotential equations see: G.E. Brown and A.D. Jackson: *The Nucleon–Nucleon Interaction*, Amsterdam : North–Holland 1976
3. J.W. Van Orden: Czech. J. Phys. **45**, 181 (1995)
4. W.W. Buck and F. Gross: Phys. Rev. D **20**, 2361 (1979)
5. F. Gross, J.W. Van Orden and K. Holinde: Phys. Rev. C **41**, R1909 (1990); Phys. Rev. C **45**, 2094 (1992)
6. R.A. Arndt and L.D. Roper: Scattering Analysis and Interactive Dial-in (SAID) program, Virginia Polytechnic Institute and State University
7. F. Gross and D.O. Riska: Phys. Rev. C **36**, 1928 (1987)
8. J.C. Ward, Phys. Rev. **78**, 182 (1950);
   Y. Takahashi, Nuovo Cimento **6**, 371 (1957)
9. F. Coester and D.O. Riska: Ann. Phys. (N.Y.) 234, 141 (1994)
10. R.E. Arnold, C.E. Carlson, and F. Gross: Phys. Rev. C **21**, 1426 (1980)
11. V.G.J. Stoks et al. : Phys. Rev. C **48**, 792 (1993)
12. E. Hummel and J.A. Tjon: Phys. Rev. Lett. **63**, 1788 (1989); Phys. Rev. C **42**, 423 (1990)
13. G. Höhler et al. : Nucl. Phys. **B114**, 505 (1976)
14. S. Galster et al. : Nucl. Phys. **B32**, 221 (1971)
15. E. Nyman and D.O. Riska: Phys. Rev. Lett. **57**, 3007 (1986); Nucl. Phys. **A468** 473 (1987);
    M. Wakamatsu and W. Weise, Nucl. Phys. **A477**, 559 (1988)
16. H. Ito and F. Gross: Phys. Rev. Lett. **71**, 2555 (1993)
17. K.L. Mitchell: Ph.D. Thesis, Kent State University (1995) (unpublished); K.L. Mitchell and P.C. Tandy, to be published
18. A. Stadler and F. Gross: Private Communication

Few-Body Systems Suppl. 9, 429–438 (1995)

Few-Body Systems

Elastic Electron-Deuteron Scattering with New Nucleon-Nucleon Potentials and Nucleon Form Factors

W. Plessas, V. Christian, R.F. Wagenbrunn

Institute for Theoretical Physics, University of Graz, Universitätsplatz 5, A-8010 Graz, Austria

Abstract. We present elastic *e-d* scattering observables obtained from the recent *N-N* interaction models proposed by the Bonn and Nijmegen groups. In particular, we discuss the pertinent results for the electric and magnetic structure functions as well as the deuteron tensor polarization and charge form factor, especially with respect to their dependence on the choice of the nucleon form factors. We find that for these new *N-N* potentials only a particular model of nucleon form factors, different from the traditional ones, allows for a comprehensive reproduction of all elastic *e-d* data at low and moderate momentum transfers.

In a recent study [1] we investigated the performance of the new versions of the *N-N* interaction models developed by the Bonn and Nijmegen groups. In particular, we presented results for elastic *e-d* scattering observables for the meson-theoretical Bonn-B [2], FULLF as well as OBEPF [3], and Nijm93 [4] potentials; beyond these we also considered the phenomenological Reid-like *N-N* potentials Nijm-I (non-local) as well as Nijm-II (local) [4], since they provide a high-quality reproduction of the *N-N* data base. In ref. [1] it was found that neither one of the *N-N* models employed was able to describe the electromagnetic structure of the deuteron really in agreement with the experimental data for momentum transfers $q^2 \lesssim 80$ fm^{-2}. Notably even at low q^2 all of the above models failed to reproduce in particular the Saclay data [5] for the electric structure function $A(q^2)$ and – at the same time – the charge form factor $F_C(q)$, especially with respect to its zero at $q_0 = 4.39 \pm 0.16$ fm^{-1} [6]. Already below $q^2 \lesssim 20$ fm^{-2}, deviations were of the order of 10%, i.e. much larger than the reported experimental uncertainties. Practically the same kind of shortcomings were observed before for previous *N-N* potentials [7].

Our approach of evaluating *e-d* observables (for more details see ref. [8]) consists in: first calculating the non-relativistic impulse approximation (IA), then adding relativistic corrections (RC), specifically the spin-orbit, Darwin-Foldy, and nuclear-motion terms, and finally summing up with the most important meson-exchange currents (MEC), the π-pair, $\rho\pi\gamma$, and π-retardation currents. Note that all further MEC are either negligible, at least in the momentum-transfer region $q^2 \lesssim 20\ \mathrm{fm}^{-2}$, or their contributions practically cancel each other. In the evaluation of the MEC contributions for the case of the meson-exchange potentials (all Bonn versions and Nijm93) we choose the vertex form factors consistently with the underlying dynamical model. For the folded-diagram Bonn potentials (FULLF and OBEBF), which rely on instantaneous meson exchanges, we leave out the retardation currents; the latter are also not included in the case of the Nijm93 potential.

We demonstrate the effect of the various additional contributions on top of the non-relativistic IA for the deuteron electric structure function $A(q^2)$ in Fig. 1, where we have taken the Bonn-B potential as a representative example. It is evident that in the momentum-transfer region shown the most important effect is provided by the π-pair current. The contributions by the $\rho\pi\gamma$ and π-retardation currents of course grow up to 20% but they largely cancel each other. The lowest-order RC remain at a few percent and thus further (higher-order) RC cannot distort the total result decisively. For momentum transfers up to $4-5\ \mathrm{fm}^{-1}$, the latter observation is confirmed, even with respect to the charge form factor $F_C(q)$, by the detailed study in ref. [9]. Consequently we may consider the framework of our calculations reliable, at least at these low momentum transfers.

In the study of ref. [1] we relied on some traditional parametrizations of the nucleon form factors. The results for the above N-N potentials were demonstrated along with the best-fit nucleon form factors of Lomon [10], usually called IJLG [11, 12]. The findings and conclusions, however, hold also in the case of other nucleon form factors, such as IJL [11] or Höhler [13]. In particular, the reproduction of the Saclay data [5] for the electric structure function $A(q^2)$, simultaneously with the charge form factor $F_C(q)$, cannot be achieved. This is made evident in Fig. 2 and Table 1 below. In addition to the results with the above nucleon form factors, we have also inserted there the predictions with an earlier parametrization by Gari and Krümpelmann (GK-85) [14], which was already discarded as inappropriate in a previous work by Mosconi and Ricci [15].

Apparently the uncertainties related to the nucleon form factors, in the first instance stem from the neutron, and there mostly from the electric form factor (Fig. 3). A few years ago Gari and Krümpelmann [19] improved their earlier parametrization and proposed a new model of nucleon form factors, which we denote as GK-92. In Fig. 3 we have included these form factors too. Especially with respect to the electric neutron form factor, GK-92 has a behaviour rather different from the other ones: it is lower at low q^2 and higher at high q^2. Such characteristics turn out advantageous to remedy the shortcomings in the reproduction of the experimental data. We have therefore repeated our study of the new N-N potentials with the GK-92 nucleon form factors [19].

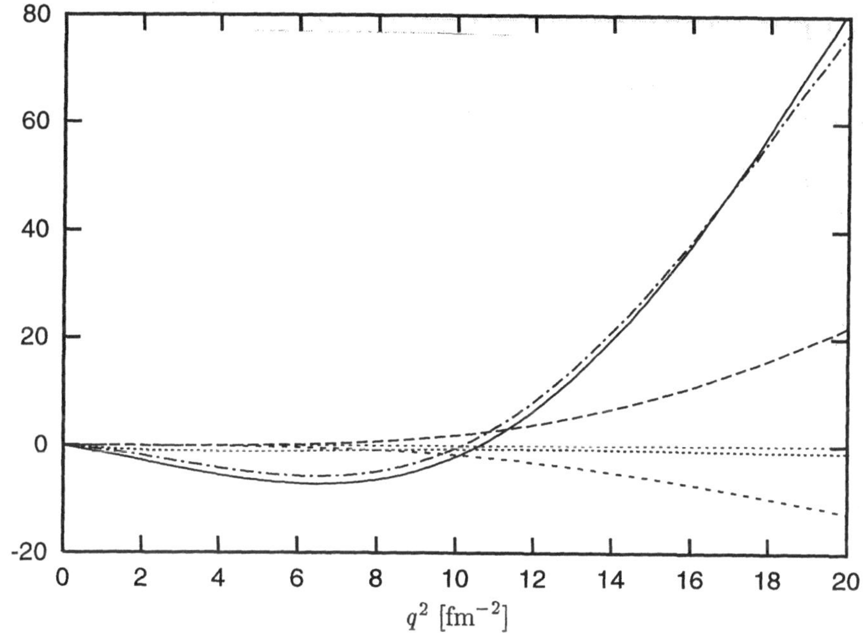

Figure 1. Percentage contributions relative to the non-relativistic IA result for the electric structure function $A(q^2)$ in case of the Bonn-B potential: RC (dotted), π-pair (dash-dotted), $\rho\pi\gamma$ (long-dashed), π-retardation (short-dashed), total RC+MEC (solid).

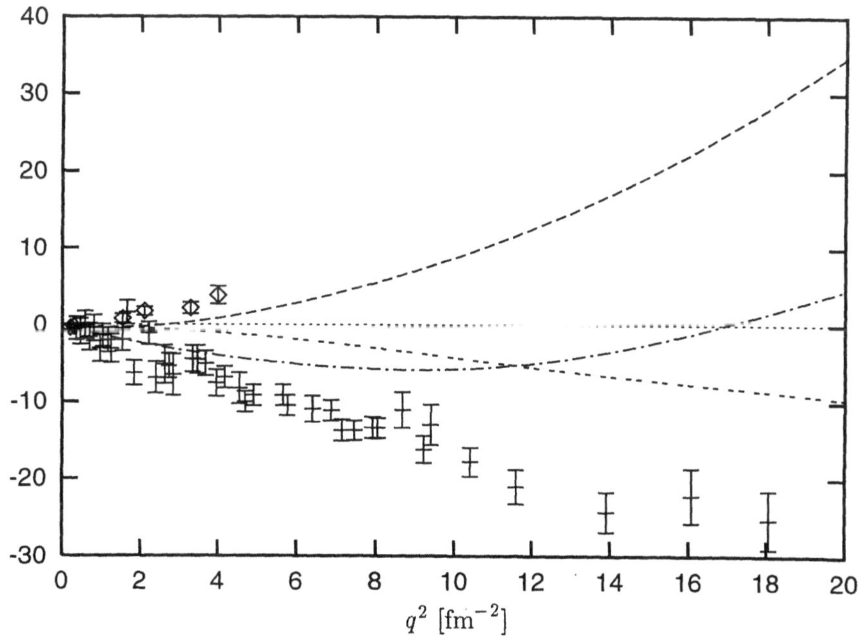

Figure 2. Percentage deviations relative to the full (IA+RC+MEC) calculation with the IJLG nucleon form factors for the electric structure function $A(q^2)$ in case of the Bonn-B potential: IJL (short-dashed), Höhler (dash-dotted), GK-85 (long-dashed). Experimental data from refs. [5] (crosses) and [16] (diamonds).

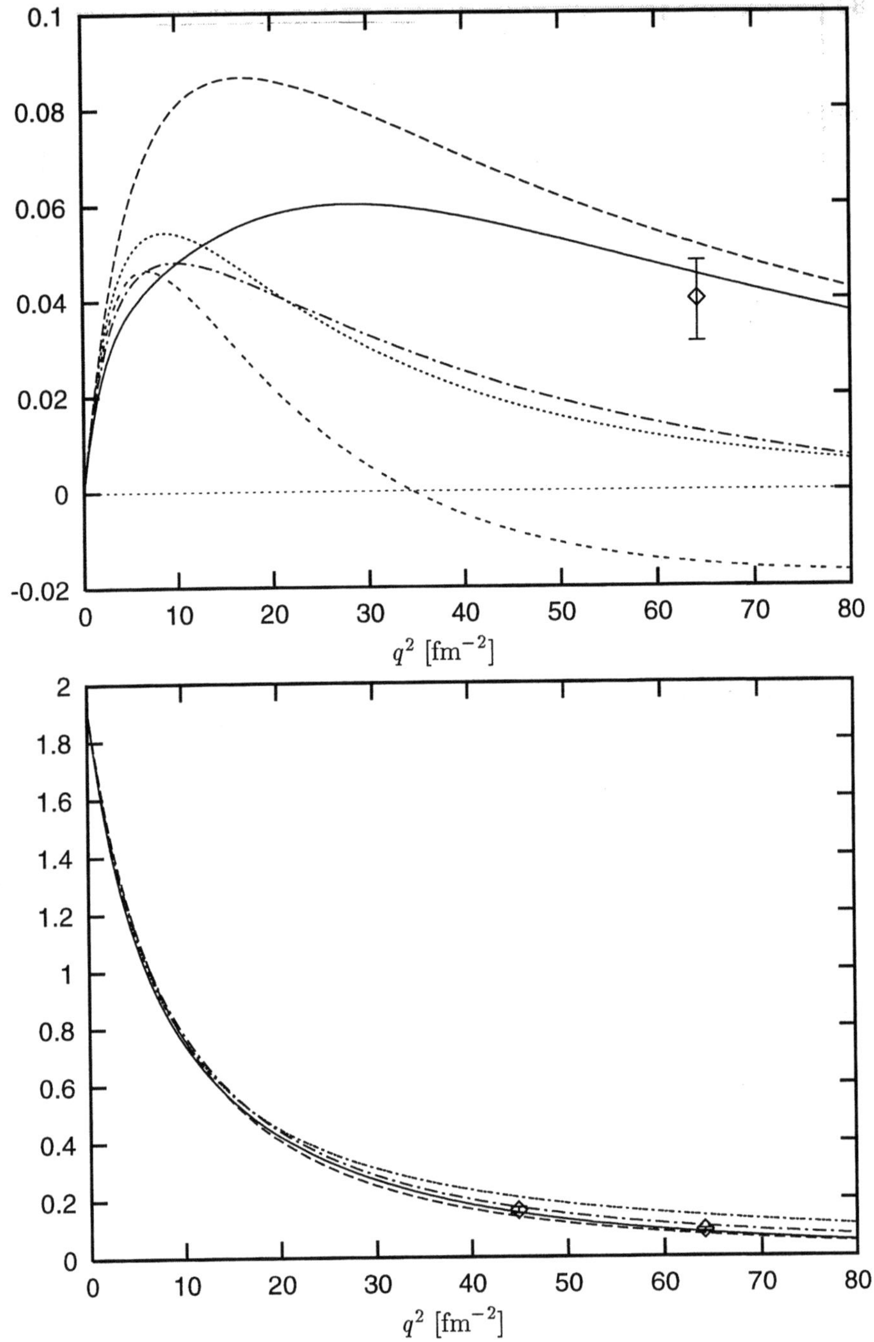

Figure 3. Different parametrizations of the neutron electric (top) and magnetic (bottom) form factors $G_{E,n}$ and $G_{M,n}$, respectively: IJLG (dotted), IJL (short-dashed), Höhler (dash-dotted), GK-85 (long-dashed), GK-92 (solid). Experimental data from refs. [17] (top) and [18] (bottom).

Table 1. Positions of the zero of the charge form factor $F_C(q)$ as resulting from the full (IA+RC+MEC) calculation in case of the Bonn-B potential for various models of nucleon form factors employed.

Nucleon Form Factor	q_0 [fm^{-1}]
IJLG [10]	4.28
GK-85 [14]	4.27
IJL [11]	4.25
Höhler [13]	4.21
Experiment [6]	4.39 ± 0.16

In what follows we show the corresponding results of all relevant e-d observables for the above-mentioned versions of the Bonn and Nijmegen potentials from our full (IA+RC+MEC) calculations. From Figs. 4 and 5 it is immediately evident that for all N-N potentials the reproduction of the Saclay data for the electric structure function $A(q^2)$ is largely improved at low q^2 (cf. Figs. 1 and 2 in ref. [1]). At the same time the results for the charge form factor $F_C(q)$ remain acceptable or have even slightly meliorated, see Figs. 8 and 12 as well as Table 2 (cf. the latter with Table 1 in ref. [1]). As a consequence also the tensor polarization $T_{20}(q)$ is now described better, practically within the experimental errors, by all potentials considered here (Figs. 9 and 13).

In Figs. 6 and 10 the electric structure functions $A(q^2)$ are shown over the whole range of momentum transfers up to $q^2 = 80$ fm^{-2}. The theoretical predictions start to deviate from the experimental data above $q^2 \approx 20$ fm^{-2}. A similar behaviour is observed for the magnetic structure functions $B(q^2)$ in Figs. 7 and 11. We are reluctant to attribute these differences to either the deuteron wave-function (half-off-shell) properties of the considered N-N models or the

Table 2. Positions of the zero of the charge form factor $F_C(q)$ in case of the GK-92 nucleon form factors [19].

Potential	q_0 [fm^{-1}]
Bonn-B	4.28
FULLF	4.26
OBEPF	4.10
Nijm93	4.18
Nijm-I	4.28
Nijm-II	4.15
Experiment [6]	4.39 ± 0.16

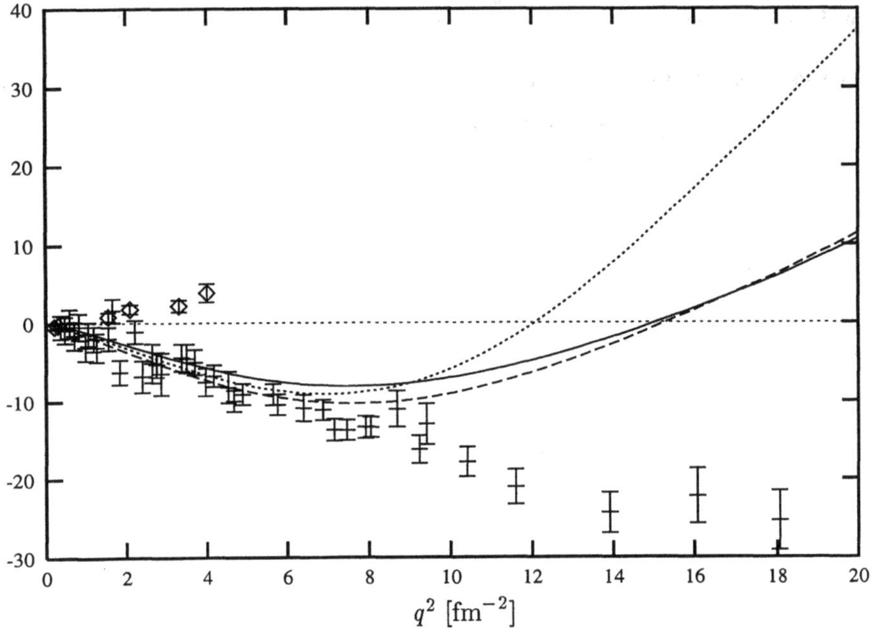

Figure 4. Relative percentage deviations of the electric structure function $A(q^2)$ in case of the GK-92 nucleon form factors for the Bonn-B (solid), FULLF (dashed), and OBEPF (dotted) potentials. As in Fig. 2 the Bonn-B calculation with the IJLG nucleon form factors was taken as reference. Experimental data as in Fig. 2.

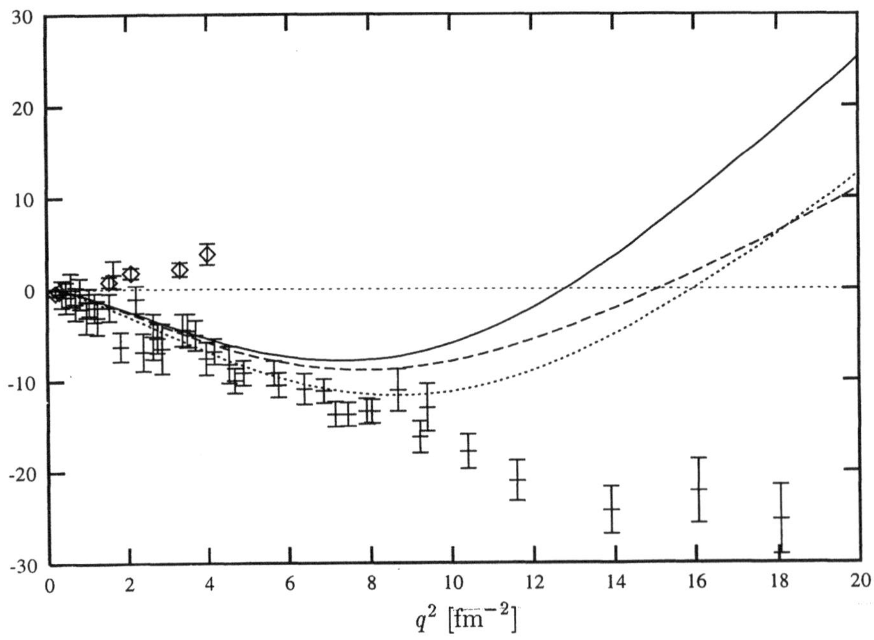

Figure 5. Same as in Fig. 4 for the Nijmegen potentials Nijm93 (solid), Nijm-I (dashed), and Nijm-II (dotted).

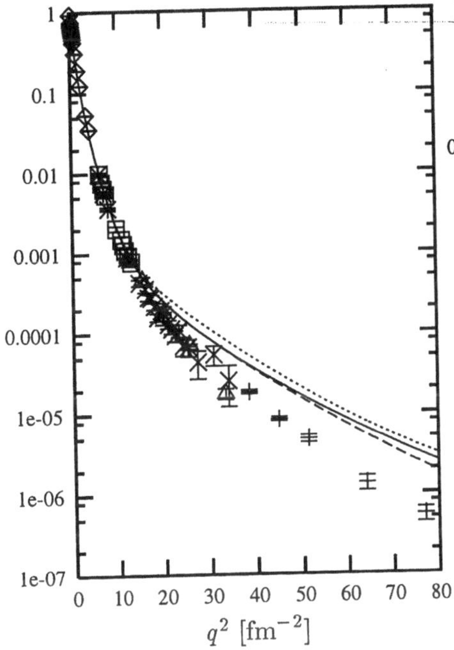

Figure 6. Electric structure function $A(q^2)$ for the Bonn potentials as in Fig. 4. Experimental data from refs. [5], [16], [20], [21], [22], and [23].

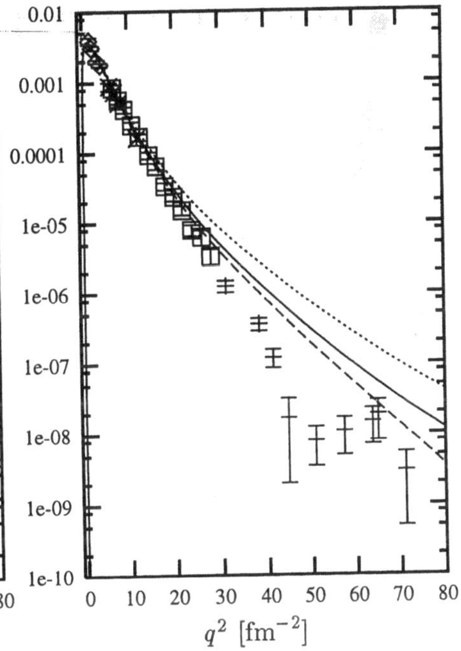

Figure 7. Magnetic structure function $B(q^2)$ for the Bonn potentials as in Fig. 4. Experimental data from refs. [16], [20], [24], and [25].

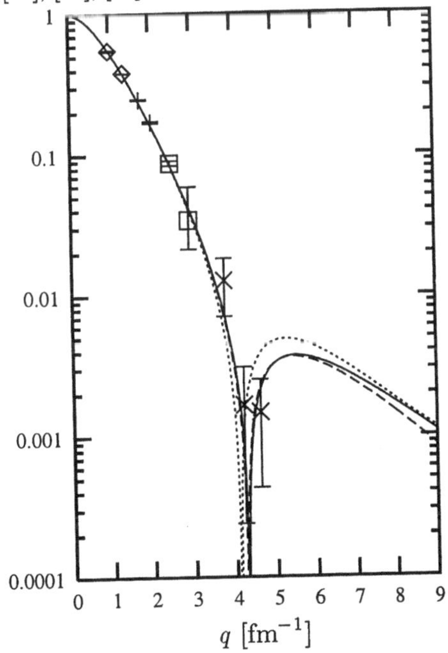

Figure 8. Deuteron charge form factor $F_C(q)$ for the Bonn potentials as in Fig. 4. Experimental data from refs. [6], [26], [27], [28], and [29].

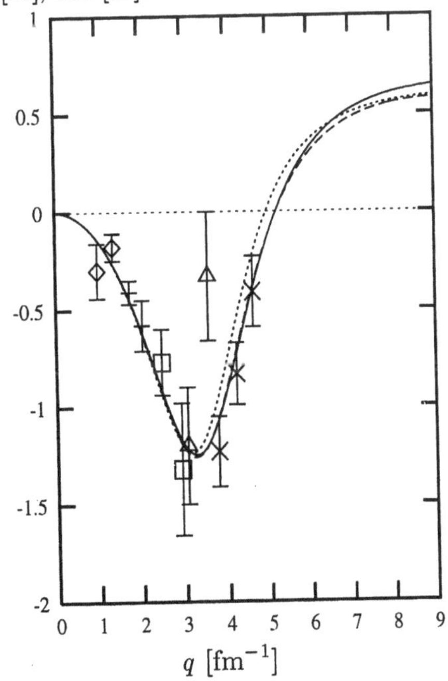

Figure 9. Deuteron tensor polarization $T_{20}(q)$ for the Bonn potentials as in Fig. 4. Experimental data from refs. [6], [26], [27], [28], [29], and [30].

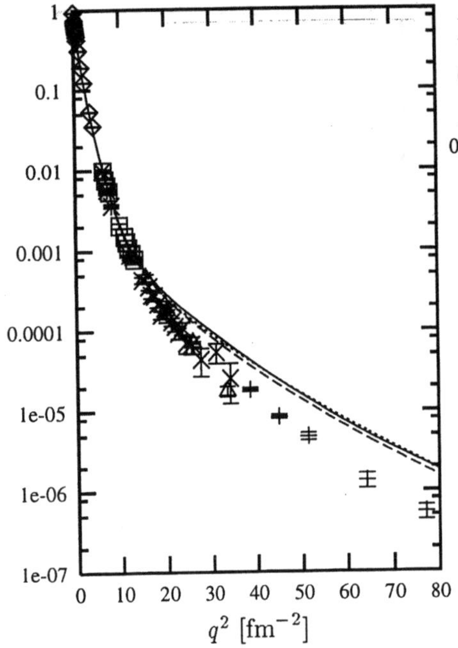

Figure 10. Electric structure function $A(q^2)$ for the Nijmegen potentials, as in Fig. 5. Experimental data as in Fig. 6.

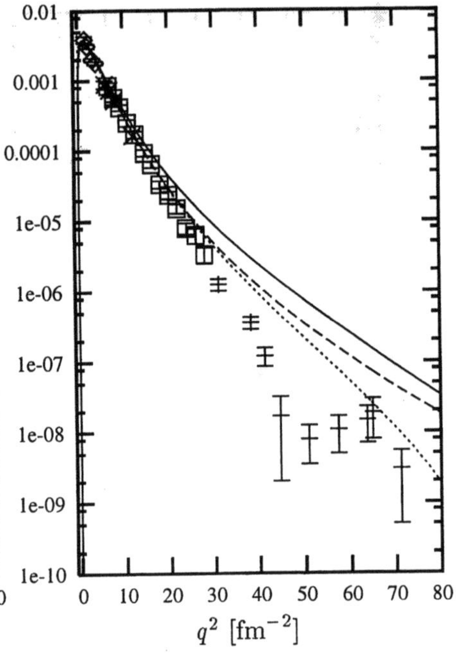

Figure 11. Magnetic structure function $B(q^2)$ for the Nijmegen potentials, as in Fig. 5. Experimental data as in Fig. 7.

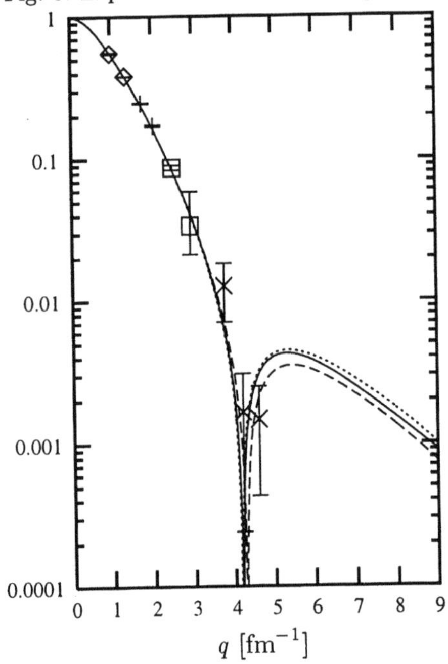

Figure 12. Deuteron charge form factor $F_C(q)$ for the Nijmegen potentials, as in Fig. 5. Experimental data as in Fig. 8.

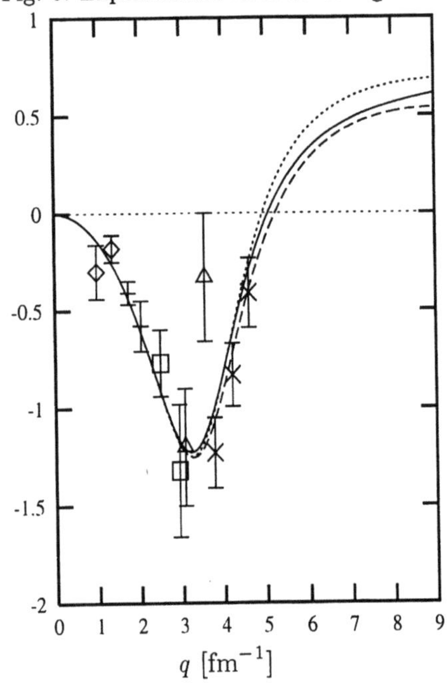

Figure 13. Deuteron tensor polarization $T_{20}(q)$ for the Nijmegen potentials, as in Fig. 5. Experimental data as in Fig. 9.

various ingredients of our calculations. As we have demonstrated above, our approach is certainly reliable up to momentum transfers of $q \approx 5 - 6$ fm^{-1}. Beyond that domain any theory is afflicted with uncertainties (stemming from RC, MEC, nucleon structure, subnuclear degrees of freedom, etc.), which can become quite large in some cases.

The elastic *e-d* observables are generally considered as a useful tool to examine the half-off-shell behaviour of *N-N* interaction models. Such tests of the deuteron wave functions must be undertaken with care, however. As we have shown, certain ingredients of the calculation, particularly the nucleon form factors, affect the results appreciably already at low momentum transfers. In this region, on the other hand, many experimental data with relatively small uncertainties have been achieved over the recent years. Any theory must first of all explain these data accurately and comprehensively before it may be put to a test towards higher momentum transfers. For the set of the new *N-N* potentials of the Bonn and Nijmegen groups a more or less acceptable description of the low-momentum transfer data is reached in case the GK-92 model of the nucleon form factors is employed.

References

1. R.F. Wagenbrunn and W. Plessas: Few-Body Systems, Suppl. **8**, to appear
2. R. Machleidt: Adv. Nucl. Phys. **19**, 189 (1989)
3. J. Haidenbauer, K. Holinde, and M.B. Johnson: Phys. Rev. **C45**, 2055 (1992)
4. V.G.J. Stoks, R.A.M. Klomp, C.P.F. Terheggen, and J.J. de Swart: Phys. Rev. **C49**, 2950 (1994)
5. S. Platchkov et al. : Nucl. Phys. **A510**, 740 (1990)
6. M. Garçon et al. : Phys. Rev. **C49**, 2516 (1994)
7. W. Plessas, Ch. Brandstätter, S. Cvijetic, J. Haidenbauer, L. Mathelitsch, P. Obersteiner, J. Pauschenwein, and R. Wagenbrunn: Few-Body Systems, Suppl. **7**, 251 (1994)
8. P. Obersteiner, W. Plessas, and J. Pauschenwein: Few-Body Systems Suppl. **5**, 140 (1992)
9. J. Adam, Jr., H. Göller, and H. Arenhövel: Phys. Rev. **C48**, 370 (1993)
10. E.L. Lomon: Ann. Phys. (N.Y.) **125**, 309 (1980)
11. F. Iachello, A.D. Jackson, and A. Landé: Phys. Lett. **43B**, 191 (1973)
12. S. Galster, H. Klein, J. Moritz, K.H. Schmidt, and D. Wegener: Nucl. Phys. **B32**, 221 (1971)

13. G. Höhler, E. Pietarinen, I. Sabba-Stefanescu, F. Borkowsky, G.G. Simon, V.H. Walther, and R.D. Wendling: Nucl. Phys. **B114**, 505 (1976)

14. M.F. Gari and W. Krümpelmann: Z. Phys. **A322**, 689 (1985)

15. B. Mosconi and P. Ricci: Few-Body Systems **6**, 63 (1989); ibid. **8**, 159 (1990)

16. G.G. Simon, Ch. Schmitt, and V.H. Walther: Nucl. Phys. **A364**, 285 (1981)

17. S. Rock et al. : Phys. Rev. Lett. **49**, 1139 (1982)

18. A. Lung et al. : Phys. Rev. Lett. **70**, 718 (1993)

19. M.F. Gari and W. Krümpelmann: Phys. Lett. **B274**, 159 (1992); ibid. **B282**, 483 (1992)

20. C.D. Buchanan and M.R. Yearian: Phys. Rev. Lett. **15**, 303 (1965)

21. J.E. Elias, J.I. Friedman, G.C. Hartmann, H.W. Kendall, P.N. Kirk, M.R. Sogard, and L.P. van Speybroeck: Phys. Rev. **177**, 2075 (1969)

22. R.G. Arnold et al. : Phys. Rev. Lett. **35**, 776 (1975); ibid. **58**, 1723 (1987)

23. R. Cramer et al. : Z. Phys. **C29**, 513 (1985)

24. S. Auffret et al. : Phys. Rev. Lett. **54**, 649 (1985)

25. P.E. Bosted et al. : Phys. Rev. **C42**, 38 (1990)

26. V.F. Dmitriev et al. : Phys. Lett. **157B**, 143 (1985)

27. B.B. Voitsekhovskii et al. : JETP Lett. **43**, 733 (1986)

28. M.E. Schulze et al. : Phys. Rev. Lett. **52**, 597 (1984)

29. R. Gilman et al. : Phys. Rev. Lett. **65**, 1733 (1990)

30. C.E. Jones et al. : Few-Body Systems, Suppl. **7**, 112 (1994)

Few-Body Systems Suppl. 9, 439–443 (1995)

Few-Body Systems

Elastic γd Scattering

M.I. Levchuk[1], A.I. L'vov[2]

[1] B.I. Stepanov Institute of Physics, Belarus Academy of Sciences, F. Skaryna prospect 70, 220602 Minsk, Belarus

[2] P.N. Lebedev Physical Institute, Russian Academy of Sciences, Leninsky prospect 53, 117924 Moscow, Russia

Abstract. Elastic γd-scattering is considered below pionic threshold with taking into account MEC and polarizability of the nucleon. Two methods of treating the two-body current and seagull are discussed: Siegert-like procedure of the minimal substitution, and a microscopic computation with one-meson exchanges of the Bonn potential.

Studies of electromagnetic interactions of hadrons and nuclei are aimed to getting a deeper insight to the structure and dynamics of hadronic systems. Matrix elements of the electromagnetic current j_μ which can be measured in exclusive nuclear photo- and electro-reactions of the lowest order in the electromagnetic coupling e give us a knowledge of effective constituents of nuclei, their wave functions, explicit mesonic and isobar effects, etc. Reactions of the next order in e, such as the Compton scattering, proceed through virtual photoexcitation of different intermediate states and are, therefore, useful for learning global properties of the electromagnetic current weighted over the whole spectrum, quite similarly to inclusive reactions of the first order in e. However, the most intriguing feature of the second-order reactions is in getting an access to the so-called seagull operator $S_{\nu\mu}(y,x) = S_{\mu\nu}(x,y)$ which is the next term in the expansion of the Hamiltonian $H[A]$ in powers of the electromagnetic potential:

$$H[A] = H_0 + \int A^\mu(x) j_\mu(x)\, dx + \tfrac{1}{2} \iint A^\nu(y) A^\mu(x) S_{\nu\mu}(y,x)\, dx\, dy + \cdots .$$

The seagull interaction is a well-known part of the non-relativistic Hamiltonian $(\boldsymbol{p} - e\boldsymbol{A})^2/2M$ and simply means that the photons involved are absorbed and emitted at close space-time points which are not distinguishable at the energy scale considered.

The gauge invariance constrains the longitudinal components of the seagull $S_{\nu\mu}$ [1] but leaves completely free its transverse part. In the momentum representation: $[j_0(k), j_\nu(-k')] = k^i S_{\nu i}(-k', k)$ and $S_{0\mu} = 0$. This equation, together with the current conservation, $[j_0(k), H_0] = k^i j_i(k)$, ensures the gauge invariance $k^\mu T_{\nu\mu} = 0$ of the Compton scattering amplitude

$$-T_{\nu\mu}(-k', k) = S_{\nu\mu}(-k', k) + \langle f|j_\nu(-k')G(E_i + \omega)j_\mu(k)|i\rangle + \text{crossed}.$$

In the long wave-length limit the above constraints unambiguously determine the "trivial" part of the current density, $j_i(0)$, and the seagull, $S_{ij}(0,0)$, in terms of the nuclear Hamiltonian H_0 and the charge density j_0. For the single nucleon the non-trivial transverse parts of the current and seagull start with the magnetic moment and electric and magnetic polarizabilities of the nucleon. A lot of efforts had been devoted to measuring and theoretical understanding of these quantities.

Non-trivial transverse two-body contribution to the current begins with the magnetic moment caused by mesonic exchanges. This contribution explains the notorious 10%-enhancement in the radiative *np*-capture and in the past triggered impressive progress in microscopic understanding of MEC. Nothing similar has been achieved yet for the non-trivial two-body contribution to the seagull determining medium corrections to the polarizabilities of the nucleon.

The present work is mainly devoted to the investigation of the seagulls in the elastic γd-scattering. We also explore the possibility of measuring the nucleon polarizabilities through this reaction. We calculate the scattering amplitude $T_{\nu\mu}$ at energies below the pionic threshold. The one-body current $j_\mu^{(1)}$ and seagull $S_{\nu\mu}^{(1)}$ are borrowed from the non-relativistic Hamiltonian keeping the magnetic moment and polarizabilities of the nucleon. As for treating the two-body effects, two approaches are used. The first one takes into account only "minimal trivial" parts of the two-body current and seagull. Technically this is done by virtue of the minimal substitution $p \to p - eA$ in the potential V of *np*-interaction.

The procedure of the minimal substitution is based on the formula [2]

$$F(p - eA(x)) = F(p) - eA_\mu(x)\frac{\partial}{\partial p_\mu}\int_0^1 F(p + sk)\,ds + O(e^2), \tag{1}$$

valid for any function $F(p)$ of the operator arguments $p_\mu = \mathrm{i}\partial/\partial x_\mu$ and the plane electromagnetic wave $A(x) \sim \exp(-\mathrm{i}k_\mu x^\mu)$, provided the non-commutating operators $\pi_\mu = p_\mu - eA_\mu$ in the l.h.s. of (1) are symmetrically ordered (like $\pi_x\pi_y + \pi_y\pi_x$). The integral over s in (1) is the price one has to pay for the non-commutating π_μ.

Considering the *np*-potential $V(\boldsymbol{p}', \boldsymbol{p})$ in the momentum space and applying the minimal substitution to the proton momenta p_p, p_p' we recover the two-body electromagnetic current

$$j_\mu^{(2)}(x) = \frac{\delta V(p - eA)}{\delta A^\mu(x)}, \quad j_\mu^{(2)}(k) = \exp(\mathrm{i}\boldsymbol{k}\boldsymbol{R})\,[\gamma_\mu V - V\gamma_\mu], \tag{2}$$

where $\boldsymbol{R} = (\boldsymbol{r}_p + \boldsymbol{r}_n)/2$, and γ_μ is a differential operator in the space of the relative momenta $\boldsymbol{p} = \frac{1}{2}(\boldsymbol{p}_p - \boldsymbol{p}_n)$:

$$\epsilon^\mu \langle \boldsymbol{p}'|\gamma_\mu V|\boldsymbol{p}\rangle = -\tfrac{1}{2}\boldsymbol{\epsilon}\boldsymbol{\nabla}_{p'} \int_0^1 \langle \boldsymbol{p}' - \tfrac{1}{2}s\boldsymbol{k}|V|\boldsymbol{p}\rangle \, ds,$$

$$\epsilon^\mu \langle \boldsymbol{p}'|V\gamma_\mu|\boldsymbol{p}\rangle = +\tfrac{1}{2}\boldsymbol{\epsilon}\boldsymbol{\nabla}_{p} \int_0^1 \langle \boldsymbol{p}'|V|\boldsymbol{p} + \tfrac{1}{2}s\boldsymbol{k}\rangle \, ds. \qquad (3)$$

Two terms in (2) correspond to minimal substitutions of p_p and p'_p. In the coordinate space $\gamma_i = -ier_i(e^a - 1)/2a$, where $\boldsymbol{r} = \boldsymbol{r}_p - \boldsymbol{r}_n$ and $a = \mathrm{i}\boldsymbol{kr}/2$. Using $k^\mu \gamma_\mu = e(e^a - 1)$ we get that the current (2) exactly satisfies the conservation low: $[j_0(k), V] = k^i j_i^{(2)}(k)$.

In the next order in A_μ the minimal substitution results in the minimal two-body seagull

$$S_{\nu\mu}^{(2)}(-k', k) = \exp\Big(\mathrm{i}(\boldsymbol{k} - \boldsymbol{k}')\boldsymbol{R}\Big) \, [\gamma'_\nu, [\gamma_\mu, V]], \qquad (4)$$

where the operator γ'_ν is γ_ν with $k \to -k'$. It exactly obeys the conservation law: $[j_0(k), j_\nu^{(2)}(-k')] = k^i S_{\nu i}^{(2)}(-k', k)$.

In the long wave-length limit the formulae (2), (4) lead to exactly the same results as Siegert's theorem. Moreover, these minimal current and seagull can be viewed as a special case of the general Siegert transformation.

In diagrammatic calculations with the minimal current and seagull [2, 3] the two-body contributions appear in the form of the so-called contact vertices. Considering the transition matrix element of the (total) current between the deuteron and the plane-wave state $|p\rangle$, we have:

$$T_\gamma = \epsilon^\mu \langle p|j_\mu|d\rangle = \epsilon^\mu \langle p|j_\mu^{(1)} + \gamma_\mu V - V\gamma_\mu|d\rangle = t_\gamma - V\psi_\gamma,$$

where the amplitude t_γ includes both the contribution of the one-body current and the contact term $\Gamma_\gamma = \epsilon^\mu \langle p|\gamma_\mu V|d\rangle = \epsilon^\mu \gamma_\mu \Gamma$, which is the derivative (3) of the dnp-vertex $\Gamma(p) = \langle p|V|d\rangle = -\frac{p^2+\alpha^2}{M}\psi(p)$, $\alpha^2/M =$ the deuteron binding energy. Also $\psi_\gamma = \epsilon^\mu \gamma_\mu \psi(p)$. When the seagull operator (4) acts to the deuteron state another contact vertex appears: $\Gamma_{\gamma\gamma'} = \epsilon'^\nu \epsilon^\mu \langle p|\gamma'_\nu \gamma_\mu V|d\rangle = \epsilon'^\nu \epsilon^\mu \gamma'_\nu \gamma_\mu \Gamma(p)$, which is the second derivative of the vertex Γ. Using the Lippmann-Schwinger relations $G = G_0 + G_0 T G_0$ and $T = V + V G_0 T$ between the potential V, the off-shell np-scattering T matrix, and the exact and free propagators G and G_0, we find that in the total amplitude $T_{\nu\mu}$ terms like $\gamma'_\nu V \gamma_\mu$ are cancelled and the amplitude gets the diagrammatic form of Fig. 1. Of course, the found amplitude is gauge invariant and satisfies low energy theorems.

Another approach we also use [4] is based on constructing two-body currents and seagulls through photon coupling to all the real and fictitious mesons included in the Bonn potential OBEPR (they are π, ρ, ω, σ, δ, η, and auxiliary heavy Λ-bosons [5] constituting monopole form factors of vertices). Also the Δ-isobar currents are included. The diagrammatic representation of the obtained Compton scattering amplitude is very similar (see Fig. 2), but now all the two-body effects get clear physical meaning.

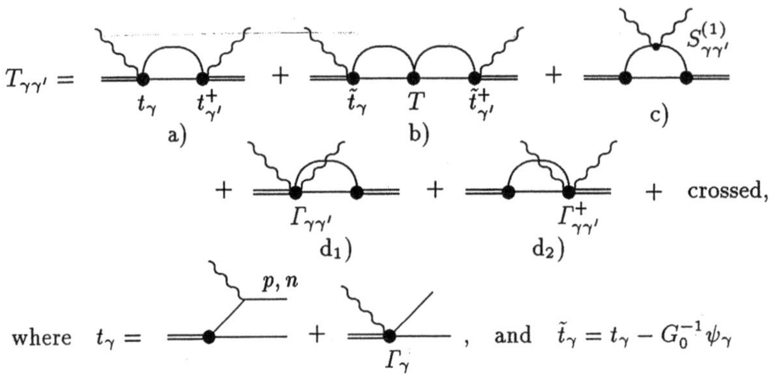

Figure 1. Diagrams of γd-scattering in the effective model of minimal substitution.

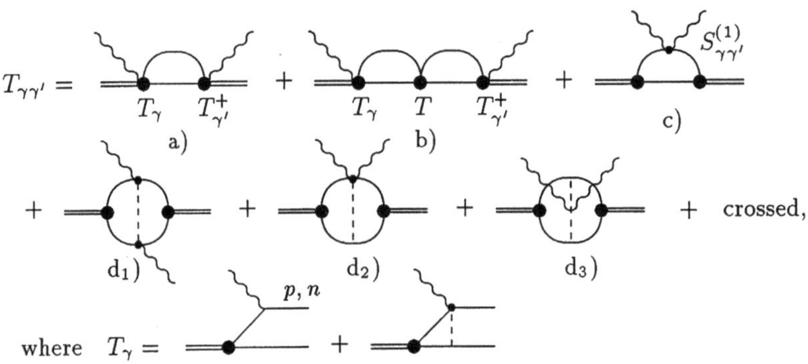

Figure 2. Diagrams of γd-scattering with explicit mesonic currents ("Bonn model").

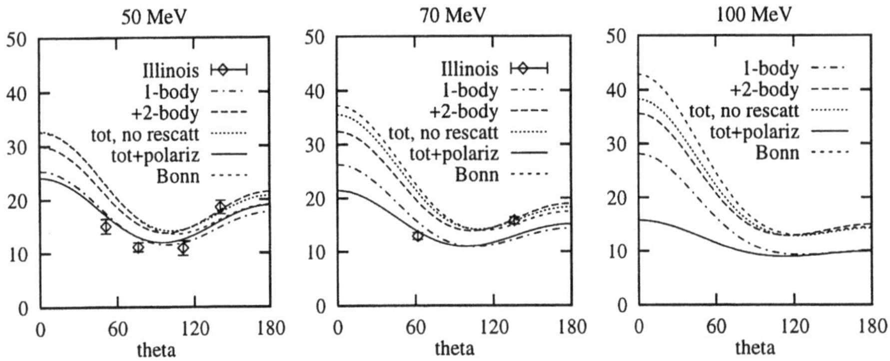

Figure 3. Differential cross section $d\sigma/d\Omega_{\mathrm{cm}}$ [nb/sr] in the minimal model (Fig. 1). (1) Dashed-dotted lines: contribution of the one-body current and seagull. (2) Dashed lines: total contribution, i.e. both one- and two-body operators are retained. (3) Dotted lines: total contribution without the rescattering term (Fig. 1b). (4) Solid lines: total contribution with adding the nucleon polarizability. In cases 1–3,5 the polarizability is not included. (5) Short-dashed lines: total contribution without the rescattering (Fig. 2b) in the "Bonn model". Experimental data are from [10].

Numerical results of the minimal model are shown in Fig. 3. We used the Paris potential V in a separable approximation [6] to simplify treating the rescattering diagram of Fig. 1b, see [7]. However, we anticipate very weak dependence of the amplitude $T_{\nu\mu}$ at energies considered on any particular choice of V. Two-body effects essentially enhance the differential cross section, whereas the rescattering contribution leads to a small decrease of $d\sigma/d\Omega$ at forward angles. These findings perfectly agree with those of [8, 9], however our cross sections at 100 MeV are lower at backward angles, we don't know why. Adding the nucleon polarizabilities (we use the nucleon-averaged $\bar{\alpha}_N = 12$ and $\bar{\beta}_N = 3 \times 10^{-4}\,\mathrm{fm}^3$) does further decrease of the cross section, in qualitative agreement with the data available [10]. One can pursuit extracting the polarizabilities from γd-scattering but both more accurate data and more careful investigation of the model dependence is needed.

As yet we have not completed evaluation of the diagram of Fig. 2b. So, in Fig. 3 we show results of the model with explicit mesons ("Bonn model") without the rescattering correction which has to decrease $d\sigma/d\Omega$. They are in a close correspondence with those of the minimal model (to avoid confusion note that Figs. 1b and 2b involve different amplitudes $\tilde{t}_\gamma$ and T_γ). The seagull contribution found in the "Bonn model" reveals more rapid energy dependence than that in the minimal model, especially in the spin-flip amplitudes. This becomes important in some polarization observables.

Acknowledgement. We are thankful to organizers of the Conference for their warm hospitality and support. This work was supported by Russian Foundation for Fundamental Research and by Advance Research Foundation of Belarus.

References

1. H. Arenhövel: In: *New Vistas in Electro-Nuclear Physics.* Plenum Press 1986
2. A.I. L'vov: Voprosi At. Nauki i Tekhniki (Kharkov Phys. Tech. Inst. Publ.), ser. Obsh. Yad. Fiz. **2/35**, 51, 53 (1986); **2/38**, 93 (1987)
3. A.I. L'vov, V.A. Petrun'kin: In: *Perspectives on Photon Interactions with Hadrons and Nuclei* (Lecture Notes in Physics, vol. 365), p.123, Berlin, Springer 1990
4. M.I. Levchuk: Few-Body Systems (in print)
5. D.O. Riska: Progr. Part. Nucl. Phys. **11**, 199 (1984)
6. J. Haidenbauer, W. Plessas: Phys. Rev. **C30**, 1822 (1984), **C32**, 1424 (1985)
7. M.I. Levchuk, A.I. L'vov, V.A. Petrun'kin: Few-Body Systems **16**, 101 (1994)
8. M. Weyrauch, H. Arenhövel: Nucl. Phys. **A408**, 425 (1983)
9. M. Weyrauch: Phys. Rev. **C41**, 880 (1990)
10. M.A. Lucas: PhD Thesis, Univ. Illinois at Urbana-Champaign 1994

Few-Body Systems Suppl. 9, 444–448 (1995)

Few-Body Systems

Isospin Violation in Low-energy Hadronic Physics

U. van Kolck *

Department of Physics, Box 351560, University of Washington, Seattle, WA 98195-1560, USA

Abstract. Isospin violation in pionic and nuclear systems is studied at energies comparable to the pion mass using the general effective chiral Lagrangian. It is shown that the smallness of observed isospin breaking arises naturally from QCD (with electromagnetic interactions included). Except for process that involve two neutral pions where one expects an $O(\frac{m_d-m_u}{m_d+m_u})$ effect, isospin violation from the quark mass difference is further suppressed by powers of the pion mass over the characteristic scale of QCD. As a result, the pion mass difference tends to be the dominant effect. Pion-pion, pion-nucleon scattering and the nucleon-nucleon interaction are considered. The implications for the latter case are: i) charge dependent forces due to the pion mass difference in one pion exchange are the largest; ii) charge asymmetric forces arise from subleading operators, which are related to meson mixing.

1 Introduction

Why is isospin such a good symmetry of low-energy hadronic physics? The answer is not immediately obvious. A measure of isospin violation compared to chiral symmetry breaking in the QCD Lagrangian is the ratio of the quark mass difference to the sum, $\varepsilon \equiv \frac{m_d-m_u}{m_d+m_u} \sim 1/3$, or 30%, using standard values (essentially those found by Weinberg long ago [1], $m_u \cong 4\text{MeV}$ and $m_d \cong 7\text{MeV}$). However, the pion masses, for example, are due to this explicit symmetry breaking plus electromagnetic effects and still, the pion mass difference is only 3% of the average pion mass. Here I want to argue, assuming only that chiral symmetry is spontaneously broken and that the dynamics of QCD is natural, that the answer is due i) generically to the constraints imposed on operators by the chiral transformation properties of the relevant fields, and ii) specifically to experimental limitations. I should note that some of these points have been made before by Weinberg either using current algebra [1] or looking at the pion chiral Lagrangian [2], but I believe this is the first time the subject is treated systematically with chiral Lagrangians involving nucleons.

* *E-mail address:* vankolck@alpher.phys.washington.edu

2 Leading Isospin Violating Operators

At energies and momenta Q small compared to its typical scale $M \sim m_\rho$, QCD is equivalent to a theory of pions and non-relativistic nucleons. The coupling constants that appear are not small, but a perturbative expansion exists in Q/M, as long as chiral symmetry is implemented correctly. For a process involving A external nucleons and any number of pions, an irreducible diagram with L loops, C separately connected pieces and V_i vertices of type i is of order Q^ν, with $\nu = 4 - A + 2(L-C) + \sum_i V_i \Delta_i$, where the index of an interaction with $d(n)$ derivatives (nucleon fields) is defined by $\Delta \equiv d + \frac{n}{2} - 2 \geq 0$ [3]. The larger the indices of the interactions of a diagram, the more powers of soft momenta will be involved, and thus, under an assumption of naturalness —that any coupling of mass dimension $-\delta$ is of order $M^{-\delta}$—, the smaller the contribution of that diagram to a given process. Chiral symmetry is crucial in ensuring that the index is not negative.

The most general isospin violating Lagrangian [4] is constructed out of operators that break isospin in the same way i) as the quark mass difference — i.e. as the 3-component of an SO(4) vector—, and ii) as four-quark interactions generated by hard photon exchange —i.e. as the 34-component of an SO(4) antisymmetric tensor. I list here the most important isospin violating operators, with the index as a superscript. Below I denote the pion decay constant by $F_\pi \simeq 190$ MeV, the axial vector coupling of the nucleon by $g_A \simeq 1.25$, the pion covariant derivative by $\vec{D}_\mu = \frac{D^{-1}}{F_\pi}\partial_\mu \vec{\pi}$, and $D = 1 + \frac{\vec{\pi}^2}{F_\pi^2}$.

From the quark mass difference we find that the most important interactions are

$$\mathcal{L}^{(1)}_{\rm qm} = \delta m_N(-\bar{N}t_3N + 2D^{-1}\frac{\pi_3\vec{\pi}}{F_\pi^2}\cdot\bar{N}\vec{t}N), \tag{1}$$

and

$$\begin{aligned}\mathcal{L}^{(2)}_{\rm qm} &= \frac{1}{2D^2}\delta m_\pi^2 \pi_3^2 + \frac{\beta_1 D^{-1}}{F_\pi}(\nabla\vec{\pi}_3 - \frac{2D^{-1}}{F_\pi^2}\pi_3\vec{\pi}\cdot\nabla\vec{\pi})\cdot\bar{N}\boldsymbol{\sigma}N \\ &\quad +\gamma_s[\bar{N}t_3N - 2D^{-1}\frac{\pi_3\vec{\pi}}{F_\pi^2}\cdot\bar{N}\vec{t}N]\bar{N}N \\ &\quad +\gamma_\sigma[\bar{N}t_3\boldsymbol{\sigma}N - 2D^{-1}\frac{\pi_3\vec{\pi}}{F_\pi^2}\cdot\bar{N}\vec{t}\boldsymbol{\sigma}N]\cdot\bar{N}\boldsymbol{\sigma}N,\end{aligned} \tag{2}$$

with $\delta m_N = O(\varepsilon\frac{m_\pi^2}{M})$, $\delta m_\pi^2 = O(\varepsilon^2\frac{m_\pi^4}{M^2})$, $\beta_1 = O(\varepsilon\frac{m_\pi^2}{M^2})$, and $\gamma_{s,\sigma} = O(\varepsilon\frac{m_\pi^2}{M^4})$.

As for terms originating from electromagnetic interaction,

$$\mathcal{L}^{(-2)}_{\rm em} = -\frac{1}{2D^2}\bar{\delta}m_\pi^2(\vec{\pi}^2 - \vec{\pi}_3^2), \tag{3}$$

and

$$\begin{aligned}\mathcal{L}^{(-1)}_{\rm em} &= \bar{\delta}m_N(-\bar{N}t_3N + \frac{2D^{-1}}{F_\pi^2}\bar{N}(\vec{\pi}^2 t_3 - \pi_3\vec{\pi}\cdot\vec{t})N) \\ &\quad +\bar{\beta}_1 F_\pi^{-2}D^{-2}(\vec{\pi}^2 - \pi_3^2)\bar{N}N\end{aligned} \tag{4}$$

with $\bar{\delta m}_\pi^2 \propto \alpha m_\rho^2$, and $\bar{\delta m}_N \sim \bar{\beta}_1 = O(\frac{\bar{\delta m}_\pi^2}{M})$.

In anticipation of the forthcoming discussion, I will not list here higher order operators.

3 Phenomenology

We can have an idea of how electromagnetic and quark mass difference effects compare by looking at the pion mass difference. To leading order in e^2 and $\frac{m_\pi}{m_\rho}$, $\Delta m_\pi^2 \equiv m_{\pi^\pm}^2 - m_{\pi^0}^2 = (\bar{\delta m}_\pi^2 - \delta m_\pi^2)$. The contribution from the quark mass difference is suppressed compared to the pion mass squared by factors of both $\frac{m_\pi^2}{m_\rho^2}$ and ε^2, and contributes only some $(8\text{MeV})^2$ to Δm_π^2. The electromagnetic contribution $\bar{\delta m}_\pi^2$ is typically bigger, since $\alpha m_\rho^2 \sim (66\text{MeV})^2$. So, in order to further estimate the size of the electromagnetic terms, I neglect δm_π^2. Experimentally, $\Delta m_\pi^2 \sim (35\text{MeV})^2$, which is *numerically* $\sim \varepsilon \frac{m_\pi}{m_\rho} m_\pi^2$. Equation (3) corresponds then to a hypothetical quark mass term in $\mathcal{L}_{\pi,qm}^{(1)}$, and one can expect that $\mathcal{L}_{em}^{(n)} \sim \mathcal{L}_{qm}^{(n+3)}$. We can now discuss the implications of this dimensional analysis to various processes.

I start with pion-pion scattering. It is trivial to use Eq. (3) to calculate the leading isospin violation: it is remarkable that it does *not* change the result of the lowest order chiral Lagrangian first obtained by Weinberg long ago [5], that the amplitude is $M_{ab,cd} = A(s,t,u)\delta_{ab}\delta_{cd} + A(t,s,u)\delta_{ac}\delta_{bd} + A(u,t,s)\delta_{ad}\delta_{bc}$ with $A(s,t,u) = s - m_\pi^2$. That is, at fixed s, t and u, there is no isospin violation [1] of order $O(\frac{\Delta m_\pi^2}{m_\pi^2}) \sim 5\%$, contrary to what one might have expected. As a consequence, isospin violation from high-energy effects will arise at two orders higher in the perturbative expansion, which includes one loop: either a hopelessly small $\frac{\Delta m_\pi^2}{m_\rho^2} \sim 0.2\%$ from quark masses, or $O(\alpha) \sim 1\%$ from soft photon exchange between pion legs [6]. I am forced to conclude that purely pionic reactions are not a good place to see isospin violation other than from photon exchange, and turn to systems with nucleons.

The leading term involving a nucleon is, according to the above discussion, given by Eq. (1). On one hand, it provides a contribution of order $\varepsilon \frac{m_\pi^2}{m_\rho} \sim m_u - m_d \sim -$(a few MeV) to the nucleon mass splitting. On the other hand, there appear interactions of an even number of pions with a nucleon, with δm_N as coupling constant. For example, the effect on the pion-nucleon scattering lengths (neglecting the fact that the thresholds themselves depend on the mass differences) is

$$\Delta a(\pi^0 N \to \pi^0 N) = \frac{1}{1+\frac{m_\pi}{m_N}} \frac{1}{\pi F_\pi^2} \delta m_N t_3^{(N)},$$

[1] Except for a "kinematic" breaking hidden in s, t and u, that appears, for example, in the scattering lengths, due to different thresholds; but this is a trivial effect from the pion masses.

$$\Delta(a(\pi^+ n \to \pi^0 p) - a(\pi^- p \to \pi^0 n)) = \frac{1}{1+\frac{m_\pi}{m_N}} \frac{\sqrt{2}}{\pi F_\pi^2} \delta m_N,$$
$$\Delta a(\pi^\pm N \to \pi^\pm N) = 0. \qquad (5)$$

For processes that involve only one π^0, the contribution is $O(\varepsilon \frac{m_\pi}{m_\rho}) \sim 5\%$ of the isosymmetric result dominated by the Tomozawa–Weinberg term [7, 5]. But for $\pi^0 N \to \pi^0 N$ which vanishes in lowest order, it should be $O(\varepsilon) \sim 30\%$, thus revealing the full isospin breaking in the QCD Lagrangian! Unfortunately, this one case of large isospin violation cannot be measured easily, and so we are led to consider the possibility of at least one virtual π^0, which brings us to systems with several nucleons, i.e. nuclear physics.

The power counting indicates that to order $\nu = 8 - 3A$ the isospin violating nuclear potential is a two-nucleon potential of the form

$$V = V_{II}\,[t_3^{(1)} t_3^{(2)} - \vec{t}^{(1)} \cdot \vec{t}^{(2)}] + V_{III}\,[t_3^{(1)} + t_3^{(2)}], \qquad (6)$$

where

$$V_{II} = -\left(\frac{2g_A}{F_\pi}\right)^2 \frac{\boldsymbol{q}\cdot\boldsymbol{\sigma}^{(1)}\boldsymbol{q}\cdot\boldsymbol{\sigma}^{(2)}}{(\boldsymbol{q}^2 + m_{\pi^0}^2)(\boldsymbol{q}^2 + m_{\pi^\pm}^2)} (\Delta m_\pi^2 + \Delta m_N^2) \qquad (7)$$

and

$$V_{III} = \frac{2g_A\beta_1}{F_\pi^2} \frac{\boldsymbol{q}\cdot\boldsymbol{\sigma}^{(1)}\boldsymbol{q}\cdot\boldsymbol{\sigma}^{(2)}}{\boldsymbol{q}^2 + m_\pi^2} - (\gamma_s + \gamma_\sigma \boldsymbol{\sigma}_1 \cdot \boldsymbol{\sigma}_2). \qquad (8)$$

Here we see that the dominant pion mass difference term contributes only to charge dependence. In the $^1S_0$ state, Eq. (7) agrees with the result of ref. [8], where it was found to account for half of the observed charge dependence in the scattering lengths, the rest being attributed to pion-photon exchange and multi-pion exchange. Charge symmetry breaking, on the other hand, comes from the smaller Eq. (8). We can use the values for the pion-nucleon coupling constants determined by the Nijmegen group [9] to find $\beta_1 = 1(8)\cdot 10^{-3}$ [10]. This is what is expected from dimensional analysis, $\varepsilon(\frac{m_\pi}{m_\rho})^2$, and from $\pi-\eta-\eta'$ mixing (using the values for couplings and mixing element in ref. [11]). Similarly, γ_s —but not γ_σ— might be viewed as originating in $\rho-\omega$ mixing, but then my estimate $\varepsilon(\frac{m_\pi}{m_\rho})^2 m_\rho^{-2}$ is somewhat smaller than what follows from $\rho-\omega$ mixing obtained on the ω mass shell. Finally, pseudovector meson exchange (in particular close-lying doublets such as $a_1 - f_1$) could contribute to the γ_σ spin-spin force [10].

In any case, we see that the symmetries of QCD naturally explain the observed pattern of isospin violation in the nuclear potential: If V_N denotes a class N force [12], one concludes that $V_{N+1} = O(\frac{m_\pi}{m_\rho})V_N$. From the two-body nature of the potential we further expect that all isospin violation that is not due to one-photon exchange (Nolen-Schiffer anomaly) is compatible with the information from the scattering lengths, as found empirically [13]. Similarly, when there are external pions or photons, impulse approximation is dominant, so any isospin violation would arise from the above elements.

4 Conclusion

As a result of a conspiracy of symmetries and experimental limitations, isospin violation in nuclear and pion physics is mostly due to trivial mass splitting effects, and is factors of $\frac{m_\pi}{m_\rho}$ smaller than it appears from the looks of the QCD Lagrangian.

Acknowledgement. I am grateful to S. Weinberg for inspiration and J. Friar for many valuable discussions. This research was supported in part by the U.S. Department of Energy grant DE-FG06-88ER40427.

References

1. S. Weinberg: Trans. N.Y. Acad. Sci. **38**, 185 (1977)
2. S. Weinberg: Physica **96A**, 327 (1979)
3. S. Weinberg: Nucl. Phys. **B363**, 3 (1991)
4. U. van Kolck: Ph.D. Dissertation. Univ. of Texas 1993 (unpublished)
5. S. Weinberg: Phys. Rev. Lett. **17**, 616 (1966)
6. F.S. Roig and A.R. Swift: Nucl. Phys. **B104**, 533 (1976)
7. Y. Tomozawa: Nuovo Cim. **46A**, 707 (1966)
8. E.M. Henley and L.K. Morrison: Phys. Rev. **141**, 1489 (1966)
9. R.A.M. Klomp, V.G.J. Stoks and J.J. de Swart: Phys. Rev. **C44**, R1258 (1991)
10. J.L. Friar, T. Goldman and U. van Kolck: In preparation
11. P. Langacker and D.A. Sparrow: Phys. Rev. **C25**, 1194 (1982)
12. E.M. Henley and G.A. Miller: In *Mesons in Nuclei*, North-Holland 1979
13. G.A. Miller, B.M.K. Nefkens and I. Slaus: Phys. Rep. **194**, 1 (1990)

Few-Body Systems Suppl. 9, 449–453 (1995)

Few-Body Systems

Radiative Pion Photoproduction from the Proton and π^+ Meson Polarizabilities*

J. Ahrens[1], V. Alekseyev[2], J. Arends[1], R. Beck[1], S. Cherepnya[2], D. Drechsel[1], L. Fil'kov[2], P. Hardy[3], V. Kashevarov[2], B. Krusche[4], C. McGeorge[3], V. Metag[4], R. Owens[3], J. Peise[1], H. Ströher[3], Th. Walcher[1]

[1] Institut für Kernphysik der Johannes-Gutenberg-Universität, Mainz, Germany

[2] Lebedev Physical Institute, Leninsky Prospect 53, 117924 Moscow, Russia

[3] Department of Physics and Astronomy, Glasgow University, Glasgow. GB

[4] II. Physikalisches Institut, Universität Giessen, Giessen, Germany

Abstract. We study the possibility of investigating radiative pion photoproduction from the proton at the microtron MAMI-B with the aim to obtain an experimental information about the π^+ meson polarizabilities. It is shown that an exposition time of about 30 days will allow to determine the π^+ meson polarizability with quite high accuracy.

The electric (α) and magnetic (β) pion polarizabilities characterize the deformation of the pion in an electromagnetic field. They depend on the rigidity of its internal structure as a composite particle and, therefore, are important quantities to test the validity of theoretical models.

Our present knowledge about the polarizability of the pion is still quite unsatisfactory. The present experimental data obtained by the Primakoff effect [1] agree with the prediction of the dispersion sum rules but are at variance with the chiral perturbation theory (ChPT). The values derived from radiative pion photoproduction [2] have very large error bars and also show the largest discrepancy with regard to the ChPT predictions. Unfortunately, the attempts to determine the polarizability from reaction $\gamma\gamma \rightarrow \pi\pi$ [3] are very model dependent.

There is no doubt that the statistical errors of radiative pion photoproduction can be considerably reduced at the new electron accelerators, and careful studies of the extrapolation to the pion pole in various kinematical situations should help to reduce the systematical errors. It is therefore the aim of this

*This work was supported by the Deutsche Forschungsgemeinschaft (SFB201)

work to investigate the possibility to carry out the experiment on the radiative π^+ meson photoproduction from the proton at the microtron MAMI-B in kinematical regions most favourable to get more correct information on the pion polarizabilities.

The elastic $\gamma\pi^+$-scattering cross section can be found by extrapolating the experimental data for radiative pion photoproduction to the pion pole [2,4,5]. Such method was first suggested in [6] and has been widely used for determination of the cross section and phase shifts of elastic $\pi\pi$-scattering from the reaction $\pi N \to \pi\pi N$. For investigations of $\gamma\pi^+$-scattering this method was first used in [7,2].

There are five independent invariant variables for this reaction:

$$s = (p_1 + k_1)^2, \quad t = (p_2 - p_1)^2, \quad t_1 = (k_2 - k_1)^2,$$
$$s_2 = (p_2 + q_2)^2, \quad s_1 = (k_2 + q_2)^2, \tag{1}$$

where k_1 (k_2) is the 4-momentum of the initial (final) photon, p_1 (p_2) is the 4-momentum of the initial (final) nucleon and q_2 is the 4-momentum of the pion. As a result of the extrapolation, the elastic $\gamma\pi$-scattering cross section is [4,5]

$$\frac{d\sigma_{\gamma\pi}}{d\Omega} = -K(s_1) \cdot \lim_{t\to\mu^2} F(t, s, s_1, t_1, s_2) \tag{2}$$

where

$$F = (t - \mu^2)^2(s - m^2)^2 \frac{d\sigma_{\gamma p\to\gamma\pi N}}{dt ds_1 d\Omega_{\gamma\gamma}}, \qquad K = \frac{4\pi}{g^2_{\pi NN}} \frac{2\pi}{\mu^2(s_1 - \mu^2)}, \tag{3}$$

m (μ) is the nucleon (pion) mass.

An extrapolation requires to fix four independent kinematical variables. The extrapolation of the function F at the fixed invariant variables s, s_1, t_1, s_2 is the simple method to avoid additional singularities on the path of extrapolation. Unfortunately, in this case the contribution of the pion polarizability is small ($\sim 5\%$) in the physical region of the process under investigation. So it is necessary to have a high experimental precision to obtain a correct value of the pion polarizability. In addition, there is a pole at $t_3 = (k_1 - q_2)^2 = \mu^2$ which is close to $t = 2\mu^2$ for backward scattering in the $\gamma\pi$ system. This pole may influence the accuracy of the extrapolation.

Let us consider the function F at the fixed s, s_1 and angles $\theta^{\rm cm}_{\gamma\gamma}$ and $\varphi^{\rm cm}_{\gamma} = \varphi_b$ (φ_b is the Treiman-Yang angle). The variables t_1 and s_2 are not fixed now and are functions of t. For example,

$$t_1 = -\frac{(s_1 - \mu^2)(s_1 - t)}{2s_1}(1 - \cos\theta^{\rm cm}_{\gamma\gamma}) \tag{4}$$

On the other hand, the contribution of the difference of the polarizabilities $(\alpha_{\pi^+} - \beta_{\pi^+})$ to the radiative photoproduction cross section is proportional to t_1. Therefore it is expected that the contribution of the polarizabilities gets larger with increasing $|\, t\,|$. Moreover, the pole $t_3 = \mu^2$ is moved to $t = s_1 > 4\mu^2$, i.e. a value above threshold for the $\gamma\gamma \to \pi\pi$ process and far away from the region of extrapolation.

However, in this case additional singularities appear on the path of the extrapolation. For example, at $\theta_{\gamma\gamma}^{cm} = 180°$ there is a pole at $t = 0$. Fortunately these singularities may be removed by integrating the function F over φ_b from 0 up to 2π and over $\theta_{\gamma\gamma}^{cm}$ from 140° up to 180°. The latter region of integration carries the biggest contribution of polarizabilities in the $\gamma\pi^+$ cross section. After integration over φ_b we have a singularity of the type $(1 - \cos\theta_{\gamma\gamma}^{cm})^{-1/2}$, which is removed by integrating over $\theta_{\gamma\gamma}^{cm}$.

It is worth noting that at $\theta_{\gamma\gamma}^{cm} = 180°$ the contribution of the background diagram with the Δ_{33} resonance in the s-channel and the photon emitted by the pion, which gives the biggest contribution to F for the case with fixed s, s_1, t_1, s_2, is equal to zero. Moreover, the integration over φ_b and $\theta_{\gamma\gamma}^{cm}$ essentially decreases the contribution of nucleon resonances from crossed channels. Therefore it is expected that in the case under consideration the contribution of the background diagram will not give rise to big modification of the results obtained by the pion pole and nucleon pole diagrams only. The calculation in the frame of the latter model shows for $\alpha_{\pi^+} = 0$ and $\alpha_{\pi^+} = 7 \times 10^{-4}\,\text{fm}^3$ that the contribution of the polarizability changes sign at small $|t|$, and then afterwards becomes larger with increasing $|t|$ reaching $\sim 20\%$ at $t = -6\mu^2$. This contribution is large enough to be measured in the experiment.

Let us describe in more detail the method to determine the pion polarizability. We shall regard the function F integrated over φ_b from 0 up to 2π and over $\cos\theta_{\gamma\gamma}^{cm}$ in the angular range $140° \leq \theta_{\gamma\gamma}^{cm} \leq 180°$. The experimental data for radiative pion photoproduction represented by the function F will be extrapolated with the help of expression:

$$F(t) = e^{A(t-\mu^2)}[A_0 + A_1(t-\mu^2) + A_2(t-\mu^2)^2 + \cdots] \tag{5}$$

The parameters $A, A_0, A_1, A_2 \ldots$ have to be determined from a fit to the experimental data for the reaction $\gamma p \to \gamma\pi N$. The coefficient $A_0(s_1)$ in this expression is connected with the π^+ meson polarizabilities in the following way:

$$A_0(s_1) = -A_0^B + (\alpha_{\pi^+} - \beta_{\pi^+})A_0', \tag{6}$$

where

$$\begin{aligned} A_0^B &= \left(\frac{e^2}{4\pi}\right)^2 \left(\frac{g_{\pi NN}^2}{4\pi}\right) \frac{\mu^2}{2s_1(s_1-\mu^2)} \{0.468(s_1^2+\mu^4)+ \\ &\quad \frac{8\mu^2 s_1^2}{s_1 + 7.547\mu^2} - \frac{8s_1\mu^2(s_1+\mu^2)}{s_1-\mu^2}L\}, \end{aligned} \tag{7}$$

$$\begin{aligned} A_0' &= \left(\frac{e^2}{4\pi}\right) \left(\frac{g_{\pi NN}^2}{4\pi}\right) \frac{\mu^5(s_1-\mu^2)^2}{4s_1^2} \left\{\frac{4s_1^2}{(s_1-\mu^2)^2}L - \right. \\ &\quad \left. 0.674622 - 0.234\frac{s_1+\mu^2}{s_1-\mu^2}\right\}, \end{aligned} \tag{8}$$

$$L = \ln\left(0.883 + 0.117\frac{s_1}{\mu^2}\right).$$

For each fixed s_1 we have

$$\alpha_{\pi^+} - \beta_{\pi^+} = \frac{A_0^{(\mathrm{exp})} - A_0^B}{A_0'}, \qquad \Delta(\alpha_{\pi^+} - \beta_{\pi^+}) = \frac{\Delta A_0^{(\mathrm{exp})}}{A_0'} \tag{9}$$

where $A_0^{(\mathrm{exp})}$ and $\Delta A_0^{(\mathrm{exp})}$ are the values of A_0 and ΔA_0 found by the extrapolation of the experimental data.

The coefficient A_0 depends on s_1 only. Obviously, we have to obtain the same values of A_0 at fixed s_1 from the extrapolation of the experimental data for different values of the initial photon energy ν_1. Therefore, such an extrapolation for different ν_1 at fixed s_1 can give information about the correctness of the extrapolation procedure.

The parameter $(\alpha_{\pi^+} - \beta_{\pi^+})$ does not depend on ν_1 and s_1. Therefore, summing the values of this parameter at different ν_1 and s_1 we shall obtain the final values of $(\alpha_{\pi^+} - \beta_{\pi^+})$ and $\Delta(\alpha_{\pi^+} - \beta_{\pi^+})$. The described procedure gives the possibility to decrease essentially the error obtained by extrapolating only one curve at fixed s_1 and ν_1.

The experiment is proposed to be performed with the tagged photon beam of the 855 MeV continuous wave electron accelerator MAMI. In combination with the Glasgow-Edinburg-Mainz tagging facility (GEM) [8], this accelerator produces quite a high intensity of the tagged photons up to $3 \cdot 10^8$ per second in the energy region $\nu_1 = 500 - 800$ MeV with a sufficient energy resolution ~ 2 MeV.

We plan to use two neutron detectors. The main neutron detector (Big TOF detector) allows to detect the neutrons in an energy region $10 - 60$ MeV with an efficiency of $30 - 50\%$ and also to determine their energy with a resolution $\sim 10\%$ using the neutron time-of-flight.

Unfortunately, the big neutron detector has a very small detection efficiency for neutrons with an energy below 10 MeV. To cover a neutron energy region between 4 and 10 MeV, it is proposed to use an additional neutron detector (Small TOF detector).

To detect the secondary photons in the angle region $\theta_{\gamma\gamma} = 60° \pm 30°$ and $\varphi_\gamma = 0° \pm 50°$ we plan to use a spectrometer based on BaF_2 crystals.

As pion detector we intend to use 4 planes of multiwire proportional chambers placed at $\overline{\theta}_{\gamma\pi^+} = 0° \pm 15°$ and $\Delta\varphi_{\pi^+} = 360°$.

The detectors were placed according to the kinematical investigation. The phase space of the process $\gamma p \to \gamma\pi^+ n$ was divided into separated cells over ν_1, s_1 and t. Then the yields of the reaction under investigation were determined for each cell using the GEANT cod.

It has been shown for the triple coincidences that an exposition time of about 30 days would allow to measure the cross section at $t \leq -1\mu^2$ with a precision better than 5% for more than 70% of the cells, and only in a few cells the precision is expected to be $6 - 7\%$. The precision of such measurement at $t > -1\mu^2$ is expected to be better than 10% as a rule.

Furthermore, the background process $\gamma p \to \pi^0 \pi^+ n$ has been analyzed. The suppression of this background is mainly ensured by locating the pion detector at $\theta_{\gamma\pi^+} = 0° - 15°$, while most of π^+ mesons from the reaction $\gamma p \to \pi^0 \pi^+ n$ are emitted at $\theta_{\gamma\pi^+} > 20°$. A further suppression of this reaction is caused by the conservation laws. As a result, we have obtained practically the full separation of this main background reaction for triple (γ, n, π^+) coincidences in the kinematical region under discussion.

Moreover, the triple coincidences allow to separate Compton scattering on the proton and π^0 meson photoproduction $(\gamma p \to \pi^0 p)$.

The errors of extrapolation depend on the shape of the curve F and the precision of the experimental data. Let us estimate the expected error for the pion polarizability. To this purpose we use the function F obtained in the frame of the theoretical model described early. Taking into account the calculated yields of radiative photoproduction for an exposition time 30 days and summing the values of $\Delta\alpha_{\pi^+}$ (if $\alpha_{\pi^+} = -\beta_{\pi^+}$) at different ν_1 and s_1 we find

$$\Delta\alpha_{\pi^+} = 1.3 \times 10^{-4}\,\mathrm{fm}^3 \tag{10}$$

This precision is high enough to check the prediction of different theoretical models.

If for some reasons (a more complicated shape of the experimental curve, a decrease of the exposition time) the extrapolation errors are increased by a factor 2, then the expected value of $\Delta\alpha_{\pi^+}$ will be equal to 1.8.

In conclusion, radiative photoproduction offers a real possibility to determine the polarizability of the π^+ meson at MAMI. An exposition time of about 30 days will be necessary to measure this important quantity with quite a high accuracy.

References

1. Yu.M. Antipov et al. : Phys. Lett. **B121**, 445 (1983)
2. T.A. Aybergenov et al. : Sov.Phys.–Lebedev Inst. Reports **6**, 32 (1984); Czech. J. Phys. **B36**, 948 (1986)
3. F. Donoghue and B. Holstein: Phys. Rev. **D48**, 137 (1993)
4. T.A. Aybergenov et al. : Proceedings of the Lebedev Phys. Inst. **186**, 169 (1988)
5. D. Drechsel and L.V. Fil'kov: Z. Phys. **A349**, 177 (1994)
6. G. Goebel: Phys. Rev. Lett. **1**, 337 (1958);
   G.F. Chew and F.E. Low: Phys. Rev. **113**, 1640 (1959)
7. T.A. Aybergenov et al. : Sov. Phys.–Lebedev Inst. Reports **5**, 28 (1982)
8. I. Anthony et al. : Nucl. Instr. and Meth. **A301**, 230 (1991)

Few-Body Systems Suppl. 9, 455–460 (1995)

Few-Body Systems

A Model for the $\gamma N \to \pi\pi N$ Reaction

J. A. Gómez Tejedor, E. Oset

Departamento de Física Teórica and IFIC, Centro Mixto Universidad de Valencia - CSIC 46100 Burjassot (Valencia), Spain

Abstract. We have studied the $\gamma N \to \pi\pi N$ reaction using a model which includes N, $\Delta(1232)$, $N^*(1440)$ and $N^*(1520)$ intermediate baryonic states and the ρ-meson as intermediate $\pi\pi$ resonance. The model reproduces fairly well experimental cross sections below $E_\gamma = 800$ MeV and invariant-mass distributions even at higher energies. One of the interesting findings of the study is that the $\gamma N \to N^*(1520) \to \Delta\pi$ process is very important and interferes strongly with the dominant Δ-Kroll-Ruderman term to produce the experimental peak of the cross section.

1 Introduction

There are three possible double pion photoproduction reactions on the proton, and three on the neutron:

$$\begin{array}{llll} (a) & \gamma p \to \pi^+\pi^- p & (d) & \gamma n \to \pi^+\pi^- n \\ (b) & \gamma p \to \pi^+\pi^0 n & (e) & \gamma n \to \pi^-\pi^0 p \\ (c) & \gamma p \to \pi^0\pi^0 p & (f) & \gamma n \to \pi^0\pi^0 n \end{array} \qquad (1)$$

These reactions have been extensively studied experimentally in the past ([1, 2, 3, 4, 5, 6, 7]) and there is abundant information on cross sections and invariant-mass distributions for the $\pi\pi$ and πN systems. New improvements in experimental techniques and facilities have reopened the study of these reactions at Mainz, with two experiments on the proton [8, 9]. From the theoretical point of view, only the reaction $1(a)$ has been studied with one early model [10] which considers only five Feynman diagrams.

One characteristic feature of the $\gamma N \to \pi\pi N$ reaction is that it requires a fairly large number of Feynman diagrams to account for it theoretically. Hence, the apparent success of the model of ref. [10] for reaction (a), which considers only 5 Feynman diagrams, has always been intriguing.

With this reaction becoming a target of new experimental study and interesting medium effects predicted for the $(\gamma, \pi^+\pi^-)$ reaction in nuclei [11], a

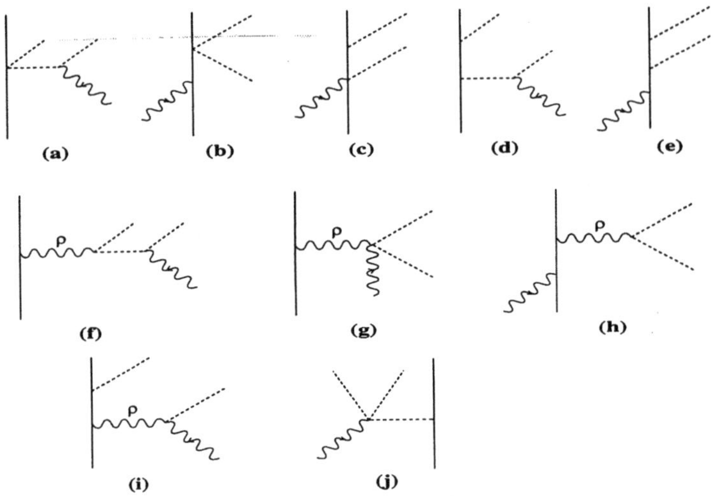

Figure 1. Classification of the Feynman diagrams into one point, two point and three point diagrams. Continuous straight lines: baryons. Dashed lines: pions. Wavy lines: photons and ρ-mesons (marked explicitly).

thorough theoretical study of the $\gamma N \rightarrow \pi\pi N$ reaction is necessary. This task has been undertaken in ref.[12].

2 The model

We classify our diagrams in one point, two points and three points diagrams, according to the number of vertices in the hadron components (see Fig.1). Our basic components are pions, nucleons and nucleonic resonances. We consider for the hadronic components N, $\Delta(1232, J^\pi = 3/2^+, I = 3/2)$, $N^*(1440, J^\pi = 1/2^+, I = 1/2)$ and $N^*(1520, J^\pi = 3/2^-, I = 1/2)$. The $N^*(1520)$ has a particularly large coupling to the photons and proves to be an important ingredient, mostly because its interference with the dominant component of the process, the $\gamma N \rightarrow \Delta\pi$ transition through the gauge Δ-Kroll-Ruderman term. Higher resonances have a weaker coupling to photons and do not interfere with the dominant term, hence their contribution is small, at least for photon energies below 800 MeV, Mainz energies, where our model is meant to work. Because of the important coupling of the ρ-meson to the two pion system and the $\gamma\pi$ system we have also considered terms involving the ρ-meson with the same organizing scheme. These terms are only relevant at high energies but show up clearly in the two pion invariant-mass distributions at these energies [1, 12].

With these considerations the basic diagrams which we consider have the structure as shown in Fig.1. In diagrams (a) and (b), the one point $NN\pi\pi$ coupling stands for the s-wave πN interaction. We consider there only the isoscalar part of the amplitude. The isovector part is mediated by ρ exchange

[13] and hence it is explicitly taken into account in diagrams (f), (g) and (h). Diagram (c) contains the gauge term $NN\pi\gamma$ or Δ-Kroll-Ruderman term. We use a pseudovector coupling for the $NN\pi$ vertex and this allows us to consider exclusively positive energy intermediate state in the hadronic propagators [14]. In the two point and three point diagrams we include nucleon and the resonances as intermediate states.

For the $\gamma p \rightarrow \pi^+\pi^- p$ reaction while all possible diagrams with N and Δ intermediate states are considered, we omit some with $N^*(1440)$ intermediate states which are very small. For the $N^*(1520)$ intermediate states we keep only the term which interferes with the dominant term of the amplitude (Δ-Kroll-Ruderman term). In addition, all different time orderings of the diagrams are considered.

For the other isospin channels we use the same model as for the $\gamma p \rightarrow \pi^+\pi^- p$ case, changing the isospin factors and introducing some terms which are only relevant in the case of neutral pions [12]. However, we have neglected some Feynman diagrams which have been found to be very small for the $\gamma p \rightarrow \pi^+\pi^- p$ channel.

The Feynman diagrams considered and detailed calculations can be found in ref.[12].

3 Results and discussion

In Figs.2 and 3 we show the total cross sections for the $\gamma p \rightarrow \pi^+\pi^- p$ and $\gamma p \rightarrow \pi^0\pi^0 p$ isospin channels, as well as the contribution to the total cross section of diagrams with *nucleon*, $\Delta(1232)$, $N^*(1440)$, $N^*(1520)$ and $\rho(770)$ as intermediate states (in ref. [12] we also showed results for the other isospin channels, as well as differential cross sections and invariant-mass distributions for the $\gamma p \rightarrow \pi^+\pi^- p$ channel). We show results up to $E_\gamma = 800$ MeV, where the new experiments at Mainz concentrate.

In Fig. 2 we observe that the $\Delta(1232)$ terms (short-dashed lines) are dominant in the $\gamma p \rightarrow \pi^+\pi^- p$ reaction. Essentially the $\Delta(1232)$-Kroll-Ruderman and $\Delta(1232)$-pion-pole terms are the most important ones. The non resonant terms (short-dash-dotted lines) are much smaller and they provide a small background which grows up moderately as a function of the energy. The $N^*(1520)$ contribution (long-dashed lines) by itself is also small compared to the $\Delta(1232)$ one, but it is essential to reproduce the total cross section due to its interference with the $\Delta(1232)$-Kroll-Ruderman term as we already remarked in ref.[12]. The ρ-terms are negligible at these energies, but they show up clearly at higher energies (see ref.[12]). Contributions from other terms are still smaller.

The inclusion of the $N^*(1520)$ terms leads to an interference with the Δ-Kroll-Ruderman terms which is responsible for the appearance of the maximum and a much better agreement with experiment. This interference occurs only between the $\Delta(1232)$-Kroll-Ruderman term and the s-wave part of the $N^*(1520)\Delta\pi$ contribution [12]. This interesting finding shows that although intuitive, it is not correct to associate the peak of the cross section to the

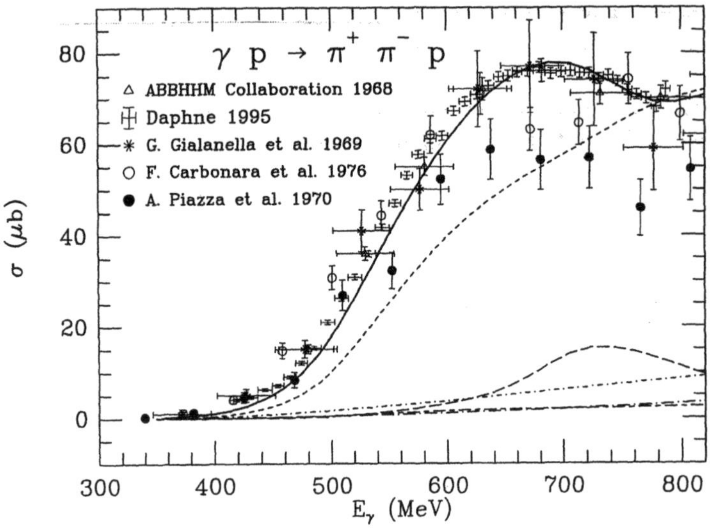

Figure 2. Total cross section for the $\gamma p \rightarrow \pi^+\pi^- p$ reaction. Continuous line: total cross section. Short-dashed line: contribution of $\Delta(1232)$-intermediate states. Long-dashed line: contribution of $N^*(1520)$-intermediate state. Short-dash-dotted line: contribution of N-intermediate states. Long-dash-dotted line: contribution of ρ-intermediate states. Short-dash-long-dashed line: rest of the diagrams. Experimental data from refs. [1, 2, 3, 4, 8].

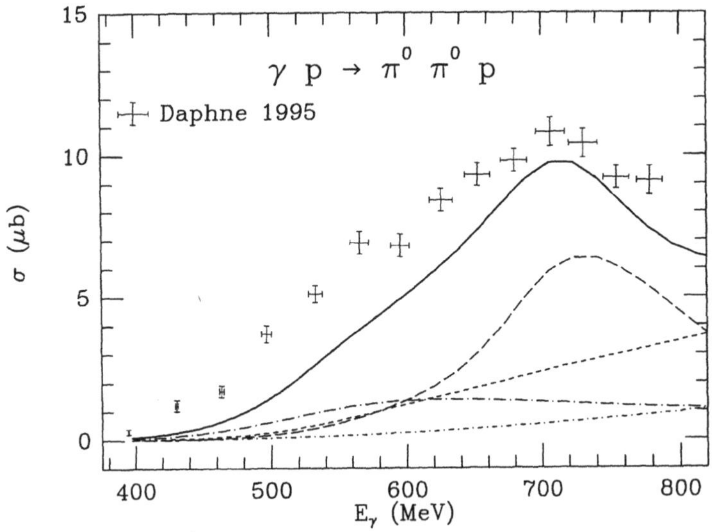

Figure 3. Same as Fig. 2 for the $\gamma p \rightarrow \pi^0\pi^0 p$ reaction. Experimental data from ref. [8].

$\Delta(1232)$ resonance. We showed that the delta terms do not lead to such a peak and it comes as an interference phenomenon.

For the $\gamma p \to \pi^0\pi^0 p$ case things are rather different (see Fig. 3). In these cases a lot of terms vanish due to the fact that the photons can not couple to neutral pions. In particular, there are no Kroll-Ruderman and pion-pole terms. Thus, the total cross section for this isospin channel is much smaller than the cross section for other channels. Furthermore, terms that in the other isospin channels are very small become important in this channel.

In Fig.3 we can see that, except for the contribution from N-intermediate states (short-dash-dotted lines) which is very small, the other contributions are all of them relevant. Below 500 MeV we can see that the $N^*(1440)$ (long-dash-dotted lines) dominate the reactions. At 500 MeV the $\Delta(1232)$ (short-dashed lines) and the $N^*(1520)$ (long-dashed lines) start to grow up, and around 600 MeV all these diagrams have similar strength. At 600 MeV the $N^*(1440)$ contribution starts to fall down, and the $N^*(1520)$ and $\Delta(1232)$ dominate the reaction. The $N^*(1520)$ contributions peaks at 720 MeV (it is responsible for the peak of the total cross sections) and from this energy on it falls down, while the $\Delta(1232)$ contribution continues growing up moderately.

It is worth noting that, in this case, the $N^*(1520)$ contribution is important by itself, and not only by its interference with other terms as it happens in the $\gamma p \to \pi^+\pi^- p$ case, and again it is essential to reproduce the peak of the cross section around 700 MeV.

In ref.[12] results are also shown for the other isospin channels. For the $\gamma p \to \pi^+\pi^0 n$ channel we have found an important discrepancy between our calculations and the experimental data. For the isospin channels on the neutron, the available experimental data are not accurate enough to extract any conclusion.

4 Conclusions

We have constructed a model for the $\gamma N \to \pi\pi N$ reaction including *nucleons*, $\Delta(1232)$, $N^*(1440)$ and $N^*(1520)$ as intermediate baryonic states as well as ρ-meson intermediate states for the $\pi\pi$ system. Our model is rather complete, but still misses terms which become relevant from $E_\gamma = 800$ MeV on. As in a previous model accounting for only a few of these diagrams [10] for the $\gamma p \to \pi^+\pi^- p$, we observe the dominance of the Δ-Kroll-Ruderman and pion-pole terms but get an appreciable contribution from other terms. In particular we found the contribution of the $N^*(1520)$ resonance very important, and essential to produce the peak which is present in the experimental cross section around $E_\gamma = 680$ MeV.

Our model reproduces quite well the experimental results of refs. [1, 8, 9] below E_γ =800 MeV for the $\gamma p \to \pi^+\pi^- p$ and $\gamma p \to \pi^0\pi^0 p$ isospin channels (we should mention that our model agrees even better with the preliminary data of TAPS collaboration [9] for the $\gamma p \to \pi^0\pi^0 p$ channel than with the DAPHNE data for the same channel), but we have found some important discrepancies for the $\gamma p \rightarrow \pi^+ \pi^0 n$ channel. For the isospin channels on the neutron we have found also some discrepancies with the old data of refs. [3, 4], but as we already mentioned in ref. [12], these experiments should be improved.

Our model is also a starting point to generate exchange currents in nuclei. For instance, by producing one pion off-shell and attaching it to a nucleon line, we generate two body contributions to the (γ, π) channel. Similarly, by producing the two pions off-shell and attaching them to two nucleons, we generate three nucleons mechanisms which contribute to photon absorption in nuclei, and so on. Such channels have been partially investigated in ref. [14] but more detailed studies would be welcome.

References

1. Aachen Berlin Bonn Hamburg Heidelberg München collaboration: Phys. Rev. **175**, 1669 (1968)
2. G. Gialanella et al.: Nuovo Cimento **LXIII A**, 892 (1969)
3. F. Carbonara et al.: Nuovo Cimento **36A**, 219 (1976)
4. A. Piazza et al.: Nuovo Cimento **III**, 403 (1970)
5. Cambridge Bubble Chamber Group: Phys. Rev. **155**, 1477 (1967); Phys. Rev. **163**, 1510 (1967)
6. H. R. Crouch et al.: Phys. Rev. Lett. **13**, 636 (1964); Phys. Rev. Lett. **13**, 640 (1964)
7. H. G. Hilpert et al.: Phys. Lett. **23**, 707 (1966)
8. A. Braghieri et al.: to be published.
9. H. Ströher: Private Communication.
10. L. Lüke and P. Söding: Springer Tracts in Modern Physics **59**, 39 (1971)
11. E. Oset and M. J. Vicente-Vacas, in *Int. Symposium on weak and electromagnetic interactions in nuclei*, Heidelberg, 1986, H. K. Klapdor Ed., p.444. Berlin: Springer-Verlag 1986; J.A. Gómez Tejedor, M.J Vicente-Vacas, E. Oset: Nucl. Phys. **A588**, 819 (1995)
12. J.A. Gómez Tejedor and E.Oset: Nucl. Phys. **A571**, 667 (1994); J. A. Gómez Tejedor and E. Oset: Preprint FTUV/95-28 (HEP-PH/9506209); J. A. Gómez Tejedor, Tesina de Licenciatura. Universidad de Valencia 1993
13. T.E.O. Ericson and W. Weise: *Pions and nuclei.* Oxford: Clarendon Press 1988
14. R. C. Carrasco and E. Oset: Nucl. Phys. **A536**, 445 (1992)

Few-Body Systems Suppl. 9, 461–465 (1995)

Few-Body Systems

The Polarized Structure Function of the Nucleon in the Constituent Quark Model

M. De Sanctis

INFN, Sezione di Roma, P.le A.Moro 2, 00185 Roma

Abstract. A study of the polarized structure function of the nucleon is presented. Relativistic corrections have been calculated for the real photon case, corresponding to the Drell-Hearn-Gerasimov sum rule. These corrections have the form of an electric dipole operator, leading to additional absorption strength when the hyperfine interaction is considered. In the case of exchange of virtual photons, a phenomenological expression for the electromagnetic current is derived assuming a sort of impulse approximation for the electron scattering on the single quark. The expected asymptotic behaviour at high momentum transfer is reproduced.

Many efforts are devoted to the study of the polarized structure function of the nucleon $I(Q^2)$, by means of reactions with electromagnetic probes at different values of the momentum transfer Q^2. The aim is to understand in some detail the spin-flavour structure of the nucleon and of its excitations. There exist definite theoretical predictions for $I(Q^2)$ both in the real photon case ($Q^2 = 0$), corresponding to the Drell-Hearn-Gerasimov (DHG) sum rule [1] and at high momentum transfer ($Q^2 \to \infty$) in the Deep Inelastic Scattering (DIS) region, corresponding to the Bjorken sum rule [2, 3]. Both the rules are based on general properties of the electromagnetic interaction. From the experimental point of view, $I(Q^2)$ is not determined up to now with sufficient accuracy [4, 5] but, due to the advent of new electron facilities with polarized beams and targets, more accurate experimental data are expected to be available in the near future [6].

We recall that the spin dependent part of the inelastic cross section of an elementary spin 1/2 particle on a nucleon may be written in terms of two invariant response functions $\bar{G}_1(Q^2, \nu)$ and $\bar{G}_2(Q^2, \nu)$. Neglecting the mass of the scattering fermion one has [2]

$$\frac{d^2\sigma^{\uparrow\uparrow}}{dQ^2 d\nu} - \frac{d^2\sigma^{\downarrow\uparrow}}{dQ^2 d\nu} = -\frac{4\alpha^2}{Q^2 E_e^2} \cdot \left[M\bar{G}_1 \cdot (E'_e \cos\vartheta + E_e) - Q^2 \bar{G}_2 \right], \qquad (1)$$

E_e, E'_e and ϑ represent respectively the energy of the incoming and outgoing fermion and its scattering angle, all measured in the laboratory reference frame;

the nucleon mass M, the invariant squared momentum transfer $Q^2 \geq 0$ and the invariant energy transfer $\nu = E_e - E'_e$ have also been introduced. The notations $\uparrow\uparrow$ and $\downarrow\uparrow$ stand for parallel and antiparallel polarization of the incoming particle with respect to the polarization of the nucleon target, respectively. Also, polarizing the target transverse to the lepton enables to obtain a different weighting of $\bar{G}_1$ and $\bar{G}_2$ facilitating the extraction of $\bar{G}_1$ and $\bar{G}_1$ from the experimental data [7].

In order to study the spin structure of the nucleon in the whole range of variability of the momentum transfer $Q^2 \epsilon (0, \infty)$, we define the transverse polarized structure function [8]

$$I_{\mathrm{T}}(Q^2) = 8\pi\alpha \int_{\nu_{\mathrm{thr}}}^{\infty} d\nu \frac{1}{\nu^2 + Q^2} \cdot [M\nu \bar{G}_1 - Q^2 \bar{G}_2] =$$

$$= -4\pi^2\alpha \int_{\nu_{\mathrm{thr}}}^{\infty} d\nu \frac{1}{\nu^2 + Q^2} \cdot [| < f|J_+|N, \frac{1}{2} > |^2 - | < f|J_-|N, \frac{1}{2} > |^2] \ . \quad (2)$$

That definition only contains the tranverse components of the current operator that are, within the available models, more reliably determined than the longitudinal one. Moreover it has the DHG and DIS limits as the conventional $I(Q^2)$. More precisely

$$I_{\mathrm{T}}(Q^2) = 8\pi^2\alpha \begin{cases} -k^2/4M^2 \ , \ Q^2 = 0 \\ Z/Q^2 \ , \ Q^2 \to \infty \end{cases} , \quad (3)$$

where k represents the anomalous part of the magnetic moment and Z is a positive constant related to the parton structure of the nucleon. The constituent quark model (CQM) predicts $|k| = 2$ plus corrections due to hyperfine interactions [9], $Z = \frac{5}{9}(0)$ for the proton (neutron) [2, 3]. In this paper we do not take into account the contributions to Z due to the "non valence" partons that are beyond the present possibility of the CQM and only focus our attention on the Q^2 depedence of I_{T}.

We now consider the problem of calculating the matrix-elements of Eq. (2). We shall use the CQM wave functions and, at $Q^2 = 0$, the effective current operators [10] with relativistic corrections up to order m^{-2}. One obtains the standard nonrelativistic current in the long wavelength approximation and a relativistic correction to the electric dipole, of the form [9, 10]

$$\boldsymbol{J}^{RC} = \mathrm{i}\omega \sum_{i=1}^{3} [\frac{1}{4m}(\frac{Z}{M} - \frac{2\mu_{ia} + e_i}{m})(\boldsymbol{\sigma}_i \times \boldsymbol{p}'_i) - \frac{Z}{2M}\{\boldsymbol{x}'_i, h_i\}] \ , \quad (4)$$

where $\boldsymbol{x}'_i, \boldsymbol{p}'_i, \boldsymbol{\sigma}_i, e_i, \mu_{ia}$ and h_i represent the intrinsic coordinate, intrinsic momentum, Pauli spin operator, charge, anomalous magnetic moment and intrinsic effective Hamiltonian of the i-th quark; Z and M stand for the total charge and mass of the nucleon. Standard calculations show that the relativistic correction of Eq.(4) is crucial in order to verify the DHG sum rule that, within the CQM, takes the form

$$I_{\mathrm{T}}(0) = -\frac{2\pi^2\alpha}{M^2}[4 - 8\, a_{\mathrm{M}}^2 - 10\, a_{\mathrm{D}}^2 - 2\, r_0\, a_{\mathrm{D}}^2] \ , \quad (5)$$

where a_D and a_M represent the standard configuration mixing cohefficients of the nucleon wave function due to the hyperfine interaction [11].

We now consider the evaluation of Eq. (2) at $Q^2 \neq 0$. We firstly note that the nonrelativistic approximation with relativistic corrections is not expected to hold as such at $Q^2 > m^2$. However, if the configuration mixing is neglected, ($a_M = a_D = 0$), only the nonrelativistic spin current is sufficient to reproduce the DHG. Also, at $Q^2 \to \infty$ the spin interaction is considered to be responsible of the Bjorken limit [3, 7]. So we consider in our model the nucleon state as pure s-wave space symmetric wave function; with the help of the angular momentum algebra, one can show that the convection current depending on spatial intrinsic operators gives no contribution to $I_T(Q^2)$. Furthermore, the recoil term is also vanishing due to transversality. The only relevant is the spin current, for which we take the form [8]

$$\boldsymbol{J}^s = \mathrm{i}qK(Q^2)\sum_{i=1}^{3} e_i(\boldsymbol{\sigma}_i \times \hat{\boldsymbol{z}})E_i \ . \tag{6}$$

Firstly, $K(Q^2)$ is a kinematic factor whose form is determined by using a sort of "impulse approximation". The virtual photon is considered to interact elastically with a free quark initially at rest; 4-momentum conservation is assumed in that interaction process. The corresponding amplitude, written in term of Dirac spinors, is then expressed in the nonrelativistic space, so that $K(Q^2)$ takes the form

$$K = [2m[(\frac{Q^4}{4m^2} + Q^2 + m^2)^{1/2} + m]]^{-1/2} \ . \tag{7}$$

Note that at $Q^2 = 0$, the correct nonrelativistic value $K \simeq \frac{1}{2m}$ is recovered.

Secondly, for the phase factor E_i we take tentatively the expression

$$E_i = e^{\mathrm{i}\sqrt{Q^2}\hat{\mathbf{q}}\mathbf{x}'_i} \ , \tag{8}$$

with the usual nonrelativistic limit $E_i \simeq e^{\mathrm{i}\mathbf{q}\mathbf{x}'_i} \simeq 1$. Finally, by using the equality $Q^2 + \nu^2 = \boldsymbol{q}^2$, one obtains

$$I_T(Q^2) = 4\pi^2\alpha K^2(Q^2) \cdot \frac{1}{9}[10(T_0 + \frac{1}{2}) - (10T_0 + 13)|\Phi|^2] \ , \tag{9}$$

with $T_0 = +(-)\frac{1}{2}$ for the proton (neutron). Also, the spatial matrix element Φ, in the CQM takes the form

$$\Phi(Q^2) = <\psi|E_3|\psi> = e^{-Q^2/6\alpha_0^2} \ . \tag{10}$$

Note that Eq. (9) satisfies the limits of Eq. (3) in the CQM without relativistic corrections. The only free parameter of the model, that is the harmonic oscillator constant $\alpha_0 = \sqrt{m\omega_0}$, can be independently determined by fitting, with the spin current of Eq. (6), the form factor of the $N - \Delta(1232)$ spin-flip transition. By using the parametrization [13], one finds $\alpha_0 = 0.32$ GeV, a

Fig. 1

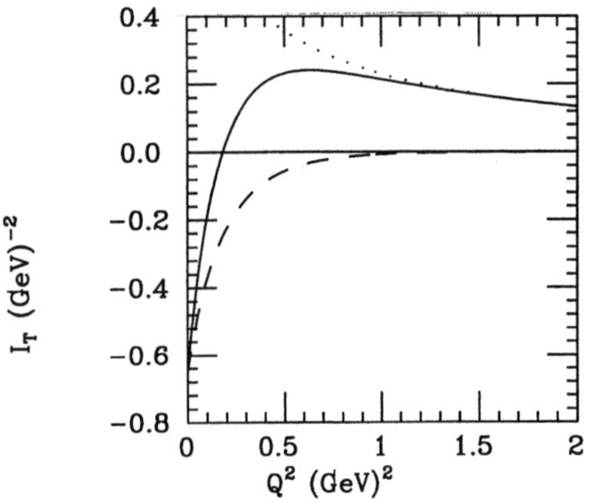

Fig. 2

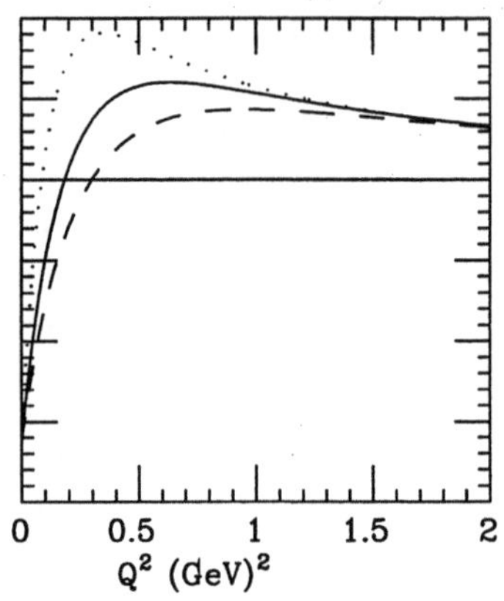

value that favourably compares with other issues of the CQM, also verifying the consistency of our model.

We now briefly comment the numerical results. In Fig. 1 the result for $I_{\rm T}(Q^2)$ obtained with $\alpha_0 = 0.32$ GeV is shown. In the proton case (full curve) $I_{\rm T}$ is zero at $Q^2 = 0.18$ GeV^2, reaches its maximum value at $I_{\rm TMAX} = 0.2405$ GeV^{-2} at $Q^2 = 0.36$ GeV^2. The derivative at the origin is given by

$$I'_{\rm T}(0) = \frac{2\pi^2\alpha}{9m^2}\left[\frac{1}{m^2} + \frac{10T_0+13}{6\alpha_0^2}\right] = \begin{cases} 6.436 \text{ GeV}^{-4} \ (proton) \\ 3.783 \text{ GeV}^{-4} \ (neutron) \end{cases} \qquad (11)$$

We note that the negative derivative at $Q^2 = 0$, found in ref. [14], should be due to the different definition of the transverse structure function. In our model the asymptotic behaviour (dotted curve) is reached at relatively low values of Q^2. At $Q^2 = 10$ GeV^2, i.e. in the range of the EMC data, we have $I_{\rm T} = 3.08 \cdot 10^{-2}$ GeV^{-2}. If the pure $1/Q^2$ dependence were taken, with $Z = 5/9$, one would have at the same $Q^2, I = 3.20 \cdot 10^{-2}$ GeV^{-2}. Due to the pure phenomenological character of our model and to the general difficulties related to the problem, we do not attempt to draw any conclusion concerning the spin crisis question [5]. We only note that our dynamic model may reproduce the main trend of $I_{\rm T}(Q)$ in a wide kinematic region. Finally in Fig. 1 we also show $I_{\rm T}(Q^2)$ of the neutron (dashed curve) that remains always negative. In Fig. 2 we show the sensitivity of our prediction with respect to the parameter α_0 by using different values ($\alpha_0 = 0.2188, 0.32, 0.41$ GeV, dotted, full and dashed curve, respectively) that are found in the literature [16]. This parameter remains a phenomenological effective constant giving rise to some uncertainties in the predictions of the CQM [17].

Further investigations are needed to clarify the problems related to the polarized structure function of the nucleon. Relevant information on the Q^2 dependence of the polarization observables can be given by the forthcoming measurements with electromagnetic probes. New theoretical models taking into account relativistic and dynamic effects are expected to interpret those results.

References

1. S. Drell and A.C. Hearn: Phys. Rev. Lett. **16**, 908 (1966); S.B. Gerasimov: Sov. J. Nucl. Phys. **2**, 430 (1966)
2. J.D. Bjorken: Phys. Rev. **148**, 1467 (1966); **D1**, 1376 (1970); J. Ellis and R. Jaffe: Phys. Rev. **D9**, 1444 (1974)
3. J. Kuti and V.F. Weisskopf: Phys. Rev. **D4**, 3418 (1971)
4. J. Karliner: Phys. Rev. **D7**, 2717 (1973)
5. J. Ashman et al. : Phys. Lett. **206B**, 364 (1988); Nucl. Phys. **B328**, 1 (1990)
6. V.D. Burkert et al. : CEBAF PR-91-023; V.D. Burkert: CEBAF-PR-94-001.
7. F.E.Close: *An Introduction to Quarks and Partons*, Chapt 13. London: Academic Press 1979.
8. M. De Sanctis: Nota Interna n. 1045, 17/1/1995, Dip. di Fisica, Università di Roma "La Sapienza" (to appear in "Il Nuovo Cimento")
9. M. De Sanctis, D. Drechsel and M.M. Giannini: Few Body Syst. **16**, 143 (1994)
10. M. De Sanctis and D. Prosperi: Nuovo Cim. **98A**, 621 (1987)
11. M.M. Giannini: Rep. Prog. Phys. **54**, 453 (1990)
12. D. Drechsel and M.M. Giannini: Few Body Syst. **15**, 99 (1993)
13. W. Albrecht et al. : Nucl. Phys. **B27**, 615 (1971); S.Galster et al. : Phys. Rev. **D5**, 519 (1972)
14. V. Burkert and Z. Li: Phys. Rev. **D47**, 46 (1993)
15. M.Anselmino, B.L. Joffe and E. Leader: Sov. J. Nucl. Phys. **49**, 136 (1989)
16. N. Isgur and G. Karl: Phys. Rev. **D18**, 4187 (1978); N. Isgur and R. Koniuk: Phys. Rev. **D21**, 1868 (1980)
17. D. Drechsel and M.M. Giannini: J. Phys. **G12**, 1165 (1986)

Few-Body Systems Suppl. 9, 466–470 (1995)

Nucleon-Nucleon Correlations and Multiquark Cluster Effects in Deep Inelastic Electron Scattering off Few-Nucleon Systems at $x > 1$

Silvano Simula *

Istituto Nazionale di Fisica Nucleare, Sezione Sanitá, Viale Regina Elena 299, I-00161 Roma, Italy

Abstract. Inclusive $A(e,e')X$ and semi-inclusive $A(e,e'N)X$ deep inelastic electron scattering processes off few-nucleon systems are investigated at $x > 1$, showing some of the relevant features of the cross section which are sensitive to the effects arising from nucleon-nucleon correlations and possible exotic multiquark cluster configurations at short internucleon separations.

The aim of this contribution is to address few relevant questions concerning inclusive $A(e,e')X$ and semi-inclusive $A(e,e'N)X$ deep inelastic scattering (DIS) of electrons off few-nucleon systems for values of the Bjorken variable $x = Q^2/2M\nu > 1$ (corresponding to kinematical regions forbidden on a free nucleon), assuming that virtual photon absorption occurs on a hadronic cluster which can be either a nucleon-nucleon (NN) correlated pair or a six-quark ($6q$) bag. The cross section for the inclusive process $A(e,e')X$ can be written as

$$\sigma^A \equiv d^2\sigma/dE_{e'}\, d\Omega_{e'} = \sigma_0^A + \sigma_1^A \tag{1}$$

where the contributions from different final nuclear states have been explicitly separated out, namely σ_0^A describes the transition to the ground and one-hole states of the $(A-1)$-nucleon system and σ_1^A the transition to more complex excited configurations (mainly $1p-2h$ states) arising from $2p-2h$ excitations generated in the target ground-state by NN short-range and tensor correlations. In what follows, the DIS contribution to σ_0^A and σ_1^A will be calculated within the impulse approximation (IA). As is well known, the IA calculation requires the knowledge of the nucleon spectral function $P^N(k,E)$, which represents the joint probability to find in a nucleus a nucleon with momentum $k \equiv |\boldsymbol{k}|$ and removal energy E. In presence of ground-state NN correlations the spectral function can be written as $P^N(k,E) = P_0^N(k,E) + P_1^N(k,E)$,

* *E-mail address:* simula@hpteo2.iss.infn.it

where the indeces 0 and 1 have the same meaning as in Eq. (1). For $A = 3$ and 4 one gets

$$P_0^N(k, E) = n_0(k)\,\delta(E - E_{\rm min}) \qquad (2)$$

where $n_0(k)$ is the nucleon momentum distribution corresponding to the ground-to-ground transition and $E_{\rm min} = 5.5,\ 20$ MeV, respectively. As for P_1^N, its momentum and removal energy structure, generated by NN correlations, can be accounted for by adopting the extended two-nucleon correlation (2NC) model of ref. [1], viz.

$$P_1^{N_1}(k, E) = \sum_{N_2=n,p} \int d\boldsymbol{k}_{\rm CM}\, n_{\rm rel}^{N_1N_2}(\boldsymbol{k} - \frac{\boldsymbol{k}_{\rm CM}}{2})\, n_{\rm CM}^{N_1N_2}(\boldsymbol{k}_{\rm CM}) \cdot$$
$$\delta[E - E_{\rm thr}^{(2)} - \frac{A-2}{2M(A-1)}(\boldsymbol{k} - \frac{A-1}{A-2}\,\boldsymbol{k}_{\rm CM})^2] \qquad (3)$$

where $n_{\rm rel}^{N_1N_2}$ ($n_{\rm CM}^{N_1N_2}$) is the momentum distribution of the relative (center-of-mass (CM)) motion of the two nucleons in a correlated pair, and $E_{\rm thr}^{(2)} = M_{A-2} + 2M - M_A$ is the two-nucleon break-up threshold. For $A = 2$ one has $P_1^N = 0$ and $P_0^N(k, E) = n(k)\,\delta(E - E_{\rm min})$, where $n(k)$ is the nucleon momentum distribution in the deuteron and $E_{\rm min} = 2.226$ MeV is the deuteron binding energy. Then, the contribution $\sigma_\alpha^{A(3q)}$ ($\alpha = 0,\ 1$), due to processes in which the struck quark belongs to a nucleon in the nucleus A, reads as follows

$$\sigma_\alpha^{A(3q)} = \sigma_{\rm Mott} \sum_{N=1}^{A} \int dW d\boldsymbol{k} dE\ \delta\left[\nu + k^0 - \sqrt{W^2 + (\boldsymbol{k}+\boldsymbol{q})^2}\right] \cdot$$
$$P_\alpha^N(k, E)\, \frac{M}{\sqrt{M^2 + k^2}}\, \frac{W}{\sqrt{W^2 + (\boldsymbol{k}+\boldsymbol{q})^2}} \sum_{j=1,2} b_j^N\, W_j^N(W, Q^2) \qquad (4)$$

where $k^0 \equiv M_A - \sqrt{(M_A + E - M)^2 + k^2}$ is the initial nucleon energy in the lab system and W_j^N is the nucleon structure function. In what follows, the free nucleon structure function, parametrized as in [2], will be adopted. In Eq. (4) $b_1^N = a_{21} + 2\tan^2(\theta_e/2)\, a_{11}$ and $b_2^N = a_{22} + 2\tan^2(\theta_e/2)\, a_{12}$, where the explicit expressions of the coefficients a_{ij} can be found in [3].

Let us now consider the possibility that the two nucleons in a correlated pair can loose their identity at short separations, so that the incoming photon can interact with a $6q$ bag structure. This is the mechanism proposed and applied to the investigation of inclusive DIS processes off the deuteron in ref. [4]. Within such a multiquark cluster picture, the inclusive cross section for the deuteron can be written as the incoherent sum of the contributions resulting from virtual photon absorption by a $3q$ and $6q$ clusters [4], viz.

$$\sigma_0^{^2{\rm H}} = P_{3q}^{^2{\rm H}}\, \sigma_0^{^2{\rm H}(3q)} + P_{6q}^{^2{\rm H}}\, \sigma^{^2{\rm H}(6q)} \quad , \quad \sigma_1^{^2{\rm H}} = 0 \qquad (5)$$

where $P_{3q}^{^2{\rm H}}$ and $P_{6q}^{^2{\rm H}}$ are the probability that the struck quark belongs to a nucleon of a NN pair or to a $6q$ bag in the deuteron, respectively ($P_{3q}^{^2{\rm H}} + P_{6q}^{^2{\rm H}} =$

1). In Eq. (5) the quantity $\sigma^{^2\mathrm{H}(6q)}$, representing the contribution of the process in which the struck quark belongs to a $6q$ bag, is given by

$$\sigma^{^2\mathrm{H}(6q)} = \sigma_{\mathrm{Mott}}\,\{W_2^{(6q)}(W_{6q}^*, Q^2) + 2\tan^2(\theta_e/2)\, W_1^{(6q)}(W_{6q}^*, Q^2)\} \qquad (6)$$

where $W_{6q}^* \equiv \sqrt{(\nu+2M)^2 - |\boldsymbol{q}|^2}$ is the invariant mass produced by virtual photon absorption by $6q$ bag at rest (with mass $2M$) and $W_i^{(6q)}$ is the inclusive structure function of a $6q$ bag. In what follows, $W_i^{(6q)}$ will be parametrized according to the the prescriptions of ref. [4] based on quark counting rules. For $A > 2$ only the correlated part σ_1^A can be affected by the presence of $6q$ bag configurations, for σ_0^A is related only to final states of the residual $(A-1)$ system belonging to its discrete spectrum. Thus, for $A > 2$ one can write

$$\sigma_0^A = \sigma_0^{A(3q)} \quad , \quad \sigma_1^A = P_{3q}^A\, \sigma_1^{A(3q)}/S_1 + P_{6q}^A\, \sigma^{A(6q)} \qquad (7)$$

where $P_{3q}^A + P_{6q}^A = S_1$, with $S_1 \equiv \int d\boldsymbol{k}dE\; P_1^N(k,E)$ being the total probability that, after the removal of a nucleon, the residual $(A-1)$ system is in any state of its continuum. In Eq. (7) the $6q$ bag contribution $\sigma^{A(6q)}$ reads as follows

$$\sigma^{A(6q)} = \frac{A}{2}\,\sigma_{\mathrm{Mott}} \sum_\beta \int dW_{6q} d\boldsymbol{k}_{\mathrm{CM}} \delta\left[\nu + k_{\mathrm{CM}}^0 - \sqrt{W_{6q}^2 + (\boldsymbol{k}_{\mathrm{CM}} + \boldsymbol{q})^2}\right] \cdot$$
$$\frac{W_{6q}}{\sqrt{W_{6q}^2 + (\boldsymbol{k}_{\mathrm{CM}} + \boldsymbol{q})^2}}\, n_{\mathrm{CM}}^\beta(\boldsymbol{k}_{\mathrm{CM}}) \sum_{j=1,2} b_j^{(\beta)}\, W_j^{(\beta)}(W_{6q}, Q^2) \qquad (8)$$

where $\beta = (u^2d^4, u^3d^3, u^4d^2) = ([nn],[np],[pp])$ identifies the type of $6q$ bag; $k_{\mathrm{CM}}^0 \equiv M_A - \sqrt{M_{A-2}^2 + \boldsymbol{k}_{\mathrm{CM}}^2}$ is the initial $6q$ bag energy in the lab system; n_{CM}^β is the CM momentum distribution of the $6q$ bag. For $A > 2$ the effects from the Fermi motion of the CM of the $6q$ bag can be estimated by adopting the extended 2NC model, i.e. by using for the CM momentum distribution of a $6q$ bag the one of a correlated NN pair (i.e., $n_{\mathrm{CM}}^\beta = n_{\mathrm{CM}}^{N_1N_2}$); this implies that the introduction of a $6q$ bag is assumed to modify only the intrinsic structure of a NN cluster at short separations. Since in Eq. (8) only low-momentum components (i.e., $k_{\mathrm{CM}} < 1.5\ \mathrm{fm}^{-1}$) have to be considered in $n_{\mathrm{CM}}^{N_1N_2}$ (cf. [1]), the coefficients $b_j^{(\beta)}$ can be safely taken equal to $b_1^{(\beta)} = 2\tan^2(\theta_e/2)$ and $b_2^{(\beta)} = 1$.

Within the extended 2NC model the inclusive process $A(e,e')X$ can be investigated for any value of A, including the contributions both from $3q$ (Eq. (4)) and $6q$ (Eq. (8)) cluster configurations. The results of the calculations, performed for the processes $^2\mathrm{H}(e,e')X$ and $^4\mathrm{He}(e,e')X$ at $Q^2 \sim 15\ (\mathrm{GeV}/c)^2$, are reported in Fig. 1. It can be seen that: i) up to $x \sim 1.5$ the DIS contribution, due to virtual photon absorption by a quark belonging to a nucleon in the nucleus, overwhelms the contamination due to quasi-elastic (QE) scattering processes; ii) the kinematical regions corresponding to $x > 1.5$ appear to be appropriate for investigating the effects of $6q$ cluster configurations in light nuclei, provided the value of P_{6q}^A is sufficiently large.

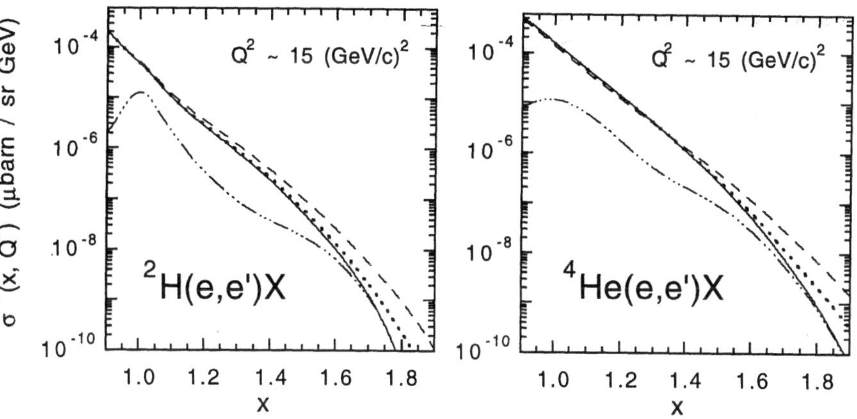

Figure 1. The cross section for the inclusive processes $^2\mathrm{H}(e,e')X$ and $^4\mathrm{He}(e,e')X$ versus x at $Q^2 \sim 15$ $(\mathrm{GeV}/c)^2$. The dash-dotted lines are the QE contribution, evaluated as in [3], whereas the solid lines include the DIS contribution calculated using the free nucleon structure function of ref. [2] in Eq. (4). The dotted (dashed) lines are the results obtained assuming $P_{6q}^{^2\mathrm{H}} = 1\%$ (5%) in Eq. (5) and $P_{6q}^{^4\mathrm{He}} = 2\%$ (10%) in Eq. (7) for the probability of a $6q$ bag admixture in ^{2}H and ^{4}He.

It should be therefore desirable to find an observable, which does not depend crucially upon the value of P_{6q}^A, not known from quark counting rules. To this end, the energy spectra of nucleons emitted in semi-inclusive DIS processes $A(\ell,\ell' N)X$ off nuclei, in which, besides the scattered lepton, a nucleon is detected in the final state, could be considered. The relevant nucleon production mechanisms in the process $A(\ell,\ell' N)X$ have been analyzed for $A > 2$ in ref. [5], where it has been shown that both at $x < 1$ and $x > 1$ the emission of nucleons in the forward hemisphere appears to be the most appropriate kinematical condition for studying multiquark cluster configurations in nuclei, provided the energy distribution of the emitted nucleons is investigated at high energy (> 0.5 GeV). However, at $x > 1$, in case of inclusive as well as semi-inclusive processes, the DIS contribution dominates over the QE scattering one only at very large values of Q^2 (> 15 $(\mathrm{GeV}/c)^2$) (cf. [3] and [5]). Therefore, it is worth noting that in case of a deuteron target the production of nucleons arising from QE scattering processes can be disentangled kinematically from the nucleon emission due to DIS events. As a matter of fact, energy and momentum conservations imply that $\nu + M_D = \sqrt{M^2 + p^2} + \sqrt{M_X^2 + (\boldsymbol{q} - \boldsymbol{p})^2}$, where ν ($\boldsymbol{q}$) is the energy (three-momentum) transfer, M_D is the deuteron mass, M_X is the invariant mass of the residual system and $\boldsymbol{p}$ is the momentum of the detected nucleon. It follows that for fixed values of ν, $\boldsymbol{q}$ and $\boldsymbol{p}$ the value of M_X is kinematically known. Therefore, it is possible to distinguish QE scattering events ($M_X = M$) from DIS ones ($M_X > M$), so that the DIS contribution to the semi-inclusive process $^2\mathrm{H}(e,e'N)X$ can be investigated at moderate values of Q^2 ($\sim 5 \div 10$ $(\mathrm{GeV}/c)^2$). A sample of the results, obtained by applying the approach of ref. [5] to the reaction $^2\mathrm{H}(e,e'p)X$, is reported in Fig. 2. It can be clearly seen that at $x > 1$ the slope of the energy distribution of forward emitted protons is almost independent of $P_{6q}^{^2\mathrm{H}}$, but still sharply sensitive to $6q$ bag effects, provided sufficiently high values of the kinetic energy of the detected proton are considered.

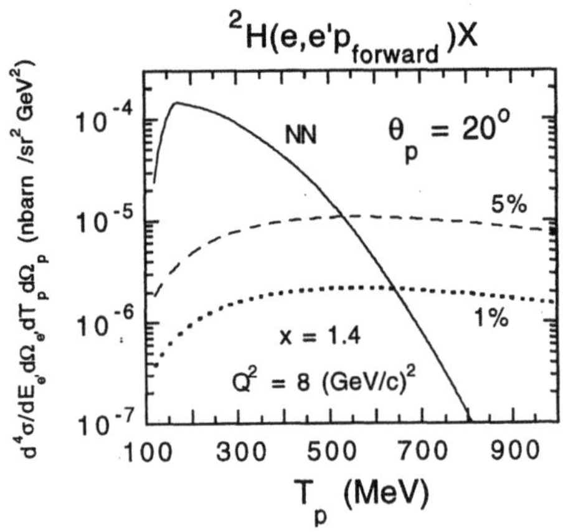

Figure 2. The DIS cross section for the semi-inclusive process $^2\mathrm{H}(e,e'p)X$ versus the kinetic energy T_p of the detected proton, emitted forward at $\theta_p = 20^o$, when $x = 1.4$ and $Q^2 = 8$ $(\mathrm{GeV}/c)^2$. The solid line is the DIS contribution due to virtual photon absorption on a NN pair, whereas the dotted (dashed) line corresponds to virtual photon absorption by a quark belonging to a $6q$ bag only, assuming $P_{6q}^{^2\mathrm{H}} = 1\%$ (5%).

In conclusion, inclusive $A(e,e')X$ and semi-inclusive $A(e,e'N)X$ processes off few-nucleon systems have been investigated at $x > 1$, showing some of the relevant features of the cross section, which are sensitive to the effects arising from nucleon-nucleon correlations as well as from the possible presence of exotic multiquark cluster configurations at short internucleon separations.

References

1. C. Ciofi degli Atti, S. Simula, L.L. Frankfurt and M.I. Strikman: Phys. Rev. **C44**, R7 (1991); C. Ciofi degli Atti and S. Simula: Preprint INFN-ISS 95/4, 1995 (submitted to Phys. Rev. **C**)

2. L.W. Whitlow, E.M. Riordan, S. Dasu, S. Rock and A. Bodek: Phys. Lett. **282B**, 475 (1992)

3. S. Simula: In: *Proceedings of the XV European Conference on Few-Body Problems in Physics*, Peniscola (Spain), June 5-9, 1995; Preprint INFN-ISS 95/7, 1995, to appear in Few-Body Systems Suppl.

4. C.E. Carlson and T.J. Havens: Phys. Rev. Lett. **51**, 261 (1983); G. Yen, J.P. Vary, A. Harindranath and H.J. Pirner: Phys. Rev. **C42**, 1665 (1990)

5. C. Ciofi degli Atti and S. Simula: Phys. Lett. **319B**, 23 (1993); C. Ciofi degli Atti and S. Simula: to appear in Few-Body Systems (1995)

Few-Body Systems Suppl. 9, 471–474 (1995)

Few-Body Systems

Spin Effects in Low-Energy Pion Scattering on ^{3}He

M.Kh. Khankhasayev, Zh.B. Kurmanov

Bogolubov Laboratory of Theoretical Physics, Joint Institute for Nuclear Research, Dubna, Moscow region, 141980 Russia

Abstract. A formalism for describing pion scattering by nuclei with spin and isospin equal to 1/2 within the unitary scattering theory is developed. This formalism is applied to the study of pion elastic scattering on ^{3}He. The results of calculations of angular distributions and spin asymmetry at low energies are presented.

1 Introduction

Pion interaction with ^{3}He is of particular interest from the point of view of studying the spin effects in pion-nucleus dynamics. There is a number of papers in which elastic $\pi\,^3$He scattering at resonance energies has been analyzed (see reviews [1,2]). In contrast to the resonance region, where pions are scattered mostly by the nucleus surface, at low energies a rather weak πN interaction permits pion to penetrate relatively deep into a nucleus. Thus, low-energy pions provide useful means to probe the nuclear interior.

Low-energy $\pi\,^3$He scattering was examined in refs. [3–6] by using a lowest-order optical potential (OP). Using the model of ref. [3], it was found that despite the overall satisfactory agreement between the calculations and experimental data, it was not possible to obtain a quantitative description of the data at both small and large scattering angles. This shows the importance of careful study of the second-order effects in the $\pi\,^3$He dynamics. In ref. [4], the calculations very similar to those of [3], but with a first-order OP constructed from a microscopic ^{3}He wave function, were presented. The Fermi-motion effect was taken into account in an exact way by calculating the complete Fermi integral. It was found that at low energies the factorization approximation usually used in calculations appeared to be unreliable.

In ref. [5], in studies of elastic $\pi\,^3$He scattering using the realistic three-body Faddeev wave functions, it was found that the first-order OP gives a good description of the low-energy data both for the elastic scattering and charge-exchange ^{3}He$(\pi^-,\pi^0)^3$H reaction.

In our previous paper [6], we analyzed the elastic $\pi\,^3$He scattering within the framework of the unitary approach. Here, we present results of our calculations of spin asymmetry in low-energy pion scattering on polarized ^{3}He.

The rather accurate experimental data for low-energy elastic $\pi^{\pm 3}$He scattering are presented in ref. [7].

2 Unitary Approach

At low energies, where pion-nucleus scattering is strongly affected by pion absorption, it is very important to use the theory consistent with the unitarity. This allows one to perform a correct separation of the pure potential effects from the nonpotential (i.e., genuine absorption) ones.

One of such unitary approaches is the Kirzhnits method based on an nonconventional formulation of quantum scattering theory, in which the evolution of a system with respect to the coupling constant is considered [8].

This method being applied to the problem of pion-nucleus scattering [9,10] allows one to construct a multiple scattering series conserving the unitarity of the scattering matrix at each step of the successive approximations and an optical-like model for the direct calculation of pion-nucleus phase shifts. At low energies, the iterative series is rapidly convergent, and the microscopic calculations of the second-order corrections show that only two first terms of the series need to be taken into account.

The unitary approach has successfully been applied to the description of low-energy pion elastic scattering by light nuclei with zero spin and isospin: ^{4}He, ^{12}C, and ^{16}O [10].

In ref. [6], we generalized this approach to the case of pion scattering from nuclei with spin and isospin 1/2 and applied the developed scheme to the consideration of $\pi\,^3$He scattering at low energies.

Let us rely here on the clear outline of the unitary approach given in ref. [10] and the generalized formalism and some results of ref. [6]. We present below our calculations and analysis of the results on differential cross sections and spin asymmetries in low-energy elastic $\pi\,^3$He scattering.

3 Results and Discussion

In Fig. 1a we present the results of our calculations of the differential cross sections for elastic π^{+3}He scattering. The difference between the solid and the dashed curves shows the role of the nucleus excitation channel. It can be seen that the role of this channel increases with the pion energy. The nuclear mean excitation-energy parameter Δ is the only free one in our approach, and in ref. [6] we fixed $\Delta = 20$ MeV. It has been found that the sensitivity to Δ is more pronounced at energies around 50 MeV and becomes weaker when the pion energy decreases. It reflects a dominance of the pion absorption correction at energies below 30 MeV in the formation of the inelasticity parameters.

The results of our calculations presented in Fig. 1b show an effect of pion

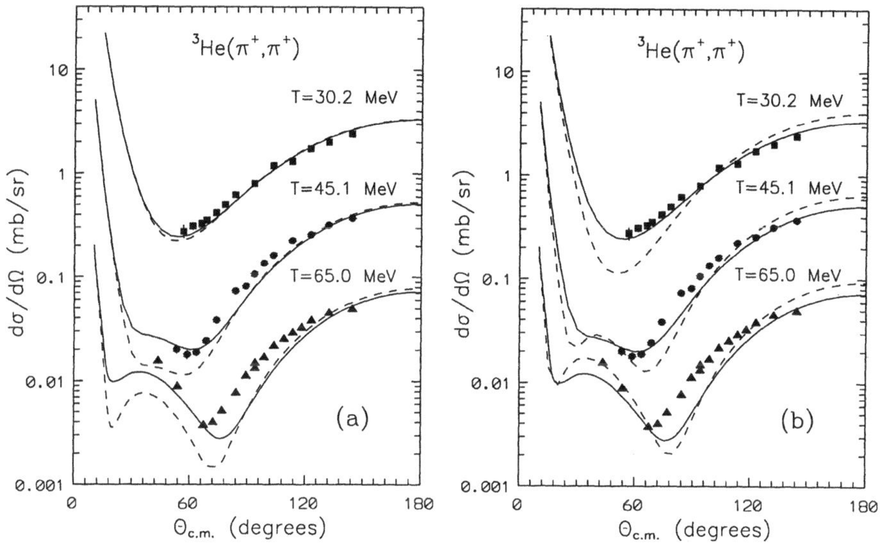

Figure 1. Differential cross sections for elastic π^{+3}He scattering at pion lab. energies $T_\pi = 30.2$, 45.1, and 65 MeV. Solid curves are for full calculations with $\Delta = 20$ MeV, short-dashed curves: (a) with excitation channel turned off, (b) with pion absorption channel turned off. The experimental data are from ref. [7]. For $T_\pi = 45.1$ MeV the data and results must be multiplied by a factor 10, and for $T_\pi = 65.0$ MeV by a factor 100.

absorption on the elastic scattering at low energies. It follows that the role of pion absorption is important for $\pi\,^3$He scattering, though it seems smaller than for $\pi\,^4$He one [10]. It was also found that the cross sections arise as a result of a strong interference between the pure potential and the absorption channels.

It can be seen that the unitary approach, being applied to elastic $\pi\,^3$He scattering at low energies, gives rather good agreement with the existing experimental data up to 50 MeV. Nevertheless, it seems interesting to study the second-order effects in a more detailed way. It would be very important especially for studying the effects of spin-asymmetry.

The pion-nucleus cross sections are sensitive mainly to the spin-independent part of the amplitudes, and the analysis of pion scattering on light nuclei shows that the differential cross sections can be reproduced very well but not the analyzing power [2]. It means that the study of spin effects on polarized nuclei can provide a much more sensitive test for the theoretical models.

In Fig. 2, we present the results of our calculations of spin asymmetry in elastic $\pi^{\pm 3}$He scattering at $T_\pi = 50$ MeV. Our calculations show that there are large spin effects at the scattering angles where the cross section has a minimum. Qualitatively the same results were obtained in ref. [4]. It will be very interesting to compare these results with experimental data. Unfortunately, there are no data on spin observables available now at low energies.

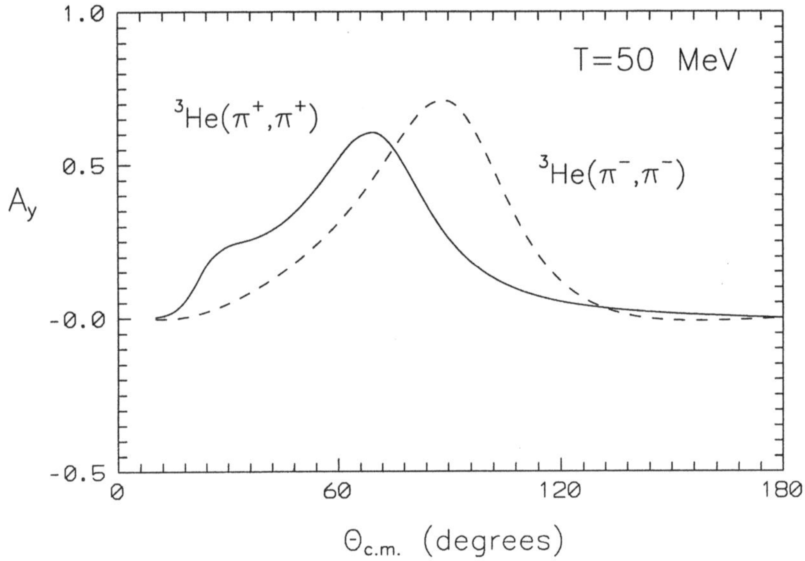

Figure 2. Spin asymmetry in elastic $\pi^{\pm 3}$He scattering at T_π = 50 MeV. Solid curve stands for π^+, dashed curve for π^- scattering.

References

1. R. Mach, M.G. Sapozhnikov, I.V. Falomkin: Sov. J. Part. Nucl. **17**, 1231 (1986)
2. E. Boschitz: In: *Mesons and Nuclei at Intermediate Energies* (Proc. Int. Conf., Dubna, May 3-7, 1994), p.282. Singapore: World Scientific 1994
3. R.H. Landau: Comp. Phys. Comm. **28**, 109 (1982)
4. F.M.M. van Geffen et al.: Nucl. Phys. **A468**, 683 (1987)
5. S.S. Kamalov, L. Tiator, C. Bennhold: Few-Body Systems **10**, 143 (1991)
6. M.Kh. Khankhasayev, Zh.B. Kurmanov: Phys. At. Nucl. **56**, 975 (1993)
7. G. Fournier et al.: Nucl. Phys. **A426**, 542 (1984)
8. D.A. Kirzhnits: Sov. J. Exp. Theor. Phys. **49**, 1544 (1965)
9. V.B. Belyaev, M.Kh. Khankhasayev: Phys. Lett. **B137**, 299 (1984)
10. M.Kh. Khankhasayev: Nucl. Phys. **A505**, 717 (1989)

Few-Body Systems Suppl. 9, 475–489 (1995)

Few-Body Systems

The Weak Decay of Hypernuclei

C. Bennhold[1], A. Parreño[2], A. Ramos[2]

[1] Center of Nuclear Studies, Department of Physics, The George Washington University, Washington DC 20052, USA

[2] Dept. ECM, Facultat de Física, Diagonal 647, 08028 Barcelona, Spain

Abstract. The nonmesonic weak decay of light Λ hypernuclei is investigated in a shell model framework. The strangeness-changing weak $\Lambda N \to NN$ transition potential is constructed by including the exchange of the pseudoscalar mesons π, K, η as well as the vector mesons ρ, ω, and K^*, whose weak coupling constants are obtained from soft meson theorems and $SU(6)_{\rm w}$. The transition matrix element includes a realistic ΛN short-range correlation function that is based on the Nijmegen YN-potential. The total decay rate and the proton asymmetry, which represents a measure of the parity-violating to parity-conserving amplitudes, are in agreement with present experimental errors while the neutron- to proton-induced ratio is underestimated by more than a factor of two. In general, the observables are found to be dominated by the pion-exchange mechanism since the contributions of heavier mesons are suppressed by form factors and short-range correlations.

1 Introduction

In single Λ hypernuclei, a Λ hyperon can occupy any orbital in the hypernucleus since it is free from the Pauli exclusion principle due to its additional quantum number Strangeness. When stable against particle emission, Λ-hypernuclei decay mainly through weak decay processes which involve the emission of pions or nucleons but are nonleptonic in nature. In the absence of strong interactions the lowest-order hamiltonian for weak quark-quark interactions is assumed to be given by the combination of V-A theory with the Cabibbo hypothesis, however, the modification of the weak force due to strong interaction corrections is not yet well understood. A free Λ-hyperon has a lifetime of 260 picoseconds and decays almost totally into a pion and a nucleon ($\Lambda \to p\pi^-$ ($\sim 64\%$), $\Lambda \to n\pi^0$ ($\sim 36\%$)), with a final momentum of about 100 MeV/c. When the Λ is embedded in the nuclear medium, this mesonic process becomes Pauli blocked and its rate is suppressed by several orders of magnitude for heavy hypernuclei such as $^{208}_{\Lambda}$Pb. Experimentally, however, one finds the lifetimes of hypernuclei to be roughly independent of A, though the data base is very poor, especially for systems with $A > 12$. Therefore, hypernuclei larger than ${}^{5}_{\Lambda}$He decay mainly through nonmesonic channels, such as the $\Lambda N \to NN$ reaction, where the Λ mass excess of 176 MeV is converted into kinetic energy of a final state of nucleons emerging with a momentum of about 400 MeV/c. As a result, this process is not Pauli blocked and its insensitivity to nuclear structure details make it a suitable channel to investigate the weak decay mechanism. The large momentum transfer indicates that the process would be sensitive to short range correlations in the initial state and may receive important contributions from heavier mesons. This nonmesonic process has received increased theoretical [1] and experimental attention in the last few years [2-5]. Recent experiments at Brookhaven [2] and KEK [3] have provided not only new measurements of the total and partial decay rates of several s- and p-shell hypernuclei, such as ${}^{12}_{\Lambda}$C, ${}^{11}_{\Lambda}$B, and ${}^{5}_{\Lambda}$He, but have also for the first time extracted the angular asymmetry of the protons from the decay of polarized hypernuclei.

The nonmesonic process resembles the weak $\Delta S=0$ nucleon-nucleon interaction that has been explored experimentally in parity-violating NN scattering measurements by measuring the asymmetry of longitudinally polarized protons. However, the $\Lambda N \to NN$ two-body decay mode contains more information since it can explore both the parity-conserving (PC) and the parity-violating (PV) sector of the $\Delta S=1$ weak baryon-baryon interaction while in the weak NN system the strong force masks the signal of the weak PC interaction.

Another possible nonmesonic decay channel, first investigated in ref. [6] and reanalyzed in refs. [7, 8], is the two-nucleon induced process $\Lambda NN \to NNN$, where the pion emitted at the weak vertex is absorbed by a pair of nucleons which are correlated through the strong force.

2 Formalism for the $\Lambda N \to NN$ Process

2.1 Decay Rates

The nonmesonic $\Lambda N \to NN$ decay rate is given by [1]

$$\Gamma_{\rm nm} = \int \frac{d^3P}{(2\pi)^3} \int \frac{d^3k}{(2\pi)^3} \overline{\sum} (2\pi)\delta(M_H - E_{\rm R} - E_1 - E_2) \mid \mathcal{M} \mid^2 \qquad (1)$$

where $\mathcal{M} = \langle \psi_{\rm R}; \boldsymbol{P}\boldsymbol{k}\; S\, M_S\; T\, M_T | \hat{O}_{\Lambda N \to NN} | _{\Lambda}A \rangle$ is the hypernuclear transition amplitude and M_H, $E_{\rm R}$, E_1 and E_2 the mass of the hypernucleus, energy of the residual $(A-2)$-particle system, and total asymptotic energies of the emitted nucleons, respectively. A transformation to the center of mass ($\boldsymbol{P}$) and relative momentum ($\boldsymbol{k}$) of the two outgoing nucleons is implied in Eq. (1). The sum $\overline{\sum}$ indicates an average over the initial hypernucleus spin projections, M_J, and sum over all quantum numbers of the residual $(A-2)$-particle system, as well as the spin and isospin projections of the exiting nucleons. Neglecting final state distortions for the moment the antisymmetric state of the two final nucleons moving with center-of-mass momentum $\boldsymbol{P}$ and relative momentum $\boldsymbol{k}$

can be written as

$$\langle \boldsymbol{R}\boldsymbol{r}|\boldsymbol{P}\boldsymbol{k}\ S\,M_S\ T\,M_T\rangle = \frac{1}{\sqrt{2}}\mathrm{e}^{\mathrm{i}\mathbf{PR}}\left(\mathrm{e}^{\mathrm{i}\mathbf{kr}} - (-1)^{S+T}\mathrm{e}^{-\mathrm{i}\mathbf{kr}}\right)\chi^S_{M_S}\chi^T_{M_T} =$$

$$\frac{1}{\sqrt{2}}\mathrm{e}^{\mathrm{i}\mathbf{PR}}\sum_{L\,M_L} 4\pi\mathrm{i}^L\left(1-(-1)^{L+S+T}\right) j_L(kr)Y^*_{L\,M_L}(\hat{\boldsymbol{k}})Y_{L\,M_L}(\hat{\boldsymbol{r}})\chi^S_{M_S}\chi^T_{M_T}, \tag{2}$$

performing a partial wave expansion for the relative plane wave motion. We follow ref. [1] by assuming a weak coupling scheme where the Λ in an orbit $\alpha_\Lambda = \{n_\Lambda, l_\Lambda, j_\Lambda, m_\Lambda\}$ couples to only the ground state wave function of the nuclear $(A-1)$ core. Employing the technique of coefficients of fractional parentage, the core wave function is further decomposed into a set of states where the nucleon in an orbit $\alpha_N = \{n_N, l_N, j_N, m_N\}$ is coupled to a residual $(A-2)$-particle state. In the present work, the single particle Λ and N orbits are taken to be solutions of a harmonic oscillator potential with parameters, $b_\Lambda = 1.87$ fm and $b_N = 1.64$ fm, that have been adjusted to experimental separation energies and the ^{12}C charge form factor. Assuming an average size parameter $b = (b_\Lambda + b_N)/2$, using Moshinski brackets and working in the LS representation, the product of the two harmonic oscillator single particle states can be transformed to a combination of relative and center-of-mass wave functions. The many body transition amplitude of Eq. (1) can then be written in terms of two body amplitudes of the type

$$\begin{aligned} t_{\Lambda N\to NN} &= \int d^3R\int d^3r\mathrm{e}^{-\mathrm{i}\mathbf{PR}}\psi^*_{\mathbf{k}}(\boldsymbol{r})\chi^{\dagger\,S}_{M_S}V(\boldsymbol{r})\Phi^{\mathrm{rel}}_{n_r\,l_r}(\boldsymbol{r};\sqrt{2}b) \\ &\qquad\times\Phi^{\mathrm{CM}}_{n_R\,l_R}(\boldsymbol{R};b/\sqrt{2})\chi^{S_0}_{M_{S_0}} \\ &= (-\mathrm{i})^{l_R}\Phi^{\mathrm{CM}}_{n_R\,l_R}(\boldsymbol{K};b/\sqrt{2})\int d^3r\psi^*_{\mathbf{k}}(\boldsymbol{r})\chi^{\dagger\,S}_{M_S}V(\boldsymbol{r})\Phi^{\mathrm{rel}}_{n_r\,l_r}(\boldsymbol{r};\sqrt{2}b)\chi^{S_0}_{M_{S_0}} \\ &= (-\mathrm{i})^{l_R}\Phi^{\mathrm{CM}}_{n_R\,l_R}(\boldsymbol{K};b/\sqrt{2})\,t_{\mathrm{rel}}\ , \end{aligned} \tag{3}$$

where isospin indices have been omitted for simplicity. The functions $\Phi_{n\,l}(\boldsymbol{a};b)$ are solutions of the 3-dimensional harmonic oscillator and $V(\boldsymbol{r})$ is the potential obtained as a Fourier transform of the nonrelativistic reduction of the corresponding Feynman amplitudes discussed later.

Introducing the operator $S_{12}(\hat{\boldsymbol{r}}) = 3\boldsymbol{\sigma}_1\hat{\boldsymbol{r}}\boldsymbol{\sigma}_2\hat{\boldsymbol{r}} - \boldsymbol{\sigma}_1\boldsymbol{\sigma}_2$, the potential can be divided into a central, a tensor and a parity violating contribution,

$$V(\boldsymbol{r}) = V_{\mathrm{C}}(r)\boldsymbol{\sigma}_1\boldsymbol{\sigma}_2 + V_{\mathrm{T}}(r)S_{12}(\hat{\boldsymbol{r}}) + V_{\mathrm{PV}}(r)\boldsymbol{\sigma}_2\hat{\boldsymbol{r}}\ . \tag{4}$$

Assuming the lambda and the nucleon to be in their lowest $l = 0$ single particle states, we obtain

$$t^{\mathrm{C}}_{\mathrm{rel}} = \delta_{S\,S_0}\delta_{M_S\,M_{S_0}}\sqrt{4\pi}\,2\left[S(S+1)-\frac{3}{2}\right]\int drr^2 j_0(kr)V_{\mathrm{C}}(r)\Phi_{100}(r;\sqrt{2}b)$$

$$t_{\rm rel}^{\rm T} = -\delta_{M_S M_{S_0}} \sqrt{4\pi}\sqrt{6}\sqrt{\frac{2S+1}{2S_0+1}} \frac{\langle S \| \left[\sigma_1 \otimes \sigma_2\right]^{J=2} \| S_0 \rangle}{\sqrt{2S+1}} \langle 20\ SM_S | S_0 M_{S_0} \rangle$$
$$\times \int dr r^2 j_2(kr) V_{\rm T}(r) \Phi_{100}(r; \sqrt{2}b)$$
$$t_{\rm rel}^{\rm PV} = -\delta_{M_S M_{S_0}} \sqrt{4\pi}\sqrt{\frac{2S+1}{2S_0+1}} \frac{\langle S \| \sigma_2 \| S_0 \rangle}{\sqrt{2S+1}} \langle 10\ SM_S | S_0 M_{S_0} \rangle$$
$$\times \int dr r^2 j_1(kr) V_{\rm PV}(r) \Phi_{100}(r; \sqrt{2}b) \ . \quad (5)$$

For nucleons in a p-shell orbit, the matrix elements are slightly more complicated [9].

A monopole form factor $F(\boldsymbol{q}^2) = (\Lambda_c^2 - m^2)/(\Lambda_c^2 + \boldsymbol{q}^2)$, with m being the meson mass, is used at each vertex, where the value of the cut-off Λ_c depends on the meson. We take the values of the Jülich YN interaction [10] ($\Lambda_\pi = 1.3$ GeV, $\Lambda_\rho = 1.4$ GeV), since the Nijmegen model distinguishes form factors only in terms of the transition channel. To account for the ΛN correlations, which are absent in the independent particle model, one has to include a correlation function, for which we take the parametrization

$$f(r) = \left(1 - e^{-r^2/a^2}\right)^n + br^2 e^{-r^2/c^2} \ , \quad (6)$$

where $r = |\ \boldsymbol{x} - \boldsymbol{y}\ |$. The values $a = 0.5$, $b = 0.25$, $c = 1.28$ and $n = 2$ provide an average spin-independent correlation function that represents a good approximation to the results obtained in microscopic G-matrix calculations in finite hypernuclei [11] using the Nijmegen YN interactions [12]. Final state interactions are taken into account via an average NN correlation function in the final state, for which we take $f_{\rm FSI}(r) = 1 - j_0(q_c r)$, with $q_c = 3.93\ {\rm fm}^{-1}$.

2.2 *Asymmetries and Hypernuclear Polarization*

At the kinematic conditions of the (π^+, K^+) reaction carried out at KEK, the hypernucleus is created with a substantial amount of polarization in the ground state. Due to the interference between the parity conserving and parity violating amplitudes, the distribution of the emitted protons in the weak decay displays an angular asymmetry with respect to the polarization axis given by

$$\sigma(\chi) = \sigma_0(1 + P_y A_p(\chi)) = \sigma_0(1 + P_y A_p \cos\chi) \ , \quad (7)$$

where P_y is the hypernuclear polarization, created by the production process, such as $(\pi^+ K^+)$ at KEK and BNL or (γ, K^+) at CEBAF [13]. A_p denotes the asymmetry parameter which is characteristic of the weak decay process. The asymmetry is defined as

$$\mathcal{A}_p = P_y A_p = p_\Lambda a_\Lambda \ , \quad (8)$$

which can be written not only in terms of hypernuclear properties,

$$A_y(\chi) = \frac{3}{J+1} \frac{Tr(\mathcal{M}\hat{S}_y\mathcal{M}^\dagger)}{Tr(\mathcal{M}\mathcal{M}^\dagger)} = \frac{3}{J+1} \frac{\sum_{M_i} \sigma(M_i) M_i}{\sum_{M_i} \sigma(M_i)} \cos\chi \tag{9}$$

but also in terms of the Λ polarization, p_Λ, and an intrinsic asymmetry parameter, a_Λ, associated with the elementary process $\Lambda N \to NN$ in the nuclear medium. In a weak coupling scheme, simple angular momentum algebra relations relate the hypernuclear polarization to the Λ polarization.

3 The Meson Exchange Potential

3.1 π-Exchange

The transition $\Lambda N \to NN$ involves a weak and a strong vertex. While there exist several strong meson-exchange potentials which, through fits to NN scattering data, provide information on the different strong NN-meson vertices, only the pion vertex is known experimentally in the weak sector

$$\mathcal{H}_{\Lambda N\pi} = \mathrm{i} G_\mathrm{F} m_\pi^2 \overline{\psi}_N (A + B\gamma_5) \boldsymbol{\tau}\boldsymbol{\phi}_\pi \psi_\Lambda \ , \tag{10}$$

where $G_\mathrm{F} m_\pi^2 = 2.21 \times 10^{-7}$ is the weak coupling constant. The empirical constants $A = 1.05$ and $B = -7.15$, adjusted to the observables of the free Λ decay, determine the strength of the parity violating and parity conserving amplitudes, respectively. The nucleon and pion fields are given by ψ_N and ϕ_π, respectively, while the lambda field, ψ_Λ, is taken as the spurion isospin state $|t\, m_t\rangle = |1/2\; -1/2\rangle$ to enforce the empirical $\Delta I = 1/2$ rule observed in the decay of a free Λ.

In order to avoid the complications present in introducing a short-range correlation function in a relativistic matrix element [14], we use a nonrelativistic transition potential, which, in the case of pion exchange, reads

$$V_\pi(\boldsymbol{q}) = -G_\mathrm{F} m_\pi^2 \frac{g_{\pi NN}}{2M} \left(A + \frac{B}{2\overline{M}} \boldsymbol{\sigma}_1 \boldsymbol{q} \right) \frac{\boldsymbol{\sigma}_2 \boldsymbol{q}}{\boldsymbol{q}^2 + m_\pi^2} (\boldsymbol{\tau}_1 \boldsymbol{\tau}_2) \tag{11}$$

where we have introduced an average mass $\overline{M} = 1/2(M_N + M_\Lambda)$. Performing a Fourier transformation one obtains the corresponding transition potential in coordinate space

$$\begin{aligned} V^\mathrm{C}(r) &= -\frac{G_\mathrm{F}\mu^2 B}{2\overline{M}} \frac{g_{\pi NN}}{2M} \frac{1}{3} \left(\mu^2 \frac{\mathrm{e}^{-\mu r}}{4\pi r} - \delta(r) \right) \\ V^\mathrm{T}(r) &= -\frac{G_\mathrm{F}\mu^2 B}{2\overline{M}} \frac{g_{\pi NN}}{2M} \frac{1}{3} \mu^2 \frac{\mathrm{e}^{-\mu r}}{4\pi r} \left(1 + \frac{3}{\mu r} + \frac{3}{(\mu r)^2} \right) \\ V^\mathrm{PV}(r) &= \mathrm{i} G_\mathrm{F}\mu^2 A \frac{g_{\pi NN}}{2M} \mu \frac{\mathrm{e}^{-\mu r}}{4\pi r} \left(1 + \frac{1}{\mu r} \right) . \end{aligned} \tag{12}$$

3.2 The Weak Coupling Constants

Previous calculations have mostly employed a one-pion-exchange mechanism to describe the reaction $\Lambda N \to NN$ which has then commonly been evaluated in nuclear matter. However, due to the difference in the rest mass between the Λ and the nucleon, the two nucleons emerge with momenta $\sim$ 400 MeV, leading to a large momentum transfer process that is sensitive to short-range effects. This suggests that heavier mesons may play an important role. We explore here the effects of adding the exchanges of the η, K, ρ, ω and K^* mesons, thus constructing a complete weak one-boson exchange potential. Relying on earlier successes of such models for the parity-violating weak NN-interaction we limit ourselves to the exchange of pseudoscalar (π, η,K) and vector (ρ, ω, K^*) mesons. The strong couplings are taken from the Nijmegen [12] or Jülich interactions while we employ the methods of ref. [15] to obtain the weak vertices. The parity-conserving weak vertices are computed by using a pole model that requires the weak meson $\to$ meson and baryon $\to$ baryon transition amplitudes as input. Writing the ΔS=1 weak nonleptonic Hamiltonian in SU(3) tensor notation

$$H_{\mathrm{W}} = \frac{G_{\mathrm{V}}}{2\sqrt{2}} \cos\Theta_{\mathrm{c}} \sin\Theta_{\mathrm{c}} \{J^2_{\mu 1}, J^{\mu 1}_3\} + h.c. \tag{13}$$

where $J^i_{\mu j} = (V_\mu + A_\mu)^i_j$ is the weak hadronic current with SU(3) indices i, j, and Θ_{c} is the Cabbibo angle. Following ref. [15] the weak vector and axial currents are expressed in terms of $\mathrm{SU}(6)_{\mathrm{W}}$ currents. The Hamiltonian is the product of two currents, each belonging to the 35 representation. From the expansion $35 \otimes 35 = 1_{\mathrm{s}} \oplus 35_{\mathrm{s}} \oplus 189_{\mathrm{s}} \oplus 405_{\mathrm{s}} \oplus 35_{\mathrm{a}} \oplus 280_{\mathrm{a}} \oplus \overline{280_{\mathrm{a}}}$ one can associate the symmetric pieces with the parity-conserving Hamiltonian and the antisymmetric pieces with the parity-violating Hamiltonian. Therefore, using $SU(6)_{\mathrm{W}}$ symmetry and enforcing the empirical ΔI=1/2 rule yields the necessary meson $\to$ meson amplitudes in terms of the $K \to \pi$ amplitude, $< M'|H_{\mathrm{W}}|M > \sim < \pi|H_{\mathrm{W}}|K >$. Employing PCAC this amplitude can be related to the physical $K \to \pi\pi$ decay rate via

$$\lim_{q\to 0} < \pi\pi|H_{\mathrm{pv}}|K > = -\frac{\mathrm{i}}{F_\pi} < \pi|[F^5_\pi, H_{\mathrm{pv}}]|K > = -\frac{\mathrm{i}}{2F_\pi} < \pi|H_{\mathrm{pc}}|K > \tag{14}$$

thus constraining the meson $\to$ meson transition amplitudes. Similarly, the baryon $\to$ baryon amplitudes can be related to the physical free lambda and sigma mesonic decay amplitudes. Applying this model to the pionic decays of the different hyperons reproduces the measured rates to within 25%.

3.3 ρ-Exchange

A number of theoretical studies in recent years have investigated the contribution of the ρ-meson to the $\Lambda N \to NN$ process [16-18]. Due to the different models employed for the weak $\Lambda N\ \rho$ vertex these calculations have yielded widely varying results. However, all works until now that have studied the ρ-meson exchange diagram have only included the tensor part of the parity-conserving ρ-exchange term. This was in part motivated by the observation that the central potential of the π-exchange term gives only a negligible contribution. Below we demonstrate by explicit calculation that the central part of the ρ-exchange is not only nonnegligible but is in fact larger than its tensor interaction. This can be traced to the fact that the ρ-exchange diagram has a much shorter range than the π-exchange potential. Our findings are independent of the particular model that is used for the weak $\Lambda N\rho$-vertex.

For the weak $\Lambda N\rho$ and strong $NN\rho$ vertices we take [16]

$$\mathcal{H}^{\mathrm{W}}_{\Lambda N\rho} = G_{\mathrm{F}}\mu^2 \overline{\Psi}_N \left(\alpha\gamma^\mu - \beta \mathrm{i}\frac{\sigma^{\mu\nu}q_\nu}{2\overline{M}} + \varepsilon\gamma^\mu\gamma_5\right) \boldsymbol{\tau}\boldsymbol{\rho}_\mu \Psi_\Lambda \tag{15}$$

$$\mathcal{H}^{\mathrm{S}}_{NN\rho} = \overline{\Psi}_N \left(F_1\gamma^\mu + \mathrm{i}\frac{F_2}{2M}\sigma^{\mu\nu}q_\nu\right) \boldsymbol{\tau}\boldsymbol{\rho}_\mu \Psi_N\ , \tag{16}$$

respectively, where the four momentum transfer q is directed towards the strong vertex. The strong and weak coupling constants of the ρ used in this paper are given in Table 3.1.

Table 3.1. Nijmegen (Jülich) strong and weak coupling constants for the ρ-exchange mechanism

Strong	F_1	3.16 (3.25)
	F_2	13.34 (19.82)
Weak	α	-3.80 (-3.91)
	β	-6.77 (-10.56)
	ϵ	1.09

The nonrelativistic potential for the ρ-meson contribution is given by

$$V_\rho(\boldsymbol{q}) = G_{\mathrm{F}} m_\pi^2 \left(F_1\alpha - \frac{(\alpha+\beta)(F_1+F_2)}{4M\overline{M}}(\boldsymbol{\sigma}_1 \times \boldsymbol{q})(\boldsymbol{\sigma}_2 \times \boldsymbol{q}) + \mathrm{i}\frac{\varepsilon(F_1+F_2)}{2M}(\boldsymbol{\sigma}_1 \times \boldsymbol{\sigma}_2)\boldsymbol{q}\right) \frac{1}{\boldsymbol{q}^2 + {m_\rho}^2}(\boldsymbol{\tau}_1\boldsymbol{\tau}_2)\ . \tag{17}$$

Using the relation $(\boldsymbol{\sigma}_1\times\boldsymbol{q})(\boldsymbol{\sigma}_2\times\boldsymbol{q}) = (\boldsymbol{\sigma}_1\boldsymbol{\sigma}_2)\boldsymbol{q}^2 - (\boldsymbol{\sigma}_1\boldsymbol{q})(\boldsymbol{\sigma}_2\boldsymbol{q})$ and performing a Fourier transform of $V_\rho(\boldsymbol{q})$ one obtains the corresponding transition potential in coordinate space, which, as for π exchange, can be divided into central, tensor and parity-violating pieces

$$V_\rho(\boldsymbol{r}) = \left({V_\rho}^{\mathrm{c}}(\boldsymbol{r}) + {V_\rho}^{\mathrm{t}}(\boldsymbol{r}) + {V_\rho}^{\mathrm{PV}}(\boldsymbol{r})\right)(\boldsymbol{\tau}_1\boldsymbol{\tau}_2)\ . \tag{18}$$

In the case of the ρ-meson, the central piece is further decomposed into a spin independent (SI) and a spin dependent (SD) part

$${V_\rho}^{\mathrm{c}}(\boldsymbol{r}) = {V_\rho}^{\mathrm{SI}}(\boldsymbol{r}) + {V_\rho}^{\mathrm{SD}}(\boldsymbol{r}) \tag{19}$$

with

$$V_\rho{}^{\mathrm{SI}}(\boldsymbol{r}) = G_{\mathrm{F}}\mu^2 F_1\alpha \frac{e^{-m_\rho r}}{4\pi r}$$

$$V_\rho{}^{\mathrm{SD}}(\boldsymbol{r}) = G_{\mathrm{F}}\mu^2 \frac{(\alpha+\beta)(F_1+F_2)}{4M\overline{M}} \frac{2}{3} \left(m_\rho{}^2 \frac{e^{-m_\rho r}}{4\pi r} - \delta(\boldsymbol{r}) \right) (\boldsymbol{\sigma}_1\boldsymbol{\sigma}_2) \quad (20)$$

The tensor and parity-violating pieces of the transition potential are given by

$$V_\rho{}^{\mathrm{t}}(\boldsymbol{r}) = -G_{\mathrm{F}}\mu^2 \frac{(\alpha+\beta)(F_1+F_2)}{4M\overline{M}} \frac{1}{3} T(m_\rho r) \left(m_\rho{}^2 \frac{e^{-m_\rho r}}{4\pi r} \right) S_{12}(\hat{\boldsymbol{r}})$$

$$V_\rho{}^{\mathrm{PV}}(\boldsymbol{r}) = -G_{\mathrm{F}}\mu^2 \frac{\epsilon(F_1+F_2)}{2M} m_\rho V(m_\rho r) \frac{e^{-m_\rho r}}{4\pi r} \hat{\boldsymbol{r}}(\boldsymbol{\sigma}_1 \times \boldsymbol{\sigma}_2) , \quad (21)$$

respectively, where the following definitions

$$T(m_\rho r) = 1 + \frac{3}{m_\rho r} + \frac{3}{(m_\rho r)^2}$$

$$V(m_\rho r) = 1 + \frac{1}{m_\rho r} \quad (22)$$

have been used. We note that, since the PC constants α, β for the ρ-meson have the same sign as the constant B for the pion, the ρ tensor potential interferes destructively with that of the pion, as in the case of the strong NN interaction.

4 Results

4.1 π-Exchange

The results of our calculations are shown in Tables 4.1-4.5 where the non-mesonic decay rate of ${}^{12}_{\Lambda}\mathrm{C}$ is given in units of the free lambda decay rate. Table 4.1 presents our results for π-exchange alone, demonstrating the effects of short-range correlations (SRC), form factors (FF), and final-state interactions (FSI) separately for the central, tensor and parity-violating potentials. The free central term is reduced dramatically by SRC, however, most of the free central potential is in fact due to the δ-function which is completely eliminated by SRC. Ignoring the δ-function, the central part is reduced by a factor of two,

Table 4.1. π-exchange contribution to the $\Lambda N \to NN$ decay rate of ${}^{12}_{\Lambda}\mathrm{C}$. The values in brackets are obtained when the δ function in the central channel is ignored.

	Free	SRC	SRC,FF	SRC,FF,FSI
C	0.288 (0.006)	0.003	0.013	0.004
T	0.818	0.739	0.598	0.645
PV	0.470	0.379	0.327	0.354
Γ	1.576 (1.294)	1.121	0.938	1.003
$\Gamma_{\text{nuclear matter}}$				1.45 [7]

from 0.006 to 0.003. Including SRC, FF, and FSI reduces the total central potential by a factor of 70, from 0.288 to 0.004. In contrast, the tensor interaction of the π-exchange diagram is only reduced by 10% by SRC and by 20% once FF and FSI are applied as well. Therefore, the contribution of the central term amounts to less than 0.5% of the total π-exchange rate. This behavior has been found and discussed by other authors as well [15-17]. We note, however, that in general nuclear matter results overpredict our results obtained in a shell model framework. Even if a Local Density Approximation is used [7] the result is still larger by about 40%. Part of this difference can be attributed to slightly different Λ wave functions (which appear explicitly even in the LDA); if the same wave function is used in ref. [7], the total rate decreases to about 1.3. Furthermore, ref. [7] uses a Landau-Migdal parameter of g'=0.52 which is taken from the description of NN correlations. If a smaller parameter of g'=0.4 is employed to account for the slightly softer ΛN correlations the nuclear matter result reduces to 1.12. The remaining difference may not be entirely surprising since the LDA is expected to work well for heavier nuclei and begins to break down for s- and p-shell nuclei. Furthermore, we point out that our PV potential yields about 30% of the π-exchange rate, at variance with older nuclear matter results that reported negligible PV rates [16].

4.2 ρ-Exchange

Our results for the ρ-meson exchange contribution are shown in Table 4.2. As noted before, the central potential can now be divided into a spin-dependent (SD) and spin-independent (SI) piece, which are shown separately. In contrast to the pion case, the factor m_ρ^2 in front of the Yukawa function in the SD central part of Eq. (20) enhances this contribution which then becomes comparable in magnitude to the piece containing the delta function. The two terms interfere destructively, as can be seen from the large value of 3.155 (without SRC,FF, or FSI), obtained when the δ-function is removed. Even with this interference, the SD central part is larger than the tensor contribution [19].

Table 4.2. ρ-exchange contribution to the $\Lambda N \to NN$ decay rate of ${}^{12}_{\Lambda}$C using the Nijmegen constants. The values in brackets are obtained when the δ function of the central channel is ignored.

	Free	SRC	SRC,FF	SRC,FF,FSI
C (SI)	0.774	0.088	0.051	0.033
C (SD)	0.234 (3.155)	0.361	0.015	0.023
C (Total)	0.669 (5.247)	0.599	0.090	0.079
T	0.156	0.065	0.022	0.023
PV	1.7×10^{-3}	4.7×10^{-4}	1.8×10^{-4}	1.7×10^{-4}
Γ	0.828 (5.406)	0.665	0.112	0.103

The results shown in Table 4.2 illustrate that SRC reduce both the central SI- and the SD-part without δ-function by almost a factor of 10, compared to a factor of 2 in the π case, again reflecting the much shorter range of the ρ-exchange diagram. Similarly, the tensor interaction of the ρ is reduced by a factor of 2.5, compared to a 10% reduction in the π case, as soon as SRC are included. The additional inclusion of FF and FSI further reduces both the central and tensor rates by substantial amounts. The final result of the central contribution exceeds the tensor term by more than a factor of three. We emphasize the relative contribution of the central and the tensor parts of the ρ-exchange; clearly, all of these potentials scale with the magnitude of the weak PC $\Lambda N\rho$-coupling constant which is model dependent. Due to the particular model we employ, the PV part of the ρ-exchange is negligible.

4.3 π- and ρ-Exchange

Finally, Table 4.3 presents our results for the π- and ρ-exchanges combined. Since both the $\Lambda N\pi$- and the $\Lambda N\rho$-couplings are obtained within the same model there is no sign ambiguity. We find a destructive interference between the two mesons, leading to a 17% reduction compared to the rate calculated with π-exchange only. Ignoring the central pieces of the ρ exchange mechanism would give a reduction of 22%. Clearly, this 5% difference is a reflection of the particular model employed for the weak $\Lambda N\rho$vertex. In fact, this effect is

Table 4.3. π and ρ exchange contribution to the $\Lambda N \to NN$ decay rate of ${}^{12}_{\Lambda}$C using the Nijmegen (Jülich) constants

	π	ρ	$\pi + \rho$
C (SI)	—	0.033 (0.037)	0.033 (0.037)
C (SD)	0.004	0.023 (0.086)	0.008 (0.053)
C (Total)	0.004	0.079 (0.169)	0.054 (0.125)
T	0.645	0.023 (0.086)	0.423 (0.273)
PV	0.354	1.7×10^{-4} (3.3×10^{-4})	0.360 (0.373)
Γ	1.003	0.103 (0.256)	0.836 (0.770)
Γ_n/Γ_p	0.134	0.026	0.157

already slightly more pronounced when we use the Jülich model for the strong coupling constants which, in turn, produce different predictions for the weak constants, as shown in Table 3.1. The result for the $\pi + \rho$ decay rate with the Jülich constants, shown in brackets in Table 4.3, is 0.770, which is 23% smaller than for π-exchange only. The reduction would amount to 35% if the central contributions were ignored. The findings reported here for the ρ-meson contribution hold for the other vector mesons, such as the K^* and the ω, as well, which can have larger couplings.

The neutron to proton induced ratio Γ_n/Γ_p, on the other hand, is increased by about 20%, from a value of 0.13 for π-exchange only to 0.16 when the ρ is included. Using the Jülich constants the $\pi + \rho$ value becomes 0.19, which represents an increase of almost 40%. While this is nowhere near an increase that would be needed to explain experimental data [2, 4], it is clearly a reflection of the fact that the tensor interaction of the ρ-exchange is not as dominant as it is in the π case.

4.4 The full weak one-meson-exchange potential

In this subsection we explore the effect of the contributions of the η, K, ω and K^* mesons, in addition to the π and ρ, on the decay observables of $^{12}_{\Lambda}$C. Although the structure of the η and K potentials is the same as that of the π, while the ω and K^* potentials are similar to that of the ρ-meson, the isospin factors are different and this can significantly alter the neutron- to proton-induced ratio.

4.4.1 Total Rates

The results in Table 4.4 demonstrate the significance of the short-range correlations and form factors in the nonmesonic decay. Clearly, without form factors

Table 4.4. Sensitivity to model ingredients of the total rate of $^{12}_{\Lambda}$C using the Nijmegen (Jülich) coupling constants. Note that for π-exchange alone the experimentally measured weak $\Lambda N\pi$-vertex is employed, thus no model dependency is present

	Free	SRC	SRC, FF	SRC, FF, FSI
π	1.580	1.121	0.938	1.003
$+\eta$	1.589 (1.590)	1.059 (1.121)	0.875 (0.938)	0.936 (1.003)
$+K$	2.076 (2.230)	1.385 (1.578)	0.999 (1.117)	1.076 (1.202)
$+\rho$	2.307 (2.541)	1.740 (2.709)	0.823 (0.794)	0.897 (0.887)
$+\omega$	6.613 (20.325)	1.511 (1.513)	0.882 (1.096)	0.921 (0.971)
$+K^*$	4.542 (14.914)	2.683 (4.439)	0.935 (1.359)	1.020 (1.345)

and SRC (column 1) the vector mesons give an appreciable contribution, i.e., adding the ω increases the total rate by more than a factor of ten in the free case when the Jülich coupling constants are used. This is due to the ωNN vector coupling which is much larger in the Jülich potential than in the Nijmegen model. However, the effect of the vector mesons is considerably suppressed by short-range effects, as shown in the second column. Note that the addition of the ω now leads to a reduction of the rate while the K^* enhances it. Adding form factors reduces the contribution of the vector mesons further by up to a factor of three. Finally, adding final-state interactions tends to increase the rate by about 10% on the average. Using the Nijmegen coupling constants the interferences between the different mesons conspire to give a final rate which is almost identical to the one obtained with pion exchange alone. In contrast, the Jülich constants lead to stronger interferences between the various mesons (last column in Table 4.4), the K and K^* both add constructively, while the ρ, as discussed in Sect. 4.3, interferes destructively. In this case, the final result is enhancement by over 30%. While both potential models give results that are in agreement with the older data [2], the newer data [4] are slightly below the prediction obtained with the Jülich constants.

4.4.2 Partial Ratios

Results for the neutron to proton induced ratio Γ_n/Γ_p are shown in Table 4.5. The neutron to proton induced ratio is, as expected, more sensitive to the isospin structure of the exchanged mesons. It has been known for a long time that pion exchange alone will produce only a small ratio [16]. The main contribution comes from the two strange mesons that, at least qualitatively, play opposing roles regardless of the choice of strong coupling constants. Including the K-exchange which interferes destructively with the pion amplitude in the neutron induced channel leads to a dramatic reduction of almost a factor of ten in the ratio, while the K^* adds constructively. However, using the Nijmegen constants leads to a final ratio that is smaller than the pion ratio by almost a factor of two, while the Jülich constants increase the final ratio by about a factor of 3 over the ratio obtained with π-exchange alone. Clearly, this observable should be well suited to delineate between different models of weak coupling constants. Even though the hope has been that the inclusion of additional mesons will dramatically increase this ratio we find a result that still greatly underestimates the newer central experimental values, even though the large experimental error bars do not permit any definite conclusions at this time. Other mechanisms that have been explored to remedy this puzzle include quark-model calculations which yield a large violation of the $\Delta I = 1/2$ rule [20, 21], and the consideration of the $3N$ emission channel ($\Lambda NN \rightarrow NNN$) as a result of the pion being absorbed on correlated $2N$ pairs. However, this channel has recently been reanalysed with the result that the discrepancy worsens [7, 8]. We note that our results differ from the nuclear matter calculation of ref. [22] even though we use the same model for the weak vertices.

4.4.3 Asymmetries

The proton asymmetry is also very sensitive to the different meson exchanges included in the model. As discussed in Sect. 2.2 the asymmetry parameter A_y, characteristic of the weak decay, must be multiplied by a theoretically determined hypernuclear polarization P_y in order to be compared to experiment. The energy resolution of the experiment measuring the decay of polarized ${}^{12}_{\Lambda}$C produced in a (π^+, K^+) reaction [3] was $5-7$ MeV which did not allow distinguishing between the first three 1^- states. Before the weak decay occurs, the two excited states will decay electromagnetically to the ground state. Therefore, in order to determine the polarization at this stage, one requires: i) the polarization of the ground and excited states, together with the corresponding formation cross sections, and ii) an attenuation coefficient to account for the loss of polarization in the transition of the excited states to the ground state. We have taken the value $P_y = -0.21$ obtained in ref. [24]. Just as for

Table 4.5. Weak decay observables for ${}^{12}_{\Lambda}$C using the Nijmegen (Jülich) coupling constants

	$\Gamma_{\rm nm}$	Γ_n/Γ_p	$\mathcal{A}(0°)$
π	1.003	0.134	−0.038
$+\eta$	0.936 (1.003)	0.209 (0.134)	−0.045 (-0.038)
$+K$	1.076 (1.202)	0.029 (0.019)	−0.020 (-0.017)
$+\rho$	0.897 (0.887)	0.029 (0.017)	−0.013 (-0.002)
$+\omega$	0.921 (0.971)	0.040 (0.077)	−0.021 (-0.024)
$+K^*$	1.020 (1.345)	0.077 (0.350)	0.013 (0.041)
EXP:	1.14 ± 0.20 [2]	$1.33^{+1.12}_{-0.81}$ [2]	-0.01 ± 0.10 [3]
	$0.93 \pm 0.17 \pm 0.01$ [4]	$1.74 \pm 0.54^{+0.23}_{-0.94}$ [4]	
		0.70 ± 0.30 [23]	
		0.52 ± 0.16 [23]	

the proton- to neutron-induced ratio, the present level of uncertainty in the experiment does not yet permit using the asymmetry as an observable that differentiates between different models for the weak vertices.

In order to avoid the need for theoretical input and access A_y directly, a new experiment at KEK [25] is measuring the decay of polarized ${}^{5}_{\Lambda}$He, extracting both the pion asymmetry from the mesonic channel, $\mathcal{A}_{\pi^-}$, and the proton asymmetry from the nonmesonic decay, $\mathcal{A}_p$. The asymmetry parameter a_{π^-} of the pionic channel has been estimated to be very similar to that of the free Λ decay [26], and, therefore, the hypernuclear polarization can now be obtained from the relation $P_y = \mathcal{A}_{\pi^-}/a_{\pi^-}$. This in turn can then be used as input, together with the measured value of $\mathcal{A}_p$, to determine the asymmetry parameter for the nonmesonic decay from the equality $A_p = \mathcal{A}_p/P_y$. This experiment will not only allow a clean extraction of the nonmesonic asymmetry parameter but will also check theoretical model predictions for the amount of hypernuclear polarization.

5 Conclusion

The $\Lambda N \rightarrow NN$ decay has been investigated in a nonrelativistic meson exchange-model which includes short range correlations, form factors and final state interactions. We have demonstrated that the central potential of the ρ-meson contribution to the $\Lambda N \rightarrow NN$ process cannot be neglected and is in fact larger that its tensor part. Due to the very different ranges of the π- and the ρ-exchange their contributions are modified differently when short-range correlations and form factors are included. We emphasize that this result is independent of the particular model chosen for the weak $\Lambda N\rho$-vertex. In view of a number of more recent theoretical efforts that increase the complexity of the $\Lambda N \rightarrow NN$ reaction mechanism by calculating correlated 2π-exchanges, through the coupling to the isoscalar σ-meson [27] or via strange $\Lambda N \rightarrow \Sigma N$

mixing [28], it is imperative that all parts of the ρ-meson (and the other vector mesons) potential be included first. The full weak meson exchange potential with either the Nijmegen or the Jülich strong coupling constants yields total rates differ by about 30% and are both close to the measurements. However, both the proton- to neutron-induced ratio and the asymmetry displayed great sensitivities to the values of the weak coupling constants. This is an indication that these observables can be used to discriminate between different theoretical models, as soon as new data with reduced statistical errors are available [25].

Our study clearly indicates that further theoretical effort must be invested to understand the dynamics of the nonmesonic weak Λ-hypernuclear decay. Within the one-meson exchange picture it would be desirable to use weak coupling constants developed with more sophisticated models. Several recent studies [20, 21] have gone beyond this conventional picture and have developed mechanisms based purely on quark degrees of freedom. One should keep in mind, however, that such models have not always been able to reproduce the experimentally measured free hyperon decays. On the experimental side, it is critical to obtain new high accuracy data soon. Even with the demise of KAON, the promising efforts at KEK with an improved measurement of the $^5_\Lambda$He decay, the continuing program at BNL, and the advent of the hypernuclear physics program (FINUDA) at DAΦNE represent excellent opportunities to obtain new valuable information that will shed some light onto the still unresolved problems of the weak decay of Λ hypernuclei.

Acknowledgement. The work of CB was supported by US-DOE grant no. DE-FG02-95-ER40907 while the work of AR was supported by DGICYT contract no. PB92-0761 (Spain). AP acknowledges support from a doctoral fellowship of the Ministerio de Educacion y Ciencia (Spain).

References

1. A. Ramos, E. van Meijgaard, C. Bennhold, B.K. Jennings: Nucl. Phys. **A544**, 703 (1992); A. Ramos, C. Bennhold: Nucl. Phys. **A577**, 287c (1994), and references listed therein.

2. J.J. Szymanski et al.: Phys. Rev. **C43**, 849 (1991)

3. S. Ajimura et al.: Phys. Lett. **B282**, 293 (1992)

4. H. Noumi et al.: Preprint KEK 95-10 1995

5. A. Zenoni for the FINUDA collaboration, *Proc. of the Second Workshop on Physics and Detectors for DAΦNE*, Frascati, Italy, April 4-7, 1995, (in press)

6. W.M. Alberico, A. De Pace, M. Ericson, A. Molinari: Phys. Lett. **B256**, 134 (1991)

7. A. Ramos, E. Oset, L.L Salcedo: Phys. Rev. **C50**, 2314 (1994)

8. A. Ramos: Contribution to this Conference

9. A. Parreño, A. Ramos, C. Bennhold: to be published

10. B. Holzenkamp, K. Holinde, J. Speth: Nucl. Phys. **A500**, 485 (1989)

11. D. Halderson: Phys. Rev. **C48**, 581 (1993)

12. M.N. Nagels, T.A. Rijken, J.J. de Swart: Phys. Rev. **D15**, 2547 (1977); P.M.M. Maessen, Th. A. Rijken, J.J. de Swart: Phys. Rev. **C40**, 2226 (1989)

13. C. Bennhold: Phys. Rev. **C43**, 775 (1991)

14. A. Parreño, A. Ramos, C. Bennhold, D. Halderson: In: *Dynamical Features of Nuclei and Finite Fermi Systems*, X. Viñas, M. Pi and A. Ramos, eds., p.318. Singapore: World Scientific 1994; A. Parreño, A. Ramos, E. Oset: Phys. Rev. **C51**, 2477 (1995); A. Parreño: Contribution to this Conference

15. J.F. Dubach: Nucl. Phys. **A450**, 71c (1986); L. de la Torre: Ph.D. Thesis, Univ. of Massachusetts 1982

16. B.H.J. McKellar, B.F. Gibson: Phys. Rev. **C30**, 322 (1984)

17. K. Takeuchi, H. Takaki, H. Bandō: Prog. Theor. Phys. **73**, 841 (1985)

18. G. Nardulli: Phys. Rev. **C38**, 832 (1988)

19. A. Parreño, A. Ramos, C. Bennhold: submitted to Phys. Rev. **C**

20. K. Maltman, M. Shmatikov: Phys. Lett. **B331**, 1 (1994)

21. T. Inoue, S. Takeuchi, M. Oka: Nucl. Phys. **A** (in print)

22. B. Holstein et al.: *Proc. of the Second Workshop on Physics and Detectors for DAΦNE*, Frascati, Italy, April 4-7, 1995, in press

23. A. Montwill et al.: Nucl. Phys. **A234**, 413 (1974)

24. K. Itonaga, T. Motoba, O. Richter, M. Sotona: Phys. Rev. **C49**, 1045 (1994)

25. T. Kishimoto: Nucl. Phys. **A585**, 205c (1995)

26. T. Motoba, K. Itonaga: Nucl. Phys. **A577**, 293c (1994)

27. K. Itonaga, T. Ueda, T. Motoba: Nucl. Phys. **A585**, 165c (1995)

28. M. Shmatikov: Nucl. Phys. **A580**, 538 (1994)

Few-Body Systems Suppl. 9, 490–494 (1995)

Few-Body Systems

Λ Decay Induced by Two Nucleons

A. Ramos[1], E. Oset[2], L.L. Salcedo[3]

[1] Departament d'Estructura i Constituents de la Matèria, Facultat de Física, Diagonal 647, 08028 Barcelona, Spain

[2] Departamento de Física Teórica and IFIC, Centro Mixto Universidad de Valencia-CSIC, 46100 Burjassot (Valencia), Spain

[3] Departamento de Física Moderna, Universidad de Granada, 18071 Granada, Spain

Abstract. The decay of Λ hypernuclei induced by two nucleons is revised, along with its implications in the experimental determination of the ratio of neutron- to proton-stimulated Λ decay.

1 Introduction

The decay of Λ hypernuclei has traditionally been assumed to proceed through the mesonic ($\Lambda \to N\pi$) and the nonmesonic ($\Lambda N \to NN$) channels. In medium and heavy hypernuclei the mesonic channel is largely supressed by Pauli blocking, even if the renormalization of the pion properties in the medium can increase the rate considerably [1-3]. The one-pion exchange mechanism for the $\Lambda N \to NN$ decay reproduces quite nicely the measured widths [1, 4, 5] but gives a neutron- to proton-induced decay ratio, Γ_n/Γ_p, between 0.1 and 0.2, while the measured value is around one [6, 7], albeit with very large experimental error bars.

In the two-nucleon induced mechanism $\Lambda NN \to NNN$, depicted in Fig. 1, the pion emerging from the weak vertex is assumed to be absorbed by a pair of correlated nucleons. This channel was first studied in ref. [8] for a Λ decaying in nuclear matter and the decay rate was found to be 30% of the total width. In this contribution we present an improved calculation of the two-nucleon induced decay channel, extending the results of ref. [8] to finite nuclei.

2 Formalism for the Λ decay

The decay of a free Λ is governed by the $\Lambda \to \pi N$ Lagrangian

$$\mathcal{L}_{\pi\Lambda N} = G\mu^2 \bar{\psi}_N (A - B\gamma_5) \boldsymbol{\tau} \boldsymbol{\phi}_\pi \psi_\Lambda + h.c. \ , \qquad (1)$$

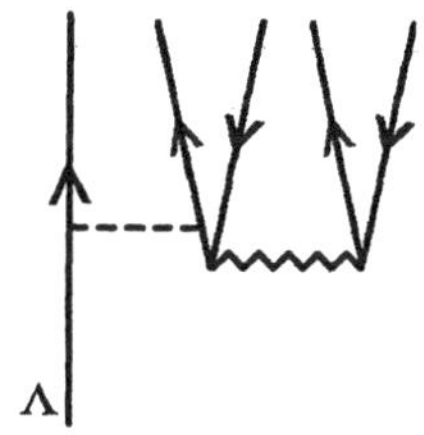

Figure 1. Mechanism for the two-nucleon induced Λ decay, where the virtual pion (dashed line) is absorbed by a pair of correlated nucleons.

where G is the weak coupling constant, $(G\mu^2)^2/8\pi = 1.945 \times 10^{-15}$, μ the pion mass, and the Λ is taken as the isospurion $\mid 1/2, -1/2\rangle$ to implement the empirical $\Delta T = 1/2$ rule. The constants $A = 1.06$ and $B = 7.10$ determine the parity violating and parity conserving transition amplitudes, respectively.

A practical way to evaluate the Λ width in nuclear matter and introduce the medium corrections is to start from the Λ self-energy

$$-\mathrm{i}\Sigma(k) = 3(G\mu^2)^2 \int \frac{d^4q}{(2\pi)^4} G(k-q) D(q) \left(A^2 + \left(\frac{B}{2M} \right)^2 \boldsymbol{q}^2 \right) \tag{2}$$

where $D(q)$ and $G(p)$ are the renomalized pion and nucleon propagators in nuclear matter, and then use the relationship $\Gamma = -2\mathrm{Im}\,\Sigma$, with the result [1, 9]

$$\begin{aligned} \Gamma(k,\rho) &= -6(G\mu^2)^2 \int \frac{d^3q}{(2\pi)^3} \theta(\mid \boldsymbol{k}-\boldsymbol{q} \mid - k_\mathrm{F})\, \theta(k^0 - E(\boldsymbol{k}-\boldsymbol{q}) - V_N) \\ &\times \left(A^2 + \left(\frac{B}{2M} \right)^2 \boldsymbol{q}^2 \right) \mathrm{Im} D(q)]_{q^0 = k^0 - E(k-q) - V_N} , \end{aligned} \tag{3}$$

where k_F is the Fermi momentum, $V_N = -50\rho/\rho_0$ MeV the nucleon potential and ρ the nuclear matter density. The inclusion of form factors and short range NN and ΛN correlations [1, 9] gives rise to a slightly more complicated expression for Γ but, for the purpose of our discussion, it is clearer to retain the simpler Eq. (3). The contributions to $\mathrm{Im}D(q)$ are proportional to the different terms of the pion self-energy, $\Pi(q)$, which, to account for the new $3N$ emission channel, must contain a piece allowing for the coupling to $2p2h$ excited states

$$\Pi = \Pi_{1p1h} + \Pi_{\Delta h} + \Pi_{2p2h} \; . \tag{4}$$

The integrand of Eq. (3) is shown in Fig. 2, for a Λ of momentum zero. As we perform the integral over $\boldsymbol{q}$ we move through momentum and energy regions where the pion couples to the different excitations. We first cross the point of the renormalized pion pole. Since there $\mathrm{Im}\Pi_{1p1h} = 0$ and $\mathrm{Im}\Pi_{\Delta h}$ is small, when we ignore Π_{2p2h} the integrand is practically a delta function shown by the dashed vertical line. Since the Pauli principle forces the momentum $\boldsymbol{q}$ to be larger than k_F (shown by the arrow), the pionic decay is forbidden in nuclear matter at normal density. For larger values of $\boldsymbol{q}$, which in turn imply smaller values of q^0, we enter the region of $1p1h$ excitations displayed by a bump around 400 MeV/c, which gives rise to the one-nucleon induced decay rate. The region of $2p2h$ excitations covers practically all points in the energy-momentum plane. Therefore, including Π_{2p2h} in the pion self-energy the shape of the integrand (full curve) spreads out as a consequence of the coupling to $2p2h$ components, and part of this strength overcomes the Pauli blocking giving rise to a genuine $3N$ emission channel.

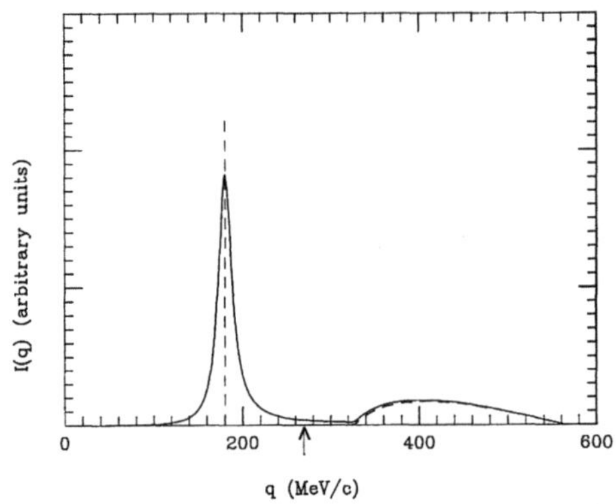

Figure 2. Integrand of Eq. (3) in arbitrary units. The arrow indicates the location of $q = k_F$, below which Pauli blocking forbids the decay.

3 The 2N-induced decay widths

The crucial ingredient for obtaining the $2N$ induced width is the piece Π_{2p2h} of the pion self-energy. In this work we use an improved input with respect to that of ref. [8]:
i) The second order potential, Π^*_{2p2h}, is obtained from a recent empirical analysis of pionic atoms [10], which is also consistent with pion elastic scattering data and total reaction cross sections. We find:

$$\Pi^*_{2p2h}(q_0 = \mu, \boldsymbol{q} = 0, \rho) = -4\pi \boldsymbol{q}^2 C_0^* \rho^2 \ , \tag{5}$$

with $C_0^* = (0.105 + i0.096)\mu^{-6}$. Our value for $\mathrm{Im}\, C_0^*$ is about one half of the one used in ref. [4], while $\mathrm{Re}\, C_0^*$ is about four times smaller.
ii) we extend $\Pi^*_{2p2h}(q^0, \boldsymbol{q})$ to new kinematical regions away from pionic atoms. The imaginary part is taken proportional to the available phase space of $2p2h$ excitations at $(q^0, \boldsymbol{q})$ and density ρ

$$\mathrm{Im}\Pi_{2p2h}(q^0, \boldsymbol{q}) = \boldsymbol{q}^2 \mathrm{Im}\overline{\Pi}_{2p2h}(q^0, \boldsymbol{q}) = \boldsymbol{q}^2 \overline{\Pi}_{2p2h}(\mu, \mathbf{o}) \frac{\mathrm{PH}(q^0, \boldsymbol{q})}{\mathrm{PH}(\mu, \mathbf{o})} \ , \tag{6}$$

while the real part is left constant.

We have also extended the work of ref. [8] to obtain the widths in finite nuclei via a local density approximation. Our results for the different decay channels are shown in Table 1, in units of the free Λ decay width, Γ_Λ. We find that $\Gamma_{\rm nm}^{(2)}$ is about 0.3 Γ_Λ, which is about one half of what was obtained in the nuclear matter calculation of ref. [8].

Table 1. Mesonic, one-body and two-body induced decay rates of Λ hypernuclei

	$\Gamma_{\rm m}$	$\Gamma_{\rm nm}^{(1)}$	$\Gamma_{\rm nm}^{(2)}$
${}^{12}_{\Lambda}$C	0.31	1.45	0.27
${}^{16}_{\Lambda}$O	0.24	1.54	0.29
${}^{20}_{\Lambda}$Ne	0.14	1.60	0.32
${}^{40}_{\Lambda}$Ca	0.03	1.76	0.32
${}^{56}_{\Lambda}$Fe	0.01	1.82	0.32
${}^{89}_{\Lambda}$Y	—	1.88	0.31
${}^{100}_{\Lambda}$Ru	—	1.89	0.31
${}^{208}_{\Lambda}$Pb	—	1.93	0.30

4 The neutron to proton ratio

Since the $2N$-induced channel amounts to about 15% of the total decay rate, it is clear that one cannot associate all the measured n and p to the nucleons emerging in the primary $\Lambda n \to nn$ and $\Lambda p \to np$ reactions and hence a reanalysis of the experimental data is needed. We find [9]:

$$\frac{\Gamma_n}{\Gamma_p} = \left(\frac{\Gamma_n^{\rm exp}}{\Gamma_p^{\rm exp}}\right) \frac{1 - \frac{1}{2}\left(\frac{\Gamma_p^{\rm exp}}{\Gamma_n^{\rm exp}} + 1\right)\frac{\Gamma_{\rm nm}^{(2)}}{\Gamma_{\rm nm}^{\rm exp}}}{1 - \left(\frac{\Gamma_n^{\rm exp}}{\Gamma_p^{\rm exp}} + 1\right)\frac{\Gamma_{\rm nm}^{(2)}}{\Gamma_{\rm nm}^{\rm exp}}}, \tag{7}$$

written in terms of the experimental values $\Gamma_{\rm nm}^{\rm exp}$ and $\Gamma_n^{\rm exp}/\Gamma_p^{\rm exp}$. It is clear from Eq. (7) that if $\Gamma_n^{\rm exp}/\Gamma_p^{\rm exp}$ is larger than 0.5, as it is the case of almost all the experimental results, the ratio Γ_n/Γ_p increases making the disagreement with theory stronger. As an example, from the data for ${}^{12}_{\Lambda}$C [6] one extracts $\Gamma_n^{\rm exp}/\Gamma_p^{\rm exp} = 1.04$, while the new analysis gives $\Gamma_n/\Gamma_p = 1.54$. Actually, the precise value for the reanalyzed ratio depends on the detection thresholds. If the nucleon emitted in the weak vertex was not detected [11], then the new ratio would decrease to a value of 0.86. However, this would imply that the emitted pion would have to be practically on-shell, and this is far from the most probable kinematical situation in which the nucleons share equally the initial energy excess.

5 Conclusions

The two-nucleon induced decay width of Λ hypernuclei has been found to be about 15% of the total decay width and, therefore, it is necessary to consider the influence of this channel in the experimental analysis. It would be very interesting to carry out improved experiments, which are able to disentangle this channel from the the other two. A possibility would be to obtain good resolution proton spectra, although the multiple scattering of the nucleons may prevent a clear separation of the channels. However, due to the different kinematics of each process, a combination with angular correlation measurements could be extremely useful. From the theoretical side, it is necessary to perform more sofisticated calculations in which the emitted nucleons are followed through the nucleus, allowing for final state interactions, to know their energy distribution and hence enabling a proper comparison with experiment. In any case, what is clear is that the consideration of the $2N$-induced mechanism is of relevance to extract the value of Γ_n/Γ_p from the data.

Acknowledgement. This work has been supported by DGICYT contracts PB92-071, AEN93-1205, PB92-0927 and by the EU contract CHRX-CT 93-0323.

References

1. E. Oset and L.L Salcedo: Nucl. Phys. **A443**, 704 (1985)
2. K. Itonaga, T. Motoba and H. Bandō: Z. Phys. **A330**, 209 (1988)
3. J. Nieves and E. Oset: Phys. Rev. **C47**, 1478 (1993)
4. J. Dubach: Nucl. Phys. **A450**, 71c (1986)
5. B.H.J. McKellar and B.F. Gibson: Phys. Rev. **C30**, 322 (1984)
6. J.J. Szymanski et al.: Phys. Rev. **C43**, 849 (1991)
7. H. Noumi et al.: Preprint 95-10 KEK 1995
8. W.M. Alberico et al.: Phys. Lett. **B256**, 134 (1991)
9. A. Ramos, E. Oset and L.L Salcedo: Phys. Rev. **C50**, 2314 (1994)
10. O. Meirav et al.: Phys. Rev. **C40**, 843 (1989)
11. A. Gal: Private Communication

Few-Body Systems Suppl. 9, 495–509 (1995)

Few-Body Systems

Hypernuclear Structure and Hyperon-Nucleon Interactions

T. Motoba

Laboratory of Physics, Osaka Electro-Communication University, Neyagawa, Osaka 572, Japan

Abstract. The hyperon-nucleon G-matrix interactions derived from the Nijmegen and Jülich potentials are shown to have remarkably different spin character which should be reflected in light hypernuclei. Various outputs of the calculations are demonstrated in connection with the recent (π^+, K^+) data from KEK. A possibility of deducing Λ spin-orbit splittings in ${}^{16}_{\Lambda}$O and in heavy hypernuclei have been also discussed.

1 Introduction

Much attention has recently been payed to the properties of interactions of hyperons not only in the $S = -1$ sector but also in the $S = -2$ sector. First this has been encouraged by the recent experimental progress and future possibilities of new proposals at the major accelerator facilities. A typical example is the result of (π^+, K^+) reactions performed at KEK, providing a series of major Λ single-particle orbits bound in heavy nuclear systems such as ${}^{89}_{\Lambda}$Y, ${}^{139}_{\Lambda}$La and ${}^{208}_{\Lambda}$Pb [1,2]. Especially it is quite interesting that reanalysis by Nagae et al. [2] show *fine* structures of the peaks due to Λ spin-orbit splittings. The merit of preferential excitation of the reaction process also discloses detailed level structures of ${}^{12}_{\Lambda}$C [1] and ${}^{10}_{\Lambda}$B, although further improvement seems still necessary. In near future we can expect higher resolution data of, for example, $(\pi^+, K^+\gamma)$, (K^-, π^0) and (γ, K^+) reactions [3].

The second motivation is based on the fact that we have already five kinds of meson theoretical potentials for hyperon interactions. It is one of the fundamental goals to understand baryon many-body structures on the basis of the properties of baryon-baryon interactions. Although this is not always a straightforward approach and there might be many phenomenological intermediate steps in this direction, it seems more and more meaningful to try to build a bridge (or bridges) between hypernuclear structure and the properties of meson theoretical interactions.

Corresponding to the theoretical $\Lambda - N$ potentials proposed by the Nijmegen group (model-D and model-F)[4], Yamamoto and Bandō [5] derived the G-matrices by solving the Bethe-Goldstone equation in nuclear matter and expressed them in terms of three-range Gaussian functions in the coordinate space for easy application to the structure calculations. These G-matrices are called YNG interactions and denoted here simply by ND and NF. Then a soft-core potential model was proposed by the same group [6], and recently two potentials (A and B versions) were presented by the Jülich group [7,8] with the same spirit as the Bonn model of $N - N$ interaction. Thus the corresponding YNG-type interactions were derived and called as NS (soft-core), JA and JB, respectively.

In this paper we demonstrate the results of structure outputs obtained in the application of these five $\Lambda - N$ interactions, and discuss the problems to be answered from both sides of a *bridge*.

2 Basis properties of meson-theoretical $\Lambda - N$ G-matrix interactions

Starting with a meson-theoretical potential having hard cores or momentum-dependence, the G-matrix is obtained by solving the coupled-channel Bethe-Goldstone equation to treat two-body correlations in nuclear matter [5,9]:

$$G_{cc_1} = v_{cc_1} + \sum_{c'} v_{cc'} \frac{Q_{y'}}{e_{yy'}} G_{c'c_1}.$$

Here c specifies a channel of $Y - N$ relative state (y, T, L, S, J) with y denoting $[\Lambda N]$ or $[\Sigma N]$ pair and $L + S = J$. The Pauli exclusion operator Q_y acts on the intermediate nucleon states and the relevant energy denominator for a starting energy $\omega_1 = \epsilon_Y + \epsilon_N$ is given by $e_{yy'} = \omega_1 - \epsilon_{Y'} - \epsilon_N + \Delta_{yy'}$ with $\Delta_{yy'} = M_y - M_{y'}$ and $M_y = M_Y + M_N$ being the channel mass. The hyperon single-particle energy ϵ_Y consists of the kinetic and potential energies as

$$\epsilon_Y(k_Y) = \frac{\hbar^2 k_Y^2}{2M_Y} + U_Y(k_Y), U_Y(k_Y) = \sum_{k_N} \langle k_Y k_N | G_{YN}(\omega) | k_Y k_N \rangle (1 - \kappa_N),$$

where $U_Y(k_Y)$ is obtained selfconsistently as a sum of G-matrix elements and $\omega = \epsilon_Y(k_Y) + \epsilon_N(k_N)$. One may refer to ref. [9] for the momentum space expressions for the Jülich potentials. The wound integral κ_N accounts for a kind of the rearrangement effect. In solving the G-matrix, the so called QTQ prescription has been adopted, where no potential terms are taken for the $Y - N$ intermediate states. For the Nijmegen soft-core model, as the QTQ result (YNG(NS0)) gives rise to too shallow potential energy U_Y, we tried to get the modified one, YNG(NS), by introducing the continuous intermediate spectrum (cf. NS0 vs. NS in Table 2).

The G-matrix interactions can be expressed very well in terms of three-range Gaussian functions in the coordinate space [5,9,10]:

$$V_{\Lambda N}^{\alpha}(r) = \sum_{i=1}^{3} (a_i + b_i k_{\rm F} + c_i k_{\rm F}^2) \exp[-r^2/\beta_i^2], \ \alpha = {\rm Central, LS, ALS, T}.$$

These ('YNG interactions') have a unique parameter k_F, the nuclear Fermi momentum to be determined for a particular hypernucleus, and the use of these local potentials is practically easy. The simulation parameters $\{a_i, b_i, c_i, \beta_i\}$ are found in refs. [5,9,10] and thus we have five kinds of the YNG $\Lambda-N$ interactions (JA, JB, ND, NF, NS (NS0)) to be tested in the structure calculations.

Table 1. The Λ single-particle energies and their distances Δ (in MeV) obtained in the DDHF calculation with five YNG interactions, $\Delta_{sp} = \epsilon_\Lambda(0p) - \epsilon_\Lambda(0s)$.

	$n_\Lambda l_\Lambda$	JA	JB	ND	NF	NS		EXP[1]
$^{91}_{\Lambda}$Zr	$0s$	-22.1	-24.3	-31.2	-25. 0	-24.1	$^{89}_{\Lambda}$Y	(-22.1 ± 0.3)
	$0p$	-14.7	-16.8	-22.9	-17.3	-16.9		(-15.8 ± 0.8)
	$0d$	-7.0	-8.8	-14.3	-9.5	-9.3		(-8.7 ± 1.0)
	Δ_{sp}	7.4	7.4	8.3	7.7	7.2		6.3 ± (1.1)
	Δ_{pd}	7.7	8.0	8.6	7.8	7.6		7.1 ± (1.8)
$^{141}_{\Lambda}$Ce	$0s$	-23.6	-25.7	-33.0	-26.8	-25.6	$^{139}_{\Lambda}$La	(-24.5 ± 1.0)
	$0p$	-17.8	-19.9	-26.8	-20.8	-20.1		(-20.4 ± 0.2)
	$0d$	-11.4	-13.3	-19.7	-14.1	-13.8		(-14.3 ± 0.2)
	Δ_{sp}	5.8	5.8	6.2	6.0	5.5		4.1 ± (1.2)
	Δ_{pd}	6.4	6.3	7.1	6.7	6.3		6.1 ± (0.4)
$^{209}_{\Lambda}$Pb	$0s$	-25.2	-27.3	-35.1	-28.5	-27.2	$^{208}_{\Lambda}$Pb	(-26.3 ± 0.5)
	$0p$	-20.3	-22.4	-29.9	-23.4	-22.5		(-21.9 ± 0.3)
	$0d$	-14.7	-16.8	-23.9	-17.7	-17.2		(-16.8 ± 0.4)
	Δ_{sp}	4.9	4.9	5.2	5.1	4.7		4.4 ± (0.8)
	Δ_{pd}	5.6	5.6	6.0	5.7	5.3		5.1 ± (0.7)

The first application of these interactions is to see the Λ single-particle energy $\epsilon_\Lambda(nlj; A)$ as a function of mass number. For this purpose we adopt sample hypernuclei which consist of double-closed core plus a Λ particle ($^{17}_{\Lambda}$O, $^{29}_{\Lambda}$Si, $^{41}_{\Lambda}$Ca, $^{49}_{\Lambda}$Ca, $^{91}_{\Lambda}$Zr, $^{141}_{\Lambda}$Ce, $^{209}_{\Lambda}$Pb), and we carried out the density-dependent Hartree-Fock (DDHF) calculations in which the k_F-dependence in the G-matrix interactions are treated within the local density approximation. For the $N-N$ interaction, Skyrme III [11] is employed. In the actual comparison with the experimental data of $^{51}_{\Lambda}$V, $^{89}_{\Lambda}$Y, $^{139}_{\Lambda}$La and $^{208}_{\Lambda}$Pb, the average interactions between Λ and excess particles and/or holes, which is found to be around 0.5 MeV, are taken into account for modification.

Table 1 lists the numerical comparison of the results of $\epsilon_\Lambda(nlj; A)$ for typical heavy systems. For a graphic illustration see refs. [10,12]. One sees clearly that ND gives rise to overbinding of the Λ particle in the finite nuclear medium. The JB, NF and NS interactions result in the similar behaviour and they are acceptable when compared with the experimental data on $\epsilon_\Lambda(0p)$ and $\epsilon_\Lambda(0d)$ while the JA interaction leads to slightly shallower Λ-bindings but it explains $\epsilon_\Lambda(0s)$ well at $A > 89$.

Table 2. Partial wave contributions to the potential energy $U_\Lambda(k_\Lambda)$ as calculated with $k_\Lambda = 0$ at the nuclear normal density $k_F = 1.35\,\text{fm}^{-1}$. All entries are in MeV.

	$^1S_0$	$^3S_1+^3D_1$	Σ_{even}	$^3P_0$	$^1P_1+^3P_1$	$^3P_2+^3F_2$	Σ_{odd}	tot
JA	-3.6	-27.2	-30.8	0.6	2.0	-0.8	1.8	-29.8
JB	-0.5	-34.4	-34.9	0.6	2.9	0.1	3.6	-32.0
ND	-7.4	-25.1	-32.5	-0.1	-2.4	-5.5	-8.0	-40.5
NF	-10.0	-20.7	-30.7	0.2	2.9	-4.0	-0.9	-31.6
NS0	-14.6	-9.2	-23.8	0.4	3.7	-3.7	0.4	-23.3
NS	-16.0	-14.8	-30.8	0.3	3.4	-3.8	-0.1	-30.8

The behaviour of $\epsilon_\Lambda(nlj;A)$ is one of the average or gross properties of the $\Lambda-N$ interactions. In spite of the similar Λ-potential depths $U_\Lambda(k_\Lambda=0)^{\text{tot}}$ obtained here (except ND and NS0), we emphasize that their partial wave contributions are very different for different interactions, as seen from Table 2. First, as the common feature, the even-state attraction is predominant and the odd-state contribution is generally very small (except ND). In the ND case the large odd-state attraction is responsible for the overestimate of U_Λ obtained above. Secondly it is revealed that, in the even-state, the shares of the spin-singlet and spin-triplet contributions are remarkably different from each other. We get the ^1E and ^3E ratios from Table 2 as

$$\frac{[S=0]}{[S=1]} = \frac{1}{7.6}(JA) : \frac{1}{68.8}(JB) : \frac{1}{3.4}(ND) : \frac{1}{2.0}(NF) : \frac{1}{0.9}(NS).$$

It should be noted that the spin-triplet attraction is remarkably strong in JA and JB, while in NS the spin-singlet interaction is strong. Therefore it is quite interesting to test these interactions in light hypernuclei where the interaction characters tend to appear more directly and clearly due to less averaging.

3 Spin character of $\Lambda - N$ interactions exposed in light hypernuclei

It is interesting to test the spin-spin interaction strength ($\boldsymbol{\sigma}_\Lambda \cdot \boldsymbol{\sigma}_N$ term) by applying these YNG interactions to the energy level calculations of $^4_\Lambda$H($^4_\Lambda$He). For both $A=4$ systems the ground state is known to have $J_{\text{GS}} = 0^+$ ($B_\Lambda = 2.04$ MeV in $^4_\Lambda$H, 2.39 MeV in $^4_\Lambda$He) from the π-decay analysis and the excited state has $J=1^+$($E_x = 1.05$ MeV, 1.15 MeV respectively). As all the four baryons are in the orbital s-state, these two energy levels correspond directly to the $\Lambda - N$ spin-singlet and spin-triplet states, respectively. For the weak Λ-binding energy, the generator coordinate method has been applied to solve the relative motion between Λ and the three-nucleon cluster core of $(0s_{1/2})^3$ with size $b_c = 0.5$ fm and a fixed $k_F = 0.8\,\text{fm}^{-1}$.

The results with the five $\Lambda - N$ interactions are compared in Fig. 1 together with the experiment. We found that JA, JB and ND do not reproduce the experimental level order of the spin-doublet $0^+ - 1^+$, from which one can conclude that the spin-triplet part of these interactions is too strong (in JB

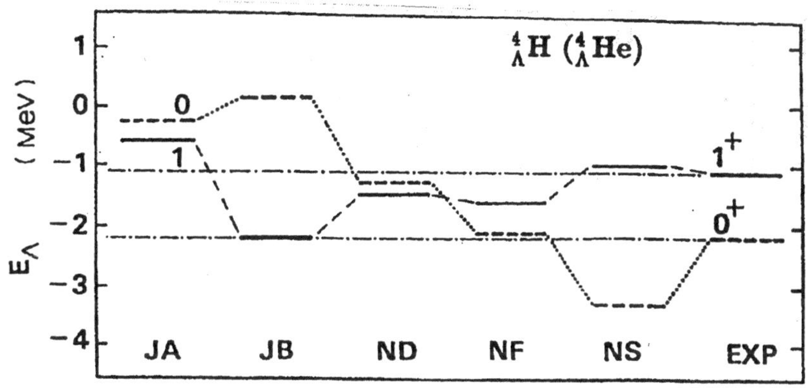

Figure 1. The calculated (GCM) and observed energies of $^4_\Lambda$H($^4_\Lambda$He) : $J = 0^+$ and 1^+.

especially). NF and NS lead to the right order. On the contrary, however, the spin-singlet attraction of NS seems too strong.

Figure 2 compares the low-lying energy levels of $^7_\Lambda$Li as calculated in the microscopic $\alpha + d + \Lambda$ three-cluster model with the YNG interactions ($k_F = 0.95\,\mathrm{fm}^{-1}$), showing again the similar spin-dependence. Noting that ^{6}Li has essentially the $\alpha - d$ cluster structure with $[l_{\alpha d} = 0 \otimes S_d = 1](J = 1^+)$ and $[l_{\alpha d} = 2 \otimes S_d = 1](J = 3^+, 2^+, 1^+)$. Accordingly in $^7_\Lambda$Li the $[l_{\alpha d} = 0 \otimes [S_d S_\Lambda]_{S=1/2,3/2}]$ coupling is responsible for $J = 1/2^+_1$ and $3/2^+_1$, while $[l_{\alpha d} = 2 \otimes [S_d S_\Lambda]_{1/2,3/2}]$ for the lowest $J = 5/2^+_1$ and $7/2^+_1$. In fact the dominant spin component has been confirmed in the wave function analysis as shown on the right side of Fig. 2. Thus the $S = 1/2$ dominant states are indicated by dashed line and the $S = 3/2$ ones by solid line. Figure 2 shows that in JB $J = 3/2^+_1$ and $7/2^+_1$ states having the $S = 3/2$ character (solid line) become lower in energy as a direct consequence of the strong spin-triplet attraction. Again NS manifests as another extreme, pushing down the spin-1/2 levels ($J = 1/2^+_1$ and $5/2^+_1$: dashed line).

In the $\alpha + x + \Lambda$ cluster model applied successfully to the light p-shell hypernuclei [13], we have an experience to use such a phenomenological $\Lambda - N$ interaction

$$V_{\Lambda N}(r) = -38.19\{1 + \eta(\sigma_\Lambda \cdot \sigma_N)\}\exp[-(r/1.034)^2],\ \eta = -0.1$$

that can reproduce the observed $0^+ - 1^+$ energy splitting in $^4_\Lambda$H($^4_\Lambda$He) as well as $B_\Lambda(^5_\Lambda\mathrm{He})$. Therefore the result of the column P suggests an acceptable case in view of the $\Lambda - N$ spin-spin property.

The above comparison for $A = 4$ and 7 suggests a possible direction of improving the $\Lambda - N$ potential models concerned.

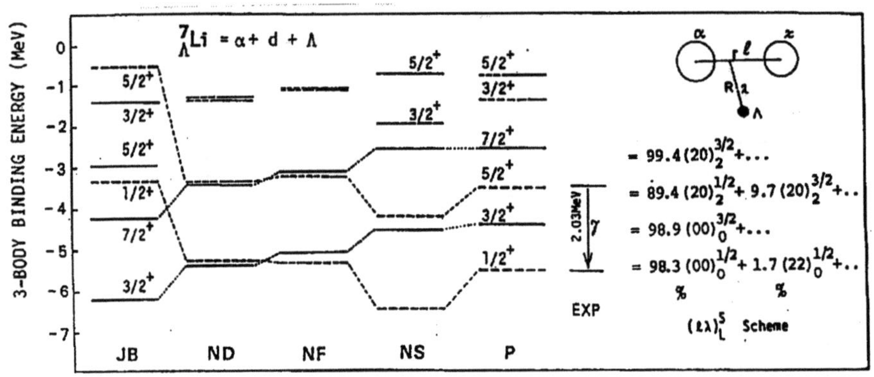

Figure 2. Comparison of the $^7_\Lambda$Li energy levels calculated with the cluster model. The phenomenological $\Lambda - N$ interaction in the text leads to the result 'P' and the spin structure of the wave functions. The experimental γ-ray energy [14] is compared.

4 Structure of $^{12}_{\Lambda}$C and the (π^+, K^+) reaction

In order to find the parameters of the effective $\Lambda - N$ potential, several shell-model calculations [15-17] have been performed for p-shell hypernuclei with relatively 'rich' data. From the meson-theoretical interaction side, the YNG effective interactions without any artificial modification were first applied in ref. [9] to test the ground state doublet structures of $^{10}_{\Lambda}$B, $^{11}_{\Lambda}$B$(J_{\rm GS} = 5/2^+)$ and $^{12}_{\Lambda}$C($^{12}_{\Lambda}$B$(J_{\rm GS} = 1^-)$). Here, in view of the new (π^+, K^+) data from KEK[1], we focus our attention to the excited states in $^{12}_{\Lambda}$C. By adopting the $[(0p_{3/2}0p_{1/2})^n_N(0s_{1/2}1s_{1/2})^1_\Lambda]$ configuration for the negative-parity states and the $[(0p_{3/2}0p_{1/2})^n_N(0p_{1/2}1p_{1/2})^1_\Lambda]$ one for the positive parity states, the shell model diagonalization was carried out, where the Hamiltonian consists of the Cohen-Kurath $N - N$ matrix elements, the effective $\Lambda - N$ interaction $v_{\Lambda N}$ and the Λ kinetic energy t_Λ with respect to the nuclear core[18]: $\mathcal{H} = H_N^{(\rm C-K)} + t_\Lambda + \sum v_{\Lambda N}$. The appropriate harmonic oscillator size parameter (b_N=1.522 fm) is used by keeping the relation $M_N b_N^2 = M_\Lambda b_\Lambda^2 = \hbar/\omega$ and $\hbar\omega = 41A^{-1/3}$.

Figure 3 compares the energy levels, and the spin character of the five G-matrix interactions are again reflected in the ground-state doublet splittings of $J_{>,<} = J_c \pm 1/2$, though they are reduced when compared with those of $^4_\Lambda$H and $^7_\Lambda$Li. One sees that the $J_> = 2^-$ partner of the ground-state doublet comes below $J_<$ in the cases of JA and JB, and among others the splitting is extremely large for JB, indicating again the strong $^3S_1$ attraction. Note that the $J_>(J_<)$ state tends to be affected dominantly by $S = 1 (S = 0)$ two-body interaction. Both ND and NF provide almost degenerate ground-state doublets (the result of ND is not shown in Fig. 3). In contrast to above cases, NS gives rise to a large splitting with the $J_< = 1^-$ ground state as a consequence of the

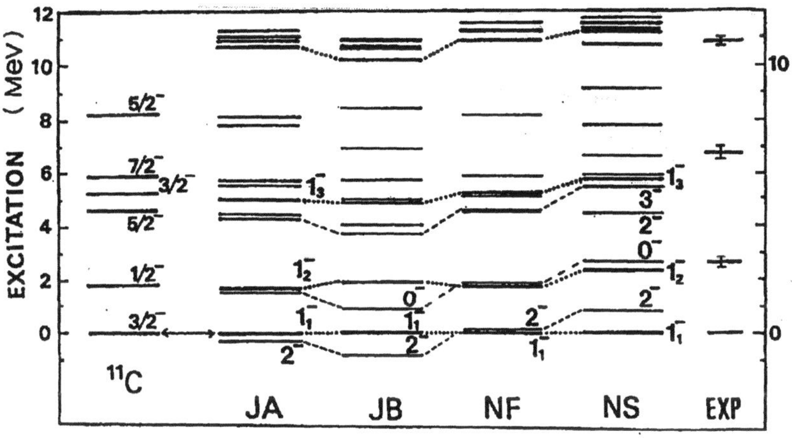

Figure 3. Calculated (ND) and experimental [1] energy levels of $^{12}_{\Lambda}$C.

strong $^1S_0$ attraction.

The theoretical $^{12}C(\pi^+, K^+)^{12}_{\Lambda}C$ cross sections, which have been calculated on the basis of the elementary amplitudes consisting of both spin-flip and spin-nonflip components, are compared in Table 3 with the values known from the BNL experiments [19,20].The basic feature of the excitation function, which has two pronounced peaks at the ground state and at 10.9 MeV excitation, can be well explained with the DWIA theory with the shell-model mixed wave functions, as shown also in Fig. 4. It is interesting that the KEK-SKS experiment [1] has confirmed new core-excited states clearly at $E_x \simeq 2.6$ MeV and 6.8 MeV for the first time. Theoretically they were predicted in the shell model [18,21], and they are explained primarily as the core excited states involving the following weak-coupling structures, respectively:

$$[^{11}C(1/2^-; 2.00\text{MeV}) \otimes s^{\Lambda}](J = 1_2^-) \text{ and } [^{11}C(3/2^-; 4.80\text{MeV}) \otimes s^{\Lambda}](J = 1_3^-).$$

The detailed comparison leads to the question, however, that the observed energy difference $E(1_2^-) - E(1_1^-) \simeq 2.6$ MeV is significantly larger than the core excitation energy observed in ^{11}C(2.00 MeV). This suggests a large amount of mixing between the lowest two bases shown above. As concerning the 6.8 MeV peak, which is much higher than any of the calculated value, there might be strong affection from other core-excited states in ^{11}C.

In order to emphasize the necessity to extend the shell-model space, we add the estimates of cross sections for these core-excited states obtained up to now within the 'standard' shell model. In Fig. 5 we show the theoretical ground-state (π^+, K^+) cross sections (solid line with open squares) calculated with five YNG interactions and, as discussed above, they are essentially in good agreement (except JB) with the observed value of $d\sigma/d\Omega(1_1^-)^{\exp}$ within possible experimental error [1, 19, 20]. It is remarkable that JB clearly fails to reproduce the value because its 1_1^- wave function deviates greatly from

Table 3. The $^{12}C(\pi^+, K^+)^{12}_{\Lambda}C$ reaction cross sections calculated with the shell-model wave functions (ND). The italic numbers denote the group* cross sections.

level	energy E_Λ [MeV]	$d\sigma/d\Omega$ [μb/sr] at $p_\pi = 1.04$ GeV/c			
EXP	CAL(J^π; E_x)	CAL(5°)	CAL(10°)	CAL(15°)	EXP
-10.8	-10.76(1_1^-;G.S.)*	12.48[*15.4*]	7.73[*10.3*]	3.36[*4.9*]	*7.2*(10°) [1]
	-10.52(2_1^-;0.14)*	0.28	0.67	0.62	*8.5*(10°) [19]
-8.3	-9.01(1_2^-;1.75)*	2.65	1.84	0.91	*10.4*(10°) [20]
-6.08	-5.86(1_3^-;4.90)	1.60	1.00	0.44	
	-0.76(2_1^+;10.0)**	9.08	5.52	2.19	
	-0.68(3_1^+;10.1)**	0.29	0.69	0.60	
0.1	-0.16(2_2^+;10.6)**	7.08[*17.6*]	4.58[*12.8*]	2.09[*5.5*]	*17.0*(5.6°) [19]
	0.10(0_1^+;10.9)**	1.10	1.03	0.65	
	1.02(2_3^+;11.8)	3.08	2.03	0.88	

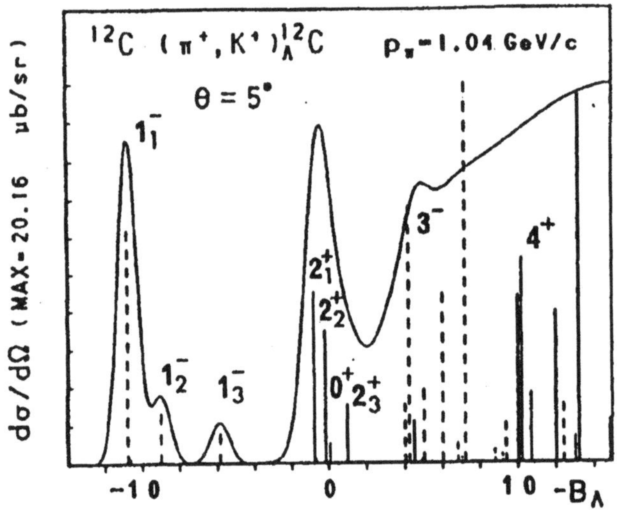

Figure 4. The (π^+, K^+) excitation spectrum at $\theta = 5°$ calculated with ND in DWIA.

the normal spin structures due to the too strong ^3E attraction. In Fig. 5 we also plot the strength ratios $R_{n1} = [d\sigma/d\Omega(1_n^-)]/[d\sigma/d\Omega(1_1^-)]$ for 1_2^- and 1_3^- with respect to the ground-state strength.Generally speaking R_{21} are possibly explained with the JA, ND, NF and NS wave functions if a slight improvement is maintained. For the third state, however, the theoretical ratios R_{31} with five interactions seem hard to explain the observed value, suggesting that it seems hard to attribute the third peak at 6.8 MeV as the 1_3^- theoretical peak.

From the above comparison new core-excited states should be further taken into account to explain the KEK data for core-excited states consistently. The candidates to be properly included are

$$[^{11}C(7/2^-; 6.48\text{MeV}) \otimes s^\Lambda](J = 3^-) \text{ and } [^{11}C(5/2^+; 6.90\text{MeV}) \otimes s^\Lambda](J = 2^+).$$

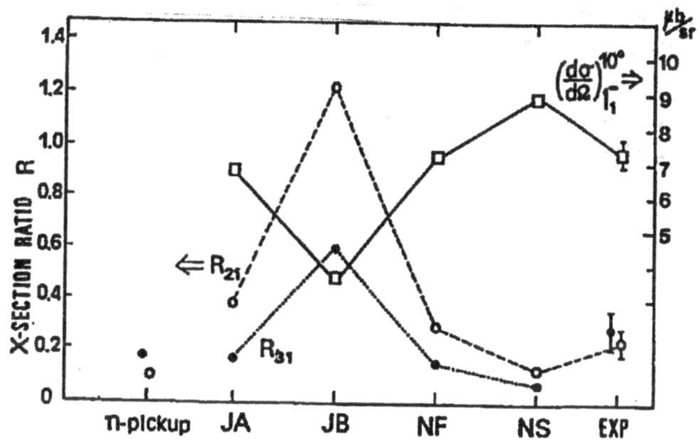

Figure 5. The ground state 1_1^- cross section (square) of $^{12}C(\pi^+, K^+)^{12}_{\Lambda}C$ at $p_\pi = 1.04$ GeV/c and $\theta = 10^o$, and the relative strengths of 1_2^- (open circle) and 1_3^- (dot). The experimental data are from ref. [1].

Both ^{11}C states are significantly excited in the n-pickup reaction on ^{12}C. Especially the latter state mixes the $[p^{-1}p^{\Lambda}]2^+$ component, and as a result the observed large formation rate might be explained [22]. The corresponding theoretical description needs further extension of the shell-model space.

In the above discussion we pointed out a necessary extension of the model space from a viewpoint of structure analysis. On the other hand, the effective $\Lambda - N$ interactions should be also improved accordingly. It is noted that, in spite of the limited experimental data, several authors [17, 23, 24] made great efforts in finding appropriate parameters of effective $\Lambda - N$ interaction in a phenomenological way. The idea of ref. [23] was to determine the radial matrix elements $\langle V_i(r) \rangle$ from the observed spectroscopic data:

$$v_{N\Lambda}(r) = V_0(r) + V_N(r)(\mathbf{s}_N \cdot \mathbf{l}_{\Lambda N}) + V_\sigma(r)(\mathbf{s}_N \cdot \mathbf{s}_\Lambda) + V_\Lambda(r)(\mathbf{s}_\Lambda \cdot \mathbf{l}_{\Lambda N}) + V_T(r)\hat{S}_{12}$$

$$\begin{aligned}\langle p^N_{j_1} s^{\Lambda}_{1/2} | v | p^N_{j_2} s^{\Lambda}_{1/2} \rangle_k &= a_V \langle V_0 \rangle + a_N \langle V_N \rangle + a_\sigma \langle V_\sigma \rangle + a_\Lambda \langle V_\Lambda \rangle + a_T \langle V_T \rangle \\ &= a_V \cdot \bar{V} \ + a_N \cdot S_N \ + a_\sigma \cdot \Delta_\sigma + \ a_\Lambda \cdot S_\Lambda + \ a_T \cdot T.\end{aligned}$$

Table 4. Three sets of parameters determined for the $N(0p) - \Lambda(0s)$ interactions.

Radial m.e.	GSD('78) [23]	MGDD('85) [24]	FMZE('91)[17]
$\bar{V} \equiv \langle V_0(r) \rangle$			
$\Delta_\sigma \equiv \langle V_\sigma(r) \rangle$	0.15	0.50	0.30
$S_N \equiv \langle V_{S_N}(r) \rangle$	-0.21	-0.08	0.10
$S_\Lambda \equiv \langle V_{S_\Lambda}(r) \rangle$	0.57	-0.04	-0.02
$T \equiv \langle V_T(r) \rangle$	0	0.04	0.02

Three sets of radial matrix elements are obtained up to now and are listed in Table 4 [17,25]. However they seem much different from each other due to the very limited data. Nevertheless it is interesting to decompose the YNG interactions into the five components so as to see their spin-tensor characters. The decompositions have been carried out by calculating the inverse matrix $A^{-1} \equiv (\tilde{a}_{ik})$ of the matrix of transformation coefficients $A \equiv (a_{ki})$:

$$\begin{pmatrix} \bar{V}_0 \\ S_N \\ \Delta_\sigma \\ S_\Lambda \\ T \end{pmatrix} = \begin{pmatrix} & & \\ & \tilde{a}_{ik} & \\ & & \end{pmatrix} \begin{pmatrix} \langle p^N_{1/2} s^\Lambda_{1/2} | v | p^N_{1/2} s^\Lambda_{1/2} \rangle_{0^-} \\ \langle \qquad |v| \qquad \rangle_{1^-} \\ \langle p^N_{3/2} s^\Lambda_{1/2} | v | p^N_{1/2} s^\Lambda_{1/2} \rangle_{1^-} \\ \langle p^N_{3/2} s^\Lambda_{1/2} | v | p^N_{3/2} s^\Lambda_{1/2} \rangle_{1^-} \\ \langle \qquad |v| \qquad \rangle_{2^-} \end{pmatrix}$$

For JA and JB we have only central interactions with us. From the radial matrix elements of the spin-spin term, Δ_σ in Table 5, one sees big difference between JB (JA) and NS which is attributed to the difference of spin-singlet/triplet ratio shown already in Sect. 2. We point out that the Nijmegen potentials have large LS component $((S_N + S_\Lambda)/2)$ and very small antisymmetric-LS component (ALS:$(S_N - S_\Lambda)/2$), while that both LS and ALS components are large in the Jülich potentials. Further analysis is still necessary to compare YNG interaction m.e. with the phenomenological values.

Table 5. Decomposition of the YNG interaction matrix elements $\langle p^N_{j_1} s^\Lambda_{1/2} | v | p^N_{j_2} s^\Lambda_{1/2} \rangle_k$ into the radial matrix elements of five components.

	JA	JB	ND	NF	NS"	J$\tilde{\text{A}}$	J$\tilde{\text{B}}$
$k_F =$	(0.97)	(1.21)	(1.31)	(1.15)	(1.07)		
Radial m.e.	Central	Central	Full	Full	Cent+LS	Full[26]	Full[26]
$\bar{V} \equiv \langle V_0(r) \rangle$	-1.355	-1.319	-1.492	-1.383	-1.485	-0.931	-1.044
$\Delta_\sigma \equiv \langle V_\sigma(r) \rangle$	0.634	2.249	0.187	-0.163	-1.213	0.106	1.301
$S_N \equiv \langle V_{S_N}(r) \rangle$	0.0	0.0	0.232	0.267	0.351	0.327	0.323
$S_\Lambda \equiv \langle V_{S_\Lambda}(r) \rangle$	0.0	0.0	0.345	0.363	0.351	0.006	-0.017
$T \equiv \langle V_T(r) \rangle$	0.0	0.0	0.0	0.0	0.0	-0.061	-0.063

5 The Λ spin-orbit splittings from ${}^{16}_{\Lambda}$O and heavy hypernuclei

The amount of the one-body Λ spin-orbit splitting has not been established precisely in spite of its importance. The analysis of two substitutional peaks in the ${}^{16}\mathrm{O}(K^-, \pi^-){}^{16}_{\Lambda}\mathrm{O}$ reaction, which are atributed to $[0p^{-1}_{1/2} 0p^\Lambda_{1/2}]J = 0^+$ and $[0p^{-1}_{3/2} 0p^\Lambda_{3/2}]J = 0^+$, leads to a very small splitting: $\delta_\Lambda = \epsilon_\Lambda(0p_{1/2}) - \epsilon_\Lambda(0p_{3/2}) =$ 0.8$\pm$0.7 MeV as deduced by Bouyssy [27]. May et al. [28] gave another empirical value of $\delta_\Lambda = 0.36 \pm 0.3$ MeV. However the energy resolutions involved were not good and, on the other hand, they simply neglected unknown details of the residual $\Lambda - N$ interactions in their analysis. Recently there appeared new experimental data: one from the high resolution analysis of emulsion data of

$^{16}O(K^- \text{stopped}, \pi^-)^{16}_{\Lambda}O$ events [29] and the other from the measurements of $^{89}Y(\pi^+, K^+)^{89}_{\Lambda}Y$ and $^{139}La(\pi^+, K^+)^{139}_{\Lambda}La$.

5.1 *Structure of* $^{16}_{\Lambda}O$

The analysis of the observed pion spectrum for the $^{16}O(K^- \text{stopped}, \pi^-)^{16}_{\Lambda}O$ events yields two peaks, large and small ones with the energy separation of 1.56 MeV. According to the theoretical estimates of the formation rate of $^{16}_{\Lambda}O(J_f)$, the $J = 2^+$ state should be produced approximately twice as large as the $J = 0^+$ state irrespective of the Kaon atomic orbitals of $3d$ or $2p$. This fact leads to:

$$\Delta E^{\exp} = E(0_1^+) - E(2_1^+) = 1.56 \pm 0.12\,\text{MeV}.$$

Based on the empirical splitting of the neutron orbits, $\delta_N = \epsilon_N(0p_{1/2}) - \epsilon_N(0p_{3/2}) = 6.18$ MeV, the eigenenergies of $J = 0_1^+$ and 2_1^+ states are calculated by adopting the following basis configurations:

$$J = 0^+ : [0p_{1/2}^{-1} 0p_{1/2}^{\Lambda}] \text{ and } [0p_{3/2}^{-1} 0p_{3/2}^{\Lambda}],$$
$$J = 2^+ : [0p_{1/2}^{-1} 0p_{3/2}^{\Lambda}],\ [0p_{3/2}^{-1} 0p_{3/2}^{\Lambda}] \text{ and } [0p_{3/2}^{-1} 0p_{1/2}^{\Lambda}].$$

The residual $p_\Lambda - h_N$ interactions are fully taken into account in the diagonalization, with the Λ spin-orbit splitting δ_Λ being changed gradually. In order to see the role of the residual interactions, first let us look at the results with no Λ spin-orbit splitting ($\delta_\Lambda = 0$), which are summarized in Table 6(a). The effect of the noncentral forces (in parentheses) appears mostly small but it seems not negligible when one wants to discuss the energies less than 0.3 MeV (see below).

Next we change δ_Λ in diagonalization procedure, which results are displayed in Fig. 6. It is interesting to find that the excitation energy $E(2^+)$ is almost independent of δ_Λ. This is because the non-diagonal $p - h$ matrix elements are small enough when they are compared with the unperturbed excitation energies given mainly by $\delta_N^{(\exp)} = 6.18$ MeV. As a result, the obtained $E(2_1^+)$ is hardly shifted from the diagonal $p - h$ interaction $V_{ph}^{(11)} \equiv V_{ph}(0p_{1/2}^{-1} 0p_{3/2}^{\Lambda}; J = 2^+)$. In other words, the 2_1^+ state can be decsribed with the lowest-energy single configuration to a very good approximation. Thus the energy of the 2_1^+ state is predicted to be rather stable: i.e. $E(2^+) = 0.57 - 0.98$ MeV above the origin $\epsilon_N(0p_{1/2}^{-1}) + \epsilon_\Lambda(0p_{3/2})$.

On the other hand, as clearly seen in Fig. 6, the 0_1^+ energy obtained in the diagonalization changes approximately linearly as a function of δ_Λ. This feature is understandable, because that the unperturbed energy itself depends directly on δ_Λ. However, the positions of $J = 0^+$ curves are more scattered than the $J = 2^+$ case, as the values of the $p - h$ matrix element $V_{ph}^{(11)} \equiv V_{ph}(0p_{1/2}^{-1} 0p_{1/2}^{\Lambda}; J = 0^+)$ are considerably different for different potentials. In any case, we emphasize that the energy separation between 0_1^+ and 2_1^+ states provides an important

Table 6. The eigenenergies, $E(0_1^+)$ and $E(2_1^+)$ in $^{16}_{\Lambda}O^*$, and particular $p-h$ interactions as calculated with YNG ΛN central potentials (in MeV). Entries in the parentheses are for the nocentral potentials. In case (b), δ_Λ is so chosen that $E(0_1^+)-E(2_1^+)$ reproduces ΔE^{exp}=1.56 MeV after diagonalization.

(a) $\delta_\Lambda = 0$	JA	JB	ND	NF	NS
$V_{ph}(0p_{1/2}^{-1}0p_{1/2}^{\Lambda};0^+)$	1.81	2.78	1.70(1.63)	0.88(1.10)	0.35(0.26)
$E_0(0_1^+)$	1.37	1.56	1.63(1.50)	0.86(1.08)	0.34(0.26)
$V_{ph}(0p_{1/2}^{-1}0p_{3/2}^{\Lambda};2^+)$	0.72	0.88	0.91(0.98)	0.65(0.65)	0.73(0.77)
$E_0(2_1^+)$	0.71	0.76	0.91(0.97)	0.65(0.57)	0.71(0.71)
$V_{ph}(0^+) - V_{ph}(2^+)$	1.09	1.90	0.79(0.65)	0.23(0.53)	- 0.38(-0.51)
$E_0(0_1^+) - E_0(2_1^+)$	0.65	0.80	0.72(0.53)	0.21(0.51)	-0.37(- 0.45)
(b) δ_Λ changed	JA	JB	ND	NF	NS
δ_Λ	0.95	0.85	0.85(1.05)	1.35(1.05)	1.90(2.00)
$H_{11}(0^+)$	2.76	3.63	2.55(2.68)	2.23(2.15)	2.25(2.26)
$E(0_1^+)$	2.26	2.32	2.46(2.53)	2.20(2.12)	2.24(2.26)
$E(2_1^+)$	0.71	0.77	0.91(0.98)	0.65(0.57)	0.71(0.72)
$E(0_1^+) - E(2_1^+)$	1.55	1.55	1.55(1.55)	1.55(1.55)	1.53(1.54)

information on δ_Λ. It is interesting to deduce the necessary Λ spin-orbit splitting δ_Λ which reproduces the experimental separation energy of 1.56 MeV. The appropriate values are all larger than 0.8 MeV:

$$\delta_\Lambda \simeq 0.95(\mathrm{JA}),\ 0.85(\mathrm{JB}),\ 0.85\text{-}1.05(\mathrm{ND}),$$
$$1.35\text{-}1.05(\mathrm{NF}),\ 1.90\text{-}2.00(\mathrm{NS}) \text{ in MeV}.$$

For the latter three potentials one sees the effect of noncentral forces.

5.2 *The Λ spin-orbit splittings from heavy hypernuclei*

It is well known that the (π^+, K^+) reaction preferentially excites a series of a high-spin natural-parity states based on the conversion of neutron in the large-l orbit at surface. The typical $1p_\Lambda - 1h_N$ states in $^{89}_{\Lambda}Y$ and $^{139}_{\Lambda}La$ are partly shown in Table 7 together with the DWIA cross section estimates. As several spin-orbit partners can be excited pronouncedly, one of the interesting aspects in heavy hypernuclei is to disclose the hyperon spin-orbit splittings for which there has been no direct measurement up to now.

Recently Nagae [2] performed the reanalysis of the KEK data, suggesting 2.5 ± 0.2 MeV for the $0f_{7/2}^{\Lambda} - 0f_{5/2}^{\Lambda}$ splitting in $^{89}_{\Lambda}Y$. This value corresponds to the strength $V_{\mathrm{LS}} = 6 - 7$ MeV of the Woods-Saxon prescription, and it is approximately twice as large as the previous understanding. The theoretical excitation functions for the $^{89}_{\Lambda}Y$ and $^{139}_{\Lambda}La$ production are in good agreement with the experimental ones [30]. The strength seems to be compatible with the $\delta_\Lambda > 0.8$ MeV obtained above for $^{16}_{\Lambda}O$. However we have to wait for the dedicated experiment planned at KEK before extracting a definite conclusion.

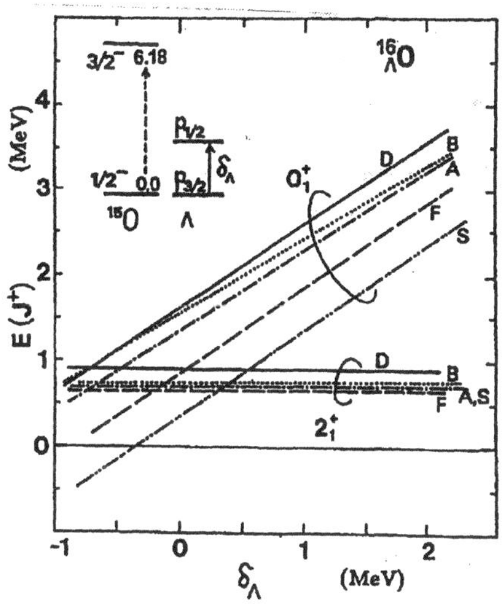

Figure 6. Eigenenergies $E(0_1^+)$ and $E(2_1^+)$ are plotted as a function of the Λ spin-orbit splitting δ_Λ.

Table 7. The DWIA cross sections $d\sigma/d\Omega(\theta = 0^o)[\mu b/sr]$ for the $[(0lj)_N^{-1}(0lj)^\Lambda]$ states excited in the (π^+, K^+) reaction. The Woods-Saxon wave functions for both N and Λ are used.

	N-hole	$0s^\Lambda_{1/2}$	$0p^\Lambda_{3/2}0p^\Lambda_{1/2}$	$0d^\Lambda_{5/2}0d^\Lambda_{3/2}$	$0f^\Lambda_{7/2}0f^\Lambda_{5/2}$	$0g^\Lambda_{9/2}0g^\Lambda_{7/2}$
	$0g^{-1}_{9/2N}$	1.88	3.79 3.73	6.07<9.73	8.80<17.03	(unbound)
$^{89}_{\Lambda}$Y	$0f^{-1}_{5/2N}$	0.95	3.49 ≫ 0.14	7.30 ≫ 0.59	11.29≫1.01	(———)
	$0h^{-1}_{11/2N}$	0.89	2.06 1.89	3.38<5.38	5.41<10.40	7.68<16.30
$^{139}_{\Lambda}$La	$0g^{-1}_{7/2N}$	0.50	2.03 ≫0.14	4.70 ≫0.44	8.26 ≫0.85	12.15≫1.40

6 Conclusion

We have demonstrated interesting aspects of hypernuclear structure outputs when we employed the YNG effective interactions derived from the meson-theoretical potentials from Nijmegen and Jülich. They are remarkably different in spin-character from each-other, which are shown to be reflected in light hypernuclear structures. We have also emphasized importance of trying to make a bridge (bridges) from both directions including phenomenological approaches.

Acknowledgement.

The present paper is based on several collaborations and helpful discussions with Professors K. Itonaga, Y. Yamamoto, L. Majling, M. Sotona, R.H. Dalitz, D.H. Davis and D.N. Tovee. The author is grateful to all of them. He also thanks to Professors R. Mach and M. Sotona for extending the hospitality at the Institute of Nuclear Physics, Czech Academy of Sciences.

References

1. T. Hasegawa: PhD. Thesis, University of Tokyo 1994; T. Hasegawa et al.: Phys. Rev. Lett. **74**, 224 (1995)

2. T. Nagae: In: *Proc. 23rd Intern. Symposium on Nuclear and Particle Physics with Meson Beams in the 1 GeV/c Region* (March 1995, Tokyo), to be published.

3. For example, see Nucl. Phys. **A585** (1995); *Proc. 23rd Intern. Symposium on Nuclear and Particle Physics with Meson Beams in the 1 GeV/c Region* (March 1995, Tokyo), to be published.

4. M.M. Nagels, T.A. Rijken, and J.J. de Swart: Phys. Rev. **D12**, 744 (1975); **D15**, 2547 (1977); **D20**, 1633 (1979)

5. Y. Yamamoto and H. Bandō: Prog. Theor. Phys. **73**, 905 (1985)

6. P.M.M. Maessen, T.A. Rijken, and J.J. de Swart: Phys. Rev. C**40**, 226 (1989); Nucl. Phys. **A547**, 245c (1992)

7. B. Holzenkamp, K. Holinde, and J. Speth: Nucl. Phys. **A500**, 485 (1989)

8. A. Reuber et al.: Czech. J. Phys. **42**, 1115 (1992) and references therein

9. Y. Yamamoto et al.: Czech. J. Phys. **42**, 1249 (1992)

10. Y. Yamamoto, T. Motoba, H. Himeno, K. Ikeda, and S. Nagata: Prog. Theor. Phys. Suppl. **117**, 361 (1994)

11. M. Beiner et al.: Nucl. Phys. **A238**, 29 (1975)

12. T. Motoba, and Y. Yamamoto: Nucl. Phys. **A585**, 29c (1995)

13. T. Motoba, H. Bandō, and K. Ikeda: Prog. Theor. Phys. **70**, 189 (1983)

14. M. May et al.: Phys. Rev. Lett. **51**, 2085 (1983)

15. A. Gal, J.M. Soper, and R.H. Dalitz: Ann. Phys.(N.Y.) **63**, 53 (1971); **72**, 445 (1972); **113**, 79 (1978)

16. R.H. Dalitz, and A. Gal: Ann. Phys.(N.Y.) **116**, 167 (1978); **131**, 314 (1981); D.J. Millener et al.: Phys. Rev. C**39**, 499 (1985)

17. V.N. Fetisov, L. Majling, J. Žofka, and R.A. Eramzhyan: Z. Phys. **A339**, 399 (1991)

18. K. Itonaga, T. Motoba, and H. Bandō: Prog. Theor. Phys. **84**, 291 (1990)

19. C. Milner et al.: Phys. Rev. Lett. **54**, 1237 (1985)

20. P.H. Pile et al.: Phys. Rev. Lett. **66**, 2585 (1991); R.E. Chrien, Nucl. Phys. **A478**, 705c (1988)

21. K. Itonaga, T. Motoba, O. Richter, and M. Sotona: Phys. Rev. C**49**, 1045 (1994)

22. This possibility has been also pointed out by A. Gal and by L. Majling (private communication)

23. A. Gal, J.M. Soper, and R.H. Dalitz: Ann. Phys.(NY) **113**, 79 (1978)

24. D.J. Millener, A. Gal, C.B. Dover, and R.H. Dalitz: Phys. Rev. C**31**, 499 (1985)

25. L. Majling, and R.A. Eramzhyan: Prog. Theor. Phys. Suppl. **117**, 55 (1994)

26. T.T.S. Kuo, and J. Hao: Prog. Theor. Phys. Suppl. **117**, 351 (1994)

27. A. Bouyssy: Phys. Lett. **84B**, 41 (1979); **91**, 15 (980)

28. M. May et al.: Phys. Rev. Lett. **47**, 1106 (1981)

29. R.H. Dalitz, D.H. Davis, T. Motoba, and D.N. Tovee: in preparation

30. T. Motoba: In: *Proc. 23rd Intern. Symposium on Nuclear and Particle Physics with Meson Beams in the 1 GeV/c Region* (March 1995, Tokyo), to be published

Few-Body Systems Suppl. 9, 510–512 (1995)

Conference Summary

R. J. Peterson*

Nuclear Physics Laboratory, University of Colorado, Boulder, CO 80309-0446, USA

This was a very advanced conference, bringing together experts in the many interlocking features of the relations between mesons and light nuclei, and the organizers put these speakers into a coherent program. New data of high quality and relevance were shown, and both old and new theoretical methods were presented, based increasingly upon fundamental processes and much less upon empirical fitting. I will not try to summarize the many results and conclusions reached, but give my general and personal observations on the skeleton of what we have been doing, and how we have come to learn what we know. Rather than highlights of the Conference, I will end with a summary of the promises that were made, and what we can expect at the next in this series of Conferences on Mesons and Light Nuclei.

Three mesons were treated, not just two as in the preceeding Conferences. The pion retains its principal role as the foundation of the strong interaction, and fundamental pion-nucleon observables are now finally under control. The best of the modern experiments and the most complete and soundly-based analyses now work together to let us state that we do know the free pion-nucleon interactions to high accuracy and good reliability over a wide range of energies. Kaon reactions on nuclei, with or without the exchange of strangeness, continue to improve, but we have seen no modern improvements or checks on the basic kaon-nucleon interactions. Since so much depends on these observables, and since we have seen great changes in our understanding of the pion-nucleon interactions, perhaps it would be wise to re-open studies of kaon-nucleon interactions, both by new experiments and through the correct fitting methods. The third meson, making a new and large impression on this Conference, is the η. There has been an explosion of new experimental results on η's, with both pion and photon beams. We now know the form of the η-nucleon coupling quite well, and η-nucleon cross sections are being inferred from production experiments on complex nuclei, where propagation of the η determines the final spectra. Reactions involving the η were shown at this Conference to violate

**E-mail address:* peterson@spectr.colorado.edu

both charge symmetry and chiral symmetry. This meson is surely destined to be an important tool, with properties sufficiently different from those of the pion or the kaon to ask new questions.

These meson-nucleon interactions are the driving terms for meson interactions with and within light nuclei. Why do we emphasize the light nuclei? The first of the reasons is the availibility of demonstrated good nuclear wave functions. An afternoon of this Conference was devoted to proving this, where these wave functions were used to generate observables to compare to very precise nucleon and electromagnetic reactions, across a wide range of momentum transfers. The excellent agreements shown were completely convincing. Very good nuclear structure wave functions are indeed available for the light nuclear systems.

Since the shell structure and spacing of light nuclei are so simple, these form the laboratory to learn baryon-baryon interactions, where modern boson exchange theories should also include interactions of hyperons with nucleons or with one another. We measure the spectra and infer the interactions responsible, where no other information can be obtained because we have no hyperon beams. This is a well-developed field for Λ's in hypernuclei, and new experiments with superior resolution are becoming available to use or test the models. The future also looks promising, for Λ hypernuclei and perhaps even for Σ and doubly-strange hypernuclei. The large theoretical effort that has grown over the years will soon be matched by equally good data.

Another reason to study light nuclei is the limited opportunity for reactions to become too complicated. Clear evidence for an explicit three-body nuclear effect was made possible by the complete description of the π^+ absorption reaction reaction on ^{3}He. We also saw the evolution of reaction complexity in the pion double charge exchange spectra from helium, with a simple story, to oxygen, where the spectra were merely statistical.

Polarized nuclear targets have made great advances in recent years, with ^{3}He in particular now available in dense, highly polarized samples. Meson reactions with polarized targets can give spin observables that are sensitive to interferences not otherwise visible. For instance, a second-order, and therefore small, effect of delta rescattering in elastic π^+ scattering from ^{3}He seems also to be found from these experiments. At a more fundamental level, highly polarized internal targets of H, D, and ^{3}He are being used to determine the microscopic features of the spins of the objects within these light systems, using high energy electron beams.

A fourth feature of light nuclei that we find attractive is the range of densities available, both as averages and from a greater relative amount of surface. This allows access of hadrons to interesting interior effects including multibody correlations (seen in absorption, for instance), meson exchange currents (affecting kaon, as well as electron, elastic scattering), and alterations of the natures of nucleons within the interior, using appropriate kinematics to sense individual nucleons. We saw evidence that cross sections for high energy K^+ and pions on nuclear nucleons may be larger than those in free space. These differences show signs of depending upon the density where the interior scattering occured, in both total and elastic scattering data.

Pion or photon beams can excite nucleons to N^*'s within nuclei, as well as in free space. The photon data show that the widths of these N^* become much greater within nuclei, but the strength to excite them is retained even in heavy nuclei. Pion total cross sections on complex nuclei show a decrease relative to free nucleons, due to distortions and absorptions that shield some nucleons from the beam, but the scanty data seem to indicate also a broadening of the N^* within nuclei. Optical model and other methods to account for the distortions give predictions below those pion data. Are the pion-nucleon cross sections therefore greater than assumed, or are distortions less than computed? Theoretical methods for high energy meson reactions were shown at this Conference to be highly reliable, so this sort of question can be addressed now. Several types of medium effects are expected, and new data are needed to note them and possibly to separate them.

There were many satifying moments at this Conference, where ideas and methods that should work well were shown to do so by comparison to sensitive experiments. We obtained a general sense that we are gaining both information and, more importantly, wisdom. Our goal is a tested understanding of how the simplest hadrons, our mesons, interact with and within the simplest of the complex nuclei. We see better experiments, closely connected to theory. Theoretical efforts are using intelligent assumptions, approximations and methods, increasingly based on fundamental and general theories of the strong interaction. It seems to be true that only at conferences such as this that all these are brought together effectively for us all to see.

What is coming next? Many speakers gave us promises of what we can expect next time at a Conference on Mesons and Light Nuclei. High resolution hypernuclear spectra from several classes of reactions will soon be available, with natural parity states emphasized using a meson beam, with spin excitations from electromagnetic reactions complementing these. Baryon-baryon interactions based on natural one boson exchange models, within a relativistic mean field theory, will be available to compare to these data. Perhaps we will even finally see usable spectra for Σ and double-Λ hypernuclei. The taste we saw for a possible H particle will be confirmed (or not) solidly very soon. The η will be the subject of new experiments, with even the current data enough to attract new theoretical work in several new directions. High energy meson scattering and reactions will give a deeper and broader base of data, with reliable theories now ready to be applied. Here, then, is the best way to study the excitations and properties of N^* within nuclei. New methods to deal with the off-shell nature of these scatterings seem to be available. Also, high energy meson beams for meson production are splendid tools to study the limits of chiral symmetry. And finally, we can expect the theoretical foundations of this field to be yet more strongly connected with fundamental constraints, of symmetries, field theories and QCD.

The next Conference on Mesons and Light Nuclei promises to show yet more progress towards the questions we asked at this one. See you there.

List of Participants

Jiří ADAM
Institute of Nuclear Physics
250 68 Řež
Czech Republic
adam@ujf.cas.cz

Jürgen AHRENS
Institut für Kernphysik
Universität Mainz
D-55099 Mainz
Germany
ahrens@vkpmzp.kph.uni-mainz.d

John R.M. ANNAND
Department of Physics and Astronomy
University of Glasgow
University Avenue
Glasgow G12 8QQ
Scotland
jrma@np.ph.gla.ac.uk

Bernd BASSALLECK
University of New Mexico
Dept. of Physics
Albuquerque, N.M. 87131
U.S.A.
bossek @ bootes.unm.edu

Mijo BATINIĆ
Rudjer Boskovic Institute
Bijenicka c. 54
P.O.Box 1016
41001 Zagreb
Croatia
batinic@olimp.irb.hr

Yuri A. BATUSOV
Joint Institute for Nuclear Research
Laboratory of Nuclear Problems
Moscow Region
141980 Dubna
RUSSIA
yuabat@nusun.jinr.dubna.su

Vladimir B. BELYAEV
Joint Institute for Nuclear Research
Lab. Theoretical Physics
141980 Dubna
Russia
belyaev@thsun1.jinr.dubna.su

Cornelius BENNHOLD
Centre of Nuclear Studies
Department of Physics
The George Washington University
Washington, DC 20052
U.S.A.
bennhold@gwuvm.gwu.edu

Barry L. BERMAN
Corcoran Hall, Suite 208
The George Washington University
Washington, D.C. 20052
U.S.A.
berman@gwuvm.gwu.edu

Edmund BOSCHITZ
Inst. für Exp. Kernphysik
Universität Karlsruhe
76021 Karlsruhe
Germany

Tancredi BOTTO
NIKHEF-K
409 Kruislaan
1009 DB AMSTERDAM
The Netherlands
tancredi@nikhefk.nikhef.nl

Jo van den BRAND
NIKHEF-K
P.O.Box 41822
1009 DB Amsterdam
The Netherlands
elise@nikhefk.nikhef.nl

Volker BURKERT
CEBAF
12000 Jefferson Avenue
Newport News
Virginia 23602
U.S.A.
burkert@cebaf.gov

Petr BYDŽOVSKÝ
Institute of Nuclear Physics
250 68 Řež
Czech Republic
bydzovsky@ujf.cas.cz

P. CAMERINI
Inst. Nazionale di Fisica Nucleare
Dip. di Fisica dell' Universita 'di
Trieste
Italy
grion@trieste.infn.it

Luciano CANTON
Sez. INFN e Dipartimento di Fisica
dell'Universita' di Padova
via F. Marzolo, n.8
35131 Padova
Italy
canton@padova.infn.it

Piergiorgio CERELLO
Instituto di Fisica Generale
Universita di Torino
Via P. Giuria 1
Torino
Italy
cerello@toux30.to.infn.it

Rene CEULENEER
Universite de Mons -Hainaut
Faculte des Sciences
Avenue Maistriau 19
B-7000 Mons
Belgium
sceule@vm1.umh.ac.be

Colston CHANDLER
Dept. of Physics and Astronomy
University of New Mexico
Albuquerque, NM 87131
U.S.A.
chandler@triton.unm.edu

Stanislaw CIECHANOWICZ
Institute of Theoretical Physics
University of Wroclaw
Pl. Maksa Borna 9
50 - 204 Wroclaw
Poland
ciechano@ift.uni.wroc.pl

Aleš CIEPLÝ
Hebrew University
Israel
ales@vms.huji.ac.il

Janusz DABROWSKI
Theor.Div.
Institute for Nuclear Studies
Hoza 69
Pl-00 681 Warsaw
Poland
dabrnucl@fuw.edu.pl

Maurizio DE SANCTIS
INFN sezione di Roma
c/o Dipartimento di Fisica
Universita' " La Sapienza"
P.le A. Moro, 2 00185 Roma
Italy
desanctis@roma1.infn.it

Jean P. DIDELEZ
Institut de Physique Nucleaire
IPN - Orsay - Bat 100
91406 Orsay - Cedex
France
didelez@ipncls.in2p3.fr

Jan DOBEŠ
Institute of Nuclear Physics
250 68 Řež
Czech Republic
dobes@ujf.cas.cz

Peter J. DORTMANS
University of Melbourne
Department of Phyics
Parkville 3052
Victoria
Australia
pjd@higgs.ph.unimelb.edu.au

David J. ERNST
Department of Physics and Astronomy
Box 180 7B
Vanderbilt University
Nashville, TN 37235
U.S.A.
ernst@compsci.cas.vanderbilt.edu

P.J. FERNANDEZ de CORDOBA
Dept. de Fisica Teorica
Univ. Valencia
46100 Burjassot (Valencia)
Spain
fernande@evalvx.ific.uv.es

Lev V. FIL'KOV
Lebedev Phys.Inst.
Leninsky Prospekt,53
Moscow
Russia
h01fil@dsyibm.desy.de

E. FRIEDMAN
TRIUMF
4004 Wesbrook Mall
Vancouver, B.C.
V6T 2A3
Canada
elifried@triumf.ca

Salvatore FRULLANI
Inst. Nazionale di Fisica Nucleare
Sezione Sanita
Viale Regina Elena 299
Italy
frullani@iss.infn.it

Walter GLÖCKLE
Institut für Theoretische Physik
Ruhr-Universität Bochum
D - 44780 Bochum
Germany
walter.gloeckle
@ruba.rz.ruhr-uni-bochum.d400.de

Stefan GMUCA
Institute of Physics
Slovak Academy of Sciences
Dubravska cesta 9
842 28 Bratislava
Slovakia
gmuca@savba.sk

Jacek GOLAK
Jagiellonian University
Institute of Physics
4 Reymonta St.
PL 30059 Krakow
Poland
ufgolak@cyf-kr.edu.pl

Inna GRACH
ITEP
B.Cheremushkinskaya 25
Moscow 117259
Russia
grach@vxitep.itep.ru

N. GRION
Inst. Nazionale di Fisica Nucleare
Dip. di Fisica dell'Universita' di
Trieste
Italy
grion@trieste.infn.it

Toru HARADA
Department of Social Information
Sapporo Gakuin University
11-Bunkyodai, Ebetsu
Hokkaido 069
Japan
harada@rikvax.riken.go.jp

Otto HÄUSSER
Simon Fraser University
Burnaby
British Columbia
Canada V5A 1S6
hausser@triumf.ca

Satoru HIRENZAKI
University of Valencia
Departamento de Fisica Teorica
46100 Burjassot (Valencia)
Spain
zaki@evalvx.ific.uv.es

Dirk HÜBER
Ruhr-Universität Bochum
D - 44780 Bochum
Germany
hbo002@zam001.zam.kfa-juelich.de

Sabit KAMALOV
Joint Institute for Nuclear Research
Bogolubov Lab. Theor. Phys.
141980 Dubna
Russia
kamalov@theor.jinrc.dubna.su

Alejandro KIEVSKY
Inst. Nazionale di Fisica Nucleare
Dipartimento di Fisica
Piazza Torricelli 2
56100 Pisa
Italy
kievsky@pisa.infn.it

Jörg KÖHLER
Universität Basel
Inst.Exp.Kern.u. Teilchenphysik
Klingelbergstr. 82
CH-4056 Basel
Schweiz
koehler1@ubaclu.unibas.ch

Ubirajara van KOLCK
Department of Physics, FM-15
University of Washington
Seattle, WA 98195
U. S. A.
vankolck@alpher.phys.washington.edu

Sergei I. KRUGLOV
B. I. Stepanov Institute of Physics
F. Skaryna prosp. 70
220602 Minsk
Belarus
levchuk%basnet.minsk.by@demos.su

Anna KRUTENKOVA
Inst. Theor. Exper. Physics
B.Cheremushkinskaya, 25
Moscow 117259
Russia
krutenkova@vxitep.itep.ru

Vladimir KUKULIN
Institute of Nuclear Physics
Moscow State University
Moscow 119899
Russia
kukulin@compnet.msu.su

Juris LUKSTINS
Joint Institute for Nuclear Research
Lab. High Energies
Dubna, Moscow region 141980
Russia
juris@lhe17.jinr.dubna.su

V. V. KULIKOV
Inst. Theor. Exper. Phys.
Moscow, 117259
Russia
kulikov@vxitep.itep.ru

A. I. L'VOV
Lebedev Physical Institute
Leninsky Prospect 53
Moscow 117924
Russia
lvova@vxdesy.desy.de

Zhanat B. KURMANOV
Joint Institute for Nuclear Research
Bogolubov Lab. Theor. Phys.
141980 Dubna
Russia
zhanat@thsun1.jinr.dubna.su

Rostislav MACH
Institute of Nuclear Physics
250 68 Řež
Czech Republic
mach@ujf.cas.cz

Jean LABARSOUQUE
Centre d'Etudes Nucleaires
de Bordeaux Gradignan
Universite Bordeaux I
F- 33175 Gradignan Cedex
France
labars@frcpn11.in2p3.fr

Ruprecht MACHLEIDT
Department of Physics
University of Idaho
Moscow
ID 83843
U. S. A.
machleid@tamaluit.phys.uidaho.edu

Dmitriy LANSKOY
Institute of Nuclear Physics
Moscow State University
119899 Moscow
Russia
lanskoy@compnet.msu.su

Lubomír MAJLING
Joint Institute for Nuclear Research
Laboratory of Theoretical Physics
Moscow Region
141980 DUBNA
Russia
majling@theor.jinrc.dubna.su

Michael LEVCHUK
B.I. Stepanov Inst.Phys. of Belarus
Academy of Sciences
F. Scaryna prospect 70
220602 Minsk
Belarus
levchuk%basnet.minsk.by@demos.su

J. MAREŠ
TRIUMF
4004 Wesbrook Mall
Vancouver, B.C.
V6T 2A3
Canada
mares@erich.triumf.ca

H. MARKUM
Institut für Kernphysik
TU-Wien
Wiedner Hauptstr. 8-10/142
A-1040 Vienna
Austria
markum@ekpds1.tuwien.ac.at

Terry MART
Institut für Kernphysik
Universität Mainz
55099 Mainz
Germany
mart@vkpmzp.kph.uni-mainz.de

Leopold MATHELITSCH
Institut für Theoretische Physik
Universität Graz
Universitätsplatz 5
A-8010 Graz
Austria
mathelitsch@edvz.uni-graz.ada.at

June L. MATTHEWS
Massachusetts Institute of Technology
Room 26-433
77 Massachusetts Avenue
Cambridge
MA 02139 U.S.A.
matthews@mitlns.mit.edu

Kazuya MIYAGAWA
Okayama University of Science
1-1 Ridai-cho
Okayama 700
Japan
hbo004@djukfa11.bitnet

Dario MORICCIANI
INFN, Sez Roma 2
Via della Ricerca Scientifica 1
I-00173, Rome
Italy
moricc@graal1.lnf.infn.it

Chris MORRIS
Physics Division
Mail Stop H846
Los Alamos National Laboratory
Los Alamos, NM 87545
U.S.A.
morris@Lampf.lanL.gov

Toshio MOTOBA
Physics Department
Osaka Electro-Commun. Univ.
Hatsu-machi, Neyagawa
Osaka 572, Japan
motoba@
jpnyitp.yukawa.kyoto-u.ac.jp

V. G. NEDOREZOV
Institute for Nuclear Research
Academy of Sciences of Russia
60-th Oct. Anniversary Prosp.,7A
Moscow 117312
Russia
nvg@ksrs.msk.su

Bernard M. K. NEFKENS
University of California
Department of Physics
Los Angeles, CA 90024
U.S.A.
bnefkens@uclapp.physics.ucla.edu

Juan M. NIEVES
Physics Department
University of Southampton
Southampton SO95NH
Great Britain
j.m.nieves@soton.ac.uk

Malte OELSNER
Institut für Theor. Physik
Universität Hannover
Appelstr. 2
D - 30167 Hannover
Germany
oelsner@itp.uni-hannover.de

Eulogio OSET
Dpto. Fisica Teorica
Universita Valencia
Avd. Dr. Moliner 50
46 100 Burjassot (Valencia)
Spain
oset@evalvx.ific.uv.es

A. PARREÑO GARCIA
Dpt. ECM
Facultat de Fisica
Diagonal 647
08028 Barcelona
Spain
assum@dirac.ecm.ub.es

Eduard PARYEV
Institute for Nuclear Research
Russian Academy of Sciences
60th October Anniversary Prosp.7a
Moscow 117312
Russia
paryev@inr.msk.su

R. J. PETERSON
Nuclear Physics Laboratory
University of Colorado
Boulder CO 80309-0446
U. S. A.
peterson%spectr@vaxf.colorado.edu

Willibald PLESSAS
Institute for Theoretical Physics
University of Graz
Universitatsplatz 5
A-8010 Graz
Austria
plessas@edvz.kfunigraz.ac.at

Angels RAMOS
Dpt. ECM
Facultat de Fisica
Diagonal 647
08028 Barcelona
Spain
ramos@fermi.ecm.ub.es

Yu Yu. L. RATIS
Samara Aerocosmic State Univ.
Moskovskaya av. 34
443086 Samara
Russia
ratis@kulini.samara.su

Michael SADLER
Abilene Christian University
Box 7963
ACU Station
Abilene, TX 79601
U. S. A.
sadler@acuvax.acu.edu

Vladimír ŠAULI
Faculty Math. Phys.
Charles University
Prague
Czech Republic

Reinhard SCHUMACHER
Department of Physics
Carnegie-Mellon University
Forbes Ave
Pittsburg, PA 15213
U.S.A.
reinhard@ernest.phys.cmu.edu

Michael SCHWAMB
Institut für Kernphysik
J. J. Becherweg 45
55099 Mainz
Germany
schwamb@vkpmza.kph.Uni-Mainz.

Shoji SHINMURA
Gifu University
Dept. Applied Mathematics
Faculty of Engineering
Yanagido 1-1, Gifu 501-11
Japan
shinmura@cc.gifu-u.ac.jp

Fedor ŠIMKOVIC
Institut für Theoretische Physik
Universität Tübingen
Auf der Morgenstelle 14
72076 Tübingen
Germany
fedor.simkovic@uni-tuebingen.de

Pia THÖRNGREN
Department of Physics
Stockholm University
Box 6730
S-113 85 Stockholm
Sweden
pia@physto.se

Silvano SIMULA
Inst. Nazionale di Fisica Nucleare
Sezione Sanita'
Viale Regina Elena 299
I-00161 Rome
Italy
simula@iss.infn.it

Lothar TIATOR
Institut für Kernphysik
Universität Mainz
55099 Mainz
Germany
tiator@vkpmzp.kph.uni-mainz.de

Jaroslav SMEJKAL
Institute of Nuclear Physics
250 68 Řež
Czech Republic
smejkal@ujf.cas.cz

R. G. E. TIMMERMANS
Kernfysisch Versneller Institut
Zernikelaan 25
NL - 9747 AA Groningen
The Netherlands
timmer@qmc.lanl.gov

Miloslav SOTONA
Institute of Nuclear Physics
250 68 Řež
Czech Republic
sotona@ujf.cas.cz

Natalia S. TOPILSKAYA
Institute for Nuclear Research
Academy of Science
60th October Anniv. prosp. 7A
117312 Moscow
Russia
topilskaya@inr.msk.su

Petr ŠTECHER
Interconex Prague
Czech Republic
stecher@ujf.cas.cz.

Emil TRUHLÍK
Institute of Nuclear Physics
250 68 Řež
Czech Republic
truhlik@ujf.cas.cz

Alfred ŠVARC
Rudjer Boskovic Institute
FEP Department
P.O.Box 1016
41001 Zagreb
Croatia
svarc@olimp.irb.hr

Tamotsu UEDA
Ehime University
Faculty of Science
Matsuyama
Ehime
Japan 790
uedatam@dpc.ehime-u.ac.jp

J. W. VAN ORDEN
CEBAF
12000 Jefferson Avenue
Newport News
Virginia 23602
U.S.A.
vanorden@cebaf.gov

Taiichi YAMADA
Kanto Gakuin University
Laboratory of Physics
Yokohama 236
Japan
yamadat@tansei1.cc.u-tokyo.ac.jp

M. J. VICENTE-VACAS
Departamento de Fisica Teorica
Univ. Valencia
46100 Burjassot (Valencia)
Spain
vicente@evalux.ific.uv.es

Yasuo YAMAMOTO
Tsuru University
Physics Section
Tahara 3-8-1, Tsuru
Yamanashi 402
Japan
yamamoto@ins.u-tokyo.ac.jp

Thomas WILBOIS
Institut für Kernphysik
Joh. Gutenberg - Universität
J. J. Becherweg 45
D - 55099 Mainz
Germany
wilbois@vkpmza.kph.uni-mainz.de

Rakhim YARMUKHAMEDOV
Institute Nuclear Physics
Ulugbek,Tashkent 702132
Uzbekistan
rakhim@pc202.suninp.tashkent.su

Slawomir WYCECH
Theory Division
Soltan Institute for Nuclear Studies
Hoza 69
PL - 00 - 681 Warsaw
Poland
wycech@fuw.edu.pl

Valentina P. ZAVARZINA
Institute for Nuclear Research
Academy of Science
60th October Anniv. prosp. 7A
117312 Moscow
Russia
topilskaya@inr.msk.su

Author Index

Springer-Verlag and the Environment

WE AT SPRINGER-VERLAG FIRMLY BELIEVE THAT AN international science publisher has a special obligation to the environment, and our corporate policies consistently reflect this conviction.

WE ALSO EXPECT OUR BUSINESS PARTNERS – PRINTERS, paper mills, packaging manufacturers, etc. – to commit themselves to using environmentally friendly materials and production processes.

THE PAPER IN THIS BOOK IS MADE FROM NO-CHLORINE pulp and is acid free, in conformance with international standards for paper permanency.

MIX
Papier aus verantwortungsvollen Quellen
Paper from responsible sources
FSC® C105338

If you have any concerns about our products,
you can contact us on
ProductSafety@springernature.com

In case Publisher is established outside the EU,
the EU authorized representative is:
Springer Nature Customer Service Center GmbH
Europaplatz 3, 69115 Heidelberg, Germany

Printed by Libri Plureos GmbH
in Hamburg, Germany